MANUFACTURING ENGINEERING

Principles for Optimization

Third Edition

DANIEL T. KOENIG

Library of Congress Cataloging-in-Publication Data

Koenig, Daniel T.
  Manufacturing engineering : principles for optimization / Daniel T. Koenig. -- 3rd ed.
    p. cm.
  ISBN 0-7918-0249-3
1. Production engineering.   I. Title.

  TS176.K625 2006
  658.5--dc22

2006026443

DEDICATION

This book is dedicated to my granddaughters Alison and Jillian, may they grow up to be wonderfully accomplished women. To my sons Michael and Alan, my daughters-in-law Donna and Cindy. And to my wife Marilyn who encouraged me to continue with this project even though it meant many hours of isolation for her. I love you all very much.

Daniel T. Koenig
Lake Worth, Florida

TABLE OF CONTENTS

Chapter 1

MANUFACTURING ENGINEERING ORGANIZATION CONCEPTS

A FABLE: THE COMPANY THAT COULD AND THE COMPANY THAT COULDN'T

Let me tell you a story about the Company That Could and the Company That Couldn't. Both of these companies were vying for a lucrative market for the same new product, which was different from anything that had been made before. The market for this product was enormous, and both companies were very eager to enter it.

The Company That Couldn't specialized in selling. They could literally sell "refrigerators to Eskimos." They did not understand manufacturing well. They were content to let equipment vendors equip their factory and were miserly about providing manufacturing management support. They were convinced that sales were of primary importance and all other functions were secondary. Their philosophy was to emphasize sales, and whatever resources were left over, which were not much, were parcelled out to finance, design, and manufacturing.

The Company That Could specialized in nothing. They believed in maintaining a balance of skills within all of their functions. Their salespeople could sell, but not "refrigerators to Eskimos." They sold to serve the needs of their customers. Their engineers designed products that their salespeople could sell and their factory could produce. Most important, they tried to determine how best to make their products and did not depend on vendors to equip their plants.

The competition to introduce the new product to the marketplace was fierce. At first, the Company That Couldn't had a substantial lead over the Company that Could. Their order books were full and their factory was swamped with production requirements. In order to produce the new product to meet customer demands, they relied on their traditional method. They asked their equipment vendors to supply them with equipment to do the job. The vendors, exceedingly happy to do so, provided the Company That Couldn't with all sorts of equipment which, they claimed, "was the ultimate in making the new product." Perhaps it was, but the Company That Couldn't failed in making the product. The price to make it was higher than the selling price, and the output did not meet demand. Why? They did not have an understanding of the equipment. It was probably too complex. They also did not have the knowledge to manage the factory. Their management staff did not know how to train their workers to make the product on the vendor-specified and vendor-designed equipment. They also did not match their product design to the equipment they bought. All of this happened because the Company That Couldn't did not pay attention to supporting a balanced organization.

The Company That Could had a much different approach. They found out what the market wanted in the new product. Then they designed the product and presented it to their marketing, finance, and manufacturing departments. Marketing reviewed it for what the customers really wanted in the product. Manufacturing reviewed it for ease of manufacturing and how much it would cost to make. They determined whether existing equipment could do the job, or whether they had to develop something new. They thoroughly explored how they would train their operators to make the product. They determined what the critical design elements were and how many units of the new product they could produce in a given time period. They also estimated the costs for development and implementation. Finance reviewed these costs and stated what the company could afford.

After several iterations of this process, the Company That Could was satisfied that it could successfully enter the market for the new product. And as a result, the Company That Could took the market away from the Company That Couldn't. They were able to produce products that were of better design and higher quality, to produce them at a lower cost, and to deliver them on time. This was possible because the Company That Could had a balanced approach. It worked as a team with all facets of its operations participating. Perhaps even more important, before it tried to sell the new product, it learned how to make it so that it could be sold at a profit.

The moral of this story is that manufacturing engineering is a vital part of a company's success. Learning how to produce at the lowest costs and still meet the constraints of the design and the marketplace is essential for any company. This fable is universal. As long as a company is engaged in providing a product or service it will require a balanced team approach to be successful, and manufacturing engineering will always be a vital team member. This book is about the techniques and philosophies of manufacturing engineering that make industrial organizations "companies that could."

THE INDUSTRIAL MATRIX

Let us first look at the makeup of a typical industrial organization to see how manufacturing engineering fits in this matrix. An industrial organization must, at a minimum, consist of sales, operations, and finance. Traditionally we think of this breakdown as: marketing, finance, design, and manufacturing; where the latter two make up the entirety of operations. These basic functions can be further refined to another level. For example, design can be broken down into research and development, applications engineering, and product service.

Now let us look at the basic responsibilities of each of the major functions.

Marketing: Define what is sellable within the charter of the organization and then sell it.
Design: Based on what is sellable, design the product in accordance with good scientific and engineering principles and produce the specifications for the product.
Finance: Raise the necessary funds for the organization and dispense them in accordance with recognized fiscal practices. Keep track of all funds to optimize their uses.
Manufacturing: Produce the product at the lowest possible cost, in the shortest possible time, and in such a way that it meets all design specifications.

The four functions are interrelated, of course. Marketing cannot obtain orders for products that manufacturing cannot produce or design cannot engineer. Finance must recognize that it is supporting a manufacturing entity; its policies, for example, cannot be based on principles developed for the successful operation of banks and insurance companies. Design cannot specify products that are beyond the scope of manufacturing to produce. Nor can it ask for nuances that improve the elegance of the design but do little to improve the saleability of products. Manufacturing must live within the budgetary levels deemed prudent by finance and must make the product within the specifications required by design. Manufacturing must also deliver finished goods in accordance with the desires of the customers as defined by marketing. Marketing must take into account the effects on manufacturing of all delivery promises made to customers.

THE MANUFACTURING MATRIX

How does manufacturing engineering fit into this organizational matrix? Manufacturing consists of two major categories: shop operations and support. It is necessary to understand the duties of these categories to properly understand the role of manufacturing engineering.

Shop operations is the producing arm of manufacturing. Here management directs the activities of people, machines, and processes in producing the product in accordance with an overall schedule.

Shop Operations receives the equipment, instructions, raw materials, and master schedule from the service groups, then applies labor to produce the product.

The service groups consist of materials, quality control (sometimes known as quality assurance), and manufacturing engineering. Their function is to provide direct support to operations in the form of raw materials and equipment to work with and information on how to use both.

Materials

Materials has the responsibility for producing a master production schedule in accordance with orders received and anticipated by marketing. The materials sub-function, then, controls and monitors the production schedule of shop operations, the transmission of specification information from design to manufacturing, and the supply of manufacturing instructions from manufacturing engineering to shop operations. It has the other major function of purchasing raw materials and supplies and ensuring that they arrive on time to support the overall master production schedule while minimizing inventory costs. It also controls all raw materials and finished goods in storage locations before the former is released for production and the latter distributed to customers. It normally does all this via management of the Manufacturing Resources Planning (MRP II) program, also known as Enterprise Resources Planning (ERP).

Manufacturing Engineering

Manufacturing engineering has the responsibility for instructing shop operations on how to make the product, the sequence, and the facilities to use. It also has the overall responsibility for planning the nature of the factory and its present and future equipment.

In addition, manufacturing engineering evaluates capacities per time frame for marketing to use in sales strategies; evaluates manufacturing capabilities for design engineering to use as constraints on product specifications; and evaluates current manufacturing performance for overall monitoring and for modifiers to capacity and capability evaluations.

Manufacturing engineering is responsible for the maintenance of current equipment and the evaluation and purchasing of new equipment. It also provides this service for non-producing facilities such as buildings, offices, vehicles, and other miscellaneous items.

Another function frequently assigned to manufacturing engineering is that of process control. Measurements of quality are continuously made during a process, usually as part of the traditional system of evaluating productivity performance. This is a relatively recent amalgamation of duties and strives to combine the requirements for high quality with those for improved productivity necessary for corporate survival.

Quality Control

Quality control has traditionally been the liaison between manufacturing and design. This function interprets design's specifications for manufacturing and develops the quality plan to be integrated into manufacturing engineering's methods and planning instructions to operations. Quality control is also responsible for recommending to management what level of manufacturing losses (cost of mistakes in producing the product) can be tolerated. This is based on the complexity of the product design, specifically the degree of preciseness necessary in tolerances. Quality control traditionally monitors manufacturing losses by setting a negative budget that is not to be exceeded, and establishes routines for measurement and corrective action. Although, now with zero defects being a defined goal, the concept of losses budgets is becoming increasingly taboo. This is unfortunate, since from a pragmatic view, a six sigma realization is fictitious (but not necessarily an erroneous objective).

Within the past decade or two, quality control has become increasingly involved with marketing and customers in establishing documentation systems to ensure guaranteed levels of product quality; e.g., managing the ISO 9000 program. This new role has led to the new title quality assurance, to differentiate it from traditional in-house quality control.

Quality assurance strives through documentation of performance and characteristics at each stage of manufacture to ensure that the product will perform at the intended level. Whereas quality control is involved directly with manufacturing operations, quality assurance is involved with the customer support responsibilities generally found within the marketing function. Many industrial organizations have chosen to establish an independent quality assurance sub-function within the manufacturing function and have placed the technical responsibilities of quality control, namely process control, within the manufacturing engineering organization.

MANUFACTURING ENGINEERING RELATIONSHIPS WITH OTHER FUNCTIONS

Manufacturing engineering interacts with the major functions of the industrial organization as well as the subfunctions within manufacturing. Manufacturing engineers are essential in future business planning activities led by marketing, where factory capabilities and know-how on optimizing costs are paramount in any strategy. Their inputs are vital to finance for planning future allocation of funds, and their definition of what is manufacturable and what is not, and the various degrees in between, greatly influences the design function and the type of design specifications produced.

MANUFACTURING ENGINEERING WITHIN THE MANUFACTURING FUNCTION

Figure 1-1 depicts the organization of the manufacturing function with a specific breakout for manufacturing engineering.

There are several alternative ways to depict the manufacturing engineering organization. For example, there could be a producibility engineering unit under advanced manufacturing engineering (AME) if there is a sufficiently strong need in a particular company's operation. Likewise, some managers choose to have separate methods engineering units if their products demand continuous redesign of workstations and fixturing. Another breakout could be that of a computer

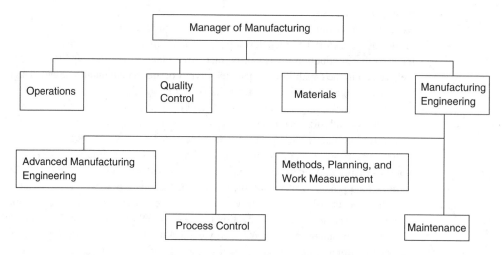

Figure 1-1. Organization of the Manufacturing Function with a Breakout for Manufacturing Engineering

integrated manufacturing (CIM) systems unit with resources derived from AME and from the methods, planning, and work measurement unit. This would be especially useful if the company was just beginning integrated computerization. Process control could very easily be a quality control unit, and in many organizations it is. It is included here as a separate unit because of its inherently technical nature compared to the trend in quality control toward nontechnical activities supporting marketing.

The point here is that manufacturing engineering organizations are not cast in concrete, but can and should be modified to fit the specific needs of individual companies. The organization shown in Fig. 1-1 is the classical manufacturing engineering organization. These units cover all phases of manufacturing engineering's responsibilities. We will now define the charters of the four manufacturing engineering units.

ADVANCED MANUFACTURING ENGINEERING

The major responsibilities of the advanced manufacturing unit are:

> Area planning
> Capacity analysis
> Capability evaluations
> New technology evaluations and needs
> Producibility engineering
> Computer-integrated manufacturing development
> Investment project management
> Long-range planning and forecasts

A separate engineer could have each or several of these responsibilities, or several engineers could share one area of responsibility. It matters little what the specific organizational structure looks like as long as all responsibilities are properly attended to. For simplicity, we will discuss each area of responsibility separately.

Area Planning

This activity determines the present and near-future shape of the factory to meet all needs for facilities. Area planners use the master production schedule of the materials subfunction to determine what the factory must produce. They match the requirements for production with the present capabilities of the factory and then develop plans for additions and modifications to the factory's equipment so that production plans can be met. Area planners are the experts on factory floor layout, well versed in the capabilities of the many types of equipment and processes. They must also have intimate knowledge of the services available and needed in conjunction with these types of equipment and processes. For example, they must be aware of the electric power, voltage, amperage, phase, AC or DC power, transformer, and filtering requirements for control circuits; availability of potable and sanitary water; purity and pressure of compressed air; pressure and condensate control of heating and process steam; hook height above the floor and lifting capacity of cranes; number of cranes on a rail; and sundry other items involved in making machines work and getting materials to and from those machines; and other technical specifications as needs arise.

Area planners work with materials schedulers, marketing analysts, design engineers, manufacturing engineers, and others to determine the needs of the factory for the near-term to medium-term future. They compare the requirements to current capacities and decide whether the factory has the necessary equipment to meet the future needs. If it does, the area planner simply verifies that fact for management. If there are shortfalls in capacity, the area planner will notify management recommend a course of action to remedy the situation.

The area planner has another function, that of optimizing present capabilities so that the factory can produce products at the lowest possible cost. This responsibility is a corollary to that of determining the shape of the factory to meet future production goals.

The work tools of area planners are the various layouts (factory floor blue-prints) and capacity evaluations for the equipment and processes. Area planners are systems engineers. Their task is to gather information from experts and determine whether a manufacturing flow exists. If it does, is the flow sufficient to meet production volume needs? If it does not, what must be done to rectify the situation? They obtain information on capacity from the capacity analysis and capability evaluation sections of advanced manufacturing engineering. Information on machine rates of speed comes from the methods, planning, and work measurements unit. The physical characteristics of the machine tool processes are obtained from the maintenance unit or the engineers responsible for new technology evaluation and needs or investment project management.

After assembling the necessary information, the area planner begins to place facilities on the layout to achieve the desired product flow. The layout covers either the entire factory floor or specific portions of it, depending on the scope of the task at hand. If the master production schedule calls for producing the same product, but more or less of it, the layout will probably cover only a portion of the factory and concentrate on optimization of current equipment, or perhaps an addition to the equipment inventory. If the plan calls for introducing a new product, a total layout will be required showing how to integrate the old and new products and how facilities are to be shared or dedicated. The flows will also have to be considered for interferences. Typically, the area planner will use templates derived from a computer software layout program to locate facilities on the monitor screen and will list the capacities and required services next to the templates. This procedure will reveal bottlenecks where there is too large a difference between facilities and capacities. After identifying bottlenecks, the area planner will determine the alternatives for solution, such as adding additional similar equipment or substituting improved or larger capacity equipment. Once a layout is produced, reviews with shop operations and other manufacturing engineering personnel are scheduled in order to obtain operational viewpoints and other technical inputs. Usually, valid points are made during these reviews which result in a need for further refinements and iterations of the layout.

As implied above, area planning is an art as well as a science. It involves many engineering disciplines, which must be artfully melded together to achieve an optimal factory plan. This melding is the task of the area planner, who is most often a senior manufacturing engineer.

Capacity Analysis

Capacity analysis is the detailed study of the amount of current product and product mixes that can be produced within a specified period of time. This information is vital for the makeup of the master production schedule and for marketing sales strategies.

Engineers in this discipline are very concerned with the product mix being imposed on the factory. They traditionally convert specific products to the basic elements of fabrication time, machine time, process time, and assembly time necessary to manufacture them. They create prototype workstation loads per product. Knowing the amount of product to be produced in a given time period and the work time per manufacturing element, it is possible to calculate capacity, which is normally expressed as capacity per workstation. The engineer can then specify which manufacturing areas have the capacity to produce the load and which areas do not. If a factory is required to make more than one product at a time, which is often the case, the job of determining the capacity becomes more complex, and as a practical matter capacity analysis is not usually done. Instead, the engineer will show limits; that is, 100% product A—0% product B, and 0% product A—100% product B, etc.. The engineer will also list the number of hours per workstation required for each product. With this information the factory limits are established. This is sufficient because the master scheduler, using the MRP II system, is mortgaging time against workstations in order to load the factory. Hence, the scheduler is interested in work elements as devised by the capacity analysis engineer.

The work of the capacity analyst is normally transmitted to the area planner for inclusion in the master plan for the specific area. In many manufacturing concerns, capacity analysis is considered part of area planning.

Capability Evaluations

Capability evaluation is the analysis of what tolerance can be created at what incremental cost, what critical processes can and cannot be performed, and what maximum size and weight can be handled in the factory. This information is critical for the design function's ability to continuously evolve the product or product lines. Design engineers must understand the limitations of the factory in order to design the products for the factory. Engineers involved in capability evaluations have the responsibility for determining the limitations and codifying the results for use by others.

These engineers require very detailed evaluations of equipment performance and an understanding of the present capabilities of equipment. They must understand the effects of deterioration of equipment on its ability to produce to close tolerances. Their primary function is to be aware of the performance level of the equipment and to periodically update the database.

Capability evaluation engineers perform another vital function, that of technically evaluating proposed products to determine whether they can be made. If it is feasible to make a proposed product, then capability evaluations engineers will develop the preliminary optimized sequence and workstation selection to be used in making the product. This is done in close consultation with the area planner for integration with the factory's master plan. Like capacity analysis, capability evaluation is often considered to be part of area planning.

New Technology Evaluation and Needs

The competitive marketplace requires that manufacturers continuously evaluate costs of production. Failure to do this invites competitors to erode away hard-earned market share and makes it more difficult to sell one's product. Just as the design function must continuously evolve a better design, so the manufacturing function must continuously evolve better ways of producing the product. The goal is always to reduce the cost of production. The process of reducing cost is done by knowing how the product is made and searching for new types of equipment to produce the product at lower cost.

Considerable effort is expended in keeping current on new ways to produce the firm's products. This is usually done by maintaining close liaison with vendors who sell equipment traditionally used by the company, attending trade shows, participating in technical societies' programming, and, of course, regularly reading pertinent trade and technical publications.

Engineers involved in new technology evaluation produce a series of reports on new technologies of interest to the firm. These reports become the research basis for long-range planning and product introductions.

Producibility Engineering

Producibility engineers ensure that designs produced by the design function are workable, that is, optimally producible in the company's manufacturing facilities. Producibility engineers deal in tolerance evaluations and changes in design to facilitate manufacture on optimal machines and facilities. Another task is to document detailed factory limitations and procedures for the understanding of design engineers. Their main goal is to lower the costs of making the product by obtaining design optimization relative to manufacturing needs. The producibility engineer is often said to be the manufacturing ambassador to the design function. They interpret manufacturing capabilities for the designers and convey the designers' real needs for product functionality to manufacturing.

The producibility engineering discipline has become more important as the need for productivity improvement in industry has increased. This discipline, once one of many in the spectrum of manufacturing engineering, has been singled out as decisive in helping industries compete in the modern high-cost era. The subject of producibility engineering and its relationship to the philosophy of concurrent engineering will be covered in considerable detail in Chapter 6.

Computer-Integrated Manufacturing Development

This activity investigates productivity gains that can be made by applying computers to all phases of manufacturing. Engineers involved in CIM development are active in all three areas of the Computer-Aided Design/Computer-Aided Manufacturing (CAD/CAM) triad: machine/process control, measurements control, and planning control. They are responsible for developing and enhancing all CIM systems, then transferring operating control to other units within the manufacturing section for actual use.

The CIM development engineers work closely with their counterparts in the design and finance functions to conceive and implement the company's CIM plan so that an integrated system can exist. The integrated system is usually based on a common database shared by all functions. The CIM development unit is charged with making sure that the integrated data base will serve the manufacturing section's needs while not being detrimental to the needs of other functions. The development and management of CIM systems will be considered at length from the viewpoint of manufacturing in later chapters.

Investment Project Management

This segment of advanced manufacturing engineering works very closely with the finance function. It is responsible for "keeping book" on all funds allocated to the capital investment program. These engineers produce the investment plan and manage the projects from inception through approval by appropriate levels of management. When the plan has been approved, the investment project management engineers place orders for the equipment, monitor the progress of vendors in producing the equipment, and finally arrange for the equipment to be installed and debugged. Once the equipment is deemed operational by the intended users, control is released to the users.

Investment project management depends heavily on inputs from the area planners. Area planners, being responsible for determining the immediate and near-future needs of manufacturing to produce the product, are the natural initiators of capital investment projects. They provide the reasons for requests and preliminary information regarding cash flow and payback periods. The investment project management engineers, using this information, prepare the detailed documentation used to obtain the appropriate levels of management approval. They also use the services of the finance function to provide support documentation for the capital expenditure request.

Having received approval to spend money for capital equipment, the investment project management engineer will place an order for the equipment with an approved vendor. Vendor appraisal and selection normally take place during the preparation stages of the request for funds so that the investment project management engineer will know how much money to request. In consultation with the new technology evaluation/needs engineer and the responsible area planner, a list of appropriate vendors is drawn up. Vendors may be asked to visit to discuss the project, or visits may be made to vendor facilities to see similar equipment. The overall purpose is to determine exactly what piece of equipment will be purchased. If many vendors are capable of supplying the necessary equipment, requests for quotes are issued to the vendors inviting them to state prices and delivery times and to specify what their product is capable of doing. A request for quote is often required even when there is only one vendor.

This entire process tends to define the project for all involved. For the financial people, it quantifies near-term expenditures of money and allows comparison to budget plans. For shop operations, knowledge of incoming equipment allows plans to be made concerning staffing logistics and product output levels. For manufacturing engineering, this information is vital for factory layouts, support services planning, obsolescence evaluation, and productivity/producibility issues.

Once an order is placed, the investment project management engineer plans for receipt of the equipment, its installation, and its final turnover to shop operations. The engineer works with the area planner on matters concerning where the facility will be located, and oversees the maintenance unit for the actual installation. Debugging of the unit is done in conjunction with the methods, planning, and work measurements unit. Various elements are involved in installing capital equipment. However, the investment project management engineer develops the schedules, determines the project's critical path, and is responsible for its successful completion.

Long-Range Planning/Forecasts

This activity develops the company's plans for technical improvement, capacity improvement, and capability improvement in accordance with its overall strategic and operating plans. Responsibilities include translating overall objectives and goals into technical and hardware plans, usually over a 5-year or longer period. These specific plans allow the finance function to develop criteria for profits or other means of funding. They allow for an evaluation of the stated objectives and goals. This is done by an iterative process between the manufacturing and finance functions. Most often the commencement of this activity is initiated by the business planning cycle, usually as part of the annual business plan development. The givens, or baseline depends on the circumstances of the company. The long-range planning/forecast engineers will start the process either from a known dollar value for investments per year, or from a known or targeted production level over the forecast period. In either case, the iterative process applies.

The process is started by gathering inputs from the marketing and design functions on quantities and types of products to be sold over the forecast period. Accurate forecasting being very difficult, marketing is normally reluctant to make specific forecasts, and whatever forecasts it makes are all too often vague and full of restrictions and disclaimers. Nevertheless, manufacturing must have a forecast and must be able to believe it. Hence, a very important assignment for the long-range planning/forecast engineers is obtaining a forecast in sufficient detail and verifying it. This may be done informally by asking marketing planning to explain how they arrived at their forecast, or formally by asking marketing to have another source, usually an external or internal consulting service, make a forecast. Whatever the source, manufacturing engineering must have confidence in the forecast before proceeding.

Given a forecast, the long-range planning/forecast engineers develop scenarios of manufacture in consultation with the area planners, and those engineers responsible for capacity analysis, capability evaluation, and new technology evaluation/needs. This will result in a facility capital investment plan for the forecast period. The plan is then reviewed with other functions and the iterative procedure begins.

METHODS, PLANNING, AND WORK MEASUREMENT

Like the advanced manufacturing engineering unit, the methods, planning, and work measurement (MP&WM) unit may have one engineer for each major activity, several for one activity, or one for several activities. Again, the personnel allocation depends on the staffing needs of the particular organization.

MP&WM is the core of the industrial engineering discipline conducted by manufacturing engineering. This unit performs the traditional industrial engineering work carried out for the factory. It develops and implements non-facility or personnel productivity improvement programs. Hence the MP&WM unit is the branch of manufacturing engineering that is closest to shop operations on a daily basis. It instructs the shop on how to do a specific job, trains the operators, determines how much time it should take to do each operation, and measures the shop's performance.

The major phases of the MP&WM unit's work are listed below and discussed in greater detail in the following sections. The activities are interrelated, but for the purpose of definition we can examine them independently.

Workstation methods
Workstation planning
Workstation and tool design
Time standards
Measurements of productivity
Variance control programs
Job rate evaluations
Operator training

Workstation Methods

Methods engineers devise the basic techniques for producing products as required by the design specifications. They mate a workable design with the specific capabilities of the factory to produce the part or assembly at an optimal cost. Methods engineers use the best combination of machine feeds and speeds or process controls to achieve the results called for in the design specifications. They specify tools to be used for cutting; proper voltages, amperages, and wire feed rates for welding processes; the best suited jigs and fixtures for assembly; and so forth for the particular manufacturing operations. They outline the chronological sequence of events necessary to do the required work, but do not arrange in sequence anything that does not have an operational prerequisite. The specific sequencing is left to workstation planning personnel, who will optimize the entire sequence in time.

Methods engineers also spend considerable time and effort determining the positioning of work for proper presentation to the machine or process equipment—for example, how the work piece should be secured and loaded to minimize tool chatter, droop, or process-induced stresses. Proper positioning helps to ensure that the finished part will have all the characteristics required by the design specification.

The specification of tooling falls within the domain of the methods engineer, who outlines what is necessary and what is to be accomplished, and then turns the task over to the tool designer for specific development. The methods engineer defines the constraints within which the tool designer operates.

Methods engineers normally are the manufacturing engineers most closely associated with factory floor operations. They have many opportunities to observe ongoing operations and evaluate how they are progressing. They are often called upon to participate in productivity improvement projects, which may range from minor tooling improvements to complete realignments of production techniques. Methods engineers are frequently people who have had long experience in a particular factory or with a particular product line.

Workstation Planning

Planners create the detailed step-by-step sequencing instructions for shop operations to follow. A properly sequenced set of workstation plans will always result in the optimum way of producing a part or assembly with repeatable results. The planners use the design specifications, engineering drawings, and methods instructions to develop a "road map" for the operators to follow. Their information output covers the exact detailed chronological sequence of tools to be used, the number of passes to make for a weld or the number of cuts to take on a specified machine tool, and the exact setup on the machine tool to be followed. Each step also includes the time allowed for the step to be completed. Good planning leaves nothing to chance. Every aspect of the job to be performed must be accounted for, leaving no room for supposition or inference.

The final function of the planner is to put the plans into the manufacturing organization's production control system. This may be a manual or computerized system. The only overall requirement is that the operator or the intelligent machine understand what is required and be able to respond effectively.

The planning function is the storehouse of the body of knowledge on how its products are manufactured. One ongoing activity is cataloguing and codifying this information for future use. Many companies use Computer-Assisted Process Planning (CAPP) for this purpose. This technology allows easier repeatability of planning for like or similar parts and assemblies by applying a group technology code—a classification of parts by shape, manufacturing sequence, or materials from which they are made. In this way families of plans can be developed, grouped, and easily recalled, and only the superior plans will be used over and over again. Thus, CAPP encourages standardization and optimization and leads to improved productivity. By using the computer to search for similar plans already on file using their group technology code number (many times inputted via a bar code scanner), the number of new plans required per time period is reduced and the quality of planning is improved. In essence, all planning is being done by the senior or more qualified planners. Still another advantage of CAPP is the reduction of cycle time required to go from engineering drawing to completed planning. The computer can search for similar planning in a fraction of the time required to do it manually.

The computer has changed and continues to change the nature of how workstation planning organizations are run. Drawings are quite frequently data inputs for Computer Numerical Control (CNC) machines, step-by-step sequences are generated from data bases rather than handwritten instructions for a specific part, and timing sequences are calculated by previously programmed systems rather than one at a time. All these abilities have increased the output of the planning function, but the output is still the same and still performs the very necessary function of telling the operator what to do and how to do it.

Workstation and Tool Design

The design of tools is an exclusive responsibility of manufacturing engineering. Tool design engineers create tooling on the basis of constraints spelled out in the specifications prepared by methods engineers. Tool designers must understand not only the classical concepts of stress analysis, but also how tools are used and what they will produce. They must have a good background in metallurgy, welding, forming, cutting, and other associated technologies. An excellent knowledge of processes are also required. While the tool design engineers specialize in the ultimate contact device between the material to be worked on and the workstation, the workstation engineers are more systems-oriented than tool designers. They are charged with designing a workstation that optimizes production time and achieves the most effective and productive utilization of space and time.

Tool design engineers work similarly to the product design engineers. They use the same techniques that go into the design of a product to be produced in the factory. The difference is that their output usually calls for a short production run and its manufacture is almost always assigned to a tool specialty shop. However, tool design engineers need considerably more than technical competence. They must understand the use of tools in manufacturing. For this reason, a lengthy on-the-job apprenticeship is usually needed to convert a theoretically trained engineer into a tool design engineer.

Workstation engineers deal with principles of motion economy, the efficiency of body movements. They must know how material will arrive at the workstation, how it should be handled there, and how the finished part should leave. They design workstations to minimize both human movement and machine movement and ensure optimal efficiency of the human–machine combination. The workstation engineer is as close to the stereotypical efficiency expert as anyone likely to be found in manufacturing.

Workstation design and tool design involve an amalgamation of unlikely skills. The broad systems thinking skills of the workstation engineer and the specific detailed thinking skills of the tool design engineer come together in the workstation. Here the workstation engineer must know what the contact tool can do and how it must fit into the scheme of the workstation. The tool design engineer must know the limitations of the workstation so that the tool design will flow within the human–machine system. The systems-oriented and detail-oriented engineers complement each other and act as consultants to each other.

Time Standards

Plans show the amount of time it takes to complete each particular segment of work at the workstation. These times are developed by time standards engineers. The classical body of knowledge known as motion and time study, or efficiency measurements, deals with the most efficient ways of using the human body; for example, a wrist motion is more efficient than a full arm motion because it uses less energy and is faster. Motion and time study also deals with the motions of machines and how much time they take to complete a cycle; it is concerned, for example, with feeds and speeds for a lathe as specified by the methods engineer. Combining times for human motions with times for machine motions makes it possible to calculate the time it should take for any manufacturing operation. The time to perform a set of operations at a workstation is called a time standard.

Time standards engineers codify and tabulate time standards of operations for use by planners. They create computer programs that the planners may use to determine how much time any particular operation should take, and to assign planned times to plans issued to shop operations.

The system for writing time standards is quite specific and borders on being tedious. Fortunately, computer systems have been devised that assist in accomplishing this task, but it still remains very detailed work. Engineers must first understand the workstation: what it contains, the parts likely to be processed there, how it is loaded and unloaded, the feeds and speeds, and the number of operators working. They must understand the workstation design, the methods to be applied, and the tools to be used. They then list in chronologic order all actions to be taken by human or machine or both together. For each item on the list they research the principles of motion economy for human actions and mechanical speeds for machine actions. Ultimately, they record the least amount of time it should take to do each item on the list. For items that are variable, they construct an equation that defines the spectrum of events. The total time of all items on the list is the time standard. Although time standard development is laborious, it is the only way of knowing what the ultimate optimal time for an operation should be. Any other method of setting planned time is usually a comparison to previous performances, good or bad, and the potential for improvement is not known.

Measurements of Productivity

Using plans and time standards, methods engineers can determine how much time, in theory, it should take for a work element. Combining all the work elements of a job reveals the time necessary to complete that job. Knowing how long it takes to do a job makes it possible mathematically to compare actual work performance to theoretical or standard performance. This mathematical ratio becomes a measurement of efficiency. If enough measures of efficiency are taken, it is possible to discern trends. If the ratio approaches unity as time progresses, then productivity is improving. If the reverse happens, productivity is decreasing.

There are other measures of productivity such as dollars of product value output per direct worker—the so-called output over input measurement. These types of measurements tend to be too large-scale, and therefore of little use in managing shop operations. Efficiency trends tend to be the most direct and easiest to measure, and a considerable body of knowledge on this measurement system has been accumulated and classified.

Manufacturing engineers involved in productivity measurement are concerned with the validity of the data, hence the accuracy of the measurement trends. They spend their professional time developing data collection systems, which typically evolve about the way work is dispatched to the workstation and how progress is recorded. In addition, most computer manufacturers now offer industrial management systems, the heart of which is usually a data collection system, which often work in conjunction with MRP II systems. The goal of these systems is to capture completions of processing steps as soon as they occur so that elapsed time can be compared to planned or theoretical time, generating an efficiency measurement on a real-time basis. These systems frequently employ user-friendly terminals at the workstation that instruct the operator in how to make entries and have a mechanism for allowing the operator to record start and stop times. With accurate information of this type, factory managers can tell the state of the operations at any time and take corrective actions immediately. They can accurately gauge their production rates and make judgments for future orders. Engineers involved in measurements of productivity play an important role in devising these systems and thereby considerably influence the information gathering and analysis necessary for success.

Variance Control Programs

Variance control is very closely associated with measurements of productivity. Variance measurement is another way of saying efficiency measurement. The variance is the amount of change or mathematical distance from the norm, which in this case is 100% efficiency, that is, doing the work in exactly the planned amount of time. Thus the variance is the difference between the planned time and the actual time.

Variance control programs are any techniques that can be applied to reduce the mathematical distance between actual and planned times for accomplishing an assigned task. For example, if the time to perform a job was 129% of the standard, then the variance would be 29%. The variance control program would be aimed at bringing the 29% variance as close to zero as possible. To do this, engineers would look for all the factors that hindered the completion of the operation within the planned time and develop plans to eliminate them or, as is usually the case, reduce them. Strategies can range broadly, from ensuring that properly sharpened tools are being used to asking producibility engineers to review the manufacturability of the product.

Variance control programs tend to become complex because many unknowns must be evaluated and simultaneously accounted for. Hence engineers working in this area tend to be senior in manufacturing experience and thoroughly understand the nature of the products being produced. Many of the projects used in variance control programs require inputs from most of the organizations of the manufacturing function. For this reason, the programs require good project management skills. Such projects normally deal with underperforming workstations; hence they are highly visible and require mature judgment for their successful conclusion.

Job Rate Evaluations

To function properly, an organization must have the proper people in the proper jobs. One of the tasks of MP&WM is setting the standard requirements for successfully carrying out specific jobs. This is job rate evaluation, or determining the value on a comparison scale of each job that must be performed for the factory to function. The engineers who do this determine the order of job importance so that management knows how to pay for the various positions.

Engineers usually do this by designing several key jobs that the particular factory needs in order to produce its products. CNC lathe operator, welder, and assembler would be examples of some of the key jobs for a factory that fabricates, machines, and assembles to make its products. Once these key jobs are designed, the responsibilities assigned, and the availability of these skills on the local market determined, the jobs are ranked in order of importance. All other jobs are then ranked relative to these key jobs.

In this process it is essential that the skills associated with all jobs be properly evaluated. Major mistakes here would seriously jeopardize the company's ability to attract qualified persons to the respective positions. Factors that are taken into account are mental and physical skills necessary to perform the function, availability of such skills in the surrounding geographic area, comfort or discomfort levels associated with the job, hazards in performing the job, and cost of operator mistakes associated with the job or degree of criticality of not making mistakes. Each of these factors is given an evaluation on a point system so that a basically subjective system can be made as objective as possible.

The job rate evaluation system can set the tone of personnel relations. If it is done correctly, the factory will have established the basis for high employee morale. If not, a factory with continuous labor strife may be created. Engineers engaged in this activity thus have a great influence on the morale and the productivity potential of the factory.

Operator Training

This phase of MP&WM responsibilities is directly associated with the shop operations work force. Methods engineers are responsible for instructing the operators on how to operate the manufacturing equipment in accordance with the prescribed methods, and are therefore usually themselves experienced in operating the factory equipment.

In training the work force, engineers use all the traditional methods: classroom, training aids, on-the-job training, pairing trainees with experienced operators, and so forth. In addition, the MP&WM unit may operate an apprentice training program that gives recipients both classical technical training and specific skills training.

Regardless of the size of the organization, all factories conduct some form of operator training. The modern industrial concern cannot exist without such programs.

MAINTENANCE

The responsibilities of the maintenance unit are:

Machine tool and equipment maintenance and upgrade
Buildings and grounds maintenance
Stationary engineering
Disposal of excess and obsolete equipment
Tool room

The maintenance unit is one of the few organizations within manufacturing engineering that has skilled operators reporting to it. It is similar to shop operations in administration of personnel, so that the operation of maintenance consists of more than directing the activity of professional engineers. It also includes several foremen and their respective skilled craftsmen. In many ways this unit resembles a multipurpose construction company.

Machine Tool and Equipment Maintenance and Upgrade

Maintenance is responsible for keeping the factory's equipment operational so that production can continue. A related responsibility is that of continually upgrading or improving equipment to allow constantly better performance. Maintenance interrelates closely with shop operations to keep the equipment functioning, and with advanced manufacturing engineering to devise better ways to use the equipment to meet continually evolving design requirements.

Engineers work with a wide span of organizations: vendors selling parts and retrofit equipment, shop operations management requiring coordination of maintenance efforts and production requirements, advanced manufacturing engineers requiring information on whether and how equipment can be modified to meet design modifications, and MP&WM for tooling efforts and tooling design and impacts of methods on equipment.

The relationships are even more pronounced for the equipment upgrade activities with AME. Here maintenance engineers are involved in projects that border on capital equipment acquisition, requiring skills closely akin to those of advanced manufacturing engineers. The dollar values are usually less—a reason upgrades are usually attractive to management—but the procedure is as intense as in major acquisition projects.

The maintenance unit employs skilled operators such as electricians, carpenters, pipe fitters, steelworkers, welders, machine repairmen, and electronics technicians. These trades are supervised by foremen knowledgeable in such skills and well versed in job estimating techniques. A major function is responding to requests to fix malfunctioning equipment. Typically, the maintenance foreman will receive a request to repair a piece of equipment that in the opinion of the requester is not performing as it should. The foreman's task is determining if this is a machine or operator problem; and if an equipment problem, the cause of the malfunction and then getting it fixed quickly. If it is an operator problem the maintenance foremen will do the initial retraining, then turn over the monitoring and iterative training tasks to the shop operations foreman. Since in a producing operation quick repair is a vital necessity, manufacturing concerns pay premiums for expert skills in this area.

In diagnosing malfunctions, the foreman can consult with the maintenance engineers. Normally, if it is a serious matter, the engineer takes over and provides the required technical expertise, while the foreman directs the repair activities based on the technical decisions of the engineer. It should be noted that this occurs for very difficult problems and is not the norm. Typically, the engineers spend their time planning and monitoring upgrade projects, not tending to breakdowns.

Engineers are also responsible for maintaining the machine tool capability and replacement log, where the history of all equipment is kept. It includes the frequency of repair, types of repairs required, cost, deterioration in ability to hold as-new tolerances, and other information critical to the equipment performance. Although engineers are the keepers of the log, information inputs are received from all manufacturing engineering functions. For example, the information on new equipment is often provided by the AME unit, while that on deterioration in performance may be provided by maintenance foremen or methods engineers. These records are the basis for calibrating current production rate capabilities as well as substantiating the purchase of replacement equipment.

Buildings and Grounds Maintenance

The engineers here are responsible for ensuring that building systems such as air-conditioning, heating, sanitary systems, and lighting work adequately. A major part of their work is preventive maintenance so that disruptions in services do not adversely affect the primary operations of the company. They become involved with lesser-dollar-value capital equipment acquisitions commonly called hotel load services.

In addition to maintenance, they are usually responsible for cleaning and grooming services within offices and factory. In this activity not only less skilled personnel but also many of the highly technical trades such as electricians, master mechanics, carpenters, and electronic technicians are involved. Air conditioning, heating, sanitary systems, and energy and lighting systems in modern structures have considerable technical complexity and require sophisticated engineering and technical support.

Stationary Engineering

If the manufacturing operation is large enough, a separate stationary engineering group is established to operate the power plant, air conditioning system, and other hotel load services. Like the other parts of the maintenance unit, it is staffed with a mixture of engineers, foremen, and skilled tradesmen.

Disposal of Excess or Obsolete Equipment

This task involves selling equipment that is no longer needed or no longer capable of performing to required specifications. It is traditionally assigned to the maintenance group because of their close working relations with used equipment dealers, which come about because maintenance engineers often look for repair parts and frequently obtain them from dealers in used or renewed or rebuilt parts.

Disposal of unneeded equipment is not a haphazard operation. Each year the manufacturing engineering organization will prepare a budget for disposal of equipment, containing the cost of and the value to be received for disposal. The net effect should be a net gain for the company to offset the cost of operations. Cost of disposal would include actual removal costs, advertisement costs, and carrying charges of inventory after the equipment is declared obsolete. In well-run companies the funds to be received by selling obsolete equipment are not simply a desirable goal, but are actually planned for. The budgets for disposal of obsolete equipment are developed by evaluating the physical plant inventory records to determine the useful life of equipment. This information, coupled with the identification of replacement equipment on the capital investment budget, yields a realistic value for the used equipment sales budget.

It is the responsibility of maintenance to develop sales plans for disposing of the equipment and maximizing the company's profitability. In a sense, maintenance engineers must develop skills of successful marketing personnel. They must research the market and contact and develop customers for their particular products.

Tool Room

Larger companies have tool rooms fully equipped and staffed to support manufacturing operations. As companies downsize and outsource production the need for tool rooms decreases. However, keep in mind the services performed by a tool room will be done either as an integral part of the company, or as a contracted service as needed. But the services provided by a tool room will still be required.

The tool room is responsible for making repair parts, jigs, and fixtures for all of manufacturing. It is also responsible for redressing and grinding all tools used in manufacturing operations. It is a service shop operated by manufacturing engineering and administered by the maintenance unit. Engineers in this segment are engaged in designing repair parts to support maintenance activities only. Tooling, jigs, and fixtures are designed by MP&WM engineers and manufactured for them in the tool room.

The tool room is usually equipped with general-purpose machines of a smaller nature and stocked with raw materials used in making repairs. A small stock of materials suitable for frequent MP&WM methods tooling work is also kept on hand. Materials for special requests may be drawn from production stocks or ordered as required.

Operators assigned to the tool room are usually senior personnel, having master craftsmen skills and experience. Tool room operations are similar to those of shop operations units. Work is received in, scheduled, and dispatched to the operators. The main difference between shop operations and the tool room is that detailed method instructions are not required. These operators, having the highest skills, do their own planning to produce the required products.

PROCESS CONTROL

The last unit within the manufacturing engineering sub-function is process control. This is a fairly recent addition to manufacturing engineering. Indeed, many classical analyses of quality control/assurance still consider process control to be a part of that subfunction. However, with the ascendancy of quality assurance over traditional quality control, there has been a trend toward placing process control within manufacturing engineering. This is due to a desire on the part of manufacturing management to concentrate technical support services within the strongest and traditional manufacturing technical support function, coupled with the fact that quality assurance is more of a certification activity with a significant part of its attention focused on other than manufacturing activities.

The responsibilities of process control are:

Quality planning
Nondestructive test
In-process inspection
Incoming material inspection
Product test
Total quality management administration

Quality Planning

This is the planning and strategy activity of process control, and is sometimes referred to as process planning or inspection planning. The engineers involved develop the plans for checking the adequacy of performance of shop operations to ensure that the final product performs as designed. Using plans and methods produced by MP&WM as a guide, engineers determine where inspections and nondestructive tests will be specified during the manufacturing process. They also specify the type of inspection or test to be conducted and, based on design engineering requirements, determine what will constitute acceptance or rejection. These quality plans are then integrated into the overall operational plans by which shop operations produces the company's products.

Normal manufacturing activities (regardless of the illusions of zero defects proposed by six sigma programs) produce a certain percentage of deviations from drawing. Some are important, some of little consequence. It is the quality planning engineers' responsibility to evaluate these deviations and determine what the proper corrective action will be. They then ensure through MP&WM that the corrective actions are factored into manufacturing planning for rework. As the arbiter of quality via the deviation from drawing procedure, quality planning has the database to evaluate performance of the various shop and support functions. A score-keeping function is possible and desirable; in this way quality planning can report to management whether quality levels are improving or declining.

Quality status can be reported by statistically evaluating the numbers of deviations and their seriousness. This leads naturally to an evaluation of the cost of doing the repair work caused by the deviation. Repair work, which constitutes manufacturing losses, is an important measurement of organizational quality levels. Manufacturing losses are a significant measure of the adequacy of attention to detail of the operators and their foremen. High losses indicate a poorly managed operation. Quality planning engineers are responsible for setting the manufacturing losses, budgets, and measurements policy.

Nondestructive Test

A significant portion of the evaluation of products during manufacture is done by using non-destructive testing techniques. Engineers assigned this responsibility have extensive technical knowledge of x-rays, magnetic particle testing, ultrasonic testing, and dye penetrant testing to name a few, and how such tests can verify product integrity. It's not unusual for test engineers to spend about half their time designing tests in accordance with the quality planning engineers' specifications, and the other half supervising or performing such tests.

Nondestructive Test (NDT) engineers must have considerable knowledge of the structure of materials so that they can properly interpret test results. Such test results are often difficult to interpret, and judgments are often made on a probabilistic basis. Thus the NDT engineer must understand the design of the product, the rigor of the intended service requirements, and how the product or component is manufactured. With this knowledge, the NDT engineer evaluates the components, making educated judgments on the adequacy of the part for its intended use.

This is not to say that NDT engineers evaluate products in a void. The interpretation of test results is often guided by extensive codes and codified interpretations of those codes. In fact, many product specifications require testing in accordance with clearly defined codes. NDT engineers must be knowledgeable about the various codes and the interpretations given to these codes by technical societies and worldwide government agencies.

NDT is by far the most abstract function of manufacturing engineering. Many books, articles, and dissertations for advanced degrees have been written on the subject, and it is probably the most thoroughly publicized manufacturing engineering technology. Still, NDT is simply one tool that good manufacturing managers use along with other methods to produce high-quality products.

In-Process Inspection

This part of process control is commonly referred to as the "policeman on the beat." The common and mostly incorrect impression is that in-process inspection inspects quality into the product or component. Inspection is done to verify parameters considered critical for proper product performance. It is not done to tell an operator whether he or she performed the function correctly, but to double check what is presumed to be proper performance by the operator. This is a subtle difference and is often misunderstood. The inspector's function is to check system conformance; operators are supposed to know that a part is correct before submitting it for inspection. Operations where 100% correctness is essential and 99.9% correctness equals failure are double-checked and sometimes triple-checked through the inspection system. Where 100% correctness is not necessary, inspections are called out in the planning to provide other important information, for example, whether the human–machine system is still performing within the expected tolerance range, and whether the process is producing as expected or designed. Quite often raw material specifications can vary, resulting in insufficient results. It is important to understand that it is not the function of in-process inspection to check the knowledge or adequacy of training of the operator; this is an MP&WM function.

Inspectors receive their instructions from the quality planning engineers. These instructions, which are contained in the quality plan, specify what the inspector will inspect, how often, and in what sequence of the operation. The inspection may be a physical check with measuring tools such as verniers, micrometers, or electronic or optical instruments, or a verification of documentation prepared initially by an operator making the part. Often an inspection is a combination of both.

Usually inspections are done when a product or component passes from one major operation to another, such as from fabrication to machining. Therefore inspectors perform another very useful activity for manufacturing management. When their work is complete, the status of the part is recorded and management then knows that the part is correct per engineering design through that specific operation. However, it should be noted the trend is more toward operator certification instead of higher cost inspection activities for all but the most critical ones.

Staffing of this segment usually consists of inspectors and foremen. Inspectors may be drawn from the ranks of the operators or may be trained initially upon hiring for the inspection job. Inspection foremen almost invariably will be promoted from the rank of inspector.

Incoming Material Inspection

Process control is responsible for verifying that raw material received for use in manufacturing is of adequate quality. This task is accomplished by measuring or verifying the material in accordance with the design specification.

Incoming materials inspection engineers work with design engineers to develop material specifications that can be met by vendors at reasonable costs. Design engineers specify materials for use in the product. Incoming material inspection engineers review the tolerance band for the specified materials in terms of what can be purchased within technical, cost, and delivery time constraints. Once agreement is reached with design engineering, they specify the types of acceptance tests to be performed on materials before release for manufacturing.

Work is also done in close collaboration with purchasing to qualify vendors. Purchasing agents will find vendors who claim they can supply or produce the needed materials. It is the incoming material inspection engineers' responsibility to determine whether the vendor can actually do so and can sustain the required level of performance over the life of the contract. Ensuring that adequate materials are received is a vital task; faulty materials lead to faulty products that, at best, require high repair costs.

Product Test

Product test can be thought of as the culmination of all process control work. It can also be thought of as a quality check of the inspection process itself. If the quality plan is adequate and carried out properly, then the product's performance should have been verified and a total test is redundant. For this reason, a test of the completed product is often nothing more than a contractual requirement that must be performed before the customer accepts the product. But product test is also more than proving that the whole is equal to the sum of the parts. It allows for gathering data that support the design theory of the product, for interpretations to be made for further improvements in design so that future products will be better than present ones, and for evaluation of design evolution toward better performance costs. In addition, it is a means of verifying design, since not all design parameters can be fully calculated or predicted.

Product test engineers work closely with design engineers to plan tests to provide useful data. They must also work in close harmony with all other phases of manufacturing engineering. Not infrequently, product testing will turn up deficiencies in design that require major revisions in manufacturing processes. This is particularly true if the company produces many prototypes and has short production runs. Therefore, manufacturing engineers are as interested in product test's results as are design engineers.

For complex products, product test becomes a very important part of the total process control function. It gives the company a high degree of confidence that the product will perform as the customer expects it to, and this is a valuable marketing tool as it helps to establish the proper reputation with the customers.

Total Quality Management Administration

The final responsibility given to process control is that of administering the quality awareness program. As I have stated previously, quality cannot be inspected into a product or process. Acceptable levels of conformance to specifications and instruction need to be achieved during the creation process itself. It is impossible for the inspection process to change nonconforming work to conforming work just by virtue of the fact that the work is undergoing an inspection. The best we can hope for during inspection is that deficiencies are discovered and corrected. It is far better that performers of value-added work are cognizant of the need to do acceptable work in the first place and not rely on the inspection function to find their deficiencies. Thus the need to correct deficiencies should be minimized to the greatest extent possible. The mechanisms of fostering this awareness and the importance of doing acceptable work always are the essence of Total Quality Management (TQM).

The members of the process control staff given the task of administering the TQM process are both proponents and planners. In many aspects they are adjuncts to the quality planning engineers in that they are devising plans to ensure that tasks are performed properly within the process flow. However, the main task they perform is to demonstrate to the entire company structure that everyone is responsible for achieving acceptable quality levels. Moreover, they are responsible for creating a mental set within the company that customer satisfaction can be achieved only through creating products and services that the customer wants. Inevitably, that means thorough attention to detail and doing defect-free work.

The task of creating this necessary awareness for TQM falls within process control. Along with creating awareness is the assignment of assuring that TQM principles are being applied. This is the audit function. Also, in order for an audit function to be of use, first techniques of how to comply have to be codified and taught. These, too, are functions given to process control. The techniques involved will be explained and demonstrated in the chapter devoted to process control.

SUMMARY

I've now introduced the complexities of the manufacturing organization. In particular, the manufacturing engineering organization has been defined to demonstrate its vast area of responsibility within the manufacturing matrix. It should now be evident in a macro sense how the various units of manufacturing engineering work and how they are related to each other. Manufacturing engineering is the primary technical resource of the manufacturing function, but its influence goes beyond the limits of manufacturing and extends to the activities of the design, marketing, and finance functions of the industrial organization.

REVIEW QUESTIONS

1. Manufacturing engineering is considered to be the core technical discipline within the manufacturing organization. Discuss the interrelationships of manufacturing engineering with the other functions; in particular, state how and in what circumstances manufacturing engineering would be the lead organization for manufacturing in these relationships.
2. Referring to Fig. 1.1 and to the list of responsibilities of process control, explain why process control is considered part of the manufacturing engineering organization rather than part of quality assurance.
3. An area planner is required to develop a manufacturing facilities plan to produce a new product. List and describe the various steps the planner would take to accomplish this task.
4. Describe the difference between workstation capacity analysis and capability evaluation.
5. The area planner in the role of investments project management engineer is responsible for conceiving, purchasing, and implementing new equipment projects. Develop a generic flowchart showing in chronological sequence the steps the area planner must take to perform this task.
6. Methods engineers are responsible for defining the sequence of operations to be followed in producing parts in a factory. Show in a generic sense the chronological order to be used for the operations, and explain why the order of the sequence is significant.
7. Explain the difference between planning and time standards.
8. What is a variance control program, and why is it important for the success of a manufacturing concern?
9. List and define the types of engineers and trades likely to be found in a maintenance unit of a typical machined parts job shop.

Chapter 2

MANUFACTURING ENGINEERING MANAGEMENT TECHNIQUES

The manufacturing function cannot just exist; it must have a stated reason for being. A manufacturing organization exists to make its company's products as specified in the company's strategic and business plan. But just making the product is not enough; it must make the product so that the company can make a profit. To do less would result in a situation where the firm itself would eventually cease to exist. Therefore stated reasons or justifications for existence must be both generic and specific. It is the definitions and explanations of these generic and specific reasons for existence that lead to the management techniques employed in all manufacturing subfunctions. In this chapter we will thoroughly explore these reasons for existence to understand how they translate into specific management techniques applicable to manufacturing engineering.

THE BUSINESS PLAN CYCLE AND THE MANUFACTURING ENGINEERING ROLE

Manufacturing engineering plays a key role in defining a company's business plan. Let's take a look at how. But first let's see what the business plan cycle is and why every successful company needs one.

The business plan cycle is an annual event that looks at internal and external factors that effects how well the company did in its previous year. Internally it compares results with previously set goals; e.g., what worked well, what didn't, and what can be done to correct under performing areas. Externally it looks at the current market situations to understand how the company's existing products can be positioned for satisfactory sales levels, and more significantly does the company's products still suit the needs of the market. If not what needs to be done to enhance them or replace them. These strategic internal and external questions then become the basis for revisions in the company's objectives and become the basis for the next set of measurable goals.

There is a distinctive pattern for accomplishing the business plan cycle, some say it's a ritualistic procedure which it is if we do not take the time to understand the reasoning for the order of events sequence. All told, a business plan is an "all hands" evolution that takes approximately 30 days to complete due to the combination of series and parallel events. The important thing to note is that no single subfunction should need to spend more than approximately 5 to 10 days doing their part. While even 5 to 10 days of concentrated work is a significant assignment of resources, particularly for smaller firms, it is time well spent and enhances the productive output of the company by at least five fold for the remainder of the year. A well-done business plan becomes a living document for how the company will proceed and have a greater chance for being prosperous in the coming year. The sequence of the business plan is shown in Fig. 2-1.

Note that the sequence starts with a reaffirmation of the reason for existence of the company Via the vision and mission statement. This is followed by a statement of the objectives. (Sometimes objectives of a company become too abstract and require interpretation. How that is done is covered later on in this chapter.) Then the sequence diverts to external factors the company must contend with to be successful, culminating with the development of the sales plan for the next year and usually the next two after that. With the sales plan, the operations and finance functions can begin to do detailed planning, converting sales figures into units of production and costs associated with them. This ultimately

1. Implementation directives and guidelines—basic instructions by the most senior management as to what the overall needs of the company are, the schedule for completing the business plan, and who is assigned to do what.

2. Vision development—what the company stands for.

3. Mission statement—how the company intends to carry out the intent of the vision.

4. Objectives—broad based statements of intent that support the meaning of the vision and mission statement.

5. Situation analysis, market assessment, market segmentation—a pragmatic assessment of the current "world" state of being that sets the stage for how the company will and can compete through implementing the vision, mission, and objectives.

6. Sales plan—a setting of desires, usually numerical ($ and units), to be achieved during the time period of the business plan (typically 3 years) that are compatible with the evaluation of the "world" from step 5. How to achieve the desires is spelled out in detail.

7. Production plan—refining the numerical data of the sales plan into units of production, which are further broken down into materials and labor, by total and per time period. Often a capital equipment purchase plan is appended to the production plan. Comparisons are made between existing labor and forecasted needs, with plans to accomplish.

8. General and administrative plan— staffing for the company, along with necessary training and recruiting plans. All other plans not specifically covered elsewhere.

9. Financial plan—budgets, pro-forma P/L, and balance sheet development per the inputs of all the plans. Also a plan for raising capital and obtaining credit to fund the business plan.

10. Measurable goals—after all plans are completed, the set of goals are developed to specify what needs to be accomplished by who and by when. (This becomes the driver for all projects.)

11. Contingency plan—the "what if?" scenarios if the projects, guesses, and hunches associated with any of the above steps do not happen as forecasted. These are the backup plans, usually done in concept format only.

Figure 2-1. The Chronological Sequence of Business Plan Development

leads to specific production plans, and supporting capital plans; both of which manufacturing engineering plays a significant role in. In fact it could be said that the AME unit of manufacturing engineering and the area planners are the significant players in developing the manufacturing plan. This is true because the nature of manufacturing engineering is planning and implementation of those plans. Now let's look at the key aspects of business planning from the viewpoint of manufacturing engineering and see how this subfunction carries out its important responsibilities.

OBJECTIVES AND GOALS

Objectives and goals as indicated by the business plan define the generic and specific reasons for any organization to exist. They are general policy statements and also statements of specific intent. A system of objectives and goals consists of three basic steps:

Objective: A broad-based generalized statement of intent.
Goal: A measurable statement of specific intent bounded by a specific period of time.
Project: A specific plan with measurable steps that lead to the accomplishment of a goal. One or a set of projects may be required for the accomplishment of a single goal.

Clearly, this system progresses from very broad-based to specific action steps that define singular items to be accomplished. Figure 2-2 illustrates this concept; but notice that the arrows shown go in both directions, from broad to specific and from specific to broad.

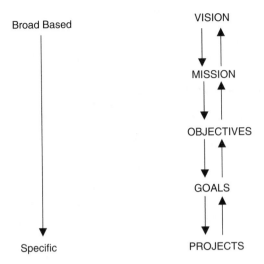

Figure 2-2. Hierarchy of Strategic Planning

Figure 2-2 implies that direction or policy is proclaimed, then actions are taken and the results are reported back to the policy authority. The concept of an objectives and goals system requires feedback of the results to allow management to evaluate the worth of the basic objectives and to determine whether or not goals were accomplished. This feedback also shows whether accomplishment of the goals supports the stated objectives.

An example of an objective would be: "Maximize profits by legal and ethical means." This is a generalized statement of intent. It tells the world that the company's policy is to maximize profits, but only by legal and ethical means. This statement sets the tone for operational procedures. It tells all personnel that they must abide by general precepts of legal and ethical conduct in their pursuit of maximized profits. However, there is nothing measurable about this statement. It does not give specific instructions to management, nor does it define legal or ethical. Therefore, this is a classical objective; it is a broad, general statement of overall policy. Implementation of that policy is left to others.

Another example of an objective is: "Improve productivity." This is an objective that would be internal to a specific function of a company and not necessarily issued by the highest level of management. Such a statement would be issued by the manufacturing function. It meets the definition of an objective. It is broad-based and generic. It simply states that the function will improve its level of productivity, but does not state how, how much, or how it is to be measured.

Objectives can be thought of as statements made by a politician campaigning for office. They sound good and are definitely the right thing to espouse. But once the campaign is over, there is no way to measure the officeholder's performance against the stated objectives. They are too broad and open to too many interpretations. In the business world such intangibles are useless for performance measurement. Therefore all policy statements, or objectives, must be quantified by supporting statements of goals.

Let us look at an example of a goal statement: "Improve profits by 15% for the next budget year compared to the performance of the current budge year." This is very specific: it refers to an amount that can be converted to dollars, and is bounded by a specific time period. It is measurable and allows virtually no room for interpretation. A goal sets a task to be accomplished and states it in such a way that it is easy to determine whether it has been accomplished.

An important characteristic of a goal is that it exists for only a certain clearly defined period of time. Once the deadline is reached, regardless of success or failure, the goal no longer exists and a new one must be proclaimed that fits the objective. The objective, of course, is the same as the one supported by the expired goal.

Goals are periodic, while objectives are ongoing and asymptotic. Figure 2-3 shows the infinity analogy of objectives. If productivity is the ratio of output of goods and services produced to

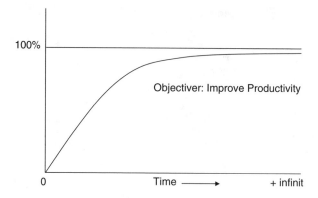

Figure 2-3. The infinity analogy of objectives

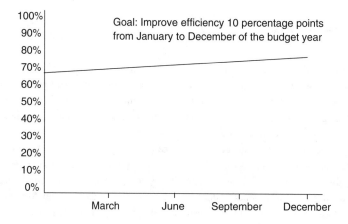

Figure 2-4. The finite analogy of goals

input of capital, material, and labor, 100% productivity corresponds to the limit of unity for this ratio. Since this is tantamount to zero losses in translating all inputs to outputs, it can occur only in a time equal to infinity. That is, an objective can never be totally achieved in practice; it can only be approached asymptotically. Who is to say that the ultimate in productivity improvement has been achieved? It is always possible that a new process or a new machine will be invented that will advance the productivity improvement horizon.

Goals are not asymptotic. They are defined parts of the total objective that allow for measurable change in a defined time period. The concept of a goal is illustrated by the goal accomplishment line in Fig. 2-4. All values to be achieved are real numbers capable of full definition and readily plotted against the goal. The actual efficiency achieved can be determined for the quarterly time periods and entered on the graph. If the actual plot coincides with or is above the goal plot, then the goal has been achieved. If it is below the goal plot at the end of the year, then the goal has not been achieved.

PROJECTS—THE MECHANISM FOR ACHIEVING GOALS

Let us look at some examples of projects that allow achievement of a goal. The goal represented in Fig. 2-4 is to raise the efficiency of the manufacturing function from 70 to 80% by December 31 of the budget year. This is a directive, an order given to the appropriate sub-function of the organization to devise plans for achieving the targeted results. Such plans in the objectives and goals system are called projects. For the efficiency goal a project could be developed to improve the attention time of operators at the workstations. Another project could involve the development of workstation tooling to allow faster feeds and speeds to be employed. A third project might involve the retraining of operators in the specified method. A fourth project could be purchasing more efficient capital equipment. The point here is that many projects may be undertaken to achieve one prescribed goal.

How does the responsible individual know whether a specific project supports a given goal? Conversely, how does the manager know what projects to select to achieve the goal? There are no absolute rule answers to these questions. However, there is an important evaluation technique or logic that can be followed, which I call the rule of project selection pragmatism. The rule is as follows: *The project that costs the least in time, money, and people resources should be selected, provided that it allows achievement of the goal.*

Often no single project activity can be launched that will allow achievement of the goal. When this happens, I recommend implementation of the series corollary to the project selection pragmatism rule. The corollary is: *Select those projects that, when executed in series or parallel, will achieve the stated goal at the lowest possible cost in terms of time, money, and people resources.*

No matter how cost-efficient a project may be, if it does not allow achievement of the goal by itself or in conjunction with other projects, then it is an unnecessary tangent and should not be pursued. Each project must enhance the cause, that is, move the organization closer to ultimate achievement of the goal. Therefore, the following questions should be asked of each project:

1. If the project is implemented, what are the likely results?
2. What is the probability of success?
3. What are the consequences of failure?
4. Will failure deter or cancel other projects?
5. What are the costs of implementing the project?
6. Are there simpler ways of achieving the same results?
7. Will the planned results allow achievement of the specified goal, or partial achievement? If partial, how much?

In practical terms, a project is a series of sequential and perhaps parallel action steps that, when completed, achieve all or part of the requirements for accomplishing the goal. Each step is a single entity with a scheduled start time and a scheduled completion time. For each step there is a simple phrase or sentence that states what is to be done during the time period assigned. Each step also has one person assigned responsibility for its completion. When all steps are completed, the project is completed. Like a goal, a project must be evaluated on completion to determine whether the planned for results have been achieved. This is vitally necessary if the results of one project lead into the start of another.

Figure 2-5 is an example of a project. Note that the elements are listed. The steps are defined in simple phrases. The time to accomplish each step is clearly shown, and the individual responsible for accomplishing each step is identified. There are also other control features such as a number for identification and a method for recording actual starts and completions for each step. What is not included on a project sheet is an explanation of the state of progress. Since a project sheet is a working schedule, it would be cumbersome to include explanations for deviations from schedule. Therefore, it is common practice to reserve deviation reports for periodic progress reports, which may be either written or oral.

Projects are very important to the success of the manufacturing engineering organization. Therefore a few thoughts on how projects can go astray are in order. Failure of projects to produce the desired results is usually caused by one or more of the following.

1. The steps are not specifically defined; they are too general. Therefore it is difficult to know what must actually be done and when the step is complete. For example, the statement

Element No.	Description	Individuals responsible	2005			2006											
			Oct.	Nov.	Dec.	Jan.	Feb.	Mar.	Apr.	May	Jun.	Jul.	Aug.	Sep.	Oct.	Nov.
1	Review history of welding problems	J.A.K.	▲	●												
2	Set up a welding council	J.A.K		Nov◄	△	○										
3	Implementation of welding manual to be prepared by Manufacturing Engineering	P.M.T.		△	▽	○ ○										
4	Set up a system to monitor weld quality	J.A.K		△			○	Mar.│								
5	Monitor weld quality	J.A.K				△										
6	Establish feedback and analysis routine	J.A.K						△	○					○		
7	Prepare and submit final cost savings report	J.A.K														△ ○

Key

△ ———— Scheduled start

△ ▽ Revised scheduled start

▲ Started

│Oct. Start before period

○ Scheduled Completion

○ ○ Revised schedule completion

● Completed

Dec. ▲│ To be completed after period

Review

Review Period (check one)
- ☐ weekly
- ☑ monthly
- ☐ quarterly
- ☐ semi-annual
- ☐ other (specify) _____

Project Title: Establish Weld Audit System   Start date: Oct.12, 2005   Finish date: _____   Project Number: _1_

(insert actual dates)

Figure 2-5. Example: Generic Project Sheet Format

"Examine the QA records for number of failures at the workstation" is very open-ended. Quality assurance records for what type of failure at the workstation? A proper statement for this step could be, "Examine the QA records for the number of failures at the workstation caused by operator error." This is much more specific and focuses on the problem to be evaluated. Examining all failures for all causes may not be relevant to the problem at hand. Another constraint on the element description would be placing a chronological bound on the search—for example, restricting it to a period of 6 months or 10 weeks. This would be helpful if it is suspected that the problem to be solved can be bounded.

2. The allowed time period for accomplishing each step is not observed. The time frame is established to do two things: allow completion of the project on time to support the time frame of the goal, and force conclusions to be drawn for the completion of each step. Frequently, manufacturing engineers will continue an investigation when the return for the effort is not worth the time and expense. Placing a time limit on each step forces evaluations to be made. It is acceptable in this type of work to draw conclusions once two-thirds of the data has been gathered and evaluated. In practical problems involving manufacturing, it is usually impractical to define 100% of the parameters and data. Strict observance of time periods is essential for efficient utilization of manpower. It is most important for those responsible for project completions to be very jealous of conserving time.

3. The project is incompletely sequenced. Either there are not enough steps to complete the project, or each step is too broad, requiring more than one item to be accomplished. For the example in item 1 above, we would not want to find all operator error failures and also evaluate the causes of these errors in one step. A better sequence would be to categorize the causes of the operator errors after the total numbers have been found and placed in the database. This is the preferable procedure because the causes in the different categories would be quantified, and the major effort could be directed to the categories with the largest number of occurrences.

Combining steps usually results in losing the sharp focus of a project. Items may be overlooked, to the future dismay of the engineer. Of course, care must be taken not to be excessively precise. Moot elements must be eliminated. People are capable of interpretation; therefore computer like sequencing is unnecessary. If the key steps are included, as understood by the engineer and the manager, then the fault of step combination will be avoided.

BASICS OF PROJECT MANAGEMENT

(Note: this section is partially based on the" Short Tutorial on Project Management" found in Chapter 3 of my book: *The Engineer Entrepreneur*, ASME Press, New York, 2003.)

One of the most important skills a manufacturing engineer needs to acquire is project management expertise. The definition of project management is:

"The art of managing a non-repeatable multistep assignment that has a beginning and an end point with desired results that are measurable and bounded by time. It is the formalization of tasks into achievable and measurable steps in order to reach a set goal."
(From Daniel T. Koenig, *The Engineer Entrepreneur*, ASME Press, New York, 2003.)

If we look at the tasks assigned to manufacturing engineers, we see that a significant proportion, if not a majority fall into this category. When engineers are given a task to accomplish it would normally have a beginning and end point, the end results being defined as goals to be achieved, and are certainly measurable with a desired time to complete. So this work definitely complies with the definition of a project. Do the work require and finishing on time definitely requires some level of thought and planning before it is undertaken. Project management is a methodology for doing that

Simple Project Steps

		Project Phase	Project Activity
(1)	Concept	Organizing	Work Breakdown Structure
(2)	Planning	Organizing/Assessing	WBS/Risk Analysis & Mgmt
(3)	Execution	Executing	Action Steps
(4)	Close-out	Executing	Action Steps

Complex Project Steps

		Project Phase	Project Activity
(1)	Concept	Organizing	Work Breakdown Structure
(2)	Definition	Organizing	Work Breakdown Structure
(3)	Design	Assessing	Risk Analysis and Management
(4)	Development or Construction	Assessing	Risk Analysis and Management
(5)	Application	Executing	Action Steps
(6)	Post Completion	Executing	Action Steps

Work Breakdown structure:
Answers the following questions
What has to be done in what order?
Who will be doing the specific tasks?
How long should it take to do each specific task?
What's the cost, both financial and in people resources?
[creates "knows out of don't knows"]

Risk Analysis and Management
Evaluates the quality of the "knows".
Points out areas where contingency planning is desireable.
Sets criteria for evaluating the effects of changes in plans.
Establishes probabilities of success.
[calculates the odds for being successful]

Action Steps
Establish schedules for accomplishing the goals of the project
Assign resources - people, equipment, and materials
Do the steps of the project in accordance with the schedule
Measure results, and take corrective actions as required
Report progress and completion to sponsors and other interested persons
[where the rubber meets the road]

Figure 2-6. Project Management Matrix (Based on data from Daniel Koenig, *The Engineer Enterpreneur*, ASME Press, New York, 2003.)

and achieving the desired end points in the most efficient and effective manner possible. Figure 2-6 illustrates the subsets of project management. Let's discuss in detail each of these subset. By understanding them we will have an excellent handle on how to successfully manage projects. In fact the remainder of this book is all about skill sets that would most likely be employed in managing projects. So in order to see how the overall system of manufacturing engineering is approached, we can think of it as an all encompassing project where the end results is optimal performance as an engineer in that subfunction.

Project Development and Application Sequence

The Project Life Cycle consist of two models, the six-phase and four-phase models. Each is essentially the same where as expected the six-phase version is used for more complex project activities than the four phase version.

Six-phase model:
a. concept
b. definition
c. design
d. development or construction
e. application
f. postcompletion

Four-phase model:
a. concept
b. planning
c. execution
d. close-out

If an engineer understands and follows either model the results more than likely will be successful. Problems occur when steps are done out of order or if the concept is used in place of definition (six phase) or planning (four phase). It is always tempting to develop a concept for managing a project without seriously testing it for credibility.

When this happens we have a project based on a foundation of cards. This leads to elaborate projects to build better buggy whips before first determining a need for one. Therefore proper project management and discipline in carrying out the sequenced steps is absolutely necessary.

Problems also occur when the project last step, close-out, is not done. And unfortunately it is often skipped because we're so relieved to have finally finished the tasks of the project that all we want to do is relax and get on with life. This is like rushing to the hotel room after a long journey and forgetting to lock the car door. The car may not be there the next day. Thus lessons learned are not transferred to the next project, and improvements in performance are minimized.

We see from Fig. 2-6 that project steps and phases and activities are linked. The steps of a project are the chronological order of accomplishment. The phases are the managerial work or process we are engaged in to do each step. The activities are the actual tasks we are performing to complete the phase and doing so in the proper chronological order. It is important to understand the project phase we are engaged in so we can emphasize the proper activity to be successful. For example in organizing the emphasis is on planning not necessarily doing things to accomplish a physical task. Here's an analogy to explain what I mean. If our task is to get five boy scouts over a ten foot wall, the emphasis in the organizing phase is considering how we are going to get ten boys over the wall, not actually accomplishing the task. It's developing the plan not executing the plan. The emphasis would shift to an action activity once we get into an execute phase.

Let's look at the three project activities in the order in which they occur.

The Work Breakdown Structure

This is a critical phase of any project. If this task is not done properly the probability of doing the project successfully and on time is severely degraded. This is the intensive planning phase of the project. Here we decide what steps to take and their most efficient sequencing. If I'm required to define the work breakdown structure in catch phrase terms here's what I would say:

"Measure twice, cut once"

Or:

"Aim, aim, aim, then shoot"

These may be fables but they are absolutely true and must be adhered with, with no exceptions. It means always take the time to plan properly, if you want success. Planning a project by specifically defining each step creates a project scope which leads to specifications for resources, e.g., materials and people to accomplish the entire work definition of the project. Add in time to accomplish the task and we have a schedule. Needless to say the more precise the planning is the more accurate the prediction of final results will be for costs and cycle time.

The project scope gives a magnitude and time frame for the project. This has to be filled in with details. The starting point is to visualize the project scope. Formats for visualization of the project scope, the work breakdown structure are:

Organization Chart,as shown in Fig. 2-7
Line indent outline,as shown in Fig. 2-8

Both are useful for identifying the pattern of how work is to be performed. Organization chart style is like a tree and branch structure, where we can see the goes into relationships between steps and also subtasks that make up the steps. The organization chart format is most useful where there are many parallel and serial steps that need to occur in proper sequence. The line indent is similar

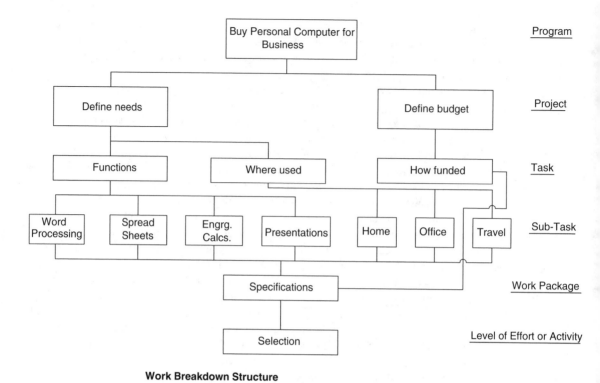

Work Breakdown Structure
example: Organization Style with Levels

Figure 2-7. Organization Chart Work Breakdown Structure

to an outline process for relating items of lesser overall importance to items of more importance. It is used where serial events predominate in accomplishing the goal of the project. The number of levels for the organization chart or line indent can and do vary considerably from project to project. The important point is each definable action that needs to be done be accounted for and that all but the first and last steps need to have both visible predecessor and successor steps. In fact the cardinal rule for work breakdown structure is that there be no loose ends. No steps that don't logically lead to the next step. When I say work breakdown structures have to be precise. I mean it. The more detail the better. There can be no ambiguity as to how a step leads to the next step. Here is an example: it may be frivolous and humorous, but it points out the emphasis of detail planning require for successful project management.

```
Define need.
        Functions
                Word processing
                Spreadsheets
                Presentations
                Engineering calculations
        Where used
                Home
                Office
                Travel
Write Specifications
        Make selection
        Make decision, based on specifications
Define budget
        Make decision—funding a lap top computer
        Make decision—buy?
        Purchase lap top computer
```

Figure 2-8. Line Indent Outline Work Breakdown Structure (Buy Laptop Computer)

EXAMPLE: The project is to supply employees with hamburgers for lunch during a working meeting. Here is a very detailed work breakdown structure to accomplish the project.

1. Assign a staff member to buy the hamburgers.
2. Determine the number of hamburgers to be purchased.
3. Select a source to buy the hamburgers.
4. Determine the approximate cost of hamburgers by telephone to the hamburger emporium.
5. Give assigned staff member cash to cover the estimated expenses
6. Have assigned staff member arrange for transportation to the hamburger emporium, by taking the company automobile.
7. Assigned staff member gets driving directions to hamburger emporium.
8. At noon time the assigned staff member gets company automobile keys from the keeper of the keys.
9. Assigned staff member leaves office, proceeds to parking lot to get company automobile.
10. Assigned staff member unlocks company automobile.
11. Assigned staff member enters company automobile.
12. Assigned staff member drives to hamburger emporium.
13. Assigned staff member parks company automobile in hamburger emporium's parking lot.
14. Assigned staff member locks company automobile.
15. Assigned staff member goes into hamburger emporium.
16. Assigned staff member places order for hamburgers.
17. Assigned staff member receives hamburgers from hamburger emporium.
18. Assigned staff member pays hamburger emporium for the hamburgers and gets receipt.
19. Assigned staff member leaves hamburger emporium.
20. Assigned staff member unlocks company automobile in hamburger emporium's parking lot.
21. Assigned staff member places hamburgers in secure position in company automobile.
22. Assigned staff member enters company automobile.
23. Assigned staff member drives to company office parking lot.
24. Assigned staff member exits company automobile.
25. Assigned staff member retrieves hamburgers from company automobile
26. Assigned staff member locks company automobile.
27. Assigned staff member proceeds to office.
28. Assigned staff member delivers the hamburgers to the person who directed the assigned staff member to buy the hamburgers.
29. Assigned staff member givers the assigner change and receipt for the hamburgers.
30. Assigned staff member returns the company automobile keys to the keeper of the keys.

While this example is trivial to the extreme, it demonstrates that a proper work breakdown structure needs to fully describe to sequence of steps necessary to gain complete coverage of the task to be done. The level of detail required is determined by the level of common knowledge that exists between the person developing the plan and the person(s) doing the actual work. If the latter is very familiar with the process then the detail break down can be more broad in scope. So we can see there will be a sliding scale of necessary detail from project to project.

When developing a work breakdown structure, keep in mind that the action of every step needs to be measurable, hence complex steps lead to more complex and intricate measurements. Also when the work breakdown schedule is converted to a schedule with the addition of step cycle time, the smaller the increments of work, the better chance we have of getting accurate times. This is a prime reason we want small incremental steps to measure, so time estimates for one step can not grossly over or underestimate the real time and resources needed.

Work breakdowns, for example, program to project to task to subtask, etc. should go down to the level sufficient to fully describe the work to be done. There can be multiple items at each level, e.g., outline and indent forms. Make sure each step from highest to lowest leads to the next without any logical breaks. If there are breaks, then a step is missing at the break points. Steps (tasks, sub tasks, etc.) should only be as large as can be properly managed, for example, the extrapolated resources and times can be reasonably certain and not lead to gross overruns.

The work breakdown structure is the initiating phase of any project. The success or failure of a project is dictated by how well it is planned. Paying attention to providing a plan with no discontinuities is absolutely required for a project to be successful. Projects without a foundation of a properly done work breakdown structure rarely succeed. So "Measure twice, cut once" and " Aim, aim, aim, then shoot" are not simply slogans, they are the absolute necessities for project success. I have been asked frequently how much planning is necessary. The answer is as much as possible so when the "go" button is engaged the project proceeds as desired and any deviations can be reacted to so the project plan can still be successfully worked and the goal achieved. In fact it is not unusual for the planning stage of a project to take the majority of the entire cycle time.

The work breakdown structure becomes a schedule when people are assigned and times defined for each step developed. Cycle time is developed for each step identified in the work breakdown structure using techniques as described later in this book. Here are some caveats to keep in mind when developing cycle times.

Calculating cycle time to do each step of a project is a need that many novice project managers ignore through ignorance. Knowing how long a project step should take to do is the key method for tracking how well a project is being executed. Any job segment can have a cycle time assigned to it. This is done by;

1. Using time standards
2. Using estimates based on similar work done previously
3. Using estimates based on a known cycle time for similar type of work

If work cycle time is not applied to a job, the results can be disastrous. The project is then at the mercy of outside forces and will be finished when they want it to be, not necessarily when the project schedule needs it to be done. Therefore any estimate of time to complete a job is better than no estimate.

Along with planning the step sequence and cycle times we need to assign personnel to each step. Who is responsible depends on the skills required for each step. For a project to be successful each step has to be lead by a competent individual with the necessary skills and training to complete the step successfully. It is the project manager's responsibility to create the project team such that each step will be done correctly, hence the choice of people is critical. Usually this is not a problem for manufacturing engineering projects because the staff presumably have been selected for the their skills and education. They all have met the requirements as described in the manufacturing engineering required skills sets. However, projects involving diverse crossfunctional organizations requires the project manager to vet each individual on the team for specific competencies and match them with the needs of the projects defined in the work breakdown structure.

Selecting people for each step is required, but knowing how long they will work and in what sequence is critical. After cycle times have been identified, specifying the people resources sequence is done. We will be look at the total hours required of each skill, profession, etc. necessary to complete the project to develop a listing of the number of people needed. The hours for each skill are put into a matrix form to see when they are needed. This allows us to schedule these resources, since in project management we will have as few permanent staff as possible to minimize staff. We tend to utilize temporary staff to minimize costs. Permanent staff are assigned to a project only if there is sufficient work available within their capabilities for the duration of the project. This means skilled people quite often are doing work on many projects over a course of a work week.

Project planning also requires cost estimates for each step. Many techniques for cost estimating and measuring are covered elsewhere in this book, so not discussed here. However, it should be noted that a project has a budget and it is the project manager's responsibility to monitor and control costs. And in many cases, also develop the project costs for senior management consideration before the project commences. Costs can be calculated for each step or for the entire project, the choice depends on the magnitude of the project

When all of this work is done: sequence steps with no discontinuities layed out, defined personnel requirements to carry out the steps, defined cycle times and costs; we have a plan. We've traveled from a project statement in support of a goal to a set of "how to" steps achieve the goal. We have in essence created "knows out of don't knows." But we're still not ready to launch the project. We need to first evaluate the risks for doing the project steps and develop contingency plans for potential problems and glitches we may encounter.

Risk Analysis and Management

Factoring in risk assessment in evaluating project management choices; using change control and uncertainty principles is a necessary step to assure that the plan developed during the work breakdown phase will be successful. Nothing is more painful and frustrating than to have done the work to create a schedule, assign people and resources, implement the project and find insurmountable roadblocks. Especially if they are surprises and there is no immediate idea as how to overcome them. Doing risk assessment and having plan B, C, D, etc. ready to go when unplanned for happenings occur makes life a lot more secure, not to mention the ability to complete the project more likely. Notice I didn't say surprises happening. Because risk analysis is a process that explores probabilities of success, thus if a step of a project has a lower than acceptable probability of success, we are keyed to have already planned a "what if?" contingency plan. Recall that the last phase of business planning is contingency planning. The same is true for project planning. We use the risk analysis exercise to determine where contingency planning is necessary, then do contingency planning for those riskier steps. Remember a major part of work breakdown structure planning is creating knows from don't knows. Contingency planning is for those steps where knows may not be entirely known.

Most of the time we are looking at risks to the project as it was planned. However, a good deal of risks, that come to be reality, occurs as a result of changes made to a project after initiation. These are caused by less than adequate evaluation of the change before it is approved. This happens primarily due to either:

- Time to plan compression factor
- Dynamic nature of change itself—not enough facts for a proper decision.

For these reasons it is important to understand what is the nature of change and how to control it. Let's look at change and change control as it effects analysis of risk, and then get on to the more general topic of Risks and Contingency planning.

Changes required for projects are the results of three factors. The first is a change in the environment necessitating a change in the goals of a project. Such things as early or late completion of supporting projects, change in direction of overriding strategies, and corrections needed for erroneous work would be examples of changes in environment. A second cause for change in a project's plan for accomplishment is erroneous assumptions made during the planning of the project. The third, and probably least considered reason for a change in project plans is personnel changes requiring re-planning to suite different skill sets. This is analogous to changing play calling for a football team if its new quarterback' prime asset is running ability vs. his predecessor's unsurpassed passing skills.

How do you evaluate whether a change should be made? Change is serious business. It means throwing out countless hours of pre-planning for a new direction and strategy that can not possibly have the same planning effort put into it as the original case. Therefore before a change can be agreed to we must evaluate its impact on the stated goal. However if the change is due to a change in goal, then regardless of its nature, the change will be made.

Here the ability to succeed with a change is not an issue for consideration. Therefore we must differentiate between a change being initiated to achieve the original goal or is it related to a changed goal.

EXAMPLE: the project illustrated previously to buy a personal computer to be used at work, at home, and travelling is for one computer for a special need. What if the sponsor requests to substitute to buy hand held computers for all members of the unit to be used as communication devices, for example, receive e-mail as the primary purpose for the devices with only minimal calculations capability, is this something the project manager should acquiesce to? Yes, because it is a change in goal not a change in procedure or methodology of achieving the goal. There is no need to make any evaluation. The sponsor has changed the goal. Therefore the change will be made.

Usually, change requests are generated for non-goal changing reasons, such as a previously planned activity is not working out as anticipated. When such change requests are made, the project manager has to evaluate the impact of the change before agreeing to it. Here are some items to evaluate before a non-goal change request should be accepted.

- How much will the change effect the project?
  - Scope
  - Cost
  - Schedule
  - Capital plan
- Will it effect adjacent, predecessor, or successor projects?
- Will it change the financial justifications? If so, are they still positive for doing the project?
- What is the justification for the change?
  - This implies a test for validity of the request.
    - It is proper only if the organization requesting the change is not higher in the chain of command
    - It is also good management practice to tell subordinates the reasons for changes

The answer for all of the evaluation questions should be positive (a better result will occur—e.g., a better chance for the goal to be achieved, or for the goal to be achieved cheaper and faster); or at least neutral. Remember—change increases risk—so the change should be more beneficial than the original plan to justify the additional risk.

It is good practice to document change requests, to have an audit trail of the project. A typical change control form is as we see in Fig. 2-9

Note that the form describes the nature of the change, the reason, then the effects on primary project considerations. Finally space for approval signatures. This last bit of bureaucracy is very necessary to ensure all members of the team involved as stakeholders in the ultimate outcome of the project agree to the change and have had the opportunity to question the wisdom of the change. In effect it creates a history of the change's genesis and why it was agreed to.

Risk and Uncertainty

As I said change creates a certain degree of risk, queasiness about the decision that's been made, and definitely some degree of uncertainty. We often use risk and uncertainty interchangeably. But as you would surmise, in engineering usages they have two different meanings. The precise definitions are:

From N. Barash & S. Kaplan, *Economic Analysis for Engineering and Managerial Decision Making,* 2nd edition, McGraw—Hill, New York, 1978.

Risk :
    A variance from average (expected) value which occurs in random chance patterns. The larger the variance the larger the risk.

Uncertainty:
    The degree of unknown caused by errors in forecasting one or more of the factors significant in determining the future values of the variables making up the variances.

Project Change Request		
Project Name:	Project Number:	Date:
Change Request No.:_____ Requested by:	 Department	Change in: (check box)
Description of requested change:		☐ scope ☐ schedule ☐ performance ☐ budget

Reason for change:

Effect on schedue:

Effect on cost:

Effect on performance:

Effect on scope:

Justification:
   non-financial:

   financial:

Approvals
   Project Manager: _____

   Customer Representative: _____

Figure 2-9.   Project Change Request Form

It is usually extremely difficult to calculate a true risk probability because we can not be sure we can foresee all the potential happenings. Statisticians make a handsome living trying to ascertain risk probabilities and it usually takes considerable effort and time to do so. Unfortunately project management is usually short on time, so other more gross but useful methods have been devised to place a relative value on the risk and uncertainty of a project being completed successfully. Let's look at some of the more practical methods that are useful for manufacturing engineering projects.

When it is impractical or impossible to use facts to calculate a risk, we need to use some sort of comparison table to help us set a value for the risk. We have all seen them many times, for example, "a certainty, most likely, somewhat likely, either likely or unlikely, somewhat unlikely, most unlikely, will not happen". The task is for the project manager to evaluate each step of the project against this arbitrary ranking system. Sort of like ranking the importance of tasks for setting priorities. This is a way to associate a broad generality of possible occurrences with some sort

Description of probability of occurrence   **Ranking**

A Certainty: It will happen, or very little chance the event will not happen.	10 9

Most Likely: Very high probability the event will happen.	8

Somewhat Likely: Should happen but a good chance it will not.	7 6

Either Likely or Unlikely Either event is at least as probable.	5

Somewhat Unlikely: The event could happen, but highier probability it will not.	4 3

Most Unlikely: Very high probability the event will not happen.	2

Will Not Happen: Very little chance the event will happen, or event will not happen.	1 0

Figure 2-10.   Probability of Occurrence Table

of explanatory verbiage. The results are usually a ranking as we see in this typical probability of occurrence table, Fig. 2-10

It is desirable to convert this ranking to a percentage. We can do this in a macro manner by simply multiplying by ten to get a percentage number. The range then would be from 10 to 100%. in 10% point intervals. This is as fine as we can get, and it is sufficient. To divide any further than that would indicate precision that doesn't exist.

Uncertainty will also influence the decision as to whether or not a project or a specific step in a project should be pursued or changed. Uncertainty looks at the degree of unknown in calculating the variance, which modifies the risk factor. It is a fact that change does increase uncertainty. The investigation for magnitude of change with respect to previous conditions has to be determined.

Calculating a value for Uncertainty is difficult. It is a complex probability analysis that is dependent on the number of variables chosen. Therefore in practice it is seldom done. The pragmatic approach is to put an arbitrary factor onto the risk factor to account for uncertainty. If there is no change in the environment, then the uncertainty factor is 1.0. If there is change in the environment (recall- change control form) then a value less than 1.0 is used. Here are some common guidelines for selecting an uncertainty factor.

- Minimal = 0.9 Uncertainty factor
- Paradigm shift = 0.1 Uncertainty factor
- Everything else is a judgment in between

There are many tables and graphs we could use for quantifying the judgments. However, they tend to be built on flimsy foundations and usually are nothing more than ratios of assumptions.

Therefore, I can not recommend their use for anything but gross estimates.

It is better for the project team to debate the various risks and uncertainties.

- List what is known and unknown.
- Make value assessments that meet the specific facts of the situation.
- Use this assessment to make a best guess estimate of the Uncertainty value.

Use this relationship to define a risk number:

Pragmatic Risk Probability = (Calculated Risk Probability)(Uncertainty Factor)

The pragmatic risk is the probability of occurrence for the risk that the project manager would use in his or her assessment decision.

Severity of the Risk

For project management we need to do more than just calculate the probability of a risk becoming an actuality. We need to determine how important the risk is to the success of the project. This is called Severity.

EXAMPLE: for a drill, the metal twists coming off the drill makes it difficult to keep it looking clean and new. The risk of the paint being abraded away and looking greasy due to oil is very high. Is this very important? No, it is cosmetic. The machine is designed to operate with metal shavings, etc. present and without paint. It will not corrode because of the oil mist it is typically covered with. Therefore the severity is low.

Severity is judged similarly to probability of occurrence of a risk. See the Severity of Effect table for a severity ranking and calculation chart as shown in Fig. 2-11. The same general comments about probability charts apply.

Detectability of a Risk

Reviewing the comments in Fig. 2-11 we can see that risks that cannot be detected pose more problems than those that can. For example: it's obvious that running a machine without oil will result in destruction of the machine. The simple preventive measures of monitoring oil levels periodically can virtually eliminate the risk from occurring. The ability to detect a risk and planning to do so can mitigate the probability of very distasteful events from happening. If the ability to detect a low oil level is taken away the probability of a seizure goes up.

So, we should be cognizant that detection can be vital in achieving a goal. Like probability and severity we can create a gross estimate of the importance of the risk to the project by using a similar rating table, but it's hardly worth the effort. It is just as effective to create your own scale of 1 to 10 with 10 being very high an 1 being very low for probability of detection estimates. Since this is so very subjective any reasonable assessment on the part of the person doing the estimating is going to be as good as any other.

Risk Probability Number

For ease of making relative ratings between different risks, project managers have traditionally combined the output of all the various tables and estimates to create one number per risk.

Risk Probability No. = (Pragmatic Risk Probability)(Severity Factor) (Detection Factor)

Setting Up Contingencies to Mitigate Risks

There are ways to mitigate and even eliminate risk. They all start with an internal assessment of the project teams resources with reference to its:

Description of severity effect   **Ranking**

A Certainty to Cause Severe Effects: It will happen without warning, or with warning, virtually unrecoverable.	10 9

Most Likely to Cause Severe Effects: Will be difficult to recover from. Take Lots of work with no guarantee of success.	8

Somewhat Likely to Cause Severe Effects: Will be difficult to recover from,but effects can be mitigated with major effort. Or effects can be mitigated with minor effort.	7 6

Either Likely or Unlikely to Cause Severe Effects: Either event requires major work to set right.	5

Somewhat Unlikely to Cause Severe Effects: Will be easy to recover from,and effects can be mitigated with major effort. Or effects can be mitigated with minor efforts.	4 3

Most Unlikely to Cause Severe Effects: Will be easy to recover from. Take Little work with high guarantee of success	2

Will Not Cause Severe Effects: Very little chance the event will cause severe effects, or event will cause no severe effects.	1 0

Figure 2-11. Severity of Effect on Project

- Strengths—what are the assets that can be employed to achieve the goals of the project?
- Weaknesses—what are the lack of capabilities that need to be protected from having to deploy?
- Opportunities—what are the areas of the project plan that can be exploited with the team's strength to optimize reaching the goal.
- Threats—what are the areas of the project plan that might require using weaker project team resource to achieve, and how can they be avoided?

The strength, weaknesses, opportunities, and threats (SWOT) evaluation is used to determine what mitigation strategy can be used. There are three generally accepted risk mitigation strategies:

- Risk avoidance
  - Used when SWOT Weakness is the dominant match up with the risk.
- Risk minimization
  - Used when SWOT Strengths is the dominant match up with the risk.
- Risk transfer
  - Used when no tactic is viable to mitigate the risk.

Risk mitigation is based on the common sense strategy of the project team not going out of its' way to find a problem coupled with a firm knowledge of the team's limitations. This means keeping the project planning within the capabilities of the organization. Never plan to do an action beyond the existing capabilities of the team. If it is necessary to go beyond the organization's capabilities, then secure the added capabilities as part of the planning process. Always apply the best resources available to every step in the process—all steps are equally important.

Contingency planning is popular in theses about business strategies and can really be done, but unfortunately after going through the rigorous process of defining the work breakdown structure and ensuing schedule many project managers tend to run out of energy or time to do contingency planning. This is a pity. Many managers take the easy way out and say "we'll wing it if a problem happens." Project teams too often succumb to the temptations to stop once the plan is thought to be complete. Too bad they don't realize that failure to do contingency planning is part of the plan, not an afterthought or a nicety to do to put "pretty" bells and whistles onto the documentation. So please consider, contingency planning is required, and here are some useful tricks of the trade to make them easy to conceive and have ready in case disaster strikes.

The first step a project manager must take is to avoid unacceptable risks. If he or she is truly successful in this aspect of project planning then contingency planning becomes nothing more than successful monitoring of the project steps and getting the actors on stage to say their lines at the required time. However, all risks are not avoidable. Businesses make large profits from strategies that accept controllable risks. For example, building a new factory based on projection of market growth is a case of taking a controlled risk. Here the projection is not "a sure thing." It is the best possible analysis of a situation. There is always a probability that the projection will not occur as forecasted. In fact forecasting the future, which all businesses have to do (business planning), is never a certainty. Businesses have to take some amount of risks. But they are controlled. In this case the controlled risk would have as part of the planning the question as to what to do if the market projections are wrong, both positive and negative. Perhaps here the contingency plan would be to build the new factory such that it can accommodate a hole host of the company's products with the idea being that at least some product lines could effectively use the space if need be, thus mitigating the risk that it would less than totally used or stand empty.

Since all risks are not avoidable we need to plan how to deal with risks that become reality. We do that with formalized contingency planning. Contingency planning answers the "What if" question. Theoretically, for every identified risk we create a contingency plan.

Supposedly the Pentagon does that for every conceived potential threat. But most companies lack the resources of the United States Department of Defense; so we have to pick and chose the risks we will apply the effort to create contingency planning for. We need to have a means of adequately selecting risks we will have contingency plans for. This is done by:

- Ranking the calculated risk probability numbers from highest to lowest.
- Establish a cutoff level that represents the highest probability of risk the company can tolerate without contingency plans.
- With the remaining ranked risk probability numbers group for commonalities where feasible.
- Establish auxiliary projects to develop group contingency plans.

I must to point out that this is a subjective evaluation for need for contingency planning. It is simply a logical way to proceed based on the level of risk becoming reality based on probability assigned.

Defining Project Success or Failure Related to Risks

Contingency plans are means of recovering from problems that have occurred. We really don't want to have to implement contingency planning. We want the project to succeed as planned. Perhaps a good way to put this in perspective is to list probable causes of success. This list can provide a good check list for avoiding failure. Let's look at those items that help a project succeed.

Emphasize teamwork. The more people feel as if they're part of a team and will reap rewards for successful conclusion of a project, the greater the chance for it to be successful. We want to create as much "ownership" for the project as we can amongst those doing the project's work.

Give the project manager authority as well as responsibility. Feeling a sense of responsibility is necessary. But to do something positive with it requires the authority to act. By giving authority management is also expressing confidence in the project manager's ability. And people tend to rise to expectation levels.

Work to continuously improve problem resolution skills. This is a learned trait, so we can improve performance by providing continuous training opportunities.

Emphasize "Communications Excellence." Thoroughly understanding each other is paramount in minimizing "assumption" mistakes. Remember the old saying:

"Those that assume make an ass out of u and me"

Set the project organization structure compatible with the work to be done. Make sure people are assigned to tasks that they are equipped and trained to do. Nothing is more demoralizing than trying to do a task that you know you do not have the level of experience or competence to do.

Use the tools:

Work Breakdown Structure
Critical Path Method/Program Evaluation and Review Technique (CPM/PERT)
Financial Analysis Evaluations

Make sure the projects support the identified goal. This seems so obvious but often over looked. The first thing any project manager and project team need to do is to evaluate what they're about to do will really achieve the intended goal. This starts with first being total knowledgable as to aspects and nuances of the goal, then making sure the work breakdown structure and ensuing schedule will get them there.

Train the project manager and team members in all skills necessary to succeed. If a team needs skills to achieve its goal then they need to acquire them. There are two ways to do this. Aquire the skills through adding the necessary skilled persons to the team. Or set about learning the skills. Often the latter is the only viable alternative. If that's the case it is quite all right to added skills learning as steps in the work breakdown structure.

Risk and its companion uncertainty define the probability of unplanned happening occurring during the course of project work. Defining the probability of risks becoming reality and then how to deal with them is the premier strategy of choice to defend against failure. So, contingency planning—how to set plans in place to recover from risks becoming reality—is the choice method employed for high-risk mitigation. Finally, we put it all in perspective by recognizing that doing good project planning and implementation is the best defense against having to implement contingency planning.

Action Steps

Finally! We've reached the point where determining what we're going to do and fretting about what happens if things do not work out as we'd like them to is over. We're ready to implement. So to speak we're launching the endeavour and pushing on to the goal. The time for heavy emphasis on planning is over, the time for the focus to shift primarily to doing is beginning.

Establishing the Schedule

The first order of business is make sure we have the schedule and the project team understands the entire plan. People quite often get confused between a schedule and the work breakdown structure. Obviously they're very closely related. You can't have a schedule without a work breakdown structure. Basically the schedule is the "complete" list of what everyone will do that has been vetted against all constraints. In essence modified to take into account realities of competing needs for resources and even the ability to obtain the first choice resources. It also has taken into account risk mitigation strategies as best as possible, and perhaps built in purposeful hold points where reviews can me made to see if any of the contingency plans need to be engaged. In essence we can say the work breakdown structure gives us:

- The sequence of events
- The time to do each event
- The resources to do the events

Scheduling puts it all together in one cohesive document howing the relationships of all 3 items as effected by the environment. But only the environment known at the time of creating the schedule. The environment consists of (but not limited to):

- The amount of time the company will work during the project's duration
- Known personal needs for time off of the people performing the work.
- The performance/ competence level of the people doing the work.
- The probability of required materials and other needed support being available when needed.
- Other major tasks that will have to be done during the life cycle of the project that will delay work on it.
- Project deadline constraints

A set of very successful "rules of thumb" to apply in scheduling projects are:

- Use the work breakdown structure to define the step dependencies and expected cycle times.
- Use a flow chart to demonstrate the relationships between steps. This is the best way to visualize parallel steps, if/then, and yes/no situations.
- First cut schedule—work from due date backwards to start date
- Use time estimates from the work breakdown structure. This validates feasibility.
- Find the critical path and make sure those items stay on schedule. (This technique will be describe in a later chapter.).
- Never use more than 75% of available person time to calculate a step cycle time. There will always be unforetold interferences preventing people from working on the project.
- Actively look for ways to do steps in parallel rather than series.
- For steps that normally follow one another, start portions of the follow-up step as soon as data from the predecessor step is available. Try to overlap as much as possible.
- Front end load as much as possible. Start quickly and keep pressure on to complete the first few steps ahead of schedule. This helps to create some more reserve time for future unknown bottlenecks.
- Order all materials and services required in the beginning of the project regardless of the true lead time to get delivery. This protects against unforeseen delays.
- Do not allow procrastination to dominate. Insist that steps get started on time.
- Remember: a schedule is a dynamic document and changes as new facts and considerations come to light. Therefore the scheduler has to be flexible.

THE GOLDEN RULE OF PROJECT SCHEDULING: *"DO IT SOONER RATHER THAN LATER."*

Implementing the project

Once more: recall that a project consists of three stages:

- Planning
- Executing
- Reporting And Close Out

We have completed the planning stage, now we have to execute. We need to implement the plan to achieve the goal of the project. The major concerns for managing projects is controlling costs and maintaining schedule. This means a prime focus of the project manager is to measure progress and be able to interpret the results of the measurements. Measurements are evaluated against baselines. If the evaluation variance is sufficiently large then, for example, implementing contingency plans if they exist for the specific deviation.

As I've emphasized before, A well planned project has a better chance of achieving its goals than one that lacks proper preparation. Schedule achievement and financial control expectations need to be pre-planned. If this is done correctly, then management of a project is more a matter of being aware of and reacting to factors outside the scope of the original planning. Here are some typical examples of factors outside the scope of the original planning.

- An unexpected mechanical failure.
- A key participant leaves the company.
- A technology that was suppose to perform to a certain level does not.
- A prime supplier can not deliver due to a labor strike.

We could say each of these could have been covered in contingency planning, and that may be correct to some extent. However, these are items that tend to be beyond the control of the project manager. Therefore they have to be monitored for. We can do this properly only if we have the proper attitude for project management. We should adhere to the philosophy:

"THOSE THAT SHOOT BEFORE AIMING RARELY HIT THE TARGET"

Here are some guidelines for effective implementation.

Have a well-defined and detailed project schedule where all steps per the work breakdown schedule are accounted for. All start dates and finish dates for all steps shown and the individuals responsible for each step are shown—principle and support personnel.

Have a budget established for the project before implementation. Make sure the budget is reasonable for the work to be accomplished. Break up the budget into sub-budgets for all personnel, for the tasks they are responsible for.

Hold a project initiation meeting. Go over the following (even if it is a review for most attendees)

- Reason for the project.
- What constitutes success, for example, what are the deliverables.
- Identify the key participants and explain what they are responsible for.
- Require that all participants be cognizant of the predecessor and successor steps to their responsibility steps.
- Require participants to start their steps on time.
- Require participants to report progress and problems on a timely basis.
- Require participants to attend scheduled progress review meetings. When conflicts arise notify the project manager and arrange for a substitute to attend for them.
- Remind participants that the project is a team effort and they should be prepared to help fellow participants in any manner necessary to achieve successful conclusion of the project.
- Distribute all pertinent information to the participants

Formally initiate the project with a memo to all participants, sponsors, and other interested personnel. This is an important part of establishing communications excellence. In many organizations project team participation is a part time thing. Therefore in order to get people fully committed to doing their assigned parts on time it is necessary to "formally" notify their supervisors that their personnel are working on this project which has now started. This helps prevent conflicts of time for participants to do their work.

Hold scheduled progress review meetings. Distribute minutes of each review meeting to all participants with action items specified with:
1. Completion date required.
2. Who is responsible for the action.

Keep the sponsor and other interested personnel informed about project progressstatus. Usually through distribution of the project review meeting minutes.

Monitor work of participants with scheduled one on one meetings. This works best when the meeting is only minimally structured. A good format I've had lots of success with is; first half of meeting the subordinate relates his progress and problems (along with proposed solution), and second half of meeting the project manager responds to the subordinate and brings up management concerns and suggestions for improvements (directives if necessary).

Monitor cash flow. Set up authorization levels for spending money at the initiation of the project in line with the project budget.

How to Make Critical Reviews of Progress

As with any activity, the only way to tell whether progress is being made is to compare the methodology of achieving the desired end point with the current level of accomplishment. Most cases this is an informal assessment. For example, when cutting the lawn on a Saturday morning we can certainly determine how well we're doing by observing how much more lawn there is to cut and comparing it with the time it has taken to get this far versus our experiences of having cut the lawn previously. Very simply we discern our rate of progress against where we were, say, last week. We're either satisfied with the results or not.

Project reviews are much the same, except they are based on a detailed schedule meticulously developed. Let's look at project reviews and see what the requirements are for them to be effective tools for project management. There are three generally accepted types of project reviews:

- Status
- Design
- Process

I should point out that project reviews share some commonality with reviews of on-going work processes that would not be considered as project work. On-going work is defined as steady state work, such as running a production line in a factory. The same types of procedures are followed for both on-going and project reviews. However, we can think of some reviews being more project than on-going, and vice versa. The point is we must be aware that techniques used for measuring and controlling project are quite often similar and even identical to those used for any activity requiring controls. The chart "Comparison of Reviews—Where Used" Fig. 2-12 demonstrates the combined usage.

Finally, before getting into specific details, we should understand that the basis for all project work is the scientific method. Any goal oriented activity employs the scientific method of investigation, proposing a solution, trying the solution, and iterating test results until a satisfactory conclusion results. So it is not surprising that critiques of project progress involves the scientific method. The scientific method is illustrated with an example in Fig. 2-13. (A detailed explanation of the scientific method will be presented later in this text in Chapter 7 on Methods, Planning, and Work Measurement.)

Project Status Review. This review tends to be a scheduled activity, with the period reoccurring daily, weekly, monthly, etc. The establishment of this review is normally done as part of the project planning step. Participants represent all functions of the company having an interest in the

Type of review	Project Activity	On-Going Activity	Primary	Secondary	Comments
Status	x	x	On-going	Project	All activities require status reviews. There are many more on-going than project tasks in any organization.
Design	x	x	Project	On-going	Design reviews occur at intervals when milestones are reached, and are usually infrequent happenings—hence project dominates.
Process	x	x	Both	Both	Process analysis for performance is required for any management activity.

Figure 2-12.   Comparison of Reviews—Where Used

successful conclusion of the project. The review has a format based on:

$$P = \text{performance level}$$
$$C = \text{cost constraints}$$
$$T = \text{Time management}$$
$$S = \text{Scope of work}$$

The script for the review is typically as follows:

- Review steps (tasks, subtasks, etc.)
- Schedule Compliance—actual vs. plan.
  - Cost compliance (where applicable)—actual vs. budget.
  - Problems encountered—severity, solution known/unknown, plan to eliminate.
- Agree on action steps to recover from any negative variances.
- Previous action variances would be reviewed within the scheduled discussion above.
- Minutes are issued showing the current status of the project with a list of open variances.
  - Correction action items assigned.

Correction actions become the basis of the next period's work activity along with the pre-planned project schedule steps due to be worked on. At times the open variance correction items overwhelm the normal work of the project. When this happens—the project is getting out of control and an entire reassessment may be required. It is an indication that the project was not planned properly or completely.

Design Review. This is a specialized form of a status review. Here, the task is very focused on a review using the Scientific Method. Designs reviews are programmed invention controls. There is a definitive technical end point to be achieved. As we can see in the simple example, the application of the scientific method is very direct.

Example:
  a. A problem is defined—*an observation*
  b. The design is proposed—*a hypothesis*

An engineer is assigned a project to find the best way to manufacturing a new design in the factory. He is given a specific cycle time goal to achieve (make observations).

Steps1 & 2

The engineer devises a sequence of events, the method, to produce the product. The method is turned over to the planning specialist to create a detailed plan with times to do each step (develop a hypothesis).

Then the plan is given to Shop Ops to try with the engineer observing the process and determining if it is adequate. (test the hypothesis).

Steps3 & 4

Having carefully observed the prototype production, the engineer modifies the method by factoring in results of the first test (make revisions to the hypothesis based on results of the test).

Once the revisions are written into the planning document, the test is done again with the engineer in attendance to observe (test the revised hypothesis).

Steps5 & 6

If the revised test meets the goal of the cycle time set for producing the product, the project concludes. If not the whole cycle is reiterated again until a satisfactory conclusion is reached. (reiterate as many times as necessary to reach an acceptable conclusion).

Figure 2-13.   Example of the Scientific Method

    c. Calculations are done—*test the hypothesis*
    d. Revised calculations done based on the test—*a revised hypothesis*
    e. Prototypes are made and evaluated—*test the revised hypothesis*
    f. Decisions are made as to suitability of the prototype for solving the problem—*iterate hypothesis/test cycle to reach a workable* conclusion.

For each item a–f we go through the same review format as with a status review, except the content is specifically focused to achieving the intent of the design.

There are two types of design review projects: new product or process design and the producibility design review. The new product or process design review tends to be pre-production oriented and primarily involves the design engineering function. It is most often aimed at solving the scientific/technical problem and only secondarily focused on economic viability. The producibility design review on the other hand is dominated with production issues. It is primarily a review process used by manufacturing engineering aimed at assuring the design is economically viable to produce.

New product or process design reviews frequently only have design engineers in attendance. It is mostly an internal exercise until the design is ready to be presented to company management for further consideration; e.g., sell and put into production.

Producibility design reviews are cross-functional activities with the main purpose being to test for viability for commercialization. The project manager normally assigns an engineer—usually out of manufacturing engineering (or someone filling that role) as the lead person for the review. While the process employed is based on the scientific method, it has been refined to allow the project manager to answer the question:

"Can the design be released in its present format into to the factory for economic production?"

In the chapter on producibility engineering we will review this technique in considerable depth, and see how the question is answered.

Process Review. The process review is basically an internal review of how well the project team is doing. The status and project reviews focus on the performance of the project in its journey toward accomplishing The assigned goal. Process reviews focus on how well the team is doing its job. It looks at the techniques of doing the job. It analyzes whether the proper managerial approaches in accomplishing the tasks are being employed.

Status and project reviews are externally focused. The only thing that counts is whether or not the goals are being met. The way which the goals are being achieved is only a secondary concern. The Process review concerns itself with techniques because it is generally agreed that proper technique will lead to a higher percentage project goal achievement on time and to budget, thus the ability to accomplish more with the same amount of resources.

A process review can be thought of as a performance review for a team vs. the traditional performance review for an individual by his or her superior in the organization. A good process review would have the following elements. There could be more, based on specific project situation. So we can say that this list is the generic version suitable for virtually any project.

- Does the project statement sufficiently define the goal to be achieved?
- Are proper management techniques being used to identify the options in achieving the goal? Explain what is being used and why.
- Are the project steps laid out in a manner showing a logical sequence of events?
- Are resources needed to complete each step being accounted for during the planning stage of the project? Personnel? Materials? Funding? Time?
- Are resources available when they are to be applied?
- Are personnel qualified/trained to do the assigned jobs?
- Is the schedule published and adequately explained to all personnel?
- Is there a mechanism for reporting back status and reporting interim results?
- Is there a mechanism for handling variances to get back on schedule?
- Is there a communications system set up to make periodic reports to the team and to all other interested parties?

This becomes a very pragmatic check list for the project manager to be constantly asking him or herself to make sure the project is being conducted in the most efficient manner possible. You can say the project manager should ask him or herself this set of question when he or she gets up in the morning and also retiring at night. Get the picture? These are the fundamentals of project management and the project manager ought to have a satisfactory answer for each question. If she does she is in very good shape vis-a-vis completing the project on time and achieving the goal.

The process review does not occur as frequently as a status or project review. Process reviews tend to occur at the start and conclusion of a project. In fact they are major components of the project close out procedures. But as discussed above the project manager has to be constantly performing a "silent-mental process review within himself or herself" in order to stay on top of the project and be successful.

A process review at the beginning of a project is done to make sure all the elements are in place to be able to accomplish the goal before the project is started. Another process review is done at the end of a project as a critique to evaluate how well it was managed, and what lessons can be learned and applied to future projects.

Setting up Measurements that Will Really Happen and Be Useful

So far we've looked at the types of project reviews. We now have an understanding of how they're set up, some insight as why they're used, and an implied knowledge that the review is a management tool for assuring that goals are being achieved. let's now look at how we can get some firm data as to whether or not the project is doing as it should—accomplishing a goal. This is really what we want to know. Here's where appropriate measurement become very useful. This means we must have definitive measurements that tell us if we're achieving of the goal. The measurements have to be related to a performance standard to know whether we're doing good, bad, of mediocre. It's very possible to do this as we shall see.

Project reviews give us broad based measurements about:

- Time—step completion actual versus schedule

- Cost—$ spent versus plan to spend, usually calibrated by the project calendar
- Assessment—whether the project will accomplish the goal.

But they need to tied to definitive measurements to be useful as an assessment tool.

EXAMPLE: A design calls for development of a 50,000-lb thrust rocket engine. Actual test is showing only 49,000-lb thrust. We know we've missed the mark, but this measurement is too broad. It doesn't even begin to assess why. It is too broad base of a measurement for control and improvement purposes. We can only solve our deficiency in engine thrust if we thoroughly understand each and every input leading up to the test. In this case we would likely have had to measure design calculations, then understand why variances occurred, so we can plan to do something about them. Perhaps measurements could have been used in the design phase to evaluate accuracy of answers. Perhaps to get ranges of values and compare to those that give acceptable results versus those that do not. Perhaps contingency planning for the design could have occurred if the range of answers were too broad, and at one end of the range, an under thrust condition would result. The point is that measurements have to be tied to the work breakdown structure such that unsatisfactory interim results can be identified and something done about it.

We can set up very specific measurements, for each step of the work breakdown structure if we thought we needed that type of data. However, we've got to be careful to not fool ourselves as to what the measurement, the numbers, are telling us. For us to really know how well we're doing we need to measure ourselves against some sort of objective standard. The following example shows how we can fool ourselves into thinking we've accomplished the goal of the project, but really didn't because an objective standard was not employed.

EXAMPLE: Our company needs to have a consistent way of having packages delivered to our plant in Europe. We set up a project to do so. The project manager Decides that a 2 week cycle to prepare the package, arrange for shipping, and having it received in the European plant is the goal. The project is set up correctly and accomplishes all the steps. Packages are now arriving in Europe 2 weeks after decision to ship is made. A successful project, right? Wrong. It is only relatively successful. It met the goal. However: the goal was wrong to start with. The project manager failed to determine the objective standard to measure success against. A relatively easy bit of research would have revealed that overnight shipment from the United States to Europe is becoming commonplace. Overnight shipment should have been the proper goal, because it is the objective standard.

The example points out that having a goal and meeting it while good is not necessarily sufficient. It is important for the goal to be set against an objective standard that is accepted as the normal performance rate for the particular topic. We can borrow from industrial engineering work measurement philosophy to set objective measurements. In work measurements we use the concept of scientific time standards to determine how much time a process should take to accomplish. (More on this later in Chapter 7 on methods, planning, and work measurements.) We need to understand the environment where the work is going to be done. Then understand the method to be followed. With this done we can determine the time to do a project step by comparing to a known standard, either direct or indirect. An example of a direct measurement where objective values for world class performance exist is key strokes/minute for data collection. Indirect objective standards for measurement are a little more complicate, but certainly doable. Here's an example of how to obtain an indirect objective measurement standard:

EXAMPLE: Installing chain link fencing around a building—know the time rate for digging fence post holes, know the time rate for attaching fence materials to the post, then we can estimate a rate for unrolling fencing material between posts. We can then compute a time it should take to do the job. Because we can get certain facts and extrapolate to a similar situation, the one we're interested in.

Setting a project time standard is very important. Many novis project managers do not realize how important it is. It is so very vital to measure against objective standards for the type of work engaged in because this is usually the "world-class" standard. If a project team can do well, measurement wise, against an objective standard it can be assured it is giving good value for the resources expended.

While the project time standard is not as accurate as a factory process time standard (usually because all the conditions can not be pre-identified), it is better than a subjective standard, and should also be the first choice. Even if it takes more effort to define it. A subjective standard is often nothing more than saying the project needs to be completed in a week, so make sure the steps fit within that time frame. This is more of a wish than a standard. It is wasteful and squanders resources by arbitrarily matching a time constraint with the work breakdown structure. The usual result is missed deadline dates because there is no relevance between the actions required and the measurement. So standards need to be set based on objective data, even if an indirect standard approach is required.

Project Close-out

This is the last phase of the project. The purpose of the close out is to tie up all loose ends and ensure that the project is really completed. By completed we mean all work is done as compared to the plan. Projects are completed when all the steps in the plan are successfully finished and the only way to tell that is to perform a process review. Since process reviews are part of the implementation plan, and done on a scheduled basis, the pending completion of a project is known well in advance.

Comparison of project plan to actual results will define when the project is complete. Sometimes the plan vs. actual have variances. The reason for the variance has to be known as well as whether an attempt needs to be made to correct it before a project can be closed.

Another reason for doing a project close out is to look for ways to improve future performance. Learning how to manage projects with greater efficiently and effectiveness is a virtue that brings good results to you and the sponsor. By applying what went right and correcting what went wrong in the performance of a project, we can successfully reduce the costs, shorten the time frame of future projects. In order to do a Lessons Learned exercise we need to:

- Review the goals set for the project
- Compare real deliverables to plan deliverables

With these facts in mind we ask the following questions.

- What was done really well?
- What could have been done better?
- What did we do poorly?
- What would you do differently if you could do it again?
- Did we understand the true needs of the project?
- Did the process reviews point out problems on time and in sufficient detail?
- Did we have the resources we needed?
  - Skills
  - Materials
  - Support
- Was the budget sufficient?

Sometimes in order to get data for these questions a feedback survey sent to the participants and the sponsor is helpful. When doing this use an easy to fill out form, otherwise busy people will probably not respond. This is not like a survey sent to a relatively large group. This is directed very precisely to people who were involved in performing the action steps and also those who were recipients of the results. I recommend a simple five-category check list to mark against a set of statements: strongly agree, agree, neither agree or disagree, disagree, strongly disagree. Typical question could be:

- Project objectives understood?
- Plan was thorough?
- Resources were adequate?
- Time to do each step was adequate?

Also, supplement "check the box" surveys with ones asking questions that require written answers. But make this optional. Doing so makes use of "reverse psychology." People will want to be helpful, but only if not forced to do so. This will get specific unique viewpoints from the participants. For each one of these questions we ask of ourselves and project participants we need an honest answer. The results of these answers then becomes the basis of the lessons learned report.

The last step of the project close out is the closure letter. A closure letter is the final report to the sponsor of the project. The recommended format is as follows:

- Summary
- Details of what was done and the results
- Supporting data
- Recommendations for future activities

A closure report creates a data base for future teams in doing similar projects and as such is a valuable resource tool for the future.

The summary is just that. An outline synopsis of what happened. Stating what the project goals and risks were, how the schedule was established to achieve the goal, and costs associated with the project. But most important the summary states "how we did" including benefits achieved and lessons learned. Most important the summary is a summary. It gets to the point immediately without any ambiguity.

The next section is the details narrative. Here all items in the summary are expanded with all the information required to explain how the project was accomplished. While the summary gives the facts of what happened, the details narrative explains how and why the results were as actually occurred. It discusses how the project was managed, with specifics of notable tasks done. It defines how the benefits were achieved and what they mean for the sponsor in the future. It also gives a detailed explanation of lessons learned.

The section following the details narrative is one with all the tables, figures, calculations, etc. This is the supporting data section which backs up the narrative with the facts that makes the narrative true.

The last section is the recommendation for future actions. This is very specific to the type of project. It could be a simple thank you for allowing us to do the work. It could go to the other extreme to be a very complex: "upon doing x, we found y existed; you should do z to eliminate the threat from y," etc.

Keep in mind that very busy people will read the first section, the summary; and then go directly to the recommendations. So these sections need to be very complete, straightforward, and free of all ambiguities.

Project management Summary

Project management as presented here is a methodology for achieving goals. It is very important that every manufacturing engineer understand the precepts and techniques of project management because virtually all work done within the manufacturing engineering function is project based. While there may be many repeated activities, such as developing a time standard for a process or purchasing capital equipment, each individual task is unique unto itself and project management procedures creates a higher probability of success at lower costs and cycle time. While all the advice and procedures shown so far are the absolute preferred way of conducting a project. Sometimes a brief summary of how to conduct a project is warranted, see Fig. 2-14. I presented this list at an ASME technical conference several years ago before writing this book. It's as valid today as then. Use it as a "check off sheet" to ensure you're covering all bases for even the simplest projects.

- Define scope of project, including what constitutes success.
- Break scope down to logical sequence steps.
- Identify the key steps to successfully complete the project scope.
  - Steps are administrative and/or mainstream for successful conclusion of the project.
- Define what constitutes successful completion of the project steps.
- Lay out project steps in sequence of completion order.
- Estimate resources needed for each step.
- Select measurements to determine status of each step and also time intervals for assessments.
- Conceptualize potential problems and their potential corrective actions.
- Put steps in chronological order referencing sequence of occurrences (predecessor steps), as modified by resources availability (people, equipment, outside influences).
- Implement project.
  - Measure.
  - Assess results.
  - Take corrective actions as required.
- Complete project.
- Prepare and issue final report.

Figure 2-14. Items to be Considered in Setting Up a Multi-task Project

SETTING UP THE OBJECTIVES AND GOALS SYSTEM

The objectives and goals system is a tool for proper management of an organization. However, the tool cannot be utilized until the specific situation of the organization is analyzed. I cannot overemphasize the need to analyze the organization's reason for being before setting out to implement an objectives and goals system.

Many manufacturing engineering organizations tend to drift with the tide, doing what is asked of them by management, but never seeming able to get ahead of the situation. They are excellent problem solvers, using techniques learned after many years of reacting to business crises. Their planning is shallow; it is characterized by Long-Range Forecasts (LRFs) that are nothing more than shopping lists of equipment they would like to have or hunches about what would help them produce their products. There are no true goals, but projects abound to solve crises and to purchase items from their LRFs. This is the situation that exists in the many manufacturing engineering organizations that take the term "service organization" too literally.

In order to truly serve its company, a manufacturing engineering organization must be an initiator instead of a reactor to stimuli. In order to initiate, it must plan ahead how to do it. Implementation of an objectives and goals system allows for continuous planning ahead, but only after identifying the objectives of the company. For manufacturing engineering that means discovering the basic objectives of the manufacturing function and then modifying them to be meaningful for manufacturing engineering activities.

Determining the Organization's Objectives

For a newly launched organization, it is relatively easy to determine what the objectives are. They may be couched in other phrases, but they are there. For example, let us say that a computer system function is established within a company. The reason given may be: "To establish our company in the forefront of the computer revolution so that we can contribute to this exciting new technology to better serve our customers." The hidden objectives are to get ahead of the competition in implementing computers and, by doing so, to reduce operating costs. Thus, the objectives that this organization would use as its overall operating policy are:

1. Implement the utilization of computers in company operations.
2. Reduce the operating costs of company operations.

Objective 1 would lead to goals such as: "Implement the XYZ computer system in the purchasing unit by April of the next budget year." Objective 2 would lead to goals such as: "Reduce the operating expense of the XYZ computer system 25% by on line processing, effective in the third quarter of the budget year."

For existing organizations, the task of finding objectives is a little more difficult. There is no preamble stating why the organization was formed. Therefore, an evaluation of what the organization does and what it should do to support overall company policy is in order. Let us look at the function of which manufacturing engineering is a part, that is, manufacturing. Understanding the core objectives of this function will certainly lead to objectives for manufacturing engineering.

Manufacturing produces the product. Therefore, an objective of manufacturing would be to maintain the capability to produce its products. However, maintaining a capability is obviously not enough; it must be usable in a cost-effective or competitive way. Therefore, another objective could involve an attempt to reduce operator costs, that is, improve productivity. An associated objective would be to minimize all types of mistakes due to misunderstandings or breakdowns in communication, which would be commonly stated as, "Improve interfaces."

Summarizing, then, the objectives of the manufacturing function would be:

1. Maintain a capability to produce the company's products.
2. Improve productivity relative to producing the products.
3. Improve the quality of the products.
4. Improve interfaces with internal and external organizations.

The same objectives could be expressed differently (and often are), but the results would be the same. For example, "Improve productivity" could just as easily have been "Reduce operating costs," because improving productivity in materials utilization, personnel performance, machine tool operations, and so forth in actuality reduces operating costs. The words used for the objectives are chosen to emphasize the needs of the organization and the importance placed on these factors of the operation. Regardless of the choice of words; these, listed above, are the four basic objectives of a manufacturing function.

Knowing the objectives of the manufacturing function, it is logical to conclude that the objectives of the manufacturing engineers sub-function must be in support. They could be the same—and, in fact, they usually are—but they could also be more specific or narrowly defined. This is illustrated in Figs. 2-15 and 2-16.

Figure 2-15 illustrates the objective of improving productivity, a broad objective that is shared with all the manufacturing sub-functions. Notice, however, that the supporting goal is a manufacturing engineering goal. Figure 2-16 illustrates a narrower objective, conserving energy. This is a subset of the broader objective, of improving productivity, and more readily defines the manufacturing engineering organization's contributions to the broader objective.

There is no rule about how many objectives an organization can have, or whether the broader objectives should be mixed with the narrower ones. The only important rule is that objectives should be identified and that they should be true objectives—ongoing, and not changing over time.

Establishing Goals

Once the objectives have been determined, it is time to determine what the goals for a particular period should be. Note that in Figs. 2-15 and 2-16 a goal is specified to support the objective. For

Objective: Improve Productivity

Goals: Write appropriation requests for the 2006 facilities plan based on the 2004 long range forecast. All ARs for major projects will be completed by the fourth quarter of 2005.

Projects:

	Completion Date	Savings ($000)	Priority
N/C chucker med. Motor	550	100	2
Test data collection equipment	548	50	1
Upgrade test equipment	548	20	1
Rebuild radial drill press	540	15	3
M bar taping machine	550	20	2
AC work center	539	150	1
DNC for frame machine center	551	45	2
Keyway broach	545	20	2
Welding robot	542	60	2
Punch press robots	548	120	1
Bar insertion system fab rotor	539	40	1
Bar shaving fab rotor	545	20	1
Building repairs	550	n/a	1
Rebuild punch press	536	40	1
Rebuild gear cutter	539	20	1
N/C mach center MM	551	100	2
Semiauto sub arc — Bldg. 60	550	40	1
Flux core weld station	548	40	2

Figure 2-15.   The Objective of Improving Productivity

simplicity, only one goal has been shown for each objective, although in actuality many goals may be required to support a single objective. The goals are not haphazard selections; rather they are the results of studies of needs done by manufacturing engineering.

The first step is to understand the overall business plan of the company, which specifies profitability levels to be reached for particular time periods. This, in turn, leads to allocations for operations, advertising, capital investment, and so on. Once the business plan is understood, its various components must be categorized under the ongoing objectives. Here there must be a 100% fit; the business plan cannot be in conflict with the objectives. Once the business plan is laid out in accordance with the objectives, the plan provides the starting point or prototype goals. Usually, these prototype goals are not identifiable with a specific function or subfunction. For example,

Objective: Conserve energy

Goals: Reduce annual steam usage by $25,000, using ultrasonic and thermal detectors to find leaking steam traps and problem areas on steam processes. Projects will be completed in October 2005.

Projects:

	Completion Date	Savings
Audit steam traps and processes for problems	2/05	
Analyze all steam applications for replacement or improvement	4/05	
Send appropriation request to Advanced Mfg. Eng.	5/05	
Mfg. Eng. as required order equipment/implement projects after appropriation request approval	9/05	
Write cost improvement	10/05	$25,000

Figure 2-16.   The Objective of Conserving Energy

a goal to make an overall $1 million profit for a product line does not automatically break down to a specific manufacturing goal. But by knowing the dollar amounts reserved for advertising, engineering, and all other non-manufacturing categories, it is possible to determine the amount of costs allocated to production. Knowing the profit level required, we can then deduce the level of productivity needed. Hence, a goal can be set to reduce the cost of manufacture to produce the desired profit. This becomes the manufacturing goal. We must now reduce it to the manufacturing engineering contribution to the overall manufacturing goal.

Manufacturing engineering now asks, "What can be done to support the manufacturing goal under the given objective?" This must be determined on the basis of the mission of the manufacturing engineering organization. If the manufacturing goal is to reduce operating costs by a specified amount, then manufacturing engineering must review the potential within all of its units to reduce costs. These reductions can be through more efficient internal operations, such as introducing an integrated computerized planning system, perhaps part of an Enterprise Resources Planning system, to improve the efficiency of each engineer. Or the reductions can be a result of a product of the engineers' work—for example, improvements in shop operation cycle time due to computerized planning giving more standardized results than manual planning. Looking for opportunities that exist within the realm of assigned responsibilities will lead to goals that contribute to the overall manufacturing goal.

In this manner a catalogue of goals can be developed in support of the overall manufacturing goal. It is the responsibility of the various levels of management to evaluate the potentials against the organization's capabilities to select goals to pursue. Some selections are quite evident, others are not.

EXAMPLE: Under a goal to improve productivity by 10% for the measurement period, the Advanced Manufacturing Engineering (AME) unit may conclude that a new machine tool to replace an older one would more than achieve the goal. Hence, justifying and purchasing the new equipment may become one of the AME unit's goals to support the overall goal of manufacturing, which in turn supports the business plan and the ongoing objective.

A not so obvious selection of a goal is illustrated in the following.

EXAMPLE: Suppose the manufacturing goal is to reduce operating expenses by 30% during the measurement period. The maintenance unit cannot as its share arbitrarily cut costs 30% to comply by reducing the number of skilled employees assigned, or cutting back on expenditures for spare parts by the dictated amount. If the maintenance unit did so, its costs would obviously be reduced substantially, but what would the effect on shop operations be? Operating expenses for shop operations would go up considerably due to delays in obtaining repair service. When service did arrive, it would be less likely that the repair part would be available. Therefore, for the maintenance unit to comply with the goal of reducing operating expenses by 30%, it must do something else. Perhaps a suitable goal in support of the overall goal would be: "Develop by the end of the fiscal period an operator dispatch and measurement system to improve the response time by 20% and the time for return of equipment to service by 10%." This goal does not mention reducing operating costs, but that certainly would be the indirect result. The machines would be running for a higher percentage of the time, which would mean more product output during a time period and less work in process inventory time, all of which would mean lower operating costs.

Selections of goals must support the overall goals of the organization. These overall goals (business plans), in turn, must support the ongoing objectives of the company. Let's now summarize the method of setting the period goals of the manufacturing engineering organization.

Receiving the period goals of the manufacturing function, the manager determines which goals pertain to manufacturing engineering. (For example, inventory reduction goals seldom concern manufacturing engineering and would not be selected.) The manager then negotiates with the manager of manufacturing to determine what percentage of the total applicable manufacturing goals will be assigned to manufacturing engineering. With this target set, the manager of manufacturing engineering further subdivides the target for each of the units within the subfunction. The unit managers then ask their engineers, project leaders, and foremen to develop specific goals and supporting projects.

What I have described above is a delegation of responsibility for planning down to the lowest possible operating level. The people who will perform the work are required to plan the work they will do to achieve the goal. Once the specific goals and supporting projects have been proposed, the procedure is reversed. At each reporting level, management evaluations are made to determine whether the specific goals and projects selected support the overall goals and can be completed within the given time period and with the available resources. When they are approved by the manager of manufacturing, the goals and projects selected become the operating plan.

The reviewing and selecting of goals is a year-round activity. It is an iterative process of evaluating current projects and making adjustments as outside conditions change. Traditionally, goals are set in the fourth quarter of the operating year, but prudent managers are evaluating and adjusting continuously so that at planning time the results need only be collated and sent out for approval.

WORK PLANNING

Setting up the objectives and goals system does not guarantee the accomplishment of goals. Work planning does. Work planning is the meticulous attention to detailed steps that makes attainment of all the assigned goals possible. Work planning makes extensive use of project management techniques discussed previously. In fact the difference between a set of well-developed work plans and a series of small projects is at times difficult to discern. One way to differentiate is to ask the questions:

- Is this a one of a kind task, uniquely different from routine tasks?
- Are the people involved in completing the task normally from different organizations?
- Is there a constant matching of skill sets to specific tasks, and are they uniquely different for each task?
- Is the lead person for each task constantly changing as the task progresses from start to finish?

If the answers to these questions are more often yes than no, then the task can be considered a project. However, in reality a small project and a work planning task are very much the same and differentiating is more of score keeping activity than a practical management activity. In fact many companies chose to use the term projects instead of work planning task, simply because they think it gives them the aura of modernity.

The basic elements of work planning (similar to the action step of project planning), which are valid for any organization assigned a task to perform in a specified time frame, are:

1. Schedules: Lists of things to accomplish, including responsibilities assigned and due dates.
2. Action plans: Very specific plans for items to be accomplished by a specified individual, by a certain date.
3. One-on-one meetings: Periodic meetings between a higher-level person and a subordinate to discuss items pertaining to all phases of assigned work.
4. Communications: Documentation, both formal and informal of the status of scheduled items.

All work planning schemes must have these four elements. The titles of the elements may vary from organization to organization and sequences may be different, but all four elements must be present for a true work planning system. Without such a system, there is no management; there are only reactions to perceived crises. Let's now examine the elements of a work planning system individually, but keep in mind that they are all underway simultaneously and interact continuously.

Schedules

If anything can be called a starting point of work planning, it would be the activity of making schedules. There is always a multitude of goals assigned to each unit of manufacturing engineering. Each goal may involve many projects, which in turn have many steps. All of these individual steps for all the projects and goals will have start and due dates. In addition to projects and goals, there may be other assigned activities that are not documented in the statements of objectives and goals but have definite starting and completion dates; an example would be financial report submittals. The scheduling step of work planning must compile all these requirements into an overall schedule of accomplishment for a particular unit or parts of a unit.

Manufacturing engineering organizations usually have computer programs to compile the start and due dates of all activities in one control chart or report. Whether or not a computer program is used, a master control chart or report is necessary for each unit. This is the only way a manager has to view the work in its entirety and see the work load distribution. Figure 2-17 illustrates a work planning schedule.

For brevity, only a few items are shown; a typical work planning schedule could run for several pages. Also note that this example is in chronological order of completion dates. Sometimes a companion chart in chronological order of starting dates is also produced. Another item that is frequently on the schedule control chart is an estimate of man-hours to achieve individual line items. This is a useful device in distributing work and determining what type of effort will be required to complete the activity according to the schedule.

Once the control chart has been constructed, it is used as a reference schedule for reviewing progress. The manager uses it to make sure activities are started on time and completed on time, or to understand why deviations in the plan have occurred. It becomes the control document for virtually all the work going on in the specific area of responsibility. If the schedule control chart is constructed properly, all items, including those that are not part of the objectives and goals, appear on the chart. One should not infer, however, that a schedule control chart made up during the fourth quarter of the previous year would be continuously adequate through the present year. Usually this is not the case. Company plans are always sensitive to the marketplace, availability of labor may vary, other activities that must happen as a precursors to the unit's plans may not occur on time, and so forth. Therefore, the schedule control chart usually goes through several modifications during the year. A manager must determine the risks in accomplishing the schedule and be ready to re-plan activities on the basis of well thought out contingencies.

Action Plans

The schedule shows the work to be accomplished by the entire unit or a larger entity. The action plan shows the work to be done by the individual. Depending on the size of the unit and the number of specific items to be accomplished, an action plan may be for several individuals or for only one person. If it is for more than one individual, then a project leader is normally assigned and is responsible for the work of all contributors.

If most of the unit's work is described in the objectives and goals statement, then the project sheets can be used as the action plan. Usually, however, an engineer will have several action plan sheets, some of which are project sheets from the specific objectives and goals statement, while the rest are action plan sheets covering miscellaneous activities.

The action plan is the individual engineer's work plan. It is what the engineer expects to be measured and evaluated against. Therefore, in drafting it, both the engineer and the manager should agree on what constitutes a satisfactory conclusion for each specific item of the plan. It must also be understood that it is the engineer or project leader's responsibility to report to the manager any obstacles that cannot be overcome in time to maintain the schedules. The traditional advice—to report back to a superior completion or reason for failure to complete—is critically important for a successfully run organization. Yet this simple rule is often neglected, leading to larger sums of money wasted in recovery efforts necessitated by lack of communication. For action plans to be

#	Activity Name	Project No.	Start Date	Finish Date	Work-hours	January	February	March	April	May	June	July	August	September	October	November	December
											2005						
1	Audit steam traps	16	1/3/05	2/25/05	28												
2	Reviw history of weld problems	22	1/3/05	2/25/05	46												
3	Set up system to monitor welding quality	22	1/10/05	2/18/05	8												
4	Set up welding council	22	2/21/05	3/18/05	18												
5	Establish feedback and analysis routine	16	3/14/05	3/18/05	9												
6	Analyze all steam use for replace/improve	16	2/28/05	05/6/05	48												
7	Implement welding manual	22	3/25/05	6/17/05	88												
8	Write appropriation requests for new steam traps	16	4/8/05	5/13/05	12												
9	Order and receive new steam traps	16	7/16/05	7/8/05	22												
10	Install new steam traps	16	7/11/05	12/9/05	178												
11	Write cost improvement, weld quality	22	12/2/05	12/16/05	6												
12	Write cost improvement, steam use savings	16	12/12/05	12/16/05	4												
						January	February	March	April	May	June	July	August	September	October	November	December

Figure 2-17. A work planning schedule

dynamic, useful management tools, managers must require timely reports from their subordinates. Managers must educate their engineers to report any deviations from the plan together with proposals for corrective actions to bring the activity back on schedule. Through this commonsense approach, they can prevent salvageable projects from becoming unsalvageable.

One-On-One Meetings

The key method for reaching understanding between a manager and an engineer relative to assigned work is a regularly scheduled (usually weekly) meeting between the two. The fact that there are no other attendees is a critical part of this element of work planning. When only two people are present, they give each other undivided attention. The purposes of the meeting are to review problems with ongoing work in a general and specific sense, to make sure the engineer understands what the manager wants, and to ensure that the manager understands the problems the engineer may be encountering.

This type of interaction, which is commonly referred to as a one-on-one meeting, is not structured. Its content depends on the current and near-term activities of the organization. The superior must come to the meeting prepared to discuss any aspect of the subordinate's work assignments, to point out good and bad performance trends, and possibly to suggest different approaches to particular problems. But the manager must be careful not to monopolize the meeting and must keep in mind that he or she should contribute a maximum of 50% of its content.

The engineer comes to the meeting with the same type of preparation, but with two additional targets: to give the manager a flavor for the work as it is progressing, and to obtain the manager's approval for any changes to current action plans that appear necessary.

The one-on-one meeting is the dynamic part of the work planning system—the part where the static plan becomes active and where policy decisions and changes are discussed and agreed to. It is the only way, except by actually doing the work, for the manager to measure the pulse of the entire organization.

It can also be thought of as a tutorial session where the superior can teach practical aspects of management to the subordinates. One of the roles of a manager is to develop skills within the employees. At the one-on-one meeting the manager can demonstrate the correct role model and instill good management principles.

Communications

Lack of adequate communications has been and will continue to be the largest single cause of failure in any organization. Subordinates always have the duty of keeping their superiors informed about work progress. Managers must always give their subordinates enough information to adequately perform their assigned tasks. The work planning system requires the establishment of such up and down communications.

This element requires formal communications in the nature of periodic written reports and oral presentations. Two types of reports that have proved to be useful are the key results report and the quarterly review.

The key results report consists of a summary of the status of all key elements of work assigned to the unit. The manager will report in graphical and/or descriptive form the status of all activities compared to the plan. Data for the report come from updating the unit's work schedule and the individual's action plans and from the notes taken by the manager during the various one-on-one meetings with the subordinates. The key results report also contains financial data related to manufacturing engineering.

I have always found it useful to report financial deviations from budget by using descriptive methods for explanations and corrective action statements. A typical key results report for manufacturing engineering would contain the following:

1. A list of projects by title under their respective goals, showing actual and planned percentages of completion. Declarative phrases explaining deviations and planned corrective action should also be included where needed.

2. A section showing current measurements versus targets, such as manufacturing losses charged to manufacturing engineering; cost improvement status; productivity measurements; actual vs. target data on critical financial accounts, such as funds spent for capital equipment purchases; and operator safety statistics.
3. One or two descriptive paragraphs outlining specific, significant accomplishments e.g., completion of an important project.
4. One or two paragraphs outlining specific concerns or directions the organization is taking on certain matters.
5. A section on personnel matters, covering morale, open positions, promotions, and so forth.

This report is sent to the manager's direct superior with copies to the subordinates and to other managers on the same reporting level within the manufacturing function. This is an easy way for a manager to keep all key associates informed and to make sure they all understand what the manager thinks is important.

The key results report is usually a monthly report, which makes it a document that can be used as a daily management tool. Since the report is due at relatively short intervals, it serves as a reminder to the manager to keep close watch on progress related to work planning schedules and action plans.

The quarterly review is an oral presentation, usually with visual aids, coordinated by the manager of manufacturing engineering but presented to the manager of manufacturing by the unit managers and key individual contributors. In other words, it is a presentation to a manager one reporting level higher than their immediate superior. The quarterly review does several things. First, it allows individuals further down the chain of command to report directly to senior management. Second, it allows the manager of manufacturing to ask questions directly of those managers and individual contributors with whom he or she would not normally have day-to-day contact. Third, it provides business exposure for junior personnel, who can get a flavor for the entire business picture by listening to the types of questions asked and other remarks made by the manufacturing manager. Fourth, it gives lower-level managers and individual contributors an opportunity to display their knowledge and ability to express themselves. Finally, it allows the manager of manufacturing to compare actual happenings with documented plans.

The quarterly review is an excellent communications technique. Since it is an oral review, the proceedings take on some aspects of a planning session. Invariably, the highest-level manager present will suggest changes, albeit small, that will affect one or more of the other three elements of work planning. Hence, this step also helps to convert a static written plan into a dynamic plan.

FINANCIAL CONTROLS

We have discussed project management, objectives and goals formulation, and work planning. These activities lead to tasks that manufacturing engineering must perform. Now it is necessary to bring in the controlling factor that determines how much manufacturing engineering can do to develop an optimally productive factory. Financial controls put the restraint of reality on all programs. They make managers assess all activities in pragmatic terms, deciding which projects should be carried out and how to husband the always limited resources to achieve the best results.

Financial controls means staying within the budgetary levels imposed by the appropriate level of management. In making a long-range forecast, manufacturing engineering has the opportunity to propose that certain programs and projects should be started or continued and that others should be terminated. This almost always requires an assessment of what the costs and paybacks will be. The long-range forecast proposals are then weighed against available funds and proposals made by the other functions. Eventually a dividing up of the available funds takes place, resulting in an operating budget for the upcoming fiscal year and an official plan for the succeeding 4 years—a 5-year plan in total (note: some companies, especially those with very dynamic product life cycles, prefer a 3-year instead of a 5-year cycle.). Usually the plan for the year after the new budget year is relatively firm and becomes less firm with each succeeding year. It is the budget year that is of primary importance for financial controls.

There are three main activities related to financial controls that must be undertaken by manufacturing engineering: (1) short-range forecasting, (2) periodic review of operations, and (3) daily accounts control. Short-range forecasting means predicting the amount of money to be spent in the next period (usually a month) and the succeeding two periods. The periodic review of operations is an analytical appraisal of how the forecast compared with the actual occurrences and determination of why there were differences. Finally, daily accounts control is a method of ensuring the specific accounts are not overspent and that the short-range forecast is attained. Let's look at these three activities in closer detail.

Short-Range Forecasting

Figure 2-18 is a typical Short-Range Forecast (SRF) Form.

It lists the accounts for which the manufacturing engineering function is responsible and has space to predict what the related expenditures will be for the forecast periods. It also has space to show how the forecast differs, if it does, from the operating budget and to give reasons for the variations. The very fact that the SRF has places to show variations from budget implies that individual accounts may vary due to unforeseen circumstances. This leads us to the major challenge for management: how to allow variations in individual accounts and still maintain the overall budget level for expenditures.

The SRF is a monthly exercise conducted by all manufacturing subfunctions. Usually the first week of the current month is established as the due date for the SRF. This discipline causes all operations to look ahead over a period of 7 to 8 weeks for immediate concerns and up to 17 weeks (one quarter of the year plus the current month) for near-term concerns. It tends to force management to be concerned about the effect of current decisions in the near future, and thus tends to prevent expedient decisions that would cause problems later on.

> **EXAMPLE:** Process control makes a concession to allow a part to proceed that is out of tolerance; that is, it takes a chance that the part will still work. This is done to make the quota for the immediate production goal. A month or two later the substandard part is used and the product fails in test. As a result, the manufacturing losses as shown in the overtime premium account are excessively high and shipment of the product to the customer is delayed. The actual expenditures are severely over the forecast, because of a decision that had only immediate short-term goals.

The SRF system tends to prevent problems such as that described above, because the manager would look at such a decision in terms of both what it might accomplish now and what the costs could be in the future.

The manufacturing engineering SRF is made up by the manager upon receipt of the SRFs from the various units. It is a system that should reach down to the project leaders, individual engineers, and foremen for inputs on their specific activities. Each manager must then critically review the inputs for practicality based on present circumstances.

> **EXAMPLE:** A forecast requiring capital expenditures at three times the level of the previous forecast must be based on a change in equipment delivery rate or some other valid reason.

Every line on the forecast must be critically appraised before it is passed up the management chain. Managers are aware that their capabilities are being constantly evaluated, and one of the most visible signs of their worth is the validity of their SRFs.

The SRF is the manager's work plan spelled out in financial terms. If it turns out to be very different from what actually occurs, then it indicates that the manager's plan was not very good. For this reason, managers must review their plans in depth to be sure that the funds are available and are

Year:_____

Department Name:_____
Department Number:_____

SHORT RANGE FORECAST

Completed By:_____
Date:

Account No.	Account Category ($000)	Next Month: Budget	Forecast	2nd Month: Budget	Forecast	3rd Month: Budget	Forecast	Reason for variance more than 10% Use back of page if more space required
	Income							
10010	Fees							
10020	Products							
10030	Reimbursed Expenses							
10040	Sales (misc.)							
10050	Services (Sales)							
10060	Uncategorized Income							
11000	TOTAL Income							
	Cost of Goods Sold							
20010	Production Materials							
20020	Product Development Mat'ls							
20030	Misc. Materials							
20040	Professional Services (DL)							
21000	TOTAL Cost Of Goods Sold							
22000	Gross Profit							
	Expense							
30010	Advertising							
30020	Automobile							
30030	Bank Service Charges							
30040	Building Materials							
30050	Contributions							
30060	Damages							
30070	Dues & Subscriptions							
30080	Equipment Purchases							
30090	Equipment Rental							
30100	Health Insurance							
30110	Liability Insurance							
30120	Workers Comp. Insurance							
30130	Interest Expenses							
30140	Late Fees							
30150	Licenses & Permits							
30160	Misc. Expenses							
30170	Office Supplies							
30180	Payroll (mgmt & non-exempt)							
30190	Postage & Delivery							
30200	Printing and Reproduction							
30210	Professional Development							
30220	Professional Fees							
30230	Recruiting							
30240	Rent							
30250	Repairs							
30260	Taxes							
30270	Telephone							
30280	Travel & Entertainment							
30290	Uncategorized Expenses							
30300	Uniforms							
30310	Utilities							
30320	Write Offs							
31000	TOTAL Expenses							
32000	Net Income							
	COGS % of Sales							
	Sales							
	Gross Income % of Sales							
	TOTAL Expenses % of Sales							
	Net Income % of Sales							
	Employees							
	Hourly—direct							
	Hourly—Indirect							
	Managerial and professional							
	General salaried							

Instructions:
1. fill in for all accounts unit has responsibility for.
2. base forecast on known or projected events occuring within the forecast period

Figure 2-18. Typical Short-Range Forecast Form

		Next Month:		2nd Month:		3rd Month:		
		Budget	Forecast	Budget	Forecast	Budget	Forecast	Reason for variance more than 10% Use back of page if more space required
Account No.	Account Category ($000)							
30250	Repairs	16.0	17.5	16.0	16.0	20.0	20.0	next mo.: 2.5 over-breakdown repair CNC lathe

Figure 2-19. One Line of a Short-range Forecast Form Filled Out

in agreement with the various unit action plans. Any good management plan requires comparison of actual with predicted expenditures. Therefore, those making predictions must expect critical and timely reviews of their performance in this area.

Figure 2-19 shows one line of an SRF filled out. The forecast shows a budgeted weekly expenditure of $3923.00, which is rounded to $4000.00 and expressed as $4.0 per week because forecasts are always in thousands of dollars. The fact that the budgeted weekly expense is $4.0 is determined by looking at the "deviations" columns and knowing that a quarter is always 13 weeks, with one of the months having 5 weeks. Adding the three monthly forecasts together, subtracting the expected deviation, and then dividing by 13 gives the budget expenditure per week. The five-week months are traditionally the first month of each quarter, i.e., January, April, July, and October. Therefore, for budget purposes, we have 5-week periods and 4-week periods. Note the deviation shown in the "next month" column. This shows that management has decided to allow an over expenditure by $2.5 for the next period. The reason for this variance is explained in the "remarks" column. If not enough space is available, a notation would be made in that column signifying an explanation made somewhere else, usually at the bottom of the form. An analysis is made for each line of the SRF, but only significant variances are explained. If the variance of the example had been $0.5 or less, no explanation would have been required.

The exact amount of variance requiring explanation will depend on the magnitude of the account and the judgment of the manager. For example, an account expending $250.00 per month would not require an explanation for a $25.00 variance. After each line has been evaluated and explanations entered where necessary, all the vertical columns are totaled and the forecasts are evaluated against budget to see how the unit of the subfunction is faring. Here, the responsible manager must judge whether the forecast is acceptable. If the forecast is acceptable, it is forwarded. If not, it is sent back to the manager's subordinates for revision. Any revisions made to the SRF will of necessity result in changes in various action plans.

The last section of the SRF is an employee forecast. Here the manager is comparing the people on the staff, plus those considered necessary to conduct the plan, against the numbers that make up the wages, salaries, and benefits budgets. There are many reasons for employee numbers to be different from budget numbers: differences from expected productivity levels that were used to predict personnel needs when long-range forecast and budgets were prepared, changes in the scope of projects, and other changes in plans are a few of the more common reasons. Personnel forecasting is required to remind management that personnel allocations are just as much a budgetary item as stationery and other accounts, and that the expected levels of performance must be reached by all employees for the organization to achieve its goals.

Periodic Review of Operations

The periodic review of operations is the measurement of the plan. It is carried out monthly by all those producing the SRF. The review starts with a comparison of the actual financial results with the month's forecast. Then, for all significant deviations, an investigation is made to determine the causes of variance. This investigation of causes yields the greatest benefit.

An SRF is an earnest attempt to predict the outcome of putting a plan into action. A simple plan leads to accurate predictions because all items or activities involved in the outcome can be anticipated and controlled. Unfortunately, most business plans are not simple, and the outcomes are not quite so predictable because a significant portion of the activity is out of the control of the project leader.

EXAMPLE: A goal to expend 1000 hours of maintenance time to fix a series of machine tools by a specified date requires a knowledge of exactly what must be done and how quickly and consistently the repair team can work. The problem of the machine tools may be precisely determined, but the rate of work can only be approximated. A key individual may not come to work for a good reason, or another project may become more critical at the time, thus drawing away the key individual. It might then be necessary to have the job done by a less experienced person, who would spend more hours on the repairs than originally forecast. The result is a missed forecast and a deviation from the actual that must be evaluated.

We evaluate results so we may learn from previous errors and not repeat those errors. A deviation from forecast properly explained may lead to beneficial modifications in plans and make it easier for the organization to achieve its goals.

Figure 2-20 is an example of a form used in a typical periodic review of operations. Note that there are two sets of columns for comparing actual with forecast: current month and year to date. It is very important to be able to view one month's results in the perspective of the longer period of time signified by the year-to-date columns. This helps to keep responsible people from taking unnecessary actions when, in fact, the variances of the monthly results may not be indicators of true trends.

Figure 2-21 is an example of a completed review of operations. Note that a variance amount is shown for each account and that the year-to-date SRF column is a summation of all the previous months. Some managers may prefer to use the previous year to date plus the current month's forecast, especially if the unit is experiencing significant variance in many accounts and forecasting errors are compounding. Note also that not all variations are explained in Fig. 2-21. The same reasoning as for the SRF deviations is applied here. If the manager fully understands the causes of variance and can take necessary corrective action, complete explanations would be a waste of time. However, explanations require some sort of investigation, and this leads to an understanding of the cause of the problem. Therefore, wherever practical, all variances from the forecast greater than 10% should be investigated.

Daily Accounts Control

Daily accounts control is tantamount to ledger management. This can be tedious and should be used only for accounts that are difficult to forecast or in which large variances would be very damaging. All accounts are monitored by Finance to a certain degree, so this activity is only used when dealing with a very critical account, or there are severe budgetary constraints requiring this tedious and time consuming activity. It is practically impossible for engineers to keep track of each entry into each account; to do so would mean the virtual abandonment of real engineering work. It is also not practical to assign a financial analyst to each person who is required to manage portions of a company's money. Therefore, a middle-ground approach is used.

First, accounts that are out of control must be determined. This does not mean that accounts with consistent variances between forecast and actual should be classified as out of control. An account such as "electric utilities" may be totaling consistently higher than the budget and forecast every month, and still be well controlled. The reason may be energy prices are rising faster than predicted, and forecasts are consistently underestimating them. Daily accounts control is not intended to correct this type of situation. A correction here would typically involve senior members of the financial function who develop economic guidelines to be used and would probably result in a request that the manufacturing function develop more vigorous energy conservation measures. This, in turn, would result in major changes in the objectives and goals statement.

Once the accounts that are truly out of control have been determined, an effective plan must be devised to bring them into control. This is the purpose of daily accounts control.

Figure 2-22 is a form used to keep track of all forecast and actual amounts during a budget year for specific accounts. Note that there is space for budget, forecast, and actual amounts. This form allows the engineer to track forecasts for the previous 3 months and then compare them to actual

REVIEW OF OPERATIONS

Year:_____   Month:_____   Completed By:_____

Department Name:_____   Date:_____

Department Number:_____

Account No.	Account Category ($000)	Budget Mo.	Forecast Mo.	Actual Mo	Difference Act./Fcst	Budget YTD	Forecast YTD	Actual YTD	Difference Act./Fcst	Reason for variance more than 10% Use back of page if more space required
	Income									
10010	Fees									
10020	Products									
10030	Reimbursed Expenses									
10040	Sales (misc.)									
10050	Services (Sales)									
10060	Uncategorized Income									
11000	TOTAL Income									
	Cost of Goods Sold									
20010	Production Materials									
20020	Production Development Mat'ls									
20030	Misc. Materials									
20040	Professional Services (DL)									
21000	TOTAL Cost Of Goods Sold									
22000	Gross Profit									
	Expense									
30010	Advertising									
30020	Automobile									
30030	Bank Service Charges									
30040	Building Materials									
30050	Contributions									
30060	Damages									
30070	Dues & Subscriptions									
30080	Equipment Purchases									
30090	Equipment Rental									
30100	Health Insurance									
30110	Liability Insurance									
30120	Workers Comp. Insurance									
30130	Interest Expenses									
30140	Late Fees									
30150	Licenses & Permits									
30160	Misc. Expenses									
30170	Office Supplies									
30180	Payroll (mgmt & non-exempt)									
30190	Postage & Delivery									
30200	Printing and Reproduction									
30210	Professional Development									
30220	Professional Fees									
30230	Recruiting									
30240	Rent									
30250	Repairs									
30260	Taxes									
30270	Telephone									
30280	Travel & Entertainment									
30290	Uncategorized Expenses									
30300	Uniforms									
30310	Utilities									
30320	Write Offs									
31000	TOTAL Expenses									
32000	Net Income									
	COGS % of Sales									
	Sales									
	Gross Income % of Sales									
	TOTAL Expenses % of Sales									
	Net Income % of Sales									
	Employees									
	Hourly—direct									
	Hourly—Indirect									
	Managerial and professional									
	General salaried									

Instructions:

1. fill in for all accounts unit has responsibility for.

2. base variance on known or projected events occuring within the forecast period

Figure 2-20. Periodic Review of Operations Form

REVIEW OF OPERATIONS

Year: __2005__
Department Name: __Mfg. Engrg.__
Department Number: __1502__

Month: __March__

Completed By: __J.J.Jones__
Date: __4/14/05__

Reason for variance more than 10%
Use back of page if more space required

Account No.	Account Category ($000)	Budget Mo.	Forecast Mo.	Actual Mo	Difference Act./Fcst	Budget YTD	Forecast YTD	Actual YTD	Difference Act./Fcst	Reason for variance
	Income									
10010	Fees									
10020	Products									
10030	Reimbursed Expenses									
10040	Sales (misc.)	0	0	2	2	0	0	2	2	consult for client
10050	Services (Sales)									
10060	Uncategorized Income									
11000	TOTAL Income	0	0	2	2	0	0	2	2	
	Cost of Goods Sold									
20010	Production Materials									
20020	Production Development Mat'ls									
20030	Misc. Materials									
20040	Professional Services (DL)	10	10	10.5	0.5	35	35	35.5	0.5	fix CNC lathe
21000	TOTAL Cost Of Goods Sold	10	10	10.5	0.5	35	35	35.5	0.5	
22000	Gross Profit	-10	-10	-8.5	1.5	-35	-35	-33.5	1.5	
	Expense									
30010	Advertising									
30020	Automobile	8	8	8	0	29.3	32	31.5	-0.5	clutch repair-truck
30030	Bank Service Charges									
30040	Building Materials	3	3.5	3.2	0.2	10.5	11	10.7	0.2	
30050	Contributions									
30060	Damages									
30070	Dues & Subscriptions	1	0	0	0	3.5	1.5	1.2	0.3	
30080	Equipment Purchases (expense)	4	1	0	1	13	16	14	2	expedited lathe delivery
30090	Equipment Rental	5	2	2	0	16.5	9	11	2	
30100	Health Insurance	13	13	11.3	-1.7	42.3	42.3	40.6	-1.7	
30110	Liability Insurance									
30120	Workers Comp. Insurance	6	6	6	0	19.5	19.5	19.5	0	
30130	Interest Expenses									
30140	Late Fees									
30150	Licenses & Permits	2	0	1	1	4	1	1	0	bldg.permit-rails
30160	Misc. Expenses(O/T Prem.)	0	1	0.3	-0.7	1	5	3.6	-1.4	
30170	Office Supplies	0.5	0.5	0.4	-0.1	1.5	1.4	1.9	0.5	
30180	Payroll (mgmt & non-exempt)	22	22	0	0	71.5	70	70	-1.5	
30190	Postage & Delivery									
30200	Printing and Reproduction									
30210	Professional Development	3	0	1	1	7	3	4	1	in-house training
30220	Professional Fees									
30230	Recruiting	0	0.2	0.2	0	0	0.2	0.2	0	replace engr.
30240	Rent	1.5	1.5	9.8	8.3	5.4	5.4	13.7	8.3	mistake-wrong charge.
30250	Repairs	16	17.5	17	-0.5	48	50	49.5	-0.5	fix CNC lathe
30260	Taxes									
30270	Telephone	1	1	1	0	3.5	3.5	3.7	0.2	
30280	Travel & Entertainment	1	1	1.3	0.3	3.5	3.5	4	0.5	
30290	Uncategorized Expenses									
30300	Uniforms	0.2	0.2	0.2	0	0.7	0.7	0.7	0	
30310	Utilities	33	28	29.5	-1.5	107.3	106	102	-4	conservation plan
30320	Write Offs									
31000	TOTAL Expenses	120.2	106.4	92.2	7.3	388	381	382.8	5.4	
32000	Net Income	-130.2	-116.4	-100.7	-5.8	-423	-416	-416.3	-3.9	

	Budget Mo.	Forecast Mo.	Actual Mo	Difference Act./Fcst	Budget YTD	Forecast YTD	Actual YTD	Difference Act./Fcst	Reason
COGS % of Sales									
Sales									
Gross Income % of Sales									
TOTAL Expenses % of Sales									
Net Income % of Sales									

Employees	Budget Mo.	Forecast Mo.	Actual Mo	Difference Act./Fcst	Budget YTD	Forecast YTD	Actual YTD	Difference Act./Fcst	Reason
Hourly—direct	8	8	8	0	8	8	8	0	
Hourly—Indirect	2	2	2	0	2	2	2	0	
Managerial and professional	12	11	11	0	12	11	11	0	resignation
General salaried									

Instructions:
1. fill in for all accounts unit has responsibility for.
2. base variance on known or projected events occuring within the forecast period

Figure 2-21. Complete Review of Operations Form

Subsection Mfg. Engrg.   Account Title: Telephone   account No. 567   Budget 20.0   Target 19.0

	Jan.	Feb.	Mar.	Apr.	May	Jun.	Jul.	Aug.	Sep.	Oct.	Nov.	Dec.
Budget monthly	1.5	1.5	1.5	1.9	1.5	1.5	1.9	1.5	1.5	1.9	1.5	2.3
Forecast cum.YTD	1.5	3.0	4.5	6.4	7.9	9.4	11.3	12.8	14.3	16.2	17.7	20.0
Jan.												
Feb.	1.5											
Mar.	1.8	2.0										
Apr.	2.0	2.0	2.1									
May.		1.6	1.6	1.5								
Jun.			1.6	1.5	1.5							
Jul.				1.6	1.6							
Aug.					1.5							
Sep.												
Oct.												
Nov.												
Dec.												
monthly actual	1.72	1.44	1.93	2.05	1.55							
cum. YTD actual	1.72	3.16	5.09	7.14	8.69							

Figure 2-22. Form Used to Follow Forecast vs. Actual Amounts During a Budget Year

and budget. We can see the entire picture or history of the account for the year on one page, and see whether there is any relationship between forecast and actual. Let us look at the month of April. Note that telephone and telegraph charges have been forecast three times in three different short-range forecasts: January, $2.0; February, $2.0; and March, $2.1. The actual amount turned out to be: April, $2.05. In this case, the actual and forecast amounts are very close. Also, the actual and budget amounts are relatively close. With the actual, the three forecasts, and the budget virtually the same, the account is clearly under control. No other measures need be instituted to track it.

Incidentally, note that in Fig. 2-22 the entries are not typed, but entered in longhand. This is acceptable because this is a control document for a specific individual, and would not normally be submitted up the chain of command. This is the type of document that an individual engineer would take along to a one-on-one meeting.

Now let's look at an account that does not pass this three-way check and is not under control, and see how we can get the forecasts closer to actuality and budget. Figure 2-23 shows an account that is out of control.

The forecast amounts are either significantly over or under actual amounts. In this case the forecast is quite useless for planning expenditures. It is evident that expenses are following a pattern quite different from that of the forecasts. This means that the individual responsible for this account must do considerably more to understand what is happening and to bring it under control. This example is typical of response accounts, that is, accounts where the manufacturing engineering organization must respond quickly to current situations. However, this does not excuse manufacturing engineering from performing on budget, since that would put the manufacturing function at the mercy of the lowest level within the organization, with policy being proclaimed by the actions of the most junior personnel.

Maintenance accounts are examples of response accounts. When equipment is malfunctioning, it must be fixed as soon as possible. Even with a chance-happening activity, it is possible to control the account and still be able to respond in an emergency. The key to successful control is to spend only the allocated funds for each month, and to spend each month of the forecast's funds simultaneously. This may sound like double talk, but it is not. It simply involves knowing each month's budget (target) and anticipating when (in what month) the bill will be received and have to be paid.

Studying Fig. 2-24 will show how this is done. Note that the budget is shown for each month. This is the amount the unit wants to have charged each month against this account. In May, $1.6

Subsection <u>Mfg. Engrg.</u> Account Title: Bldg Maint account No. 587 Budget 64.0 Target 55.0

		Jan.	Feb.	Mar.	Apr.	May	Jun.	Jul.	Aug.	Sep.	Oct.	Nov.	Dec.
Budget	monthly	6.0	5.0	5.0	6.0	5.0	5.0	6.0	5.0	5.0	6.0	5.0	5.0
Forecast	cum. YTD	6.0	11.0	16.0	22.0	27.0	32.0	38.0	43.0	48.0	54.0	59.0	64.0
Jan.													
Feb.			5.0										
Mar.			5.0	5.0									
Apr.			6.0	6.0	7.5								
May.				5.0	5.5	6.0							
Jun.					5.5	6.0	5.0						
Jul.						8.0	7.5	7.5					
Aug.							5.0	5.0	6.0				
Sep.								5.0	5.5	6.0			
Oct.									8.0	8.5	8.0		
Nov.										6.0	5.5		
Dec.											5.5		
monthly actual		2.6	9.8	8.2	5.9	3.4	7.7	8.3	2.4				
cum. YTD actual		12.4	22.2	30.4	36.3	39.7	47.4	55.7	58.1				

Figure 2-23. Forecast vs. Actual Amounts for an Account that Is Out of Control

less than anticipated was charged to this account. That means that the remaining $1.6 will have to be paid at a later time. This then becomes the rollover or anticipated maximum charge against the next month's allocated funds. The column headed "special shop orders" is a list of all building maintenance activities that have occurred through June 21, the date of this report. This would include labor and material charges through June 21. Additional charges on succeeding days would be entered in later editions of this daily report. Labor will invariably be charged in the month when it occurs because it is usually manufacturing engineering's own people doing the work, hence the billing should have minimal bookkeeping delays. Material is booked and paid for only after it arrives and the bill is received. One of the main purposes of the format is to determine the month when material will have to be paid for.

It is evident in Fig. 2-24 that Purchase Orders (POs) 3, 4, 6, 7, 11, and 12 are for material purchases; and they are placed in the month when the bill is expected to be paid. Usually the date promised by the vendor is used to place the future bill in the correct month. The total through June 21 is shown for each month. Figure 2-24 shows that June has reached its allocated fund level, hence no more work should be scheduled to be performed in June. Also note that the next 3 months already have charges booked against them. For July, $2.4 of the $6.0 budget has already been expended and the month has not yet started. A similar situation exists for August and September.

This system requires meticulously recording what has been spent and managing the operation to make sure that the results will be within the budget. This means that the project's actions should be keyed to the budget or, which is more correct, the budget should be set to match the project schedules.

Of course, Fig. 2-24 as explained above represents an ideal case. Note that no contingency is shown. A maintenance account should always have a reserve fund to take care of emergency items, since all work cannot stop when the allocated money runs out. If the roof leaks on June 22, no prudent manager will wait until July to fix it. Therefore, in addition to the rollover, the manager will hold back a percentage of the funds for emergencies. This is done either by using a lower amount for the budget, or by entering a dummy purchase order in the contingency amount. In Fig. 2-24, PO 1 for $0.5 could be considered the dummy purchase order—a typical 10% contingency amount held in reserve. Also, in practice, the entire rollover never comes due for payment the next month. Therefore the contingency is somewhat greater than $0.5.

Account No.587

Account Title: Building Maintenance

Date: June 21

($000)	Current Month June	Next Month July	Second Next Month August	Third Next Month September
Budget	5.0	6.0	5.0	5.0
Rollover from previous month	1.6			
Special shop orders				
PO no.1	0.5			
PO no.2	0.2			
PO no.3		0.3		
PO no.4		0.5		
PO no.5	0.4			
PO no.6			2.2	
PO no.7				1.6
PO no.8	2.0			
PO no.9	0.1			
PO no.10	0.3			
PO no.11		0.4		
PO no.12		1.2		
Total	5.1	2.4	2.2	1.6

Figure 2-24. Purchase Order Payment Predictions

The format used in Fig. 2-24 is similar to that of the SRF. Hence, the figures for this specific account and all others tracked in this way can be used directly as the basis of the forecast. This technique, if followed daily, will result in response activities being controlled within approximately 10 to 15%, which is excellent control for this type of account and acceptable for modern manufacturing management.

SUMMARY

All manufacturing engineering activities must be controlled for results to be achieved as planned and not as a result of chance. In this chapter we have discussed the basic techniques used to plan for activities to occur and to track progress. We have also introduced the necessary financial controls.

To control manufacturing engineering activities, each activity must be scheduled—preferably before the fact, but in all cases scheduled. In this way a database is created showing the time and resources required to handle a variety of assignments. With such a database management can establish productivity targets based on real needs as constrained by time and funds, and can use the manufacturing engineering organization to study and implement projects in a tightly focused effort.

By controlling all operations with a plan, personnel needs and funds are accurately forecast and the company's business plan and targets are more likely to be achieved. If manufacturing engineering activities are permitted to exist in a reaction mode, there are never enough people to handle all the problems that occur, nor is there enough money to cover all expenses. By planning ahead, the cost of manufacturing engineering is reduced and its achievements increased.

REVIEW QUESTIONS

1. An organization wishes to upgrade the level of education of its staff by in-house training. It intends to have 65% of its 250 people take a blueprint-reading course and 30% take a machine tool familiarization course. Finally it is considered necessary for at least 15% of the staff to complete both courses. The manager hopes to have the education plan completed within 12 months. Write an objective and supporting goals for this program.

2. Define the philosophical differences between an objective and a goal as related to manufacturing engineering.

3. Explain why adherence to the rule of project selection pragmatism and the series corollary are important to a profit-making organization.

4. The following is an excerpt from a press release issued by a new company. Determine what the objectives of the company are. "We have combined resources so that we can better serve the public in discerning the practicality and feasibility of mass transit systems."

5. A company decides that it must reduce its costs of manufacturing by 5% over the course of the next year. Determine what the objective is; then establish suitable goals statements for manufacturing engineering to support that objective. Include at least one goal each for advanced manufacturing engineering, methods planning and work measurements, and process control. A sum of $1,250,000 is available for investment; shop operations employs 250 people.

6. Refer to the schedule control chart in Fig. 2-17. Because of changes in personnel availability, the following items have had revisions to the completion dates. Construct a new schedule control chart showing the changes caused by these revisions: (a) Audit steam traps—completion delayed 6 weeks; (b) Set up system to monitor weld quality—completion advanced 3 weeks.

7. Define an action plan and explain how it would be used by an engineer and how the same action plan would be used by a manager.

8. A manager of manufacturing finds that a project to purchase a horizontal boring mill and a weld positioner are $12,000 over budget and $9000 under budget, respectively. Outline the items of information that should be included in the manager's key results report in addition to the basic over-budget and under-budget facts.

9. In reference to a short-range forecast, discuss why it is important to investigate all over-budget and under-budget forecast line items, even though the total forecast is under or equal to budget.

10. Explain how changes in the schedule control chart (question 6) will affect the short-range forecast and periodic review of operations.

11. List three typical account categories that would probably require daily accounts control. Give your reasons for each one.

12. Define the difference between line management and project management.

13. List the rules for creating a work breakdown schedule and describe how it applies to manufacturing engineering projects.

14. Show how elements of risk are important for manufacturing engineering projects selection.

15. Discuss the difference between work planning and project management.

16. Show how a short range forecast and a review of operations fits within project management procedures.

Chapter *3*

FACTORY CAPACITY AND LOADING TECHNIQUES

Factory capacity and loading evaluations are critical for determining exactly what the factory can produce in its current alignment. This knowledge is essential for the company to be able to decide what products it can sell and when it will be able to deliver them. The task of determining the factory's capacity is assigned to the advanced manufacturing engineering unit. The task of loading the shop is assigned to the materials function's master scheduler for compliance with overall business plans and to production control, for the specific or short-interval scheduling. Manufacturing engineering plays a very important role in verifying that the shop loading is in accordance with the capacity evaluations. Therefore, capacity and loading are studied together, since the former leads directly to the latter.

TECHNICAL CAPACITY

The ability to accurately determine the technical capacity of the factory is worth a considerable amount of money to a company. Consider what happens if a company bids and wins a contract to produce a complex product, and then finds out that it does not have the technical capacity to produce it. What are its options? First, the company could give back the contract, admitting that it made a mistake. This would be a rarity indeed; not many companies are willing to admit they are incompetent, and very few customers would be willing to give that company additional contracts, even if they were for different products. Second, the company could buy the technology or sub-contract to another firm that has the necessary know-how. This might solve the problem with the customer but it would be frightfully expensive, especially considering that the companies having the correct technology were probably competitors for the contract in the first place. These are two very pragmatic reasons for accurately knowing your company's technical capacity.

When speaking of technical capacity, I am referring primarily to manufacturing capability, comprising the proper facilities and the technical people to oversee and maintain them. The secondary meaning of technical capacity concerns the design engineering function—whether it has the proper resources to design the product, so that the product can be produced by the company's own factory. This part of capacity analysis is the domain of the producibility engineer and will be explored in Chapter 6.

PHYSICAL CAPACITY

Besides technical capacity we must consider physical capacity: how much product can be produced in a given period of time. From an economic viewpoint, it is vital to be able to accurately determine this facet of capacity. Firms that habitually produce and deliver products later than promised do not accurately know the capacity of their factories. This costs the producing company money, because the company is usually not paid until the product is delivered, or if progress payments are allowed, not until certain benchmark production phases have been completed. In addition, if making the product is falling behind schedule, more of the company's funds are tied up in inventory and not earning interest, which is like putting money in non-interest-bearing bank accounts. This also means that the company may have to make up for poor cash flow, probably by borrowing funds to keep operating until it can collect for sale of its products.

The ability to accurately determine capacity is important for other reasons as well. One is to avoid penalty clauses in contracts for non-deliverance by a specified date. This type of contract is fairly common for large capital goods, and if the capacity is only vaguely known, the company is at considerable risk of falling into an adverse penalty situation. Finally, accurately knowing the physical capacity of the factory is a factor in preserving the company's good reputation and avoiding the situation where other companies are awarded contracts because the customer cannot chance late delivery from a firm that has a reputation for erratic deliveries.

TECHNIQUES FOR DEVELOPING WORKSTATION CAPACITIES

Unless a factory is made up of all skilled craftsmen using hand tools, it must be considered an amalgamation of many human–machine units. Therefore when we talk about determining workstation capacities, we must take into account the designed capability of the machine tool as well as the capability of the operator to run the workstation.

Because people are not machines or robots, and their performance varies from day to day, we cannot expect exact repetition of effort every time a work-piece goes across the workstation. Hence, determining workstation capacity depends on finding the lower bound of the effort range to be expected of the average person. This is a complex task. It may be done by complex work measurements for the most analytically correct answer, by estimation by knowledgeable people for a much quicker answer. The result is an efficiency factor by which the theoretical design capability of the workstation is modified downward to obtain a capacity factor for the workstation.

> **EXAMPLE:** Let us use a semiautomatic welding station utilizing a submerged arc continuous-feed welding process and a positioner to rotate the parts to be welded. In this example, we are producing circumferential welds with each weld being applied in the vertical downward position. The basic outline of the workstation is shown in Fig. 3-1.

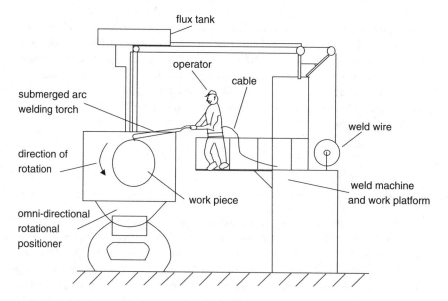

Figure 3-1. Semiautomatic Welding Work station

The known parameters of the workstation are (1) the speed and direction of rotation of the positioner, and (2) the deposition rate of the weld process. Therefore, it is possible to calculate the deposition rate on the work piece. Also knowing the volumetric cross section of the weld, it is possible to determine how long it should take to make the completed circumferential weld. This is the theoretical capacity of the weld machine workstation. Unfortunately, it is useless in developing workstation capacity.

The machine can work forever, but it needs human intelligence to control and guide it. For this example we will state that the machine is not computer controlled in whole or part. Hence, the problem of workstation capacity ultimately reduces to how much time out of the 24 hours in a day an operator will guide the machine. Let us assume that this particular submerged arc process can deposit 30 pounds of weld per hour, or 720 lb/day. This is the ultimate theoretical capacity of this workstation. To determine a practical capacity, the following factors must be taken into account:

1. *Setup time*. The work piece must be set up on the positioner. Experience dictates that roughly 10% of the time to do any job is the setup time. This is equivalent to a reduction of 3 pounds per hour in the deposition rate.
2. *Personal needs*. An operator cannot work continuously for an entire 8-hour shift because of fatigue, the need to consume food, and the need to take care of personal hygiene. As a rule, people need 30 minutes for a lunch break and another 30 minutes per 8-hour shift for rest and personal hygiene. Therefore, out of 8 hours we must subtract 1 hour for personal time, or 12.5% of the work time available per shift. This corresponds to a reduction of 12.5% of the theoretical capacity, or 3.75 pounds per hour.
3. *Absenteeism*. The usual absenteeism rate in a job shop industry is about 2%. This means that the typical worker will not show up for work approximately 5 days per year because of illness or personal needs. The absenteeism factor of 2% equates to a 0.6 lb/hr reduction in deposition rate.
4. *Interference time*. This is non-productive time when the operator is available to work but cannot. For a positioner weld workstation, interferences might be due to refilling the flux tank, replacing an expired weld wire reel, performing machine maintenance, performing in-process inspection checks, reading process instructions, and fixing the machine when it breaks down. The normal practice is to assign 10% of available time for the interference factor. In terms of deposition rates, this amounts to 3 lb/hr.
5. *Mistakes*. Operator mistakes happen and must be accounted for in any realistic capacity evaluation. Typically, the reduction in capacity allowed for operator mistakes is 3%, which in this example is equivalent to a 0.90 lb/hr reduction in deposition rate.
6. *Efficiency*. This is a measure of how the operator performs the task compared to how the task was designed to be done. Continuous performance to how the task was designed to be done. Continuous performance at 100% efficiency would require all the operator's moves and the placement of all tools to be continuous exactly as originally planned by methods engineering. This is virtually impossible for job shop applications and very difficult to achieve for continuous manufacturing operations. In semi-automated welding, an efficiency of 80% could be expected. This means that 20% of the time is nonproductive, or a reduction of 6 pounds per hour in deposition rate.

Now let us see what a practical capacity would be compared to a theoretical capacity. Table 3-1 shows the computation for the example.

In rating the capacity of the workstation in the above example, we would have to use 12.75 lb/hr for the weld deposition rate instead of the 30 lb/hr theoretically possible. This is the figure that will be used whenever we are calculating the amount of work going across this particular workstation and determining how long it will take to perform it.

Efficiency is the largest of the deduction factors. For the purpose of capacity analysis, the advanced manufacturing engineering unit must know the level of efficiency the factory segments

TABLE 3-1. Computation of Practical Capacity for a Submerged Arc Positioner Circumferential Welding Process

Theoretical capacity	30 lb/hr
Deductions	
1. Setup time	−3.00 lb/hr
2. Personal time	−3.75 lb/hr
3. Absenteeism	−0.60 lb/hr
4. Interference time	−3.00 lb/hr
5. Manufacturing losses	−0.90 lb/hr
6. Efficiency	−6.00 lb/hr
Total deductions	−17.25 lb/hr
Practical capacity	12.75 lb/hr

are working at. This information is obtained from the normal measurements of efficiency, which is continuously being evaluated. It does not matter what the basis of the measurement is, as long as it is consistent. If it is consistent, the efficiency ratings will be usable as deduction factors to obtain practical capacity values for each workstation. As the productivity performance of the factory improves, its ultimate capacity improves. Thus, the exercise of determining capacity is dynamic, and factory capacity evaluations must be an ongoing activity.

Once we have a file of workstation practical capacities, we can begin to construct a matrix of factory capacities that can be used to load the shop. For a single product line this is relatively simple.

Figure 3-2 illustrates the sequence of operations used to produce a clutch plate. There are three basic operations and one inspection that must be completed. If we know the cycle time for each operation, we can determine the capacity to make clutch plates in the factory. These operations are listed in Table 3-2, where the factored times shown are practical times; that is, all adjustments have been made to go from ideal time to practical time.

Table 3-2 gives a great deal of useful information. It shows that the complete cycle time for making a clutch plate is ½ hour (time factor), while the longest cycle time is for the Numerically Controlled (CNC) lathe and the shortest for the forge operation. Thus, in order to have no queuing of parts, we need the equivalent of 1.8 lathes, 1.2 Horizontal Boring Mills (HBMs), and

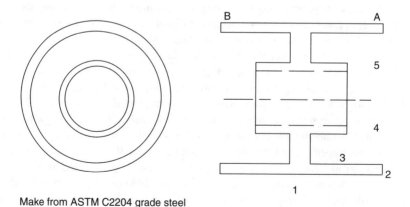

Make from ASTM C2204 grade steel

Step 1: forge basic shape

Step 2: use CNC chucking lathe to machine surfaces
1, 2, 3, 4, 5 on side A, turn around, and
machine side B

Step 3: use horizontal boring mill to mill bore

Step 4: inspect for dimensional accuracy

Figure 3-2. Operations Used to Produce a Clutch Plate

TABLE 3-2. Clutch Plate Operations Analysis

Factored Time to Perform the Operation (hr)	Number of Operators	Man-hour
Forge 0.10	2	0.20
CNC chucker lathe 0.18	1	0.18
Horizontal boring mill 0.12	1	0.12
Inspection 0.10	1	0.10
Total 0.50[a]	5[b]	0.60[c]

[a] Time factor.
[b] People required.
[c] Cost factor.

1.0 inspectors for each forge. These are the ratios of workstations required if a continuous-flow process is desired. Otherwise, a batch processing mode would be used. The decision would depend entirely on how many clutch plates are to be made. If the number is only a few thousand and another product will then be made with the same equipment, the batch mode would be chosen. If the number is in the hundreds of thousands or more, the continuous mode would be necessary.

Other useful information is obtained from this example. For instance, we can determine how many people are required (people factor) and the labor cost to produce the product (cost factor). The labor cost is determined by simply multiplying the labor rate in dollars per hour by the cost factor to obtain labor cost per part.

Now what is the capacity to make clutch plates? It takes ½ hour to make one if we have only one set of equipment. Since we already have factored times, then we know in an 8-hour shift 16 clutch plates will be produced, or a maximum of 48 per day. Deducting vacation and holiday times, the average factory operates for 46 weeks per year. Therefore we have 230 productive days per year at 5 days per week, which corresponds to a maximum capacity of 11,040 clutch plates per year. With this information, the master scheduler can now load the shop and give promise dates for parts to be delivered.

The example above is for a part that is made with machines dedicated to making that part. When we have many different parts of varying complexity to make, the technique must be extrapolated and expanded. If the parts are many and varied, and especially if they are unfamiliar, the first thing the advanced manufacturing engineering unit must do is determine whether the part can be made in the factory as presently constituted. This leads us to the basic definitions and relations between capability and capacity:

Capability: The physical limitations on what can be produced in the factory including size, weight, processes, and materials

Capacity: The number of items that the factory has the capability to produce that can be produced in a given period of time

Before we can determine factory capacity in a multi-product or component shop, it is necessary to determine whether the factory has the capability to make the product. Figure 3.2 outlined the steps necessary to make a clutch plate. If the factory does not possess any of this equipment—forge, CNC chucker lathe, horizontal boring mill—then it does not have the capability to produce the product. The capability review then becomes a matter of investigating whether the product can be made.

When the factory knows exactly what it is going to make, the capability review is a review of fitting equipment to the known product. Often a factory is set up to make a range of similar products or products made by similar manufacturing techniques. In this case capability is defined by the ranges the equipment can handle. Figure 3-3 shows a capability matrix for a company that manufactures special-purpose motors. Under each category the largest sizes and tightest dimensions the

SPECIALTY MOTORS
MANUFACURING CAPABILITIES & CAPACITIES

Item	Physical characteristics range				Processes available for mfg.	Equip. capacity shifts	coments
	Dimensions	Weight	Tolerances	Materials			
Rotors							
Shafts	18" x36" dia.	30.0 tons	+/-0.002"	Alloy steel	Lathe, slotter	3	
Punchings	0.025" thick	30 lbs	+/-0.0001"	Silicon steel	Punch press	3	min.size = 0.014" thick.
Coils	2" x 2" cross section	-	+/-0.0625"	Copper	Form, mold, tape	2	max. length = 48"
Bars	2" x 2" cross section	-	+/-0.0625"	Copper	Form, mold, tape	2	max. length = 62"
Rotor bodies	7" x 54"O.D. x 20" I.D.	25.0 tons	+/-0.001"	Alloy steel	Lathe, slotter	3	laminated construction
End rings	4" x 1" cross section 60" O.D.	-	+/-0.003"	Alum.,alloy steel	Roll form	3	
End Plates	8" x 0.25", 60" O.D.	-	+/-0.003"	Alum.,alloy steel	Mill	3	
Amort rings	2" x 2" cross section 60" wide	-	+/-0.003"	Copper, brass	Roll form	3	
Poles	7" long, 8" high, 6" wide	1.5 tons	+/-0.010"	Steel, copper	P/P, positioner wind	2	
Exciters	2" x 22" dia.	0.4 tuns	+/-0.004"	Steel, copper	P/P, hand piled	2	
Stators							
Frames, fabricated	14' x 8' x 6' x 0.75" plate	7.5 tons	+/-0.005"	Steel	Submerged arc, HBM. VBM	3	
Frames, double shell	12' x 6' x 4'	7.5 tons	+/-0.005"	Steel	Cast, HBM, VBM	3	
Top hats	14' x 8' x 4' x 0.25" plate	3.0 tons	+/-0.030"	Steel	Brake, drill	2	
Bearings							
Sleeve	24" x 25"	-	+/-0.0005"	Steel, tin, lead	Spin cast, lathe, HBM	2	
Cylindrical	14" x 15"	-	+/-0.0005"	Steel, tin, lead	Spin cast, lathe, HBM	2	
Spherical	14" x 15"	-	+/-0.0005"	Steel, tin, lead	Spin cast, lathe	2	
Ball	14" x 15"	-	+/-0.0005"	Steel, tin, lead	Purchase	2	Purchse to spec.
Roller	14" x 15"	-	+/-0.0005"	Steel, tin, lead	Purchase	2	Purchse to spec.
Coils							
Coils	2" x 2" cross section	-	+/-0.0625"	Copper	Form, mold, tape	2	
VPI	19' x 160"dia.	50.0 tons	-	Epoxy resin	Pressure tank, bake oven	3	
Varnish	-	-	-	Epoxy resin	Dip tank, bake oven	3	
Punchings	0.019" thick	0.5 lbs.	+/-0.0001"	Silicon steel	Punch press	3	Min. size = 0.014 thick
Cores	8" x 20" I.D. x 70" O.D.	45.0 ton	+/-0.001"	Steel, silicon steel	Stack press	3	
End shields	6" thick x 72" O.D.	2.0 tons	+/-0.002"	Steel	Cast, fabricate, VBM, HBM	3	

Figure 3-3. A Capability Matrix

factory can produce are listed. The processes available and the materials the factory is prepared to use are also shown. The capability matrix would be used by design engineering so they do not design a motor that cannot be made in the factory, and by marketing as a guide as to what they can sell to customers.

THEORY OF CONSTRAINTS

I bring up the Theory of Constraints (TOC) here because capacity and loading as I described may give the impression that it is a specifically defined level for a given facility and set of circumstances, and once calculated the number is cast in stone forever to be that way. This is not so. We shall see that as we go through this chapter that process improvement is at least implied and that capacity and capability can be changed, hopefully for the better. There are many ways of doing that. Manufacturing engineering is charged with continuously improving the ability of a factory to improve through engineered solutions. This means manufacturing engineering must always have on its agenda the goal of finding better ways of setting up and facilitating the factory. TOC is a philosophy that has promise in enhancing the ability to achieve that goal.

The Theory of Constraints is a methodology of identifying those limitation in manufacturing brought about by finite amounts of physical capability and capacity. Then knowing these limitations using a specific methodology of getting around these limitations, in essence creating improvements. TOC has been promoted by its creator, Dr. Eli Goldratt in his highly popular and common sense books starting with *The Goal* and several others since then, including *The Haystack Syndrome*, *It's Not Luck*, *Critical Chain*, and *Theory of Constraints*, all published by North River Press, Inc., Great Barrington, MA. They offer great insight into how to focus on what's important; e.g., making money by understanding the roadblocks in the process, the constraints, and fixing them. Dr. Goldratt offers

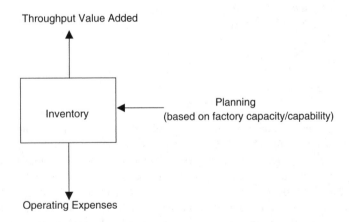

Calculations of factory capacity and capability effect the basic fundamental measurements

Three dependent variables:
        Throughput value added
        Inventory
        Operating expenses

One independent variable:
        Planning

Figure 3-4. Based on Fig 1.4 "The Three Fundamental System Measurement" (From *Introduction to the Theory of Constraints Management System* by Thomas B. McMullen, Jr.,; St. Lucie Press, Boca Raton, FL, 1998.)

a prescription which is compatible with the traditional scientific method for solving problems. It's not possible to go into sufficient detail as to what entails all of Dr. Goldratt's theory in this text. For that I refer you to his very readable and enlightening books. However I will summarize the technique and then discuss how that impacts capacity and capability decisions.

One of Dr. Goldratt's colleagues, Thomas B. McMullen, Jr. has written a book, *Introduction to the Theory of Constraints (TOC) Management System*, St. Lucie Press/APICS series on constraints management, Boca Raton, FL, 1998, in which he elaborates on the basic Goldratt theory showing a simple systems diagram depicting throughput inventory and operating expenses as the three prime factors in a manufacturing production system. McMullen describes throughput as the output we desire in sellable finished product, inventory is the material we have to work with to make the throughput and operating expenses as all the costs associated with doing so. He goes on to say that our (the manufacturer in this case) task is to find the best ways of optimizing the process by finding the core or root cause constraints to doing a better job. And either eliminate them or minimize their effect on throughput. The TOC requires the use of logic trees, elaborate cause and effect and if/then type analyses to find the best path to get to optimum throughput for the given situations. Its main tool for defining what to do is what I believe to be a specific application of the traditional scientific method. McMullen goes on to show the five step approach proposed by Goldratt:

> *TOC Five-Step Focusing Process*
> 1. Identify system constraints
> 2. Decide how to exploit the constraints
> 3. Subordinate everything else to the above decision
> 4. Elevate the constraints
> 5. Return to step 1. Don't let inertia become the new constraint

My list of defined steps for the Scientific Method are:

> *The Scientific Method*
> 1. Make observation
> 2. Develop a hypothesis
> 3. Test the hypothesis
> 4. Make revisions to the hypothesis based on the test
> 5. Test the revised hypothesis
> 6. Reach a workable conclusion

This process compares virtually identical with the traditional scientific method in that it defines what the problem to be solved is, then focuses on how to achieve a successful outcome, and to test to make sure it really does. We can see if the TOC method is applied it could logically result in achieving an improvement (for more on the scientific method see Chapter 7 on Methods, Planning, and Work Measurements). As we can see if we can find the limitations in capacity, looking at any and perhaps several if not all of the six deductions, and lift the bar so to speak, we can squeeze out considerable more product output than previously thought. Therefore we can see that capacity and capability is indeed a dynamic value, not static and cast in stone. So we use the technique of the six steps to find the deductions of capacity then can use the Theory of Constraint or other valid processes to gain improvements.

EVALUATING FOR PRACTICAL CAPACITY

Now that we have discussed the techniques of evaluating individual workstation and individual product capacities, it is possible to consider the techniques for evaluating total factory capacity. It is interesting to note here that the only independent variable is the capability of the factory

TABLE 3-3. Workstation Capacity Per Year

Workstation	Product							
	1	2	3	4	5	6	7	8
Forge	19,872	628	387	12,082	—	11,683	—	—
CNC lathe	11,040	721	172	9,212	10,563	12,692	2,163	3,812
Horizontal boring mill	16,560	823	279	—	9,732	14,678	—	1,962
Vertical boring mill	—	562	202	8,961	9,978	—	1,672	3,177
Inspection	18,065	670	311	9,743	9,853	12,672	1,833	2,468
Total	11,040	628	172	8,961	9,732	11,683	1,672	1,962

to produce, that is, the physical and process limitations identified in the capability matrix. The capability matrix shows what can be made but does not attempt to quantify how much of each known product can be made. It is very important, therefore, to be certain that a contemplated mix of products is not to be in violation of the bounds prescribed by the capability matrix. If it were, it would necessitate either investments for new equipment and personnel training, or subcontracting of phases of manufacturing to other companies.

Evaluating for practical capacity in a multi-product company requires predicting how much of the resources will be devoted to each product. This is an all but impossible task for manufacturing engineering. In fact, the marketing function is the only segment of the company that has the proper resources for estimating the market. For this reason manufacturing engineering does not try to determine market characteristics and trends, but develops basic data showing a specific mix of products with percentage quantities that can be made. Often this is not acceptable to all functions of the business because it specifies a fixed percentage for each product. Clearly, when market opportunities change, the set percentage does not remain useful. Therefore, the most recent trend in capacity analyses is to develop a matrix of all capacities and then let the user plot the effect on capacities if the set of percentages are varied. To do this, manufacturing engineering develops capacities per product as if no other product existed, and then drafts capacity charts to enable the marketing or master scheduling personnel to determine how much capacity is still available after a number of specific products is booked into the schedule.

This is a mortgaging of facilities activities. If we have 10 hours available at workstation A, and product X requires 1 hour, product Y requires 2 hours, and product Z requires 5 hours, the number of each product to go through workstation A will depend primarily on the priorities given to the products, with the result that there will be a large number of possible variations. In practice, the number of product variations to go through the workstation will depend on customer preference and orders on the book or potential orders to be received. This information comes from the master planner via the marketing function. Hence, very little depends on mathematical manipulations. An example of a capacity chart prepared by advanced manufacturing engineering is shown in Table 3-3.

The analysis for each product is done independently in the manner described above. Each workstation is evaluated for practical capacity to make each specific product as if no other product existed. Each workstation will consist of itself alone or in total if multiples of the workstation exist. Note that the total yearly capacity is always equal to that of the lowest-capacity workstation. This means that the lowest-capacity workstation is 100% booked for that product while the other workstations, having a higher capacity to make the product, are less than 100% booked.

EXAMPLE: For product 1, the yearly capacity is 11,040 pieces, while the forge can make 19,872 pieces. Therefore, only 55.6% of the yearly capacity of the forge is required to make product 1, and 44.4% of capacity is available for other products. (Product 1 is the same product used in the example illustrated by Table 3-2.)

A practical way to evaluate capacity for a proposed multi-product run would be to look at the critical workstations only. If capacity exists at the critical workstations, the likelihood of making the products at the other workstations is virtually 100%. Therefore, by focusing on the critical workstations, it is possible to reduce capacity evaluation exercises significantly. In this case, the term critical workstation is synonymous with lowest-output workstation, sometimes called the bottleneck workstation.

Some very important but perhaps unquantifiable data must be subjectively evaluated by the area planners in determining capacities. These data are often critical in factoring the theoretical capacity to a pragmatic capacity. Unquantifiable data must be rated in importance and empirical correction factors used to modify the mathematical results the area planner produces as the capacity per product. Among the hosts of unquantifiable items that must be considered in determining capacities, some of the more common ones are:

1. *Key operator retirement.* Assuming that the key operator is the most important factor in determining workstation efficiency, his or her retirement will have a detrimental effect on capacity. Usually a 5 to 10% drop in efficiency for 3 months occurs when this happens. Therefore, in the weld example (Fig. 3-1 and Table 3.1) an additional 3 pounds per hour or 10% reductions could result for a full operating quarter.

2. *Union contract due for renewal.* If the labor climate is favorable, this will have no effect. If it is one of continuous adversarial combat, a serious reduction in output will occur in all six areas of reductions of theoretical capacity.

3. *New equipment installed.* A learning curve must be anticipated going from lower to higher (designed) output. The length of the learning curve must be estimated on the basis of previous experiences, the expected competency level of on the job instructions and the complexity of the new equipment compared to the old.

4. *Tolerance changes.* Any tolerance change will affect capacity. The closer the tolerance is to the practical capability of the machine, the less product will be produced in the given time period. My experience shows that a reduction of 3:1 in output occurs whenever the required tolerance in machining goes below ±0.015 in.

5. *Weather factors.* The more uncomfortable it is to work in the factory, the higher the manufacturing losses become. I have observed 1.5:1 increases in losses in foundry-type operations from winter to summer in a north-eastern U.S. location.

6. *Workstation maintenance.* Lack of maintenance not only results in more machines being down, it also affects daily operations through failure to attend to minor items. This causes operators to make adjustments that may or may not be correct; that is, the "Kentucky windage"* applied to the engage handle on a worn gear train may not have the result the operator expects. This leads to increases in manufacturing losses and lower productivity. The area planner should use a little Kentucky windage to determine what this means in terms of capacity.

7. *Excessive planned overtime.* Rarely does a continuous 6 day schedule equal 6 days output. Usually, through fatigue, we end up with perhaps 5½ days output instead. Causes are excess absenteeism and lower efficiency.

8. *Design changes in materials specification.* This is one of the most far-reaching types of changes that can affect capacity. If the change results from a manufacturing request, it is usually done to improve manufacturability of the product; hence capacity increases. If it results from new design engineering requirements, the results tend to be materials that are more difficult to machine; hence capacity decreases. The area planner must thoroughly research any pending changes in materials.

There are countless more unquantifiable or intangible factors related to people, equipment, and material that affect practical capacities. The important point here is that the pertinent unquantifiable

---

* A sense of the correct adjustment that an individual gains after many repetitions of the same thing. Originally used to refer to a windage correction made by aiming a firearm to the left or right of a target rather than adjusting the sights.

items should be identified by advanced manufacturing engineering so that empirical factors can be assigned to subjectively adjust practical capacities upward or downward.

EFFECT OF LEVEL LOADING ON CAPACITY ANALYSIS

Strictly speaking, level loading is a production control function. Manufacturing engineering becomes involved only to ensure that all technical considerations are observed. For example, to level load for a factory to run at 90% of designed theoretical capacity for an extended period, greater than 3 months, is pragmatically impossible. I have never come across a factory that it can run continuously at a rate greater than 67% of design theoretical capacity. Hence, it is important for manufacturing engineering to evaluate the level load plans to ensure that no inherent damage to the factory can occur, and where production output plans are doomed to fail even before commencing.

Level loading is a technique used to smooth out production, that is, to produce the same amount of product over each measurement period, usually a week or a month. It calls for the plant to work at a certain level of output regardless of the amount of finished product required for any given period of time. If a workstation can make 12 units per month and the requirements for the next 3 months are 9 units, 16 units, and 11 units, respectively, then more than required will be made in the first and third months and less than required will be made in the second month. The total produced will be 36 for the 3 months, equaling the requirement for the period. The problems for the business caused by this level loading are that carrying costs for inventory are increased in months one and three, and there is a late shipment by one unit in month two. These are typical of the considerations the manager of manufacturing must make when deciding whether to use level loading.

In a more complex system, more than one product would go through a workstation or a series of workstations. The loads at the workstations are the sums of the various products. Figure 3-5 illustrates a level loading chart for a 3-month period at a multiproduct workstation or center (common workstation). In this case we can make 9,000 units, assuming that products 1, 2, and 3 require the same operation time at the workstation. The problem is that the requirement is 9,900 units, or 10% more than can be produced. Level loading would ensure that 3,000 units are produced in all 3 months. This means that in May we would be short 800 units; in June short 300 units for the month, 1100

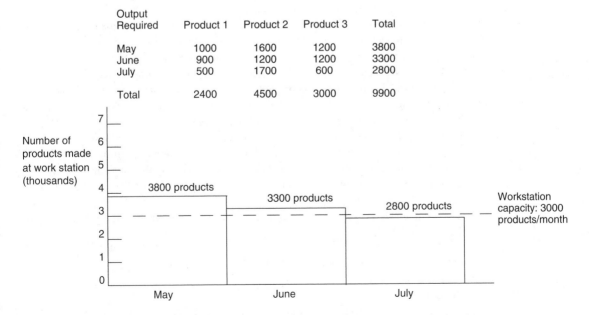

Output Required	Product 1	Product 2	Product 3	Total
May	1000	1600	1200	3800
June	900	1200	1200	3300
July	500	1700	600	2800
Total	2400	4500	3000	9900

Figure 3-5. Level Loading Chart for a Multiproduct Workstation

units cumulative; and in July produce an excess of 200 units for the month, resulting in a cumulative shortage of 900 units. Therefore, if the factory could produce 10% more for each of the 3 months, the total of 9900 units could be produced. In May, trying to achieve 3,800 units (800 units above capacity) may be impossible, but 3300 units could probably be made. If this can be done in May, it should be possible to do it in June and July. The master scheduler must decide whether it is possible. Manufacturing engineering will have to be able to answer the technical part of this question, but before we deal with the technical part, let us look further into the philosophy of level loading.

Level loading is a way of achieving a steady state. No business wants crisis after crisis, or to go from highs to lows. This fatigues people and wears out machines, which leads to mistakes and a lower life expectancy for equipment. In order to achieve success in an environment of highs and lows, a corporation must have enormous reserves to deploy at the apogee and then place them in storage at the perigee. Industrial concerns do not have such unlimited reserves, and even if they did, employing them in such a manner would be wasteful and nonoptimal. In a company with a profit motive, every dollar spent in a nonoptimal mode increases overhead and thus decreases the profit margin. The simple cost equation is

$$\text{Total cost} = \text{direct labor cost} + \text{material cost} + \text{overhead cost}$$

Since overhead costs are usually associated with control costs (in this case maintaining an un-optimized reserve to handle highs and lows), the more the variance from a level operation, the higher the overhead costs. Level loading, then, is the most common way to bring order to a business and to lower overhead costs.

Level loading lowers overhead costs by allowing routines to be established. In essence, routines are solutions that have been significantly optimized, and they tend to be the most efficient way of doing things. Routines depend on predictability; hence, random variables such as large swings in production demands must be minimized for routines to exist.

In reference to manufacturing, the basic goal of level loading is to create a smoothly running factory, with machines and processes running at the most efficient load rate, inventories of raw material at as low a level as possible, and the labor force stable. Therefore, true level loading requires a balance between demands for equipment, labor, and material. This affects manufacturing engineering primarily in the equipment areas, and secondarily in the labor and material areas.

Manufacturing engineering, through its objectives and goals statement, must consider the needs for equipment to maintain a level load. Such considerations as cost to buy, floor space for setup, and utilization (or load) must be continuously evaluated.

Considerations made by shop operations management to maintain level-loaded production are of interest to manufacturing engineering. Shop operations will consider such factors as retaining trained people, morale of the work force, effects of cyclic hiring and layoff on productivity, availability of labor to hire, and overtime. Each of these items can affect the workload of manufacturing engineering in providing technical support to the shop floor, particularly the need for training and the detail of instructions necessary in the planning. A changing work force usually has a detrimental effect on efficiency, which in turn affects workstation capacity.

Materials considerations also affect manufacturing engineering operations. Increases in materials require additional handling capability and storage locations. Manufacturing engineering must deal with these factors as it evaluates factory floor layouts. Having the machine capacity for the desired level of production is worthless if the raw materials cannot reach the workstations on time per schedule.

The ways in which level loading is accomplished have already been alluded to in the examples. Specifically, it is done by:

1. Manipulating schedules through
   a. Early starts
   b. Delayed starts
2. Making supply or stock parts during slack periods

3. Having plant shutdowns (vacations) during the slack season
4. Applying more labor during peak periods through
   a. Overtime for existing personnel
   b. Adding personnel to existing shifts and starting up idle equipment
   c. Adding shifts (i.e., if only one shift is employed, starting up a second)

The overall aim is to maintain a steady output per hour per workstation, and to regulate the total output by how long the workstation is run. This is supplemented by adding additional workstations only as a last resort. As a rule, additional workstations are added for a short duration by utilizing idle, less than optimum equipment and for a long duration by purchasing additional efficient equipment.

Now let us look at the technical portion of the answer for the problem posed in Fig. 3-5. The factory wants to run at a 10% overload for 3 months. Manufacturing engineering will have to determine whether the physical plant can handle the overload. They will do that through the following steps:

1. Recognize that no factory can run above 67% of theoretical capacity for more than short periods of time. This usually means up to 3 months.
2. Determine what facilities will be required to run at or above the 67% theoretical capacity.
   a. Review the additional maintenance needs to do so.
   b. Evaluate the spare parts inventory available and reaction times needed to obtain additional spares if necessary.
3. Evaluate the cost of maintaining the factory in an accelerated mode.
4. Evaluate the labor costs, considering
   a. Availability, vacation plans, and so on;
   b. Effect of the fatigue factor on efficiency of maintenance and repairs.
5. Determine what factors in the capacity analysis for each workstation can be optimized for a 3-month period to produce 10% more output.
   a. Reschedule vacations so that key operators are available, ensuring better efficiency, attendance, and less manufacturing losses.
   b. Consider designing new setup fixtures and improving methods to reduce setup time.
   c. Consider assigning manufacturing engineers to reducing interference times by giving specific attention to workstations that are potential bottlenecks.

These are common items that the advanced manufacturing engineering unit would look into to answer the technical question of whether a 10% increase in output can be achieved temporarily for 3 months. The answer, of course, would consist of facts and figures detailing what could be done and how that would minimize the six deduction factors used to derive practical capacity from theoretical capacity. The Advanced Manufacturing Engineering (AME) unit would also include their evaluation of the costs of such an exercise and how long they thought such an accelerated program could continue. It would then be up to the manager of manufacturing to decide whether the risks are worth the potential benefits.

BOTTLENECK IDENTIFICATION AND ELIMINATION

A discussion of factory capacity and loading would not be complete without devoting a section to the subject of bottlenecks. Bottlenecks, or choke points, in the manufacturing flow are the bedevilment of management. They cause countless frustrations, delays in production, extra costs, and conceivably lost orders. Simply stated, bottlenecks need to be identified and eliminated. If bottlenecks go undiscovered, they can and often do wreak havoc with the capacity analysis.

Bottleneck Definition

What is a bottleneck? The simplest definition is that a bottleneck is a point in the production flow where the workstation is working at capacity, yet workload volume grows faster than the completed work leaves the workstation. The situation is analogous to a three-lane highway being restricted to two lanes at a time when two lanes are inadequate to handle all the traffic. Hence we have a backup of vehicles before the restriction. Recall that when one encounters this situation vehicular velocity decreases considerably. The same happens in a factory. When a bottleneck occurs, the flow of parts traversing the factory also slows down. As in the highway analogy, bottlenecks are not permanent situations. They come and go depending on the dynamics of the situation. Let's now explore how a bottleneck can be identified and then how it is minimized and often eliminated completely.

Identifying Bottlenecks

Rates of production reveal bottlenecks. A bottleneck will be uncovered if and only if the overall rate of production is beyond the capacity of the workstation to maintain. This means that if the capacity built into the workstation is never exceeded, that workstation will not become a bottleneck. This in turn means that bottlenecks are not permanent. They exist only at specific workstations at specific rates of production. They arise and disappear as rates of production increase or decrease, if left to their own devices and not interfered with by the factory's controlling function. The manufacturing engineer's role vis-à-vis bottlenecks are predicting where bottlenecks will occur at what levels of production volume. This is done via the capacity and capability determination techniques previously discussed in this chapter.

A bottleneck, when it appears, will have the following characteristics:

1. There will be a large amount of work queued in front of the workstation waiting to be processed.
2. The workstation will be working at its maximum capacity, as presently defined.
3. Succeeding workstations will be underutilized.
4. Factory output will most likely be less than the desired amount per time period.

Possible Solution Alternatives for Eliminating Bottlenecks

Several methods are available for relieving bottlenecks. Let's examine them. In the section on level loading we explored ways to balance production over longer periods of time. We saw that it was possible to get production through a factory that had bottleneck conditions as described above, although we didn't call them bottlenecks. They were, and it becomes apparent that bottlenecks are not unusual situations but, unfortunately, common. The methods of elimination resolve down to three:

1. Schedule manipulation
2. Methods improvement
3. Facilities addition

Schedule Manipulation. The basic techniques for accomplishing balance of production were discussed in the level loading section of this chapter. Essentially, we manoeuvred the schedule so that the threshold for creating a bottleneck was raised. This meant there was no bottleneck because we made sure all workstations were working at a level that did not cause a bottleneck. Work flowed freely from one workstation to another at the desired rate. Admittedly, some workstations may have had to work longer hours than others for enough work to flow at the proper rate throughout the factory. But this is the heart of this bottleneck elimination technique. The terminology for creating this situation is "creating a buffer."

A buffer is material waiting to be worked on that is stocked beyond the bottleneck in order to feed the workstations down line that work at higher rates. Faster workstations consume inventory faster than bottleneck workstations—like a Gatling gun consuming ammunition faster than a muzzle-loading rifle. We purposely create a large inventory queue in front of these faster workstations so that they can work at their most efficient rates and still have an adequate production line balance (e.g., the fire bucket brigade synchronous flow). The level loading techniques create buffers as needed. By working overtime, for example, we are creating a buffer for the faster but idle workstations. Thus, when all the workstations are working they can work at their individual optimum speeds (which need not be the same) and the factory is level loaded. It's interesting that this technique of bottleneck elimination revolves about working the bottleneck workstation longer by use of overtime, additional shifts, and perhaps working on days the factory is normally closed. We do this to create the downstream buffer, but we never attempt to relieve the root cause of the bottleneck. The next two bottleneck solution techniques do.

Methods Improvement.   Bottlenecks are caused by workstations not being able to keep up, flow-wise, with faster workstations ahead of them. The schedule manipulation technique solves the flow problem by making sure there are adequate supplies of inventory after the bottleneck (a post-bottleneck buffer, so to speak, but one that forms naturally) so that it can work when the faster stations are tactically idle, thus creating a suitably large buffer for them to work on at superior rates when they do work. If we can get the bottleneck to work at the same rate as the downstream and upstream faster workstations, we do not have to employ the schedule manipulation technique. Instead we have solved the root cause, not manoeuvred around it.

One technique for doing this is the methods improvement technique. Here we examine how work is carried out at the workstations, look for ways to eliminate any wasteful activities, and also look for applications of better jigs and fixturing. These are the stuff of the Just In Time (JIT) procedures, which will be covered in a later chapter. Techniques for improving methods will also be presented in the chapter on Methods, Planning, and Work Measurements.

Facilities Addition.   A second way to eliminate bottlenecks, and not just bypass them, is through facilities additions. If methods improvements cannot get the bottleneck workstation up to sufficient flow rates, the company may want to purchase improved or simply duplicate (or other multiple) workstation equipment in order to produce output at a rate that eliminates the bottleneck designation.

> **EXAMPLE:** We need to deposit 22.5 lb/hr in a submerged arc positioner circumferential welding process. This will allow us to supply valve bodies in sufficient quantity to balance the downstream Computer Numerical Control (CNC) machining centers. With one head we have a practical capacity of only 12.75 lb/hr (see Table 3.1). If another head of equal capacity is added to the weld positioner, the deposition rate capacity will be 25.5 lb/hr. This is a higher capacity than needed, thus the bottleneck is eliminated.

Typically this is the most expensive and longest cycle time technique for eliminating bottlenecks. Consequently, it is the last technique to be considered. Effective approaches for purchasing capital equipment are discussed in the chapters on capital equipment programs and machine tool and equipment selection and implementation.

USE OF COMPUTERS FOR CAPACITY AND LOADING SIMULATIONS

Experience has taught me the folly of trying to develop and use a computer simulation for capacity and loading independent of total manufacturing needs. A system developed specifically for manufacturing engineering usually lacks items that can affect actual capacity. For one thing, manufacturing engineering systems usually disregard the materials aspect of capacity; they tend to make the erroneous assumption that materials will always be available and in accordance with ordered specifications. They also tend not to factor in lead times for replacement materials to arrive. Finally, independent manufacturing engineering capacity and loading systems rarely include all the workstations. A judgment is often made to cull down the volume of workstations to a number that is easier to work with, which is euphemistically called the set of critical workstations. A manufacturing engineering capacity and loading system lacks expert judgments and inputs from the other functions and subfunctions. Hence, it should be abandoned in favor of a system developed for the whole company, usually called the Manufacturing Production Control System (MPCS) which is typically part of an Manufacturing Resources Planning computer system (MRP II). Figure 3.6 shows an MPCS. By studying the various elements of this system we can understand the flow of a useful computer system and, more important, understand how manufacturing engineering's responsibilities for determining factory capacity and loading parameters fit into the overall manufacturing pattern.

The MPCS system shows the flows of information in a typical factory. This flow then becomes the basis of the Manufacturing Resources Planning (MRP II) system. MRP II takes the logic of the MPCS system and explodes it in chronological need sequence for each and every individual part and subassembly. This becomes the chronological factory load that has to be factored for workstation capacity. By running an MRP II scenario, manufacturing engineers can simulate the capacity of the factory to make the intended numbers of products, by type and volume.

By following the central core of Fig. 3-6 from top to bottom, we can see that it is a simple chronological sequence from receiving a customer's order to dispatching the work to the proper workstation. Along the line of the central core, we further refine the data from the customer's broad-based request for a finished product to the necessary discrete parts breakdown and planning required to make the parts at the various workstations.

Let us now look at each block of the flowchart to understand what is accomplished there. The basic flow diagram is presented so that the interrelations between the various manufacturing subfunctions and other functions can be appreciated. However, by understanding the nature of the work for each block we can visualize what the programmer will have to express in mathematical terms to create a useful MRP II system.

1. *Customer requisition.* The customer's order defined in terms the factory can understand, that is, by model number, drawing number, rating of output, manufacturing procedure number, and so on. This is a responsibility of the marketing function.

2. *Master schedule.* The basic sequencing of orders for the shop to produce. Here the customer's order is placed in the queue for sequence of production. Time slots for the various operations that must occur are reserved or mortgaged. This means that the design engineering cycle, the materials procurement cycle, and the major manufacturing cycles are fitted into an overall time schedule. This is a responsibility of the materials subfunction.

3. *Order to shop by start date.* The basic output to shop operations showing when major manufacturing operations must commence. It eventually triggers the production control schedule once specific planning has been completed. This is a responsibility of the materials sub-function.

4. *Shop prototype planning.* The preplanning of common models of company product offerings done to minimize the need for detailed planning for each customer requisition. Prototype planning is used to simulate capacities and hold places in the shop schedule when an order is expected but not yet received. It is also used to aid in the mechanization and reduce the time necessary to complete specific order planning. This is accomplished by storing the prototype planning in a database and making changes only for exceptions related to the specific order. This activity is a responsibility of the AME and methods, planning, and work measurement (MP&WM) units of manufacturing engineering.

5. *Shop planning.* The breakdown of the order into its component parts and the sequences of manufacture of those parts. Shop planning details the exact workstations to be used and how long it will take to complete each specific part. It is based on prototypes extracted from the database (step 4). At this stage it is ideal planning, not factored for current efficiencies at the various workstations.

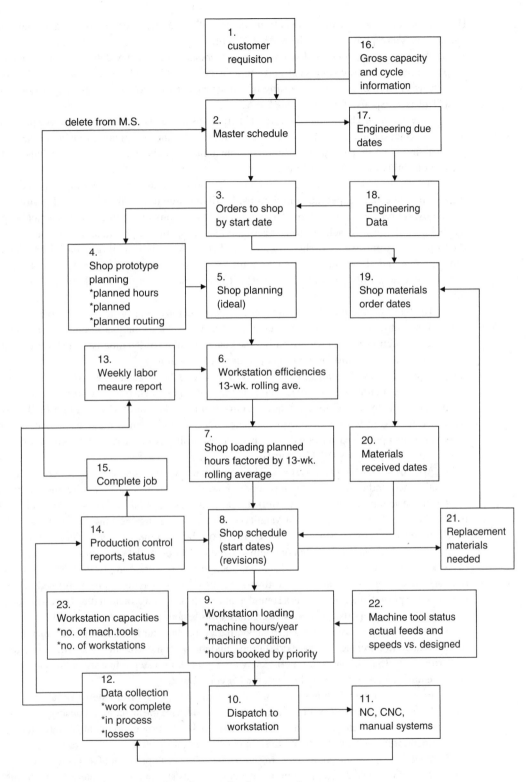

Figure 3-6. Manufacturing production control system

However, all the other five deductions from theoretical to practical workstation capacity have been made. This is a responsibility of the MP&WM unit.

6. *Workstation efficiencies.* The collating of workstation efficiency data in order to factor the shop planning by the efficiencies currently being experienced in the factory. In order to smooth out week-to-week variations in the reporting of efficiencies, a 13-week rolling average is commonly used. This is a responsibility of the MP&WM unit.

7. *Shop loading.* The planned hours and instructions for each operation that must be performed for all the components, assemblies, and tests that make up the completed product. These are the practical times that shop operations should be able to meet, barring unforeseen circumstances. This is a responsibility of the MP&WM unit.

8. *Shop schedule.* The detailed production control scheduling of the workstations, showing starting dates and expected completions. Revisions due to less than expected performance caused by poorer than anticipated efficiencies and variations in the other five practical capacity deductions are also scheduled at this step. This constitutes the actual promise date that shop operations publishes. It should be noted that work station loading is based on practical capacity as noted above and care is taken not to overload the work station beyond its current backlog and capability for additional work. It is a responsibility of the shop operations subfunction.

9. *Workstation loading.* The assignment of specific jobs to specific workstations. Here the production control specialists make reservations for workstation time based on machine tool and equipment conditions and the hours during the year when the workstation is available for production. This is a responsibility of the shop operations subfunction.

10. *Dispatch to workstations.* A time-delayed release to the designated workstations of work to be performed in chronological order. The planning packages consisting of instruction sequences, cycle times, drawings, N/C tapes, quality control forms, and so on; are stored in a file based on scheduled release to workstation sequence dates. Dispatches are then made to the workstations based on current promised starting dates. This is a responsibility of the materials subfunction. This commonly is done via the MRP II program.

11. *N/C, CNC, and manual systems.* The actual workstations where work is performed. The designation N/C and CNC refer to computer-automated subsystems that may be required to initiate and terminate actions at those types of workstations. The design of N/C and CNC subsystems is the responsibility of the AME and MP&WM units of manufacturing engineering. Operations at the N/C, CNC, and manual systems workstations are the responsibility of the shop operations subfunction.

12. *Data collection.* The manual or computerized collection of data reflecting the current status of work being performed at each workstation. Such things as work completed, work still in progress with percentage complete, and extra work having to be (or having been) performed to correct mistakes is reported on. Since this is a time-dependent step, the information may be readily used to compute efficiency reports. This is a responsibility of the shop operations subfunction.

13. *Weekly labor measure report.* The compilation of the data from step 12 into management reports related to efficiency of the workstations, and the percentage of the total work done to correct errors. The data on workstation efficiency are fed back into step 6 to correct the shop loading and scheduling. This is a responsibility of the MP&WM unit.

14. *Production control reports, status.* The compilation of data from step 12 that updates the shop schedule, step 8. In step 8 these data will be used to revise the schedule, if necessary, for early completion or incomplete work. This also forms the basis for production control status reports to management. It is a responsibility of the shop operations subfunction.

15. *Complete job.* A special input from step 14 when the last operation of a customer requisition subcomponent is finished. Completion of such components as a stator for a large AC motor or the stop valves of a steam turbine would normally require payment by the customer. Hence, a report of their completion is required. Output from this step is fed into step 2 to delete the item from the master schedule. This is a responsibility of the materials subfunction.

16. *Gross capacity and cycle information.* This is the workstation capacity data developed by AME as described in previous sections of this chapter. The master scheduler to determine

manufacturing cycle time for scheduling use of the critical equipment uses these data. This is a responsibility of the AME unit.

17. *Engineering due dates.* The schedule imposed on design engineering by the master scheduler in order to meet customer commitments. Determining the due dates is an iterative process based on marketing function needs, the complexity of the project, customer requirements, and manufacturing cycle time requirements. This is a responsibility of the materials subfunction.

18. *Engineering data.* The design data used by MP&WM to produce the planning for the specific customer order. This is a responsibility of the design engineering subfunction.

19. *Shop material order date.* The date by which material must be ordered to support the shop schedule (step 8). Material order due dates are based on design engineering requirements as expressed in step 18. Purchasing develops lead times based on inquiries and current economic conditions. MRP II computer systems typically do this. This is a responsibility of the materials subfunction.

20. *Material received date.* A confirmation that material is available to support production. This is a vital input by purchasing to the preparation of the shop schedule (step 8), and is usually accomplished via the MRP II computer system. They are very extensive purchasing, inventory control, vendor performance, and receiving computer systems that have been developed to assist in managing this important function. No material means no production, and late material means at best compressed cycles to meet completion dates. Hence, there is a significant risk of increasing costs through mistakes and inefficiency. This is a responsibility of the materials sub-function.

21. *Replacement material needed.* A result of either defective material being supplied or manufacturing mistakes causing scrapping of otherwise good material. The information is generated at the data collection stage (step 12), and is fed back to the shop schedule (step 8), where an order for replacement material is generated. The replacement material step, in turn, generates a requirement for a shop material order data (step 19) to reorder the material. This is a responsibility of the materials subfunction.

22. *Machine tool status.* Information on the current capability status vis-à-vis the actual feeds and speeds of the production equipment versus the designed capability. It is a moderating factor in workstation loading (step 9), and would be a legitimate reason for shop operations not to meet the times specified by the shop schedule (step 8). The Maintenance unit keeps very close tabs on equipment capability as compared to its original "as new" capability. This information is vital for AME data to justify new equipment. It is a responsibility of the maintenance unit of manufacturing engineering.

23. *Workstation capacities.* The modifications to the gross capacities as tabulated in the gross capacity and cycle information (step 16). They are the intangibles described in an earlier section of this chapter that temporarily affect the capacity of individual workstations to produce. This information is fed into the workstation loading (step 9), so that modifications to mortgaging of time on a workstation can take place accurately. This, too, is a legitimate reason for shop operations to miss a scheduled completion date. It is a responsibility of the AME unit.

These are the 23 steps of a typical manufacturing production control system. Such a system is a powerful integration of most functions of a business and certainly all of the direct sub-functions of the manufacturing function. It is also interesting to note that AME can easily perform capacity simulations by inputting its data to this system, and making assumptions based on experience for other function, sub-function, and unit inputs. This is inherently better and more believable by management than a capacity simulation program developed and run independently by AME.

SUMMARY

In this chapter the concepts of capacity, capability, and level loading have been defined and discussed. Companies must concern themselves with two types of capacities: technical capacity to offer a product line, and physical capacity to determine how much product could be produced in a given period of time. Level loading has been defined as a means of using capacity as a tool for optimum production.

Applying the six deductions from theoretical capacity to compute practical capacity is an important aspect of setting factory output expectations. The physical capacity has been used to explain critical workstations, a monitoring tool useful for level loading. Level loading philosophy and the work of the manufacturing engineer demonstrate the integrated nature of manufacturing. The concept of bottleneck identification and elimination is closely associated with level loading. Here we are analyzing causes that prevent successful level loading. Finally, the concept of using computers as a tool to assist in capacity analysis has been introduced.

Manufacturing engineering only plays a role, albeit a significant one, in factory capacity and loading decisions. This chapter points out the synergism between manufacturing engineering, the rest of the manufacturing function, and the other functions of the industrial organization.

REVIEW QUESTIONS

1. Discuss the differences between technical capacity and physical capacity with reference to a manufacturing facility.
2. What is meant by the term "lower bound of the typical effort range to be expected of an average worker"? How is it determined?
3. A machine tool has a theoretical capacity to produce 620 units per week. Using the values for deductions employed in Table 3.1, calculate the practical capacity of the machine tool.
4. In question 3, the setup time is improved by 15% and the interference time by 5%, but the manufacturing losses increase by 8%. What is the practical capacity of the machine tool now?
5. The following is an incomplete analysis of a manufacturing operation. (a) Fill in the missing information.

Operation	Factored time (hours) to perform per 100 feet of wire	Number of operators	Man-hours
Draw wire	0.23	2	—
Rotational bend wire	122°	4	0.32
Anneal wire	0.12	—	0.36
Spool	0.05	2	—

(b) What is the complete cycle time for the operation? (c) What is the most likely bottleneck operation that would have to be improved in order to have no queuing of product. (d) What would the direct labor cost be if the hourly rate is $8.50 per hour for the draw wire operation and $8.25 per hour for all other operations?
6. Based on the definitions of capability and capacity, give reasons why the manufacturing operation of question 5 should be considered incapable or capable of performing if the following would have to be assigned to another factory: (a) draw wire, (b) rotational bend wire 122°, (c) anneal wire, and (d) spool.
7. Eight unquantifiable items have been given in this chapter as examples of intangibles that affect times to perform work. Using the four operations of question 5, determine whether the eight items are pluses or minuses for those operations.
8. Give reasons why a factory cannot sustain a 90% capacity operating condition for a long period of time. What is considered a steady-state running condition?
9. Discuss the plus and minus factors from a management viewpoint if level loading is achieved.
10. The following data are for a factory output plan for 3 months: September, 12,830 units; October, 4,620 units; and November, 8,240 units. The standard capacity is 8000 units per month. Determine whether the factory can make the units over the 3-month period and

list the specific recommendations manufacturing engineering would make to ensure that the factory has the best chance to produce the desired quantity and yet preserve the factory's equipment.

11. Referring to Table 3-2, suppose the factored time to perform (hr) for all four operations were 0.18 hr. Does a bottleneck exist in this situation? Explain why or why not.

12. Explain how the Theory of Constraints philosophy is compatible with the manufacturing production control system depicted in Fig. 3-6.

13. Explain why schedule manipulation tends to be the primary method for relieving bottlenecks even though it normally does not eliminate the root cause of the bottleneck.

Chapter 4

CAPITAL EQUIPMENT PROGRAMS

Capital equipment programs are the most visible manifestation of manufacturing engineering activities to senior management. They represent a portion of a corporation's investment funds directed toward the future. Hence they require a precise analysis of needs coupled with the always present constraint of funds available to meet those needs.

In this chapter we will investigate the process of establishing the facilities program, the process of justifying the choices, the development of the budget for purchase of the equipment, and the management of capital equipment projects.

FACILITIES PROGRAMS

A facilities program cannot exist in a vacuum. It must be closely coordinated with the objectives and goals, the business plan, and the product scope of the company. In fact, there can be no facilities program until these items are fully explored and understood. It is an ongoing task of the Advanced Manufacturing Engineering (AME) unit to make sure that the facilities program they develop and pursue is compatible with the desires of management. Of course, the desires of management can and should be influenced by knowledgeable AME engineers.

We will start this discussion of facilities programs at the point of having an approved business plan with a fully detailed objectives and goals statement for manufacturing engineering. However, it must be pointed out that objectives and goals, business plans, product scope evaluations, and facilities plans are not strictly linear in their development. For convenience, we define them as linear, but in actuality they exist in a state of flux. For example, the objectives and goals statement may have been prepared on the basis of a previous strategic plan and may now have to be changed. Or the business plan may change due to market pressures or perceived new opportunities. A facilities plan developed on the basis of an objectives and goals statement that is now out of date may no longer be valid. To further complicate things, the new business plan may require a quick reaction affecting facilities. This could mean cancellation of previously reserved but not yet purchased equipment from a vendor or placement of a hold order on a vendor's production schedule, all major changes made before thoroughly modifying the objectives and goals statement. Consequently, the system is nonlinear even though it is described as if step A must occur before step B.

Most AME units have someone assigned to the secretary/librarian function of updating and correcting the objectives and goals statement to agree with the current version of the business plan. This keeps the facilities plan consistent with its higher-order management documents. The frequency of such changes can vary from once per quarter to once per year, depending on the volatility of the businesses the company is engaged in.

A facilities plan is a well-thought-out procedure for purchasing either new equipment or rebuilt equipment in order to meet the capability or capacity requirements of the company. In the objectives and goals system the facilities plan is a specialized action plan for the AME engineers who are assigned the task of finding, selecting, and purchasing the needed equipment. Simple facilities plans are the same as the objectives and goals project statements. For more complex facilities plans the project statements constitute only the current year's portion of the plan. The remainder of the plan would be extracted from the current 5-year or long-range forecasts of equipment needs to meet the defined capability/capacity statements.

A facilities plan could also be defined as the extraction of capital equipment purchases from the current objectives and goals statements and the long-range forecasts arranged in chronological order of intended purchase, with action steps outlined to accomplish the purchase on time and to meet the needs requirements. This definition certainly implies a need for purposeful evaluation of equipment. Without such a step we could never be sure that the equipment purchased would perform its intended function. Therefore, selection of the equipment from a long-range forecast is not a first step, but a summation step. The equipment must be on the current projects and long-range forecast document as the result of forecasts of needs made by the area planners for their particular areas of responsibility. These are the capability/capacity needs determined by analysis to meet customer demands and produce desired new products. The compilations of all entries pertaining to capital equipment by all the area planners constitute the list from which the facilities plan is drawn.

The steps to be taken in developing the facilities plan are:

1. Consolidate facilities needs lists.
2. Evaluate estimated expenditures compared to budget for each budget period.
3. Rank equipment needs.
4. Place equipment in year-of-purchase groups.
5. Establish due dates for major action items through implementation of the equipment.
6. Establish a second-tier list of equipment projects to be pursued if additional funds become available.

Let us now explore these six activities in greater detail.

Consolidate Facilities Needs Lists

In this step the AME engineer collects data on needs from all the area planners and other cognizant personnel and drafts a chronological list of equipment needs by purchase date. This gives advanced manufacturing engineering a total listing of all equipment necessary to support the company's business plan justified by the objectives and goals statement.

Evaluate Estimated Expenditures Compared to Budget for Each Budget Period

Normally, a budget for capital expenditures is established after all operating costs have been firmed up and profitably analyzed. What is left over becomes the capital available for investment in new equipment. This is true even if the firm expects to borrow funds to purchase equipment; the money is then allocated for loan repayment. Unfortunately, money allocated for capital equipment may not be sufficient to purchase the equipment because an investment budget is rarely made up after the equipment needs are identified. This means that the worth of specific equipment purchase projects must be evaluated against that of other projects. But the severity of the evaluation cannot be set until the total funds shortfall is known. Therefore, during this step, equipment is listed by year of intended purchase and expected funds available for that year, the results being shortfall figures that will require judgments of how severe the comparison between projects must be. The result of this step is a breakdown of the list of needed facilities into years of purchase matched against expected costs of each project versus expected funds available.

One thing to be aware of during this step is the level of accuracy of the expected cost of each project and the expected funds available. Since engineers want their projects to succeed and therefore want them to appear as attractive as possible to management, they invariably underestimate the cost of a project and overestimate the funds to be available. It is incumbent upon the AME manager to ensure that costs are assessed correctly. This can be done by requiring engineers to obtain cost estimates from potential vendors and by comparing costs with previous, similar programs.

The AME manager must also assess whether the engineers have thoroughly researched the project and uncovered all possible cost areas. On top of this, the prudent AME manager will also add a contingency of 10 to 20% to the project cost for unknowns. The amount of the contingency is a judgment factor depending on the type of project. For projects similar in kind to those recently accomplished, it will be closer to 10%. Finally, the AME manager will never utilize 100% of the budget available for facilities list projects. Usually only 90% will be used, leaving 10% for minor projects that are necessary for the operation of the company but not necessarily documented in the objectives and goals statement.

The problem of not having firm budgets, especially for the long-range forecast period, is one of ongoing concern for manufacturing engineering. The farther the future year is from the present, the less precise the budget figure for capital investment will be. Nevertheless, so that planning will be possible, manufacturing engineering must operate as if the budget projections are real and will be maintained. To aid senior management it must make sure the facilities plan is tied precisely to a point in time, that is, date the plan and identify it to the current version of the current year budget and long-range forecast.

Rank Equipment Needs

Ranking is a subjective activity that must be done as objectively as possible. Before it is attempted, advanced manufacturing engineering must know how severe the constraints will have to be. Can the company do projects with priorities down to 8 out of 10, or must it restrict itself to projects ranked 1 and 2 out of 10? This funding level constraint comes from step 2. Table 4-1 is a typical listing by year of equipment purchases, cost estimates, and the budget estimate prepared by AME during step 2. Clearly, the budget estimate for capital equipment investment during the budget year will not support all eight facilities projects. We must assume that the eight projects have survived the objectives and goals selection process and support the current business plan. Therefore, all the projects are desirable and would benefit the company. A choice must be made as to what will and will not be allowed to proceed.

EXAMPLE: Using data from Table 4-1, let's determine the severity level that the ranking must satisfy. The budget estimate is $1000K but, using the 90% rule, the AME manager would use $900K for the calculations, preserving the remaining $100K for unforeseen and miscellaneous investment needs. This means that the severity level will be 900/1579 for 0.569. Therefore 56.9% of the total number of projects can be funded, or 4-5 projects. Another way

TABLE 4-1. 2006 Facilities List

Equipment	Cost Estimate ($000)	2006 Budget Estimate
1. Vertical machining center, CNC	750.0	
24 × 60 in. table		
3 in. spindle		
x travel = 60 in., y travel = 24 in., z travel = 24 in.		
18 place tool changer		
2. Phoenix-type horizontal turret lathe	300.0	
40 in. swing, automatic control		
3. Notching press, 10 ton, manual control	85.0	
4. Notching press, 20 ton, automatic control	110.0	
5. Blanking press, 90 ton, manual control	225.0	
6. Electric lift truck	10.0	
7. Pick and place electronic programmable robot	85.0	
8. 10 ton panel truck (used), automatic transmission	14.0	
Total	1579.0	1000.0
Projected shortfall	579.0	

of looking at this would be: any number of projects totaling $900K can be funded. The number 4-5 may sound illogical, but it means that a project may be partially funded in 2 years, usually successive.

Knowing the severity level is necessary in order to select projects, let's look at a useful scheme for ranking them. Like so many other activities within the overall manufacturing discipline, there is no best way of accomplishing this task. The method proposed here is one that I know is effective.

All the projects listed supposedly do benefit the company. First, make sure this is true; that is, make sure that the benefit does not turn out to be detrimental to another segment of the company.

EXAMPLE: A new drawing coding system may save design engineering 10 to 15% filing time, but requires additions to the manufacturing engineering planning staff to continue to find drawings.

Second, put every proposed facility project to the following test. Can the company survive—that is, literally still be in existence—if the project is deferred? If the answer is no, then the project must be done.

EXAMPLE: The requirement for a bearing finish tolerance is beyond the capability of the equipment presently owned by the company. If the equipment is not purchased, the company cannot compete in the present market and must leave the market, possibly forfeiting current contracts.

Third, for every facilities project, ask whether its completion is highly desirable, moderately desirable, or of low desirability vis-à-vis improving the profitability potential of the company. This is a subjective evaluation, but it can be made fairly objective by comparing the engineering factors of the facilities project with those of its predecessor.

EXAMPLE: If a new lathe is four times faster than the previous way of making a product and 4,000 parts are to be made, it is more desirable than a vertical boring mill that is six times faster than an older VBM used in making 600 parts of equivalent value per part. The lathe is worth more than the VBM to the company because more parts go through that particular process.

Logic similar to this should be used to make this subjective assessment as objective as possible.

Knowing the rankings of the facilities lists projects by year group, it is now possible to assign funds to these projects until the funds are exhausted. Obviously, the essential projects will be funded first. The number of projects to be funded depends on the severity level calculated. In Table 4-1 we can surmise that none of the projects are essential for company survival. Not knowing the particulars of the company involved, it would be difficult to determine the levels of desirability, except to say that projects exhibiting trends toward newer technologies would probably be higher in desirability. Projects 1 and 7 would be in this category; they cost $835K, which leaves only $65K. The notching press, project 3, might be selected for a total investment of $920K; this is only $20K higher than the goal of $900K, which is a reasonably good match.

To summarize, the steps in ranking are:

1. Calculate the severity factor.
2. Ensure that the project listings are universally beneficial.
3. Determine what projects, if any, are essential.
4. Rate the desirability of the remaining projects.
5. Match projects with available funds according to the severity factor.
6. Determine which projects will not be funded.

Place Equipment in Year-of-Purchase Groups

Once the ranking is complete we've made the basic decision about which projects from the facilities lists will proceed and which will be deferred. Commonly, good projects may have to be deferred or canceled outright because of lack of funds. Sometimes excellent projects are simply moved 1 or 2 years into the future, when additional funds may be available. Therefore, not only items that survive the ranking process described above but also projects that did not meet the previous years's cutoff criteria are considered. Often the AME manager will add projects that were not funded to the next year's total list, and will consider the deferred project against projects initially considered for that budget year. This is not an automatic listing of all projects that were initially not funded; AME will first consider whether the project is still valid in a later year group.

Once the manipulations of non-funded projects from 1 year to later years are complete, we end up with a listing of funded projects by year-of-purchase group. This becomes the official project listing, hence the action plan to be accomplished by the AME unit. At this stage unfunded projects are removed from the objectives and goals statement and placed in a reserve category.

Establish Due Dates for Major Action Items through Implementation of the Equipment

In this step the due dates for all the important steps of a project are assigned in order to ensure successful completion of the project. The AME manager usually establishes these dates in consultation with the engineers and area planners, as described in Chapter 2. This becomes the action plan to be monitored during work planning sessions and other reviews.

Figure 4-1 is a typical weekly report showing the steps that are monitored during a facilities project. The steps must be taken in the order shown as the project progresses from a generalized concept to a specifically tailored application of equipment. Let us take one line of Fig. 4-1 and go through each item reported on. Figure 4-2 is an extraction of the heading from the total project list depicted by Fig. 4-1. We will use the lowercase letters under each category in Fig. 4-2 to explain the purpose of the specific columns.

- a. *Status.* This indicates whether a project has obtained final approval for purchase of the equipment. "AP" would indicate approved, while a blank or "pending" would indicate that the purchase of the equipment has not yet been approved.
- b. *PAR no.* The PAR (plant appropriation request) number is an index by which all functions of the company can identify the particular facilities project. A PAR number is usually assigned when the first draft of the PAR is written.
- c. *Description.* This is the English word description or title of the project.
- d. *Plan by.* The initials of the AME engineer or area planner responsible for the project are filled in here.
- e. *Estimated costs and savings.* These three columns show the financial summary of the project. The investment dollars are the funds that will be spent for capital equipment—the depreciable funds. The expense dollars are those which are associated with the investment but cannot be classified as capital investment; for example, they would include costs for moving existing equipment out of the way to make room for installation of the new equipment. These are non-depreciable expenses related to depreciable funds. Savings are the calculated project savings on an annualized basis. The savings area will come under the closest scrutiny at the various levels of approval.
- f. *Area specifications.* These are the scope specifications usually prepared by the area planner, outlining the type of equipment that should be acquired to meet a specific business need. The area planner does not specify detailed equipment, but rather what the equipment should be able to do.

PAR APPOVAL STATUS

PAR Approval Control - 2004/2005 Expenditure

Status	PAR No.	Description	Plan By	Budget Invest	Dollar Invest. Related Expense	$(000) Savings	Area Specifications T/A	Specifications T/A	Quotes T/A	NOI T/A	Write PAR T/A	ME Mgr Approval T/A	Finance Approval T/A	Mfg Mgr Approval T/A	Gen Mgr Approval T/A	Div Mgr Approval T/A	Order Equipme T/A
P	228-84	Service contract-decontroller	RR	0.0	5.8	8.1	408/408	409/408	410/409	na	411/411	412/412	412/412	412/412	413/412	na	413/412
P	228-94	Pendant cables	JM	0.0	18.0	0.0	408/408	408/408	414/420	na	415/421	416/421	417/421	418/422	419/423	na	420/424
P	228-99	Parts for toshiba	RR	23.0	0.0	13.4	412/413	412/413	416/416	na	418/418	419/419	419/419	420/420	421/421	na	422/423
P	252-81	Back face cutters	FJ	14.0	0.0	62.7	411/413	411/413	417/417	na	418/419	419/420	419/420	420/420	421/421	na	422/422
P	251-80	Repair roof	JM	0.0	30.0	0.0	410/410	410/410	410/410	na	410/411	410/411	410/411	410/411	410/411	na	411/411
P	252-84	Repair to shot blast	JM	0.0	4.4	0.0	413/412	413/412	413/413	na	414/414	414/414	na	na	na	na	414/414
P	229-91	Install BXLX	DS	4.9	0.0	0.0	420/419	420/419	na	na	421/419	421/419	na	na	na	na	421/419
AP	228-81	Service contract-Bendix cont.	RR	0.0	5.0	2.3	408/408	408/408	410/410	na	411/411	412/412	412/412	412/412	413/412	na	413/0
AP	228-86	Fire hydrants	JG	0.0	9.5	0.0	424/424	424/424	428/429	na	429/430	430/430	430/431	431/431	432/433	na	433/0
W	229-83	LM-2500 facilities	FJ	22.4	0.0	0.0	426/426	426/426	431/432	na	433/433	434/0	434/0	435/0	436/0	na	437/0
	229-88	Toshiba MS mill	FJ	162.0	30.6	105.0	440/0	441/0	450/0	501/0	505/0	506/0	506/0	507/0	508/0	510/0	511/0
	229-86	Steel racks	RC	8.6	14.4	12.0	441/0	441/0	443/0	na	444/0	445/0	445/0	446/0	447/0	na	448/0

Key

T/A - target and actual dates for action step accomplishment
(0 in action space indicates open action item)
AP - appropriation approved
W - appropriation drafted
P - purchase order placed

Figure 4-1. Steps Monitored During a Facilities Project

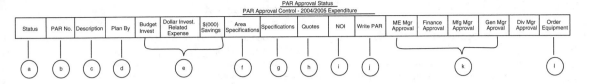

Figure 4-2. Headings from Project List in Fig. 4-1

EXAMPLE: The area planner specifies the type of equipment, such as a CNC chucking-type lathe capable of producing 1,000 items per day.

The heading T/A indicates the target date for completion of the step versus the actual date when the step was completed. Dates are shown in a code where the first digit indicates year and the second and third digits indicate fiscal week. Therefore 408 on the target line means year 2004, fiscal week eight. If the step was completed in the tenth week of 2004, then 410 would appear under A.

- g. *Specifications*. In a large AME organization the area planner now turns the project over to an AME engineer. In a smaller organization the area planner and the AME engineer may be the same person. During this step detailed specifications are prepared so that quotes can be solicited from equipment vendors.

EXAMPLE: For the CNC chucking lathe the AME engineer would specify such things as bed length, area of swing, type of controls deemed acceptable, physical size limitations, power requirements, types of tooling to be supplied with the machine, protective screens required, safety requirements, required ranges for feeds and speeds, required tolerance capabilities, required surface finish capabilities, and so on.

All requirements necessary to ensure that the equipment will perform in accordance with the requirements set forth by the area planner during step f. are included.

- h. *Quotes*. At this point vendor quotes are solicited and received on the basis of step g. The date shown is the date when the quotes are to be received; this is always specified in the invitation to quote.
- i. *NOI*. A Notification of Intent (NOI) is sometimes required by companies to let everyone in the approval process know that a plant appropriation request is coming. Whether an NOI is required usually depends on the dollar level and/or complexity of the PAR. The purpose of the NOI is to secure speedy and thorough review. If a double zero, 0/0, appears in the schedule, as it does frequently in Fig. 4-1, no NOI is required for the particular project. Notifications of intent are usually short summarized versions of the appropriation requests stating in one or two paragraphs what the project is, why it is being implemented, how much it will cost, and what the savings will be.
- j. *Write the PAR*. The drafting of a PAR consists of thoroughly describing the project so that the reviewing agencies thoroughly understand its purpose and can judge whether to approve or disapprove based strictly on the merits of the project. PARs can range in length from a single page to a large volume, depending on the complexity of the project. Regardless of length, they all contain the following information in distinct sections: (1) overall description and justification of the project, (2) savings to be achieved, (3) costs to implement, and (4) implementation schedule.

Figures 4-3 and 4-4 are illustrations of typical PARs, one very simple and the other more complex. However, PARs can take many forms other than these. PARs are selling documents; they are

Plant Appropriation Request   No. <u>228-89</u>

1. Department: <u>Power Generation</u>   Location: <u>Small Town, U.S.A.</u>

Product Line: <u>Turbine & Generators</u>

2. SUMMARY DESCRIPTION OF PROPOSED PROJECT:

Relocation of Pandijiris Manipulator for sub-merged weld arc boom machine
and positioner from Bay #2 to Bay #3, Column C31.

Category: <u>V</u>

3. AMOUNT OF APPROPRIATION

	This request	Previously approved	Future Request	Total project
Investment				
Expense	3,550			3,550
Total				3,550

Starting costs (as a memo)

Amount in included in business plan - Investment_____Expense   0

4. DATE OF INITIAL EXPENDITURE   May 2005

   DATE OF COMPLETION   May 2005

5. STATUS OF PLANNING   <u>80%</u>   complete

6. PERIOD OF EXPENDITURE

MM/YY	Total	Investment	Expense	Qtr.&Yr.	Total	Invest.	Exp.
05/05	3,200		3,200	2-05	3,200		3,200
	Total					350	350
	Contingencies						
	GrandTotal				3,550	3,550	

7. SUMMARY OF ESTIMATED CHANGES IN ANNUAL COSTS

	Proposed facility	Present facility	Reduction (increase)
Variable costs			
Direct labor (including benefits)			6,650
Direct material			
Indirect labor (including benefits)			
Maintenance of equipment			
Expense tools and suppies			
Special tools (one year amortization)			
Transportation relocation			(3,550)
Fixed costs			
Taxes and insurance			
Normal depreciation			
Product Engineering costs and exps.			
Total			3,100

Figures 4-3. A Simple Plant Appropriation Request

8. PROJECT DESCRIPTION. REASONS FOR EXPENDITURE AND RESULTS TO BE ACCOMPLISHED

Expenditure of $3,550 is requested to relocate the Pandijiris Manipulator for submerged\arc boom machine and a positioner from Bay #2 to Bay #3, col. 31. The Pandijiris manipulator was dismantled 2 yrs. ago because of plant layout rearrangement and never used. This proposal is to reinstall it together with a positioner at the west end of Bay#3 to weld Air Lock shells and LM 2500 short beams.

Savings are calculated as follows:

Based on 8 shells of Air Locks produced/yr., total length weld seams = 505 ft. Use submerged welding process w/boom machine it will take 100 hrs. With other methods it will take 400 hr including crane service and idle time waiting for crane lift.
Defect rate will be 85 hrs greater.

Savings of labor = (400 − 100) × $10.788 = $3,236
Savings in rework = 85 × $10.788 = $917

Total labor savings = $3,236 + $917 = $4,153
add 60% reduction in idle time = $2,492

Yearly savings = $6,650

9. ADMINISTRATION AND APPROVAL

Estimates and data prepared by _____

Appropriation approved by _____
Manager _____ Section

Appropriation recommended by _____
_____
_____
_____

_____
Mgr. Finance

_____

Dept. Gen. Mgr.
_____
Div. Gen. Mgr.

Appropriation approved and data certified by _____
Mgr._____ Accountiing

Date approved for implementation _____

Figures 4-3. Continued

used by various levels of management to convince other levels of management that a particular project has significant merit and should be approved. Hence they emphasize whatever the authors believe to be the pertinent or strong points, or the points that the senior reviewer likes to see on requests for capital investment funds. Nevertheless, the four points mentioned above must be covered at the level of detail outlined below:

1. *Overall description and justification of the project.* This spells out the equipment selected, the alternatives not selected and the reasons for not selecting them, and the reason for going ahead with the project.
2. *Savings to be achieved.* This section contains the calculations that lead up to and support the savings claimed. Here a very clear and logical build-up of data to calculate the claimed savings is required. This section also states when savings will commence and reach maturity.

Plant Appropriation Request   No. 229-80

1. Department: Power Generation   Location: Small Town, U.S.A.
                                          Product Line: Turbine & Generators

2. SUMMARY DESCRIPTION OF PROPOSED PROJECT:
   To purchase and install a CNC Vertical machining center (OKK MCV 800)
   and the necessary tooling

                                                        Category: V

3. AMOUNT OF APPROPRIATION

	This request	Previously approved	Future Request	Total project
Investment	351,800			351,800
Expense	18,200			18,200
Total	370,000			370,000

Starting costs (as a memo)

Amount in included in business plan - Investment $420,000 Expense $10,000

4. DATE OF INITIAL EXPENDITURE   August 2005, amount in 2005 budget

   DATE OF COMPLETION   December 31, 2005

5. STATUS OF PLANNING   75%   complete

6. PERIOD OF EXPENDITURE

MM/YY	Total	Investment	Expense	Qtr.&Yr.	Total	Invest.	Exp.
08/05	33,000	33,000	0	3-05	33,000	33,000	0
09/05	206,500	190,000	16,500	4-05	206,500	190,000	16,500
10/05	110,000	110,000	0	4-05	110,000	111,000	0
	Total				349,500	333,000	16,500
(5.5% invest. 10% exp.)	Contingencies				20,500	18,800	1,700
	GrandTotal				370,000	351,800	18,200

7. SUMMARY OF ESTIMATED CHANGES IN ANNUAL COSTS

	Proposed facility	Present facility	Reduction (increase)
Variable costs			
Direct labor (including benefits)			
Direct material	(95,452)		(95,452)
Indirect labor (including benefits)			
Maintenance of equipment			
Expense tools and suppies			
Special tools (one year amortization)			
Transportation relocation			
Fixed costs			
Taxes and insurance			
Normal depreciation			
Product Engineering costs and exps.			
Subcontracts cost		223,177	223,177
Total			127,725

Figure 4-4. A More Complicated Plant Appropriation Request

8. PROJECT DESCRIPTION. REASONS FOR EXPENDITURE AND RESULTS TO BE
   ACCOMPLISHED

Permission is requested to purchase and install a CNC vertical machining center
(OKK-MCV-800) and the necessary tooling.

This new machine tool is required to provide continuous support to the operation as
defined in the 2003 Long Range Forecaste. With the purchase of this machine tool,
three old worn-out milling machines will be scrapped. This machine will relieve the
overload situation on the existing OKK-MCV-500 N/c machine as well as handle the
work now placed on the TOS horizontal miller if additional vertical capacity is not
achieved.

See Appendix A for a description of the machine and a summary of proposed
expenditures.

Project Necessities:

1) To replace the old and worn-out machine tools. The plant at present time has the
   following small milling machines.
   a) #4705 Kearney & Trecker (horiz.) - 27 yrs. old, to be scrapped.
   b) #4801 Cinncinnati (vert.) - 34 yrs. old, to be scrapped.
   c) #4503 Milwaukee (horiz.) - 28 yrs. old, to be scrapped.
   d) #4584 TOS (horiz.) - 14 yrs. old, operable.
   e) #3050 OKK (N/C vert.) - 1 yr. old, excellent.

These machine tools, with the exception of #4584 and #3050, are in extremely poor
operation condition, to the point that they are considered hazardous.
Based on a recent survey there is approximately 10,000 non-N/C hours of small
vertical milling annual load on these machines. This load will have to be transferred
to other machines after these three are scrapped. The only machines available are
#4584 (TOS) and #3050 (OKK). #3050 is already overloaded heavily and #34584 can
take up only a small portion of the 10,000 hours.

It will be necessary to subcontract work to outside vendors if no replacement machine
is purchased.

Alternatives:

1) Continue as is. The retirement of #4801, #4705, and #4503 is way overdue. They
   should be scrapped. Without the machines, and with the projected overload
   condition of #3050 (OKK) and the vulnerable condition of #4584 (TOS) if it has to
   absorb most of this load,it is necessary to subcontract work to outside vendors if
   the production schedule is be met. This will result in high subcontract cost plus extra
   lead time. For this reason this alternative is rejected.

2) Buy a conventional machine. This alternative is rejected because a conventional
   machine will not allow for only a minimal productivity gain.

3) Rebuilding the old machine tools. The machines have an average age of 30 years.
   It is not worthwhile to commit to a major overhaul on these machines, especially
   since almost no productivity improvement will result and it is not feasible to add
   N/C controls. In fact the conditions of the machines are so bad that they are limited
   to very restricted type of operations or the safety of the operators will be jeopardized.
   The plan is to scrap these machines as soon as replacement capacity is obtained.
   For this reason, this altenative is rejected.

Figure 4-4. Continued

Major Results to be achieved:

1) Achieve a payback of 2.7 years and an anual savings in excess of $10,000.00.

2) Improve flexibility of production control because of the availability of a machine with capabilities similar to the existing OKK MCV-500. It will also serve as a back-up machine when maintenance service is required for th OKK MCV-500. Quality of product will also improve.

3. Achieve part of the plan to upgrade CAD/CAM applications of mfg. technology.

See attached appendices for references:

        Appendix A - The proposed machine and expenditures.

        Appendix B - Loads and savings summary

        Appendix C - Cash flow analysis

9. ADMINISTRATION AND APPROVAL

Estimates and data
prepared by _____

Appropriation
recommended by _____
_____
_____
_____

Appropriation approved and
data certified by _____
        Mgr._____ Accountiing

Date approved for implementation   _____

Appropriation
approved by
Manager _____
    _____ Section

_____
    Mgr. Finance

_____
    Dept. Gen. Mgr.
_____
    Div. Gen. Mgr.

APPENDIX A

Proposed Machine

The proposed machine is an OKK_MCV-800 made by Osaka Kiko Co. Ltd. of Japan. The machine fulfills all of our requirements.

One advantage of MCV-800 over other candidates is that experience gained from the MCV-500 could easily be carried over to this machine. This will include experiences in N/C programming, operation, and maintenance. This will drastically facilitate the installation, start-up, and operators training on the machine. Another advantage is the universality of tooling, trained operators, and spare parts. Past experience on the MCV-500 has been very satisfactory with respect to its features and performance.

Proposed Expenditure

	Investment	Expense	Total
1) CNC Machining Center,Foundation Transportation, & insurance	303,000		303,000
2) Installation	2,000		2,000
3) N/C Post-Processor		5,000	5,000
4) Set-up Equipment	3,000		3,000

Figure 4-4. Continued

5) Tooling: Cabinets	5,000		5,000
Durable Tooling	20,000		20,000
Non-durable tooling		5,000	5,000
6) Area Preparation		2,000	2,000
7) Training: Operators		2000	2,000
Maintenance		2,000	2,000
Programmer		500	500
Sub Total	333,000	16,500	349,500
Contingency (5.65% invest,10.3% exp.)	18,800	1,700	20.5
TOTAL	$351,800	$18,200	$370,000

APPENDIX B

Savings Calculations

Savings are based on comparing the cost of subcontracting the load to oustide suppliers vs. costs of manufacturing using the proposed machine

Cost of Subcontracting	05	06	07	08	09	10	
Load (N/C hr.)	546	1638	1638	1638	1638	1638	
N/C rate ($/hr)	$49.5	$54.5	$59.9	$65.8	$72.3	$79.7	
Cost of subcontracting to N/C Vendors	$27,027	$89,271	$98,116	$107,780	$118,427	$130,549	
Load (conv. hr)	1363*	4095	4095	4095	4095	4095	
Conv. rate ($/hr)	$29.7	$32.7	$35.9	$39.5	$43.4	$47.8	
Cost of subcontracting to Non-N/C Vendors	$40,481	$133,906	$147,010	$161,753	$177,723	$195,741	
Total Subcontract Costs	$67,508	$223,177	$245,126	$269,533	$296,150	$326,290	

Costs of Manufacturing							
Programming Costs	$5,000	$16,500	$18,150	$19,950	$21,900	$24,150	
Operating hours	1090	3276	3276	3276	3276	3276	
Operating Rate ($/hr)	$21.9	$24.1	$26.5	$29.1	$32.1	$35.2	
Operating Costs	$23,871	$78,952	$86,814	$95,659	$105,160	$115,315	
Total Mfg. Costs	$24,961	$95,452	$104,964	$115,609	$127,060	$139,465	
Net Savings	$42,547	$127,725	$140,162	$153,924	$169,090	$186,825	

The work load is derived from a survey of work station load samples over a period of 12 to 20 weeks and extrapolated over a a year (47 weeks) to obtain the annual load

machine	status	period surveyed	load in coventional hrs	Annual load in conv. hrs.
#4705	scrap-2005	12 weeks	114	445
#4801	scrap-2005	12 weeks	470	1840
#4503	scrap-2005	12 weeks	1169	4580
#4584(TOS)	operable	12 weeks	274	1073
#3050(OKK)	excellent	20 weeks	900	216
	Total load 2004			8154
	Projected load 2006			10053

Figure 4-4. Continued

APPENDIX B (continued)

Note: (1) Load from #3050 (OKK) is the overflow of 360 CNC hours, converted to
conventional hours by applying an improvement factor of 2.5:1.

    (2) The load from #4584(TOS) is the load that should be processed on the
proposed machine to achieve a cost improvement. It does not represent the
complete load on #4584 over the period surveyed.

    (3) For #4503, #4705, #4801, the load represents the complete load on the
machine surveyed in the 12 weeks period.

Anticipated load beyond 2004 will be adjusted by a projection factor based on the
Long Range Forecast as flollows

LRF	04	05	06	07	08
Total DL hrs. (000)	375	390	317	327	396
Load projection factor using 2004 as base year	1	1.04	0.85	0.88	1.06
Projected load on proposed Machine in conv. hrs.	3351*	10455	8545	8847	10656

*Adjusted for partial year

Machine capability based on 2 shift operation

        46.8 wks/yr × 35hr wk ×2 shifts = 3276 CNC hours

                = 8190 Non-CNC hours

Since the available load is greater than machine capacity, 8190 non-CNC hours will be
used in savings calculations. (8190 non-CNC hrs. is equivalent to 3276 CNC hrs. when
a 2.5:1 improvement ratio is applied)

Cost of Subcontracting

When subcontracting, the inital attempt is to find vendors with CNC capability to get the
lowest possible costs. This is not possible because of reasons of availability. It is
estimated that the load will be split 50/50 to CNC and Non-CNC vendors. An improve.
factor or 2.5:1 is used to convert non-CNC hrs. to CNC hrs. Also, a 10%/yr. inflation
rate is applied to the rates. Subcontract rates are based on quotes from vendor in our
local area. For 2004, $45/hr. is used for CNC rate and $27/hr. for non-CNC rate. A 10%
handling cost is added to the rates.

Cost of Manufacturing

Programming costs based on past experience is estimated to be half a CNC
Programmer at $30,000/yr./Programmer with a 10% inflation factor.

Operating costs are based on 2004 rate of $ 9.738/hr. plus 125% burden rate (1/2 full
rate) with 10% inflation/yr. Operating hrs. are based on 2 shifts operation.

Figure 4-4. Continued

The return on investment, payback period, and present worth and future worth of calculated
funds are also presented in this section.

3. *Costs to implement.* This is the schedule of expenditures by investment and expense categories, usually showing dates of payments.

4. *Implementation schedule.* This detailed schedule of purchase through turnover to operations
is an important section. If the implementation schedule is considered reasonable, then confidence is gained that the savings are achievable in the years claimed.

Plant Appropriation Request (pg.8)   No. 229-80

APPENDIX C

SHORT FORM CASH FLOW

MONTH	YEAR	INVESTMENT	EXPENSE	DEPRECIATION
Aug	2005	-33,000	0	20% book value
Oct.	2005	-318,800	-18,200	20% book value
Dec.	2005	92,222	0	Residual investment

Tax Rate   28.0%   2005- 2010

Year	Book Value	Invent, Red.	Savings	Net income	Deferred tax
2005	281,440	0	42,547	-25,767	0
2006	225,152	0	127,775	40,005	0
2007	180,122	0	140,162	53,274	0
2008	144,097	0	153,924	66,024	0
2009	115,278	0	169,090	78,552	0
2010	92,222	0	186,825	91,711	0

Year	cash flow	
	Net	Cumulative
2005	-307,207	-307,207
2006	96,293	-210,915
2007	98,304	-112,610
2008	102,048	-10,562
2009	107,371	96,809
2010	114,766	211,575

Zero Interest Breakeven Point: 3.43 elapsed years

Discounted Rate of Return based on 6 year life
per B.A.B. No. 414 (Yield method): 28.8%

Average Increm, Return on Increm. Investment per B.A.B. no. 413
22.42% based on 5year life.

Figure 4-4. Continued

In addition to these four generic parts, more complex PARs include an executive summary. This is normally an NOI appended to the front of the appropriations request stating what the request is about. It enables senior managers to decide whether or not it is necessary to review the PAR in detail in order to approve or disapprove it.

k. *Approvals.* Usually the minimum level for approval of a PAR is at the general manager level. All companies set approval level policies based on dollar expenditures involved. The AME manager must know who will be responsible for the ultimate approval and have the PAR written with that level in mind. In general, the greater the managerial spread between the AME manager and the ultimate approver, the more financial detail and justification will be required. The AME manager must keep in mind that when very high levels of approval are required, managers will not be familiar with the working details of the operation, nor do they need to know them. At the higher levels, the most important thing is to choose investments wisely to optimize the company's profitability. Therefore, good financial justification becomes more critical as the level of the investment increases.

l. *Order Equipment.* At this point, the AME manager has shepherded the project from generalized concept through the PAR documentation and through the approval of various diverse management authorities. Now the funds are approved and the equipment can be ordered. This is the beginning of the intensive job of ensuring that the equipment is delivered on time and to specification. Essentially, it is the end of the planning stage and the beginning of the implementation stage. (The implementation activity will be discussed in Chapter 5.) Since obtaining equipment is time-dependent, the prudent AME manager will place the order with the successful bidder as soon as approval is received to do so; hence the short time cycle shown between approval and placement order. This is possible because it is common to work out the contractual details with the vendor while the project is going through the approval stages.

Establish a Second-Tier List of Equipment Projects to be Pursued if Funds Become Available

Since AME will almost always have more worthwhile projects than funds available to implement them, it is prudent to have second-tier projects ready to go through the project planning phases described in the preceding section at very short notice. Budgets are dynamic entities, and it is not uncommon to have additional funds released part of the way through a budget year. Hence many AME managers will take several of these second-tier projects through the PAR writing stage and hold them there waiting for additional funds. This is one aspect of the manager's responsibility to equip the shop optimally to produce its products.

FINANCIAL JUSTIFICATION TECHNIQUES

Capital equipment purchases usually involve large sums of money; therefore, management must quantify the decision process to justify its affirmative or negative actions. In a business environment all actions affect the company's ability to make a profit, and all management decisions are intended to result in increased profit and not the reverse. Most decisions are based on common sense and experience, and this is usually sufficient. However, capital equipment decisions are complex and very costly, and a wrong decision can cost the company a great deal for many years to come. For this reason, quantitative decision-making (or justification) techniques have evolved for capital equipment projects as an aid in making those decisions.

Before we discuss these techniques, a word of caution is due. Like all other mathematical analysis techniques, financial justification techniques are just tools. They use as inputs the best possible data that the manufacturing engineer can gather. But they are only as good as the data used. The results of a financial justification analysis must be evaluated for their quality, that is, how well they fit the real-world situation. If the data clearly define all the variables, then the solution will have merit and the analysis will be believable. If compiling the data required considerable assumptions, the analysis will be only as good as those assumptions and may not be worth much. Engineers should not use financial justification techniques as a substitute for exercising professional judgment.

Three methods commonly used for evaluating the financial worth of a capital equipment project are:

1. Payback period method
2. Return on investment method
3. Present worth method

They all have in common that they attempt to quantify capital equipment project decisions. The theory is that if we can objectively place all projects on a selected rating scale, we will pick the best ones. The drawback is that it is usually not possible to quantify all aspects of the decision.

EXAMPLE: A decision to purchase a CNC chucking lathe depends on being able to make the most parts in a given period of time. The production rate of the machine can be quantified.

However, the cost to maintain it versus that of a manual lathe cannot be quantified. Estimates of maintenance costs can be made but hard data are not available. Therefore, to quantify maintenance costs we are forced to make an assumption, which may or may not be accurate.

Unfortunately, all financial justification techniques treat assumptions as if they were facts. As can be seen, the possibility of an erroneous analysis is not trivial. Therefore, it must be reiterated that financial justification techniques are only tools. In the hands of an experienced, competent engineer, they can be very useful, but when used by someone unfamiliar with the company's operations, they can easily be misused. Let's now consider these three techniques.

Payback Period Method

This is the simplest method, hence the most frequently used. However, since the advent of stronger financial controls, it is most often used in conjunction with the other methods. Referring to Fig. 4-4, under the heading "Major Results to Be Achieved," it is stated that a payback of 2.7 years will be achieved. This payback is determined by dividing the cost of the project by the expected annual savings, which gives the number of years it will take to reach the break-even point. The formula for determining payback period is:

$$\text{Payback period (years)} = \frac{\text{cost of project}}{\text{incremental project savings/year}}$$

EXAMPLE: Figure 4-4 shows that the cost of the project is $370. OK while the average savings over the life of the equipment is $136.7K. Therefore the payback period is 2.7 years.

This number has no significance unless the company has a specific goal in mind for the time it is willing to allow until a project returns a profit. Usually, during prosperous times a 3.0 to 4.0 year payback period is acceptable, while in recessionary times 2.0 years is often required. This calculation can give an indication of the merit of a project compared to the company's business plan parameters.

Return on Investment Method

This technique takes the payback period method one step further. In the payback period method the effective life of the equipment is not mentioned; it is assumed that the equipment has eternal life and that there will be a payback break-even period. However, if the equipment lasts only 18 months, or the project is only operational for 18 months, the 2.7 year payback period means that break-even can never be achieved. To take into account the effective useful life of capital equipment, the Return On Investment (ROI) calculation was developed. The formula for calculating the ROI is:

$$\text{ROI} = \frac{\text{average savings/year} - \text{cost of project/years of life}}{\text{cost}}$$

EXAMPLE: In Fig. 4-4 the average savings per year is $155.5K for a full 5 years. The fractional year savings of $42.5K was not used because we are using a 5-year life rather than 5½ years. The cost of the project remains $370.OK. Using the ROI formula, we calculate a value of 22.0%. Note that in Appendix C of Fig. 4-4, the average incremental return on incremental investment is 22.42%. This is the same calculation modified for tax write-offs.

The ROI claims to measure investments. An ROI of 22.0% can be compared to any other type of investment that earns interest. In this case, to find an equivalent investment the company would have to find one that would gain interest of 22.0% per year.

Present Worth Method

This is the time value of money method. It is also sometimes called the Discounted Cash Flow (DCF), Internal Rate of Return (IRR), or Net Present Value (NPV) method. The concept of the method is that all projects should be compared with an alternative financial investment yield at anticipated interest rates. That is, if we invested a sum of money equal to the cost of the project, would it yield less than, as much as, or more than the incremental savings generated by the project? If an investment at currently available interest rates yields a greater return than the proposed capital equipment project, then the project is unacceptable and should not be pursued. That is the theory of the method. In practice, we are comparing dissimilarities and are not taking into account intangible benefits that may accrue to the company.

If we have a project that yields $200.0K and the comparison investment yields $250.0K, both over the same time period, the present worth method concludes that the project should not be approved. But what if the project yielding $200.0K is an entry into a new market or expands opportunities for present markets? Then if the company settled for an additional $50.0K profit, it might miss a significantly larger opportunity. Therefore, depending on the present worth method to reach a decision can be very short-sighted, and such short-sightedness has resulted in the demise of many companies. What they did not use was good manufacturing engineering and business judgment.

This method uses the classic time value of money formula,

$$P = \frac{F}{(1+i)^n}$$

where P is the present value of money, F the future value of money after n interest periods, n the number of interest periods, usually years, and i the interest rate. We can now use this equation to start with the expected savings over the life of the project and calculate its equivalent present worth today. This tells us how much money we would have to invest today to earn as much money as we would if the capital equipment project was completed. We will use the example of the project shown in Fig. 4-4.

> **EXAMPLE:** The projected 5-year savings for the capital investment project is $155.5K average per year or $777.5K. What would the present worth of an investment have to be to earn that much money if the interest rate available for an investment is 10%?
>
> $$P = \frac{F}{(1+i)^n} = \frac{\$777.5K}{(1+0.10)^5} = \$482.8K$$

Once we have an alternative investment value—that is, how much it would cost in the financial markets to earn as much money as the capital equipment investment—we can use the present worth method to see whether the project is acceptable.

> **EXAMPLE:** Data from Fig. 4-4 show a cost of $370.0K to earn $777.5K over 5 years. The alternative is to invest $482.8K at 10% interest to earn $777.5K. Since only $370.0K must be invested through the capital equipment project to earn the same amount of money, the present worth method shows that it is less expensive to invest in the equipment than the alternative. Thus the capital equipment project should be approved.

MANUFACTURING ENGINEERING BUDGETS

The facilities program description touches on many aspects classified as budget matters and project management matters. Now let us look more closely at the budget aspect of capital equipment programs; in the next section we will consider project management.

The budget represents the funds allocated for a specific purpose, in our case for the purchase of capital equipment. It also shows by its size and adequacy the priority level given by senior management to this aspect of the business. A budget that is set low by comparison to previous budgets indicates that management is not interested in business expansion, but will be satisfied with maintaining the company's current position vis-à-vis market penetration and profits. A budget that is set high indicates that management wants to take market share away from competitors and place the company in position to become more profitable. Therefore, by evaluating the size of the budget as compared previous years, manufacturing engineering can easily gauge the intent of senior management and act accordingly. This is always true and provides guidelines for how manufacturing engineering should act once the budget is presented.

However, this is only half the story. Manufacturing engineering has considerable opportunity to influence management in deciding whether to hold the line or to set an aggressive budget. As the technical arm of manufacturing, it must make sure that senior management understands what funds are required for the company to maintain its present capability or capacity, and what would be required in addition to improve capability or capacity. Therefore, it must translate matters from a technical sense to a financial sense, that is, the cost versus opportunity relationship, keeping in mind that technical improvements must result in improved profit-making opportunities to be of use to an industrial enterprise. The algorithm the manager of manufacturing engineering must keep in mind is that technical improvement equals opportunity for profit minus costs:

$$T = OP - C$$

and that opportunity for profit minus costs must be greater than zero:

$$OP - C > 0$$

With these principles in mind, manufacturing engineering can proceed to put together proposed budgets, for both the current period and the forecast period. Even such items as upkeep of plant and facility—for example, painting offices—must satisfy the algorithm; that is, walls should not be painted unless the benefits outweigh the costs. Therefore manufacturing engineering must go beyond the reasoning that "it would be a nice thing to do" to justify a proposed budget item. In the case of painting offices, a justification that can be related to improved opportunity for profit must be shown. A reason to paint a sales office—to influence potential customers favorably—would not be sufficient for a manufacturing floor dispatch office. Perhaps the dispatch office should be painted to alleviate a safety problem that, if not solved, could cost the company funds in lost production, fines, and so forth. These mundane examples point out that manufacturing engineering must be aware that every dollar they propose to be included in the capital equipment budget must relate directly to improved profit opportunity.

When manufacturing engineering goes through the objectives and goals process, it is automatically going through a capital budget proposal process. Each project finally approved by the objectives and goals system satisfies the algorithms presented above. Whether a specific project reaches the approved budget depends largely on the degree of positiveness with which it satisfies the expression $OP - C > 0$. In the budget process manufacturing engineering is competing with all other company functions and subfunctions for limited funds. Therefore, it must strive to place only projects with large positive values of $OP - C$ calculations in its submittal of a proposed budget. Of course, it must not lose sight of the fact that projects must be consistent with overall goals. If manufacturing engineering follows the simple procedure of developing capital expenditure projects as part of the objectives and goals procedure, and then assuring that the profitability algorithms are adhered to, it will have substantial success in achieving funding for its programs.

In making a facilities program budget it is desirable to classify projects in accordance with the major category to be accomplished by implementing the project. The classifications show what the major thrust of the company is for any given budget period, that is, whether the company is attempting to enlarge its market share or trying to maintain the status quo.

Classifications vary among companies, but as a minimum, for an adequate description of all capital equipment projects, they should include the following categories:

1. Rebuild—used for projects concerned with taking existing equipment out of service and rebuilding it as new or with slightly enhanced capabilities.
2. Replace—used for direct replacement projects, replacing one piece of equipment with another, usually of the same type and capability.
3. Cost improvement—addition of equipment that allows the company to produce its products at lower costs, hence improved productivity.
4. Capacity addition—addition of equipment that allows more product to be produced than before or a new type of product to be produced.
5. Safety and environment—equipment purchased primarily to meet company, industry, or government regulations.
6. Miscellaneous—used for capital equipment projects that do not fall into any other categories.

Classification of all projects helps senior management to visualize where funds will be spent. It also makes the intent of the company visible. If the company states in its business plan that it wants to increase its market share, and all of its capital equipment projects are in the replace and rebuild categories, it is clear that the actual plans and the strategy are not in agreement. These types of mismatches occur frequently, and the check afforded by classifying capital equipment projects brings them into the open and may lead to either a change in funding levels for capital equipment projects or an honest assessment that the strategy cannot be accomplished and should be changed. Classification, then, can be thought of as a check on the system. If the majority of the equipment project classifications agree with the stated strategy as reflected in the objectives and goals, then the company's planning systems are working well and one would tend to have confidence in its ability to achieve the stated goals. If the converse is true, there is a management problem that should be addressed.

The final budget matter that concerns manufacturing engineering is the question of how much to propose. A proposal for spending at a level that is impossible to achieve will be totally discounted, and vital projects may inadvertently be discarded. To maintain its credibility, manufacturing engineering must know what level of expenditure is reasonable and what is unreasonable.

What is reasonable, of course, depends on the situation. Fortunately, there is a way to check whether a budget proposal is in the realm of reasonableness. Assuming that a company intends to stay in business and intends at least to maintain its current market level, it is fair to say that the capital equipment investment budget should be equal to or greater than the depreciation amount. This means that to stay healthy over a 3- to 5-year period, a company should invest as much money in capital equipment as it depreciates for bookkeeping and tax purposes. If it does less, the company is shrinking in size and its net worth is decreasing.

With this fact in mind the manager of manufacturing engineering can set budget targets. If the company intends to attempt to increase its market share for its product line, then the manager would be justified in proposing a budget that is higher than depreciation. I believe that the increment should be a one-for-one increase based on the amount of market penetration increase aimed at. At the other end of the spectrum, the manager of manufacturing engineering should protest if investments are kept below depreciation for the entire long-range forecast period. Such a situation is a de facto slow exiting from the business, and it is the manager's duty to point this out. Therefore the general budget target should be equal to capital equipment depreciation plus or minus amounts corresponding to current year situations, with concern for consecutive negative years.

Depreciation is a very simple yardstick for budget proposals. Depreciation accounts are calculable well in advance, since they are based entirely on existing capital equipment and firm depreciation rules. This makes it easy for manufacturing engineering to obtain the figures and set visible targets for budget preparation.

CAPITAL EQUIPMENT PROJECT MANAGEMENT

The key factor here is due date control. Figures 4-1 and 4-2 show the elements needed to ensure that capital equipment projects reach fruition by establishing dates for all activities. The schedule shows when each activity must be completed in order to meet the end date, in this case the placement of a purchase order with the selected vendor.

The purpose of capital equipment project management is to make sure that the schedule is adhered to. It involves good management techniques as previously described, including work planning, action plans, and so forth. We will describe here the aspects of project management that apply specifically to capital equipment programs.

The key management tool for capital investment projects is the PAR status meeting. This is a specialized production control meeting where the product is the plant appropriation request, its writing and its approval through the authorization to purchase stage. The PAR status meeting is a weekly meeting held by the manager of manufacturing engineering but presided over by the AME manager and attended by all the area planners, AME engineers, and other manufacturing engineering unit managers. The purpose of the meeting is twofold: (1) to review the status of all capital equipment projects, and (2) to inform the manager of manufacturing engineering of the status and solicit the manager's support and help in approval routines. The latter purpose is important because the manufacturing engineering manager is the first managerial approver and will be instrumental in having complex projects approved at higher levels of management.

The meeting is conducted by the AME manager, who reviews the status of all projects. For each project the subordinates must state whether or not a target date is met, explain why if the date is not met, and state what is being done to achieve completion of the steps. If the target date is not being met, the effect on the overall project and what can be done to get it back on schedule are also discussed. One of the main functions of the meeting is to impart a sense of urgency to the organization. Since virtually all the target dates are the result of inputs from the area planners and the AME engineers, date compliance is not an unreasonable burden.

The AME manager will also preview PARs that will be submitted to the manager of manufacturing engineering for approval in the near future, pointing out any unusual or interesting facts that the manager of manufacturing engineering should be aware of. This, of course, is the primary reason for the manager of manufacturing engineering to attend the meeting. In virtually all industrial organizations it is expected that the manager of manufacturing engineering will personally shepherd PARs of significance through the approval procedure.

At the PAR status meeting the manager of manufacturing engineering will want to be assured that the financial justifications for PARs are being adequately attended to, since this will most likely be the point of contention in any approval problems. Most approval agencies are willing to abide by the manufacturing engineering decisions pertaining to technical content but consider it within their prerogative to question the financial justification. Therefore, for capital equipment projects the manager of manufacturing engineering must make sure the financial justification is as sound as the technical justification.

One way to ensure sound financial justification is to invite early participation by the finance function. Most companies recognizing this need will assign a financial analyst to the manufacturing function to assist in this. The prudent manufacturing engineering manager will make use of this resource.

The final outcome of the PAR status meeting is the updated PAR approval status report (Fig. 4-1). This report should be distributed to all who have a need to know—that is, all approval agencies. Those who are behind schedule are officially advised of their lack of performance, and those who are doing well are officially recognized. In addition, the report tells the approval agencies when they can expect to receive PARs for their disposition. This document then becomes a control document as well as an information document. It serves to tie the entire capital equipment program together and give it substance.

SUMMARY

We have examined the mechanics and philosophies of capital equipment programs from inception to purchase of equipment. This activity is in the forefront of senior management activity and is vital to the success of the company. The range of activities involved in capital equipment programs has been covered from the development of the facilities plan and writing PARs, through financial analysis methods and project management. Of overriding importance, however, is the need for excellence in management, which makes the difference between success and failure.

REVIEW QUESTIONS

1. A facilities plan is a specialized action plan for purchasing capital equipment. Explain why a 5-year plan is desirable for this activity.
2. What is the difference between the facilities list and the facilities plan?
3. Discuss the advantages of preparing an investment budget after the equipment purchase needs are identified. Compare this method with that of allocating funds after all operating costs and profit levels have been determined.
4. Referring to Table 4-1, calculate the severity level if the budget estimate is $1050K and the majority of the projects listed are considered to represent new technologies for the company.
5. Explain the benefits to management of requiring establishment of due dates for action items through the implementation of the equipment in the factory.
6. For the vertical machining center, CNC, shown in Table 4-1, develop a set of specifications, as the area planner would for this capital equipment project. Briefly justify each major entry of the specification.
7. A PAR is to be written for the equipment in question 6. (a) Prepare an overall description and justification of the project. (b) Demonstrate a suitable method for calculating project savings. (c) Prepare a cost to implement schedule. Use 80% of the funds for purchase of equipment and 20% for installation. A three-phase payment to the vendor may be used. Recommend progress accomplishments required before each payment is made. (d) Develop a suitable implementation schedule.
8. Write an executive summary for the PAR in question 7. Include the pertinent facts that the senior operations manager would need to know to approve or disapprove the PAR. Discuss the different information that the chief financial officer would require.
9. A company wishes to increase its market share from 10 to 20% of the total for its basic product within 3 years. Explain the kind of capital equipment budget strategy it would pursue.
10. Using the algorithms $T = OP - C$ and $OP - C > 0$, justify the following projects: (a) Purchase a fleet of cars for sales personnel rather than rent or lease cars. (b) Purchase a document copier machine rather than rent or lease one. (c) Paint the plant cafeteria. (d) Install a central air conditioning system for the perishables warehouse.
11. Explain why the classification by categories system for projects is a useful management tool.
12. A project to purchase a new CNC horizontal boring mill has been proposed. It will cost $475,000.00 and save $820,000.00 over the next 6 years of operation. The company is operating in a down-turning economy and competition is intense. One of their concerns is that if they cannot improve this particular plant's productivity, they may be forced to close it down. There is also a possibility of entering a new product line in addition to their current offerings. This new line will contribute an additional $100,000.00 per year to the company's profit. Calculate: (a) The payback period. (b) Return on investment. (c) Present worth financial justifications. Based on the data given: (d) Determine whether this project should be approved. Assume all figures are after-tax figures. The company can invest at 10.5% interest.

Chapter 5

MACHINE TOOL AND EQUIPMENT SELECTION AND IMPLEMENTATION

Once the plant appropriation request has been approved, the facilities program enters into an implementation phase. We leave the domain of strategic planning and enter that of equipment purchase, including negotiation, procurement, and installation of the equipment to perform the intended task. Also, once the PAR has been approved the facilities project ceases to involve all aspects of the company. It continues to have high priority only within manufacturing engineering and, secondarily, within shop operations. To the other functions of the company the project is history; to manufacturing engineering, it is just beginning.

In this chapter we will explore the many facets of a capital equipment project that lead up to successful implementation. Some of the items occur before PAR approval, some after. The sequence described is the optimum way of approaching the accomplishment of capital equipment projects.

MAKING THE PRELIMINARY LAYOUT

As soon as a facilities project is initiated and a new piece of equipment is agreed upon, an equipment floor plan known as a preliminary layout is prepared. The purpose of this layout is to determine whether the contemplated piece of equipment will fit into the current factory situation and, if it will not, to obtain enough information to estimate the renovations that will be required to make it fit. A preliminary layout is a major factor in determining whether a proposed project is feasible or whether it is too ambitious—that is, whether the changes in the factory floor plan would be too expensive for the project to be accomplished. In this section we will look at macro requirements for preliminary layouts. Then in the next section I will introduce the layout checklist and show how it is used create the layout based on the macro considerations that have been identified.

Let us now look at a typical preliminary layout and discuss the major aspects. Figure 5-1 represents the block layout of a machining bay for relatively heavy equipment as it exists before the initiation of a facilities project to add a Computer Numerical Control (CNC) horizontal machining center with a multi-tool turret. The project is designed to increase the capacity for making large steam valve bodies from four per week to eight per week. The existing manufacturing procedure is to receive a forged body, make a machine layout for it, machine the basic interior and exterior shapes on the Vertical Boring Mill (VBM), do the intricate interior work on either the Horizontal Boring Mills (HBMs) or the lathe, depending on the inside diameter of the particular valve body, drill the various holes on the drill presses, and then inspect and ship the product. The project is to add a CNC horizontal machining center capable of doing the HBM or lathe work and the drill work at one workstation, thus saving queue time, setup time, and machining time. The goal for the new piece of equipment to be able to do eight units per week is set because it would match up well with the VBM capacity of ten units per week.

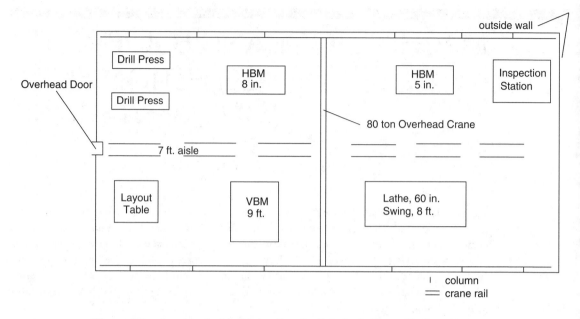

Figure 5-1. Layout of a Machining Bay for Relatively Heavy Equipment

The problem has been that each HBM and each drill press has a capacity of two units per week. Therefore, with the proposed new machining center, an additional capacity of up to 10 units per week would be available, if the older HBMs and drill presses were kept in service. The limitation in output now becomes the VBM. It was determined by the area planner, after consultation with the marketing function, that eight units per week would be adequate for the company's needs with an occasional peak load of up to ten units per week. The purpose of the preliminary layout is to see how the CNC horizontal machining center will fit into the production flow and to determine whether it will work in this factory.

Figure 5-2 shows the layout flow with the existing method of manufacture. The flow is essentially U-shaped with one double back to the inspection station. The new layout should at least not make the flow worse.

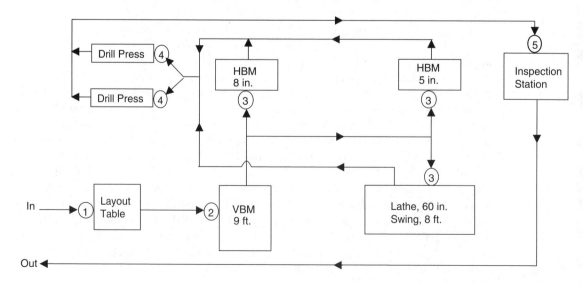

Figure 5-2. Original Flow Diagram for the Machining Bay in Fig. 5.1

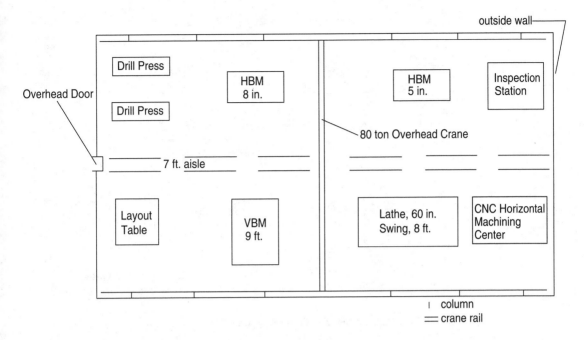

Figure 5-3. Layout of a Machining Bay in Fig. 5.1 with a CNC Horizontal Machining Center Added

An evaluation of available CNC horizontal machining centers shows that the geometry of the machine will allow placement in the factory to the right of the lathe in Fig. 5-1. However, the area planner must also be concerned with supplying electric power and compressed air to the proposed facility and making sure that it can be serviced by the overhead crane, the only method of transporting the work piece in this factory. To do this the planner reviews the architect's drawings of the building to ensure that the crane rails extend far enough to accommodate the proposed facility. This type of machine tool will require a below-grade reinforced concrete foundation, which leads to another factor to be evaluated. The area planner will be looking for any underground impediments or other constraints to putting in the required foundation. These and other questions of significance are reviewed and answered only after the preliminary layout is made; an effort to answer them beforehand is not warranted because the answer would not be accurate enough. Figure 5-3 shows the present layout with the CNC horizontal machining center added.

It appears that the proposed addition fits the space. Let us also assume that there are no underground impediments to installing the foundation. Note that the new machine is next to the lathe; therefore the area planner could expect to extend electric power and compressed air from that facility to the proposed facility. However, the preliminary layout does not indicate whether sufficient power or air exists to service an additional machine tool. This will have to be evaluated by investigating the electrical network drawings and the compressed air capacity. The layout does, however, confirm that crane coverage is adequate. If the crane rails did not extend far enough, it might be proposed to extend the rails or to install a jib crane. Fortunately, these options are not necessary.

Knowing that the new machine tool can be physically accommodated, we should now evaluate the product flow again to ensure that it is at least as satisfactory as the original flow. If the flow becomes worse, it will be necessary to make another preliminary layout. Figure 5-4 is the flow path resulting from the proposed preliminary layout.

The flow portrayed by the solid line in Fig. 5-4 is essentially a straight-line flow with finished product doubling back out the door. However, this is a flow for producing eight units per week. What will happen when 10 units per week are required? In that case we can utilize the older units, as shown by the dashed-line flow in Fig. 5-4, which is the same as the original flow shown in Fig. 5-2. In this case we only need one HBM and one drill press to produce two units per week. This

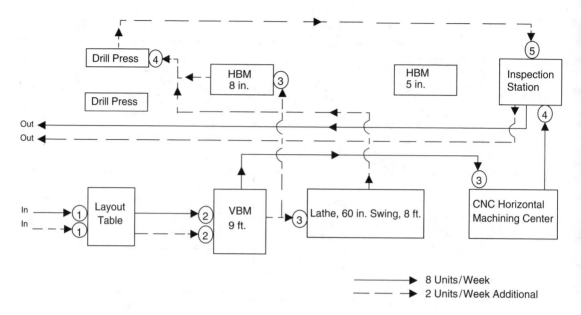

Figure 5-4. Flow Path for the Machining Bay in Fig. 5.3

gives us an opportunity to improve the layout further. Note that the additional units require the U-shaped flow and a double back to the inspection station. A review of the layout shows that if the unused drill press were removed, we could relocate the inspection station to its location and end up with a U-shaped flow for both the base load of eight units per week and the peak load of 10 units per week. Exchanging the unused drill press and the inspection station has another advantage; it places the two unused pieces of equipment adjacent to each other. They can then be used for other product projects or disposed of to make the area available for other uses. The preliminary layout opens up additional opportunities that would probably not have been uncovered if decisions had been made without a preliminary layout. Figure 5-5 shows the final preliminary layout of the CNC horizontal machining center project. This now becomes the working layout that will be used to develop the detailed plans for implementation on the factory floor.

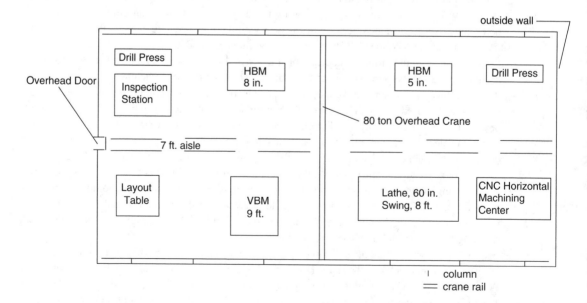

Figure 5-5. Final Preliminary Layout for the CNC Horizontal Machining Center Project

THE LAYOUT CHECKLIST AND HOW TO TIE INTO PROJECT MANAGEMENT

Are the macro considerations sufficient to proceed? We probably have a good idea that it may be, but not 100% sure. Many companies are satisfied with the macro considerations and go on from there. They base this on experience and the need to be only, say 75% sure. But if a company is in a very competitive business where cost control is tantamount to business survival, or we are looking at major changes or additions, 75% is not good enough. In that case the layout checklist process is a necessary step.

As we can see constructing a facilities layout is a linear iterative process. Therefore it leads itself to classic project management techniques. The primary tool is the work breakdown structure (WBS). In order to create a facilities layout we need to amass a significant amount of data then structure what we're going to do with it. This can be a very iterative process, but it helps to create a chronological flow relating to data and what's to be done with it. The WBS fits the bill precisely for this activity. Once we have the chronological list it is a much more straight forward task to then delineate who will obtain the data, who will do what with it and the time allowed for each step. In essence we are creating a project in the classic sense and it makes good management practice to pursue the layout as if it is an independent project. Let's look at a typical layout checklist, explore its contents, and see how it becomes first a WBS then a project.

TABLE 5-1. Layout Checklist

Do the following in sequential order:

- Create detailed method steps list of the entire process
    - Make sure there is a continuous sequence
    - Leave no action steps out no matter how trivial
- Match action steps to work stations/work areas (e.g., value added areas and non-value-added areas).
- Identify times to do each action step (the more accurate the better). Do by
    - Scientific time studies
    - Stop watch measurements of observations, or
    - Estimates based on previous experiences
- Calculate total time for each work station/work area
- Create line balance for the total of all work stations/work areas.
    - Rule of thumb:
        - Time at each work station/work area should be within ±10% with the average of all work stations/work areas for job shops;
        - Within ±2% with the average of all work stations/ work areas for flow manufacturing.
    - Balance methods action steps between work stations, by
        - Adding addition identical work stations/work areas, or
        - Transferring work to preceding or succeeding work stations/work areas.
    - Do bottleneck analysis to assure line balance is sufficient.
- Identify ergonomic space considerations for all of the line balanced work station/work areas.
    - Comply with principles of motion economy
        - Find the lowest human energy utilization method of doing work, set up work stations to favor in the following order
            o Using hands before hands and arms
            o Using hands and arms before hands, arms, shoulders
            o Using hands, arms and shoulders before hands, arms, shoulders, and back
            o Minimize use of back at any work station
        - Keep work within arm's reach as much as possible
        - Minimize distance operators travel during work
        - Keep work at waste level as much as possible
        - Prefer operator sits rather than stands
    - Layout tools and work in an arc, radius equal to average length of human arm, rather than in a straight line.
    - Keep work in sight of operators at all times, never behind them.
    - Provide proper task lighting.
- Identify square foot area and shape required for each work station/work area.
- Create a block diagram layout for entire method process action step sequence based on the required square foot area and shape for each activity.
    - Identify process flow vectors. Acceptable are:
        - Straight Line
        - Serpentine

(Continued)

TABLE 5-1 (Continued)

- – U-shape
- – Beware of and eliminate
  - o Crossovers
  - o Double backs
- ■ Set aside space for material queue at each work station/work area based on manufacturing policy.
- ■ Identify space requirements for direct support activities;
  - – QC inspection
  - – Testing
  - – Activities specific to the particular product
- • Create a final block layout for the entire process with all support activities included.
- • Identify all utilities required to support all operations
  - ■ Air
  - ■ Power
  - ■ Water
  - ■ HVAC
  - ■ Additional specific resources
- • Use final block layout as a total perimeter and sub-perimeters for the Specific Architectural Layout, which includes:
  - ■ Actual footprints of all work stations
  - ■ Utilities runs
- • Outline process flow on the Specific Architectural Layout
- • Define total costs for
  - ■ Acquisition of materials
  - ■ Acquisition of work station equipment
  - ■ Labor to install utilities, equipment, and do moves
  - ■ Downtime lost production costs
- • Review the Specific Architectural Layout with all interested parties for correctness and practicality. Modify as necessary.

Notice the layout begins with first defining the methods steps for the entire process so a cycle time can then be established for the products to be manufactured and/or assembled in the facility. This is a required step even if with a machine tool placement we are only incorporating an additional or changed piece of equipment into an established manufacturing line. Whenever a significant change is made the dynamics of the method, hence cycle time will change. So we need to understand these data in order to make sure the pieces of the puzzle fit. Imagine if the new piece of equipment required material inputs from several different locations that didn't exist before. In that case a different queuing area may be required, or more space for the original. In both cases the layout would need major modifications.

Also, imagine if the new piece of equipment produced output at a rate significantly different than the existing facilities, either faster or slower. This would result in a mismatch of parts flowing between work stations that would have to be managed. Sort of like a traffic jam on a highway that needs to be moderated somehow. This would require a line balance evaluation before the layout is even put down on a piece of paper. We see these examples of why methods and time standards need to be part of the layout WBS.

After this work is done we can begin to do the geometric work associated with the layout. The methods work sets the sequence ideally of how we want each piece of equipment to be geometrically related. With this sequence in mind we can set out how to fit the equipment into the space we have available. Ideally we want the sequence and flow layout to be 100% compatible. Unfortunately that's hardly the case and compromises have to be made. Unless this is a green field layout, for example, scoping a brand new soon to be occupied or constructed factory, then there are orders of magnitude fewer constraints as to how the machines are aligned and sequence and flow can come closer to a complete match.

Once we have the placement order established we need to look at ergonomic considerations before the real "footprints" of the layout can be established. Footprint is the square area the specific workstation will occupy. Ergonomic considerations can make or break the successful application of a layout. All too often the placement of auxiliary equipment necessary for efficient workstation operation is neglected and the planned improvements associated with the project are not fully realized. What is unfortunate is that this is something that should not happen because it is so simple to avoid. As can be seen in the checklist, there are only a few items to be considered and worked into the layout. It's basically applications of the principles of motion economy used to define time standards; for example, the most energy conserving motions results in the most efficient workstation operations. This consideration, along with common sense decisions regarding tool location and lighting are components of ergonomics necessary for a successful layout. These decisions must be made for any facilities layout, not just for manufacturing.

Once we have ergonomics factored into space requirements we can create a block layout as described in the previous section. We take into considerations what acceptable layout types are to be used based on space constraints, as defined in the checklist. Also, we assure in an iterative manner that there is space available for queue, inspection, and any special testing requirements as defined by the previously determined methods requirements. This would also include a macro scoping of the "hotel load," air, power, water, HVAC, and additional specific resources.

This final block layout can then be transferred to the architectural drawings to create a specific layout showing precisely how the layout matches and integrates with the building itself. At this point we are able to fully scope costs for achieving the physicality of the layout, which includes all materials and labor necessary to do the job. It will also include any lost production time, if any, associated with installing the layout in the factory.

What I've described above is the layout project in word format. Which is sufficient to understand what to do, considering that the check list is available. However, we want 100% successful accomplishment and we do not want to rely only on memory and experience to do all that has to be done. To help do this I recommend creating the layout project. Consider the check list as the WBS. Then add names, the assigned person to each item along with an estimated time to expect completion of the step. Also add resources necessary, if they are out of the ordinary for an Advanced Manufacturing Engineering unit. Take the WBS, assignment list, and timing and input into a gantt chart format that was discussed in Chapter 2 for project tracking. As an example I've done that for the checklist and the CNC horizontal machining center example. This is shown in Fig. 5-6.

WORKING WITH VENDORS

Let us now diverge from the chronological path of equipment implementation projects and discuss relations with vendors—a very important subject for the successful practice of manufacturing engineering. Improper relations invariably lead to problems of extra cost, late delivery, and poor equipment performance. There is no need for such problems to develop if the manufacturing engineer understands proper relations with vendors.

First, we must understand that the vendor is the supplier of equipment that allows manufacturing engineering to achieve its stated goal. The manufacturing engineer's project is not to purchase equipment, but to accomplish the stated goal. In the CNC HBM example, the stated goal is to increase weekly output from four to eight units per week with an additional capacity of two units per week for peak periods. Notice that the goal is not to purchase a CNC horizontal machining center. It is important that the manufacturing engineer purchase only what is absolutely necessary and continually keep in mind that the vendor's goal and his goal are not the same. The vendor wants to sell as much as possible. It is not necessarily in the vendor's interest to satisfy the manufacturing engineer's goal.

With these basic facts in mind, the manufacturing engineer can devise a proper relationship with vendors. The engineer must be scrupulously fair with all vendors who have the capability to satisfy the equipment needs of the project. This means giving basic information about requirements equally to all who wish to supply equipment and are deemed capable of supplying it. It is also correct to listen to vendors extol the merits of their equipment and how they would use it to satisfy the company's goal and to incorporate good ideas from vendors in the plans to achieve the goal.

ID	Task Name
1	Create detailed method steps list of the work performed on CNC mach. Ctr.
2	Match action steps to work stations/work areas (e.g.; value added areas and non-value added areas).
3	Identify times to do each action step. Investigate which time ident. Process to use
4	Scientific time studies
5	Stop watch measurements of observations, or
6	Estimates based on previous experiences
7	Review steps 4, 5, & 6 Select most appropriate for use.
8	Calculate total time for each work station/work area
9	Create line balance for the total of all work stations/work areas.
10	Do bottleneck analysis to assure line balance is sufficient.
11	Identify ergonomic space considerations for all of the line balanced work station/work areas.
12	Identify square foot area and shape required for each work station/work area.
13	Create a block diagram layout for entire method process action step sequence based on the required square for
14	Identify process flow vectors. Acceptable are: Straight Line, Serpentine, U-Shape
15	Set aside space for material queue at each work station/work area based on manufacturing policy.
16	Identify space requirements for direct support activities;
17	QC-inspection
18	Testing
19	Activities specific to Valve bodies
20	Create a final block layout for the entire process with all support activities included.
21	Identify all utilities required to support all operations
22	Air
23	Power
24	Water
25	HVAC
26	Additional specific resources for value machining
27	Use final block layout as a total perimeter and sub-perimeters for the Specific Architectural Layout, which includes:
28	Actual foot prints of all work stations
29	Utilities runs
30	Outline process flow on the Specific Architectural Layout
31	Define total costs for
32	Acquisition of materials
33	Acquisition of work station equipment
34	Labor to install utilities, equipment, and do moves
35	Down time lost production costs
36	Review the Specific Architectural Layout with all interested parties for correctness and practicality. Modify as necessary.

Project: Figure 5.6, Proj. Sht CNC Mac
Date: Wed 8/16/06

Task | Progress | Summary | External Tasks | Deadline
Split | Milestone | Project Summary | External Milestone

Figure 5-6. Project Sheet for the CNC Horizontal Machining Center Layout

Failure to keep in mind manufacturing engineering's true goal can lead to problems. The manufacturing engineer who becomes enamoured with the nuances and elegance of a design may end up specifying and purchasing equipment that does considerably more than is actually required. In this case the manufacturing engineer has fallen into the trap of being swayed through apparent friendship, small favors, technical wizardry, and so forth to spend more of the company's money than required.

It is simple not to fall into this trap. If manufacturing engineers follow a few rules, they can be assured that they will always achieve their company's goal and not the vendor's goal. The rules are as follows:

1. Always keep in mind the goal of the project. For each product presented by vendors, ask whether it is absolutely necessary for the optimum success of the project. If not, reject it.
2. Never accept a favor from a vendor such that the vendor will expect a favor in return. Remember that the only favor the manufacturing engineer can give a vendor that is meaningful is to purchase the vendor's equipment.
3. Never accept gifts of value from a vendor. This puts the manufacturing engineer in a compromising situation when evaluations must be made between competing products. It is legitimate to accept social invitations as long as the value is within common bounds of doing business, and not of such a value that the manufacturing engineer could not conceivably pay for it.
4. Never accept engineering services from vendors other than detailed information necessary to evaluate whether the product offering would fit into achieving the goal. Vendor engineering services could result in a system that would not accept other vendors' products, which is usually detrimental.
5. Always strive to obtain at least two competitive quotes for all equipment purchases. This will ensure that the company is purchasing the equipment at the lowest possible cost.

These rules amount to the exercise of common sense in dealing with persons determined to sell something. By following these rules, manufacturing engineers can pursue their assigned goal in an optimal manner. Whether the sale of equipment is optimal to the vendor is of no concern. Manufacturing engineers must be concerned only with the success of their own company and can do that only if they concentrate on optimizing achievement of their assigned goals.

SELECTING THE EQUIPMENT

Concurrent with the preparation of the preliminary layout, work starts to specify the equipment to be purchased. In specifying equipment, the more specific or detailed the manufacturing engineer can be, the better the ultimate results will be. This means that the manufacturing engineer must thoroughly understand the purpose of the project, including the technical levels to be achieved as well as the budget limitations. Once a project is close to the PAR writing stage, the area planner will hold a technical briefing session with the assigned manufacturing engineer, representatives from shop operation, the methods engineer, and the process control engineer. Specifically, the overall goal to be achieved and the options identified to achieve that goal will be discussed. During this meeting the area planner will solicit comments from those representing the other areas to ensure that the best possible solution is identified. The manufacturing engineer, on the basis of this information, will begin to determine the specifications for the equipment.

Specifications are prepared in two phases: a preliminary estimate and a detailed purchase phase. The preliminary estimate phase has the main purpose of determining the cost levels involved in the project. Usually this is done by researching current equipment catalogs from various vendors, comparing this project with previous projects, and perhaps contacting vendors to clear up technical points. This information, once compiled, gives the area planner an estimate of what the project would cost if pursued. The estimate is compared with the original budget estimate, and a decision must be made as to whether the project is still attractive and whether the company can afford to go ahead with it. This decision is usually made by the manager of manufacturing engineering, based on recommendations from the AME manager. If the decision is to continue the project, we progress to the next phase.

The detailed specification phase entails sufficient documentation to solicit firm quotes from the interested vendors. It also includes factors to ensure that the company will receive the proper value for the money spent. Therefore, detailed specifications contain requirements for performance of the equipment and requirements to make sure the equipment is manufactured to what are considered satisfactory quality standards. The performance requirements are tied directly to the goals of the project and can be determined by evaluations of engineering needs. The quality requirements are usually designed to ensure that the piece of equipment purchased will not fail in service.

The performance requirement specifications should be straightforward and set the range of what should be paid for the piece of equipment. If a machine tool is to perform to very tight tolerances—for instance, to be accurate within ±0.0001 in.—it will cost more for the vendor to produce than a machine capable of maintaining ±0.0005 in. In these specifications it is important not to overspecify. The manufacturing engineer must be certain to require only what is necessary to perform the task. Another mistake the engineer must be careful to avoid at this point is that of specifying so that a particular vendor's offering will receive favorable consideration. This is called rigging the specification, and it frequently means that the company buys something it does not need. The only certain winner here is the vendor. Performance specifications must be a true reflection of the company's need to achieve its goal and nothing more.

The quality requirements are insurance policies to make sure that the equipment works correctly and does not take extended time to debug and put into useful production. They take the form of compliances that the vendor must meet in manufacturing and testing the equipment. Usually they are tests and inspections that must be satisfactorily completed before the customer accepts the product and pays for it. Typical requirements would be: the materials used in manufacture are of the proper grade, that welding, machining, and assembly are in accordance with the engineering drawings, and that the machine performs in accordance with the specifications. How specific and tight these requirements are will depend on the criticality of the equipment in its intended use. For example, if the equipment is a simple transfer device, the quality requirements would not be as strict as if it were a safety device for a furnace fuel system.

There is a good reason for the manufacturing engineer not to insist on overly stringent quality requirements, and that is cost. The more complex and tight a quality specification is, the more the vendor will charge, so that it is possible for insurance policy to cost more than a conceivable failure of the equipment. It is important to weigh the cost of tight quality requirements against the cost of failure that the requirements protect against. If failure can destroy the entire investment or significantly hinder production, then high quality requirements costs are justified.

Another consideration in determining whether to place stringent quality requirements in the specification, is whether it is traditional in the industry to do so. If such requirements are entirely alien, it will be difficult to get vendors to quote, and if they do, the costs will be very high.

To summarize, the rules that must be followed in developing detailed specifications are:

A. Performance Requirements
   1. State explicitly what the piece of equipment must do. In our example of the CNC horizontal machining center, the manufacturing engineer would have translated the eight valve bodies per week into machining metal removal rates, that is, feeds and speeds based on acceptable depth of cuts and selection of cutting tools. In addition, the engineer would specify the types of operation the machine must be capable of performing, such as face milling, boring, and drilling, and the dimensional tolerances required for each operation. If the queue times and transfer of material from workstation to workstation are of concern, the engineer would specify limitations in times to perform these functions.

2. State the physical attributes of the machine, that is, geometric size and weight. It is important that the piece of equipment purchased fits in the space available for it. If the machine is too heavy, a more costly foundation may be required; if it is too big, it will cost considerably more to make space available for it, or it could end up in a location that is insufficient for economic production of the company's products.

3. State the required power sources. The machine must be able to work on the electrical system the company employs. For example, equipment manufactured in Europe is usually designed to run on electricity at 50 hertz, while the North American standard is 60 hertz. It would be embarrassing if a piece of equipment manufactured in Europe could not run in the company's factory until the extra cost of a power conversion was approved. Therefore, the manufacturing engineer must carefully research the company's power situation and write specifications in accordance with it.

4. State other electrical requirements, for instance, for electrical fixtures, controls, switches, and motors. These are usually based on a desire to standardize in order to minimize maintenance costs.

5. State environmental and safety requirements. The purchaser, not the vendor, is responsible for compliance with the various government and insurance requirements. Therefore, it is prudent to state these requirements to be met in the specifications. An example would be hand guards for punch presses or dead-man switches for rotating equipment.

B. Quality Requirements

1. State performance acceptance testing requirements. The manufacturing engineer must make sure that the piece of equipment will meet all performance requirements before accepting the equipment from the vendor. Even though the vendor is obligated to meet the specifications put forth in the performance requirements, the engineer is in a much stronger position if the machine does not perform as required by the quality requirements during the trials. If the equipment is delivered and paid for prior to actual testing in the company's factory, the vendor can claim that instructions for operation are not being complied with. If the vendor must demonstrate conformance before being paid, the customer is better able to obtain what he contracted for. In addition, the debugging and minor corrections always needed by complex machinery will be on the vendor's account. It is important that the specification include a clause stating exactly what will constitute acceptance of the equipment for delivery and payment. This is usually a test run in the vendor's factory in which the vendor must demonstrate compliance with the equipment specifications, for example, by machining a certain number of parts within a specified period of time under the observation of the customer.

2. State quality requirements during manufacture of the equipment. This usually consists of a set of certifications by the vendor's quality assurance personnel and perhaps key events witnessed by the customer. The exact quality requirements are placed in the specification and become part of the contract between the vendor and the customer.

3. State compliance with code requirements that the vendor must meet. In some cases the vendor must be able to assure the customer that the workers engaged in the manufacture of the equipment are qualified under industry standard or government codes.

Once the specification is prepared, the manufacturing engineer turns it over to the purchasing unit of the materials sub-function to obtain quotes from qualified vendors. Purchasing is responsible for qualifying vendors for all commodities and equipment the company purchases. However, it is common for manufacturing engineers to let the purchasing agents know who they think would be qualified to supply the equipment described in their specifications. In deciding who would be qualified vendors, manufacturing engineers usually confine their judgment to technical matters. They are interested in knowing that the vendor has built similar equipment before and that the vendor can demonstrate successful uses of the equipment. They should not consider non-technical aspects of vendor qualification with the exception of whether the vendor would be capable of delivering the product on time.

IMPLEMENTATION SCHEDULE DURING MANUFACTURING OF EQUIPMENT

This is a very busy period for manufacturing engineers. After quotes are solicited and received, the engineers will be called on to make the technical evaluations of quotes and recommend a vendor and will then become involved in placing orders with the successful bidders and monitoring progress through delivery of the equipment. Let us look at the various phases that follow the receipt of bids:

1. Technical evaluation of bids and recommendations
2. Establishment of monitoring schedules with the vendor
3. Vendor surveillance during manufacture
4. Vendor runoff and acceptance

Technical Evaluation of Bids and Recommendations

Usually vendors will reply to a request for a quote with the same level of detail used in the specification they received. Hence the amount of bid evaluation required will depend on the complexity of the specification.

Before receiving bids, the manufacturing engineer will make out a checklist containing "musts" and "wants." The former are those items of the specification that have to be met. For instance, if a tolerance for a bore inside diameter of 0.001 in., with a concentricity of 0.0001 in., is specified to achieve design engineering requirements, that tolerance capability is essential. A prospective vendor who cannot meet that requirement is automatically excluded from further consideration. "Wants," on the other hand, are desirable rather than critical traits, and inability to provide them does not mean automatic rejection. The manufacturing engineer can establish a point scale for all such items and then rate each bid received on this quasi-objective scale. It is important that every item in the specification be declared either a "must" or a "want" and a rating factor (point value) assigned to each item in the second category.

Once bids are received, they are evaluated first for compliance with the "musts." Bids that pass this test are then evaluated for the desirable but non-critical traits they can provide. The manufacturing engineer will then assign point values for the "wants" that are satisfied, giving partial values for items that are only partially satisfied. After all specification "wants" are reviewed against the bid, the manufacturing engineer will compile a score for that particular vendor.

Notice that no credit is given to items in the bid that are not asked for in the specification. Such items represent an attempt to sell something more than required at a higher selling price. It must be the practice of the manufacturing engineering organization to ignore these extras and actually let the vendor penalize himself by having a bid offering at a higher cost than his competitors.Occasionally, a vendor who has a superior offering for all the required items in the specification may be allowed to rebid without the unsolicited extras. However, care must be taken not to engage in an unethical practice vis-à-vis other vendors.

During this evaluation phase it is common for manufacturing engineers to be in contact with the various vendors. This is done to have points clarified so that the engineer thoroughly understands the bid offerings. In each case the manufacturing engineer must obtain written versions of the clarifications, which then become part of the vendor's bid, so that there will be no doubt about the vendor's intentions.

After all bids are evaluated, the manufacturing engineer will make a technical recommendation. If the technical recommendation coincides with the low bidder, everything is simple. The low bidder is awarded the contract. If the technical recommendation is given to a vendor who is not low bidder, the manager of manufacturing engineering will be called upon to decide whether the low bidder or other qualified bidders with lower prices than the technical winner should be awarded the contract. Remember that at this stage the only bidders being evaluated are those who satisfy all the "musts."

There are many rationales that can be used to determine whether a technical winner who is not the low bidder should be awarded a contract. Difference in price is a major consideration. Another consideration will be how many of the high-value "wants" are satisfied by the low bidder. In all cases the contract award problem must be resolved to the benefit of the company, not the vendor.

Normally in equipment projects conflicts between the low bidder and the technical winner are resolved in favor of the technical winner. This is true because capital equipment projects are indeed technical projects and are not treated as commodity purchases. Treating capital equipment purchases as commodities leads to more frequent equipment failures and lost profits, and is a decision that is usually regretted.

Establishing Monitoring Schedules with the Vendor

It is important that manufacturing engineers detail their expectations for monitoring requirements to the successful bidder. They should specify the kind of progress charts the vendor will be expected to produce and review with the customer. However, care must be taken not to become too involved in the vendor's internal procedures. The progress charts are designed to ensure that the equipment is produced to meet the contract date and that the vendor complies with the quality and performance requirements of the contract. Any other requests to the vendor are usually unproductive and may be a factor in late delivery to the customer.

An example of an excessive monitoring item would be one involving review of the vendor's manufacturing drawings. Too many manufacturing engineers fall into the trap of overzealousness in this regard. The engineer may wish to review the engineering drawings to make sure that the design is adequate or that codes are being complied with. However, the bid review should have shown whether the vendor is competent to design the equipment, including code conformance, and the manufacturing engineer should not waste time reviewing drawings. In addition, if reviewing the vendor's drawings becomes part of the monitoring schedule, the engineer is accepting responsibility for the design and in effect voiding the vendor's responsibility to produce a competent design. If the design should prove faulty, the vendor can then claim that the customer approved the design and refuse to accept liability for damages. The manufacturing engineer should adhere closely to schedule and contract performance items only. A typical monitoring schedule is shown in Fig. 5-7.

The monitoring schedule is produced by the vendor from points spelled out in the specification, but the witness points are established by the manufacturing engineer with the concurrence of the vendor. As can be seen from Fig. 5-7, the manufacturing engineer has placed the contractual witnesses on the schedule; at a minimum this must be done. Optional witness points have also been placed on the schedule so that the engineer can be present at critical manufacturing points. But notice that there are no mandatory holds placed on the vendor that could be used as an excuse for failure to meet the delivery date requirement.

Vendor Surveillance During Manufacture

The primary objective of vendor surveillance is to ensure that the vendor is doing everything possible to deliver the ordered equipment on time. The manufacturing engineer will be satisfied only if the vendor adheres to the end date of the schedule, that is, delivers the ordered equipment on or before the schedule date. Besides delivery on time, the manufacturing engineer must also be assured that the quality requirements and performance requirements are met. By using the customer monitoring schedule, the engineer can be in a position to do both.

The customer monitoring schedule shows when certain phases of manufacturing, including testing, are to be accomplished. As a minimum, the manufacturing engineer must check with the vendor prior to these scheduled accomplishment dates to ensure that these commitments will

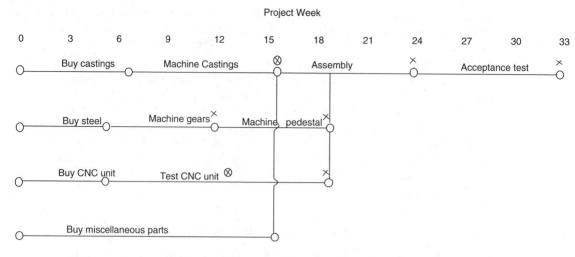

Figure 5-7. Customer Monitoring Schedule for the CNC Horizontal Machining Center

be met. If the engineer is confident that the vendor is performing adequately to schedule, nothing else need be done. If it does not appear that the vendor will meet the dates, it becomes necessary to urge the vendor to perform better. Procedures for doing this range from a simple telephone call to express concern, to a high-level management visit with implied threats of no future business if the vendor's performance does not improve. The main point is to communicate displeasure for lack of adequate performance. The vendor will favor customers who are attentive and insistent on timely performance, especially if he is experiencing delivery difficulties. Silence on the part of the customer implies acceptance of the vendor's actions.

Another method of checking a vendor's progress would be through a contract containing provisions for progress payments. On expensive capital investment projects it is common to have provisions in the contract for periodic payments to the vendor. These payments are usually tied to the vendor's achieving agreed to milestones in progressing from the start of the contract through delivery of the equipment–that is, they are tied to the customer monitoring schedule. With this system, a vendor who is late in reaching a milestone is also late in receiving the progress payment. This is a powerful tool because the vendor is probably counting on the progress payment for operating funds, especially in periods of high-interest borrowing costs.

Performance specification and quality specification surveillance are slightly different. Here the manufacturing engineer uses the customer monitoring schedule to determine when to be present to review or observe tests. As in monitoring progress to ensure delivery on time, the customer must know where the equipment stands against the schedule. However, in this case the reason is to ensure that the manufacturing engineer is present when the tests are to be performed, or data are to be approved, before the vendor moves on to the next phase of manufacturing. It is essential that the manufacturing engineer be available to perform this task, as dictated by the contract, when the vendor requests it. Failure to do so would entitle the vendor to delay delivery of the equipment by at least as much time as the delay caused by the manufacturing engineer. Often this type of delay can be multiplied by factors of two, three, or even five times if it leads to loss of priority in the vendor's sequence of manufacturing flow, so that the required facilities are no longer available because they are now being used on work for other contracts. In addition, if the engineer is not available when required, the vendor can legitimately claim the progress payment due at the milestone, which was not achieved because of a customer-caused delay.

At the opposite end of the spectrum is the case of the vendor trying to eliminate or modify tests in order to make delivery on time. This allows the vendor (1) to reduce manufacturing costs by not having to do the test, and (2) to get back on schedule and transfer the testing to the customer's factory. By delaying the test to the final installation at the customer's site, the vendor has a greater opportunity to blame any lack of performance on improper installation and operator error.

For these reasons, it is never in the manufacturing engineer's interest to waive tests. Remember, the tests were placed in the specification to ensure that the manufacturing engineer's project goal would be successfully achieved. Waiving tests should only be considered in an extreme situation and only after a thorough review of the possible consequences. The only extreme situation that would warrant considering this would be one concerned with severe economic loss to the manufacturing engineer's company if the equipment were not installed on time or within an acceptable late period. Only the manager of manufacturing engineering can adequately judge whether such an extreme situation exists.

Vendor Runoff and Acceptance

In this phase of the project the equipment will be put through its operations to see whether it does produce as the specifications require. As the company's representative, the manufacturing engineer plays a vital role in the acceptance test. Let us review the sequence of the runoff and discuss the manufacturing engineer's function.

Assembly and Lineup

Virtually all equipment used in factories is made up of subassemblies. Putting together the subassemblies is critical for successful operation of the equipment. Here's some common examples. For bake ovens and furnaces, the walls and doors must fit snugly to avoid unacceptable heat loss. For machine tools, the line-up of shafts and bearings must be precise to prevent rapid bearing wear and ensure accuracy of machined parts, and the lead screws and parts that constitute the transfer roadway of the cutting tool must be plumb and square with each other to very close tolerances, again to ensure accuracy of machined parts. Each type of equipment has specific critical parts that require precise assembly, and for a runoff or acceptance test it is important that the piece of equipment be assembled correctly.

The manufacturing engineer should be present during critical phases of assembly for two reasons. First, the engineer must make sure that the parts constituting the equipment have been made correctly, that is, that they assemble properly. This simple observation indicates that the results of the runoff will be valid, that the machine has been assembled in accordance with design, and that any lack of performance is probably inherent in the design. The second reason for being present during the assembly stage is to learn how the machine goes together and what the difficult phases are. Since the vendor's factory personnel are not likely to be available to assemble the equipment at the customer's factory, it is vital that a customer representative be familiar with the installation procedure for adequate future operation and maintenance of the equipment.

Test

This is the actual operation phase. In most runoffs the vendor runs the equipment in as close a simulation as possible of the customer's situation. During this phase the manufacturing engineer can observe the characteristics of the machine and determine whether it meets the contract specifications.

The test to which the piece of equipment is subjected is specified in the purchase contract. It is the manufacturing engineer's responsibility to see that the vendor conducts the test as required. The engineer must also keep in mind that the vendor cannot be expected to do more than is required by contract. Nuances of interpretation are important here, and the engineer must be certain that the vendor understands the requirements in the same way that he does.

Whenever controversy exists concerning whether equipment meets acceptable test criteria, there are no winners. The vendor's payment may be delayed and his reputation tarnished. The customer is in danger of receiving a piece of equipment that does not perform in accordance with specification, which can result in nonachievement of a goal that depends on the purchased equipment. Such controversy can be avoided by a pre-runoff meeting to review the specification for the acceptance test and ensure that there is complete agreement on its meaning.

The manufacturing engineer must explicitly tell the vendor what will constitute passing the test. This is done simply by reviewing the specification and determining the numbers involved in meeting it. In the case of the CNC horizontal machining center of our example, it might mean producing a fully machined valve in the specified time, or preferably two valve bodies in twice that time, since a runoff should demonstrate repeatability. In addition, the specification may require a demonstration of range of capability. In reviewing the details of the test with the vendor to make sure there is mutual understanding, there is always a phase of verbal communication. However, to prevent any misunderstanding it is recommended that this be followed by a written phase, where the manufacturing engineer produces minutes of the meeting and delivers a copy to the vendor. By receiving a copy of the minutes and not disputing any points, the vendor signifies concurrence with the interpretation of what constitutes a successful test.

During the actual running of the tests it is important that the manufacturing engineer observe all key points. If the equipment does not meet specification, the engineer will have to determine with the vendor what can be done to correct the problem. The engineer should review the corrections proposed by the vendor and determine whether they will permanently correct the deficiency and whether the time frame to do so is acceptable. In any event, the manufacturing engineer should not accept equipment that does not meet the specifications.

Occasionally, no matter what a vendor does, a piece of equipment does not meet specification. In this case the manager of manufacturing engineering must decide what should be done regarding acceptance. Several options may be considered:

1. Accept the equipment as is.
2. Negotiate paybacks by the vendor, and then accept it.
3. Reject the equipment and require the vendor to build a product that meets the specification.
4. Reject the equipment, recover progress payments if any, and place an order with another vendor.
5. Reject the equipment, recover progress payments if any, and exercise another option for achieving the goal related to the equipment purchase project.

The option selected depends on the company's situation and the degree to which the vendor misses the specification. If the mismatch is slight or not of primary significance, then option 1 or possibly option 2 is selected. If the mismatch is significant and the company can afford the time delay, option 3 or 4 may be chosen. Finally, if the company's goal must be achieved by a specified time and this time precludes another try by the same vendor, option 5 may have to be selected. In practice, option 5 is rarely used because the purchase of the equipment is typically the only way to achieve the stated goal. Therefore if a piece of equipment fails runoff, options 1 through 4 constitute the practical choices. Most equipment eventually passes the runoff test. Then the data collected by the manufacturing engineer establishes the benchmark for successful operation of the equipment and is used to develop methods and time standards. In the very rare case where equipment fails repeated runoff tests, the data will probably be used as evidence against the vendor for a law suit or settlement.

Disassemble and pack for shipment

The primary concern of the manufacturing engineer at this point is that the equipment is not damaged during disassembly and is adequately protected for shipment to the company's factory. After successfully completing a runoff, the vendor will naturally wish to deliver the equipment and get on with other business. One way of assuring that the vendor pays proper attention to the disassembly, packing, and shipping is to have this work documented in the specification. This means that manufacturing engineering must be knowledgeable about the construction and assembly procedures for the product prior to purchase, which is usually not the case. Therefore, another method must be found.

The simplest way to ensure that the vendor is not careless during disassembly, packing, and shipping is to insert a clause in the specification stating that the equipment must arrive defect free at the customer's plant. This means the vendor is responsible for the equipment until a receipt is received for it at the customer's factory. In essence, the transfer of title does not occur until the equipment arrives at the customer's site. By not accepting title at the vendor's factory, the manufacturing engineer has only to exercise concerned interest during the teardown procedure and does not have to manage the operation. Often, if the title is transferred immediately after successful completion of the test, the manufacturing engineer is put in the position of being the foreman, planner, and chief expediter over the vendor's hourly work force in order to get the purchase on its way to the company's factory. This potential frustration and confusion is eliminated if the original bid specification states F.O.B. customer's factory for the title transfer and final payment.

IMPLEMENTING THE EQUIPMENT ON THE FACTORY FLOOR

After this activity is completed, the equipment is installed and ready for operation. The company then can proceed toward achievement of the stated goal. Let us follow our example goal of producing the eight valve bodies per week and examine the steps necessary in implementing the newly arrived CNC horizontal machining center.

Once the contract for the CNC horizontal machining center is let to the successful bidder, work commences to ready the area for acceptance of the equipment. The final preliminary layout (Fig. 5-5) is used as the basis for producing architectural drawings and detailed construction drawings. At this stage of the project we know only the space required for the new machine tool. We do not know its actual shape or where the electrics, hydraulics, air, and so forth will be connected to the equipment. However, we do know what their values will be because they are part of the specification for the machine tool bid. Therefore, work on the architectural drawings can commence; and in fact, unless we are building a new factory, we are only making additions to existing drawings. These additions will constitute bringing the services to the area designated for placement of the new machine tool. After the work on the drawings begins, the locations of the services will be transmitted to the vendor, who needs to know where to place the hook-ups for the machine. Usually this is an iterative process in which the customer and vendor reach a mutually satisfactory conclusion based on the service location choices dictated by the machine's design and the site situation.

Near the end of the design phase by the vendor, the specifics of the foundation for the equipment usually become known. The vendor supplies this information to the customer, who then completes the architectural drawing by adding in the foundation outline. This architectural drawing then leads to the development of the construction drawings for emplacement of the foundation, completing the design phase of the project. The vendor has designed the equipment and the manufacturing engineering organization has designed the factory to accept it.

The construction phase begins next. The same manufacturing engineer who is responsible for vendor surveillance will usually also be responsible for the installation phase. Here the maximum use is made of project control charts. For very complex projects networking or Critical Path Method

(CPM) charts may be employed. For less complex projects gantt-type charts are used. Whatever technique is employed, it is important that it be detailed enough so that all activities are scheduled and can be monitored effectively.

Figure 5-8 is an example of a networking project control chart that is compatible with a project of the magnitude of installing a CNC horizontal machining center. Note that all the basic phases are included: work that must be done by the purchaser, the equipment vendor, and the construction vendors. This chart can be used to track all activities of all organizations involved. It may be as simple as the version shown in Fig. 5-8 or significantly more detailed, as long as it is sufficient for adequate control. Note also that the line corresponding to manufacture of the CNC horizontal machine is void of detail. For our example, this was done for clarity. In actuality that line would be replaced with the detail shown in Fig. 5-7. Hence we would have all the information required to carefully monitor all phases of the project on one chart.

Let us look at the construction phase in chronological order. During the drawing preparation, the manufacturing engineer must make sure that enough information is presented for adequate quotes to be prepared by the contractors. The term contractors can also include in-house staff if the company wishes. Lack of adequate information in drawings usually leads to cost overruns and delays. Therefore, it is important to make sure that architectural drawings are correct, particularly with respect to obstructions. Obstructions are any items that for any reason cause contractors to use more materials than anticipated, which typically leads to extensions to obtain more materials and always adds to the bill. No contractor will allow a client to pass on such hidden costs.

Construction bid activities are very similar to those explained previously in this chapter. They tend to be less complex because they involve a shorter term and a greater direct-work content. The format of construction bids usually follows standard local practices and the legal commercial code will apply. When working with in-house staff there will be no bid, but the manager of that function, usually the manager of maintenance, will be required to estimate the in-house costs involved, consisting of the time to complete the work and the cost of materials. Based on the time requirement, the manager of manufacturing will determine at an early stage whether to authorize overtime to the contractor, either in-house or external. Once contracts are let, the manufacturing engineer will monitor the work in the factory, paying particular attention to whether performance of the work is at a proper pace and to proper standards.

While the foundation is being built and services installed, the CNC horizontal machining center is going through its acceptance and runoff procedure. This activity will be repeated in the

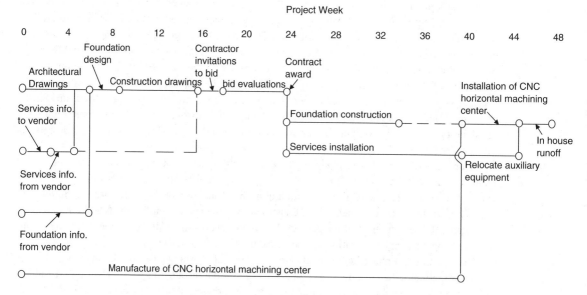

Figure 5-8. Installation Schedule for the CNC Horizontal Machining Center

manufacturing engineer's factory after assembly of the machine on its foundation. During assembly the manufacturing engineer will pay close attention to ensure that the machine is being assembled as it was at the vendor's facility. This is necessary to ensure that the equipment operates successfully and that the vendor's warranty is not voided. Usually the vendor's installation service representative will help in this activity. Virtually all large capital equipment contracts include funds for this service, which is very helpful in achieving proper installation. All reputable vendors maintain a staff of experts to assist their customers in installing machines such as a CNC horizontal machining center. Having the machine run correctly is the best form of advertising for the vendor's product.

The culmination of the installation is another runoff. This runoff or acceptance test is quite different from that held at the vendor's location, where the question was whether the machine would work as designed. At the second runoff the only things in question are whether the machine has been assembled correctly and whether it has suffered any damage during shipment. This runoff is also a fine tuning, since there is likely to be some dynamic difference between running on a solid foundation and performing on the vendor's test block foundation. During the in-house runoff, the manufacturing engineer must be cognizant of minor differences that may or may not affect long-term performance. Usually at this runoff, things such as shaft balance, hold-down bolt torques, and horizontal and vertical alignment are looked at very critically to ensure that the machine tool is set up in the most stress-free position possible.

The second runoff does not have a limit to the number of pieces to be produced before the equipment is accepted. Manufacturing engineering wants to complete this acceptance test and turn the equipment over to shop operations. But there is no high sense of urgency to get the runoff done quickly. Keep in mind that every piece now produced will be used to satisfy the production schedule needs. The only question is how many units per week will be produced and what the rate of increase will be to reach the production level in the stated goal. Also, during this second runoff the manufacturing engineer along with engineers from methods, planning, and work measurements and shop operations supervisors and operators will be fine-tuning the methods to achieve an optimum performance capability.

Now let us look in more detail at the activities involved in transferring the machine from manufacturing engineering control to shop operations control. This is the last phase in the capital equipment purchase project.

The first consideration is training of operators. Operators are assigned to the new machine tool as soon as it arrives in the factory, and it is advisable to have them participate in the assembly of the machine and be able to converse frequently with the vendor's technical support people. It is an old axiom, particularly true in job shops, that good operators know their machine and can work wonders with it. This is a good reason for introducing the operators to the new machine at the earliest possible time.

Training to operate the machine consists of learning how the new piece of equipment will produce the factory's product. It begins with assisting in the assembly and reviewing the operating manuals with the vendor's technical representative and the manufacturing engineer. The training will also cover the method to be followed in operating the equipment, tool holder positions, setup techniques, and how to utilize the CNC controls. Most of this will be accomplished through discussions between the manufacturing engineer and the operators, and some consultation with the vendor's representative. The manufacturing engineer will be keeping a record of the salient points of these sessions to be used later in the methods documentation. The ultimate goal of the training activity is for the operators to be ready to run the machine under guidance of the vendor's representative and the manufacturing engineer during the runoff, and to enable them to progress quickly to the point where they can perform correctly without close supervision.

One of the major tasks of the manufacturing engineer during the installation and runoff phase is to prepare a methods document to be turned over to the methods, planning, and work measurements unit for refinement and implementation within the planning system. This document states the sequence of operations to be followed to machine the valve bodies, such as setup procedure, sequence of tools to be used, feeds and speeds, operator performed maintenance, and how to operate the machine. Development of this document begins during acceptance testing at the vendor's location, and at least a draft document should exist by the time the second runoff is complete.

During the installation and runoff period, the manufacturing engineer also introduces all key personnel besides the operators to the machine. The engineer familiarizes shop operations management with the machine, paying particular attention to differences from their previous experiences in the factory. Similarly, the engineer briefs process control personnel and maintenance personnel. The purpose of these briefings is to reduce the time required for key personnel to become proficient with this equipment as much as possible.

Finally, we have reached the point of turning the new equipment over to shop operations. Recall that the example of producing eight valve bodies a week plus an overload capacity of 10 valve bodies required relocation of the inspection station and a drill press (see Fig. 5-5). In order to turn over the new facilities to shop operations, these other moves will also have to take place. The equipment relocations are relatively simple. It is considerations involving people that must be carefully planned. At worst, people may be relocated and not be capable of performing their jobs. With lack of attention to details, the little things that make factories work smoothly and efficiently will go

Machine/Process _____

		Review Complete
Operators by Shift		
First	_____ _____ _____	_____
Second	_____ _____ _____	_____
Third	_____ _____ _____	_____

Items Discussed	Complete
O Shift start and finish times	_____
O Notification in case of absence	_____
1st shift Foreman_____ tel. no._____	_____
2nd shift Foreman_____ tel. no._____	_____
3rd shift Foreman_____ tel. no._____	_____
O Break times	_____
O Work rules (general)	_____
O Locker assigned and moved	_____
O Work station instructions	_____
O Work station tooling	_____
O Building and facilities orientation	
Tool crib	_____
Material storage	_____
Material handling service	_____
Lubrication storage	_____
Dispatch cage, record keeping routines	_____
Inspectors	_____
Foreman's desk	_____
Rest rooms	_____
Canteen facilities	_____
Emergency telephones	_____
Infirmary/First Aid supplies	_____
Time clocks	_____
Parking facilities	_____

Foreman Signature_____

Date Complete _____

Figure 5-9. Checklist for Relocated Equipment

wrong. Such things as not knowing where the dispatch cage or the tooling lockers are located are reasons for increasing idle time, resulting in lost productivity.

It must be recognized that anything that constitutes change will result in unfamiliarity, hence lost productivity. It is management's responsibility to minimize the level of unfamiliarity. One way to do this involves making a checklist of anticipated problems, such as that shown in Fig. 5-9. A checklist should be made up for each workstation that is being relocated. It identifies the operators and requires considerable logistics information to be given to the operators by the foremen. There are places for recording that each step has been covered, so there is an attempt to require management to communicate with the people involved prior to any move. When checklists such as that illustrated in Fig. 5-9 are used, moves usually involve a minimum of disruption.

CAPITAL EQUIPMENT PROJECT CONTROL CHARTS

Most manufacturing engineers use charts of one type or another to visualize activities to be accomplished in order to achieve their stated goals. Figures 5-7 and 5.8 are examples of networking charts. Charts that take the form of bar graphs are commonly called gantt charts, after the pioneer industrial engineer Henry L. Gantt, who introduced such charts for production control and project control. Regardless of the type of chart used, their purpose is to show the activities necessary to complete a project in a manner that portrays the relationships of the steps to each other. Control charts make it significantly easier to understand the interactions between the various steps. Let us now explain the theory behind these two techniques and demonstrate how they are used.

Bar Graphs (Gantt Charts)

Bar graphs are a combination of verbal and visual information charts. The technique is to list operations to be performed in sequential order, then use a bar or line next to the items to indicate commencement and completion dates. The length of the bar indicates the time in relation to other steps. We can convert Fig. 5-7, the customer monitoring schedule for the CNC horizontal machining center, into a gantt chart to illustrate this technique. To make this conversion, we (1) list all activities by their starting dates along the left-hand margin, with the first item to be completed at the top and the last at the bottom; (2) construct a time scale along the top of the page; (3) draw a horizontal line to scale, beside each specific activity listed for starting date and completion date; and (4) use a simple key to indicate the actual starting and completion date. Figure 5-10 is the result of this task. In the figure it is assumed that the project is already under way. The filled triangles and circles indicate that the activity has been accomplished. Note that in items 1 and 6 the filled symbols do not coincide with the open ones. This indicates that the activity did not start or end on the scheduled dates.

Now compare Fig. 5-10 with Fig. 5-7. They contain the same information, and both of them show the steps in sequential order and show the relationships of the steps to each other. Which of them better visualizes the project is a matter of opinion. Perhaps the networking chart shows the dependence of later steps on earlier steps more effectively, but again, that is a subjective opinion. Both formats work; the choice of which to use is based on personal preference.

Networking Charts (Critical Path Method)

For most projects encountered by manufacturing engineers the complexity is such that there is no preferred choice between gantt charts and networking charts. However, occasionally a project is commissioned that is very complex and has a large number of incremental, interrelated steps. If this is the case, a gantt chart may become very long, so that it is difficult to visualize the thrust of the project. For simpler projects, the engineer can visualize the abstract relationships between the steps in a gantt chart and instinctively spot the critical path, or main road, from start to finish. When we have many activities taking place at the same time, as in a very complex project, we must resort to a networking chart.

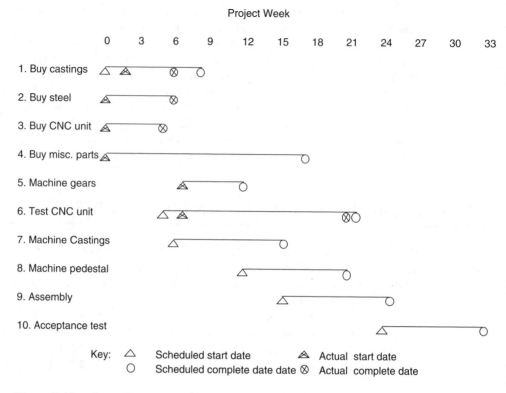

Figure 5-10. Gantt Chart: Customer Monitoring Schedule for the CNC Horizontal Machining Center

A networking chart allows us to visually describe the critical path of the project, so that the manufacturing engineer can concentrate on achieving the elements that make up the critical path. Successful conclusion of the critical path elements, which are the items with the longest lead times, means that all other elements are likely to be completed on time.

The most common form of the networking chart is the CPM chart. To illustrate the networking technique, we will develop a CPM chart for the customer monitoring schedule for the CNC horizontal machining center and end up with Fig. 5-7.

First, we may define a critical path as a sequence of chronological events selected out of a universe of events such that sequence takes the longest time to accomplish. All other sequences will be complete before the critical path sequence.

The logic of the CPM is simple. If we can find the main stream of events and make sure we accomplish them, all the tributary events will be finished on time because they have shorter cycle times. This means that we can tolerate much greater delays for the tributary items than for the main stream or CPM items. The CPM forces manufacturing engineers to focus their primary efforts on the mainstream events and not diffuse their energy on comparatively minor items.

The CPM networking technique works, but like every other tool available to manufacturing engineers it must be used with caution. Many projects will not exhibit a clear-cut critical path. One path may be a little longer than all the others and may be designated as the main stream. However, if the tributaries are almost as long, focusing on the critical path may lead to unacceptable delays of the tributaries. When this happens the late tributary becomes the new critical path and a shift in emphasis is required.

The steps for developing a critical path are as follows:

1. List and number the events of the project in chronological order.
2. Determine the immediate event precursor for each event.
3. Determine the time required to accomplish each event, using the same time scale (e.g., weeks and weeks, not weeks and days).

TABLE 5-2. Work Sheet for CPM Network: Customer Monitoring Schedule CNC Horizontal Matching Center

Node	Event No.	Description	Immediate Precursor Event	Required Time (Weeks) to Accomplish
1.,0	1	start		0
2.,7	2	buy castings	1	7
3.,5	3	buy steel	1	5
4.,5	4	buy CNC unit	1	5
5.,16	5	buy misc. parts	1	16
6.,7	6	machine gears	3	7
7.,15	7	test CNC unit	4	15
8.,9	8	machine casting	2	9
9.,8	9	machine pedestal	6	8
10.,8	10	assembly	5,7,8,9	8
11.,9	11	acceptance test	10	9
12.,0	12	finish	11	0
item 1	item 1	item 1	item 2	item 3

4. Link all the precursor and successor events with lines.
5. Add the times of all events in each of the lines to determine the longest chronological time to complete them. This is the critical path; all other paths require less time.
6. Put the network on a time scale chart similar to a gantt chart. This makes it easier to read and shows the dates when events must occur to maintain the schedule.

Now let us go through the process of constructing a CPM networking chart. Table 5-2 is the CPM work chart with data filled in representing the items pertaining to the CNC horizontal machining center example.

1. List and number the events of the project in chronological order. This list, except for the arbitrary start and finish items, is identical to the left-hand column of the gantt chart (Fig. 5-10).
2. List the immediate precursor for each event. If more than one event must occur before an item is started, then the event item will have more than one precursor. Event 10 in Table 5-2 is an example of this.
3. Determine the time required to accomplish each event. In the CNC horizontal machining center example the time scale will be weeks. This information can be found on the gantt chart.
4. Link all the precursor and successor events together with lines. For clarity, we will use the node convention, where the node is an ellipse containing a shorthand code identifying the event and showing the time it will take to accomplish it. For example, event 4 ("buy CNC unit") is estimated to take 5 weeks to complete; therefore the node would be 4.,5. Nodes for all the events are shown in Table 5.2. The linking of all the nodes is shown in Fig. 5-11.

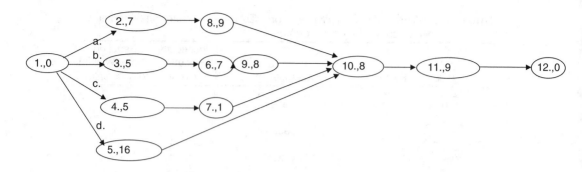

Figure 5-11.　CPM Diagram: Customer Monitoring Schedule for the CNC Horizontal Machining Center

5. Add all event time in each of the lines to determine the longest completion time. This is the critical path. We now add the times of the four paths to find the critical paths.

Path a:　(1.,0) + (2.,7)　(8.,9) + (10.,8) + (11.,9) + (12.,0) = 33 weeks

Path b:　(1.,0) + (3.,5)　(6.,7) + (9.,8) + (10.,8) + (11.,9) + (12.,0) = 37 weeks

　　　　　　　　　　　　　　　　　　　　　　　　　　　　　　　　　　critical path

Path c:　(1.,0) + (4.,5)　(7.,15) + (10.,8) + (11.,9) + (12.,0) = 37 weeks

Path d:　(1.,0) + (5.,16)　(10.,8) + (11.,9) + (12.,0) = 33 weeks

From these additions we find two critical paths and two tributary paths with time values very close to the critical paths. Unfortunately, it is typical of manufacturing engineering projects that the critical path is not very different from the tributary paths.

6. Put the network on a time scale chart similar to a gantt chart. The CPM diagram is not very revealing; only the engineer who constructed it knows what the node numbers refer to. Therefore, prudent manufacturing engineers will take the next step of constructing a time scale networking chart, such as the one shown in Fig. 5-7, so that all who are interested can follow the progress of the project. Sometimes the critical path will be superimposed on this chart, but this usually occurs only for complex projects. Figure 5-12 is Fig. 5-7 with the two critical paths shown (most often projects of this complexity would not require this further refinement).

Note that the critical path line intersects the assembly line beyond the assembly start point. This is allowable if it is understood that the assembly operation is all-encompassing. In this case it is understood that the assembly of the gears and pedestals cannot begin until the main structure, the casting, has been assembled. The same is true of the CNC unit.

COMMERCIAL SOFTWARE FOR NETWORKING AND GANTT CHARTS

There are many software offerings that manipulate project data to assist in creating gantt and CPM charts. Most of these software packages run on personal computers, so their costs are minimal. Being low in cost, they definitely are worthy of investigation for use in capital equipment programs. But be wary of their abilities to do the total project management task.

Project Week

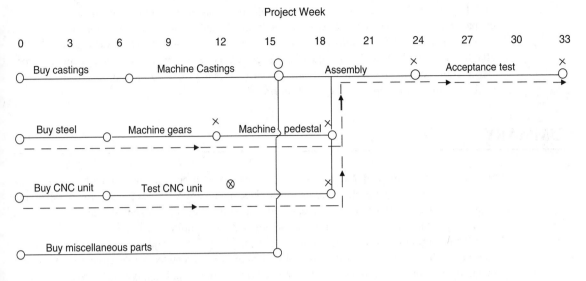

Notes

1. × = customer witness required by contract
2. ⊗ = customer witness allowed by contract
3. Vendor to give customer 72 hour notice prior each contractual witness. Absence of customer witness constitutes customer waiver of witness requirement.
4. — — ➔ = Critical path

Figure 5-12. Customer Monitoring Schedule for the CNC Horizontal Machining Center. Critical Path Superimposed on Network Chart (Fig. 5.7).

Software programs reputed to be project management programs can be very useful in curtailing the clerical effort needed to create and use the networking and gantt chart techniques—if they work. This means it is essential that the manufacturing engineer carefully investigate the capabilities of the proposed program versus the needs to control the project before attempting to use the program. Software programs that are of insufficient capacity to handle the total data can cause considerable havoc. At worst they can truncate data and give false results, such as erroneously defining the true critical path. It is therefore prudent to take the necessary time to investigate software thoroughly before employing it in a manufacturing engineering project.

Selecting the proper software is nothing more than a "musts" and "wants" exercise, similar to that discussed in the Technical Evaluation of Bids and Recommendations section of this chapter. The engineer has to determine what the "musts" and "wants" are for the intended application. At a minimum the software must be able to construct bar charts. It should also be capable of doing the steps necessary to define the critical path with minimal manipulation by the user. For example, as a minimum, in the steps for developing a critical path, as shown in this chapter, data inputs should be required only for the first three steps. If manual or off-line data manipulation is required, then the software program is probably not sufficient for a manufacturing engineering type of project.

Another point to consider is the number of individual data points the proposed software program can accept. Keep in mind that a networking activity for a large project can have hundreds or even thousands of individual nodes. These data have to be manipulated and gantt charts produced, which could require a significant amount of computer capacity. Hence, the software has to be able to handle the required amount of transactions in a reasonable time frame. This capacity will vary depending on the types of projects the manufacturing engineer will be running. Obviously, constructing a factory to produce valve bodies will require software with a larger capability than that used to design and purchase jigs and fixtures for this chapter's example CNC horizontal boring mill. In general, then, selecting software for program management requires the "musts" to do critical path planning as stated above plus a set of "wants" rated for degrees of desirability.

Commercially available software does make the job of managing large projects easier. But like all other tools available to the manufacturing engineer, it does not supplant the judgment and experience of the engineer. Computer software primarily does the clerical function. The manufacturing engineer still has to supply the creative process.

SUMMARY

We have discussed the major activities for achieving goals through the purchase of capital equipment. Capital equipment projects are complex ones, with many activities going on in parallel, and require adequate management coverage. Coordination is necessary to maintain adequate coverage and can be achieved only through proper communications techniques.

Manufacturing engineering is a pragmatic technology. This chapter describes approaches to working with vendors so that the company's interests are protected while ethical relations are maintained. Another important part of manufacturing engineering is applying good business practice based on sound engineering principles. This blend of engineering principles and business sense has been further explained through the use of project control charts. These are powerful tools for the manufacturing engineer, but professional judgment is necessary for their successful use. The pragmatic approach to business problems has also been demonstrated through the lengthy plant layout description, which shows that successful implementation of capital equipment projects requires attention to practical details as well as theory.

REVIEW QUESTIONS

1. A preliminary layout is often called a tollgate layout. With reference to the CNC horizontal machining center, discuss what the "go" or "no go" aspects might be.
2. With reference to Fig. 5-2., note the material flow. Determine where the congestion points are that would be areas of opportunity for corrective action for the new layout required by the addition of the CNC horizontal machining center. Explain why these would be areas of opportunity.
3. List in chronological order the major steps required to construct and verify the adequacy of a factory floor layout.
4. Discuss with examples how a vendor's goal and a manufacturing engineer's goal can be complementary with respect to capital equipment purchase.
5. Explain why the work breakdown structure (WBS) process is necessary for creating a layout for the addition of capital equipment machine tools to an existing factory.
6. For the CNC horizontal machining center of Fig. 5-5, prepare a list of items that a manufacturing engineer is likely to include in the detailed specifications, and explain how these are related to the CNC horizontal machining center project.
7. For the following items, determine whether the quality requirements would be specific or general and state reasons for your answers. (a) Machine tool electric drive system; (b) layout bench; (c) operator vertical lift (allows the operator to be at same level as the rotating cutting tool of an HBM); (d) machine column horizontal translating rails.
8. Make a list of "musts" and "wants" to evaluate bids for the CNC horizontal machining center. Include rating factors for the "wants" and reasons for the priority selection sequences.
9. The following vendors meet all the "must" requirements for a capital equipment project. The following data represent a summarization of the bid evaluations. The budget allocated for this project is $512K. Delivery is desired by April 15 with an absolute date of June 1. Make a recommendation with reasons for vendor selection.

	Vendor A	Vendor B	Vendor C	Vendor D
Cost	$490K	$565K	$505K	$510K
Delivery date	May 20	April 10	May 1	May 15
"Wants" score	58	68	56	66
Compatible with existing equipment	No	Yes	Yes	Yes

10. Referring to the example of the CNC horizontal machining center, the following is a list of monitoring schedule items. Explain why each item is proper or excessive. (a) Review material certifications to specified code; (b) determine progress to manufacturing schedule; (c) review design stress calculations for rotating cutting tool shaft; (d) review tolerance selections for tool holder; (e) review performance of machine tool at runoff.

11. Discuss the pros and cons of entering into a progress payment contract for capital equipment.

12. A manufacturing engineer receives a request from the vendor to waive a test pertaining to adequate performance of bearing parts. This is the fourth of five identical machines the vendor is to supply. In each of the previous three, the test was 100% satisfactory. In addition, any defect could be detected during runoff. What should the manufacturing engineer's response to the request be? Give reasons for your decision.

13. A manufacturing engineer is purchasing a CNC lathe to make studs for valve bodies. The specification requires a diametric tolerance of 0.0005 in. and a concentricity of 0.0002 in. TIR (total indicator readout). Develop a runoff acceptance test plan to meet these criteria.

14. Discuss the differences and similarities between invitations to bid for construction services and equipment purchases for projects such as the CNC horizontal machining center.

15. With reference to question 12, what differences, if any, would there be between the runoff acceptance test to be performed after installation and the runoff test after initial assembly at the vendor's plant?

16. Develop a strategy for training operators and familiarizing other personnel with the CNC horizontal machining center so that time is optimized from receipt of the machine in the factory to turning it over to shop operations.

17. Construct a gantt chart, using the data contained in Fig. 5-8.

18. Based on the data of Fig. 5-8, (a) develop the work sheet for constructing a CPM network, (b) construct the CPM network, and (c) determine the critical path.

Chapter 6

PRODUCIBILITY ENGINEERING

Producibility engineering is a coordinative discipline with the design function of the industrial organization. Through this coordinative approach we strive to optimize the process of producing the company's products. We cannot, however, simply isolate the manufacturing engineering subset called producibility engineering and discuss it and its techniques without introducing the concept of concurrent engineering (sometimes referred to as design for manufacturability). The work of a producibility engineer is in many aspects an inter-functional one. It deals with finding the pragmatic limits of feasibility for manufacturing and still meeting the intent of the design. In this case, the relationship is between manufacturing engineering and design engineering. When we expand beyond the manufacturing and design engineering relationship to include the entire industrial organization, we enter the domain of concurrent engineering.

The concurrent engineering concept unites all functions of a company into a team for conceiving, designing, manufacturing, marketing, and distributing a product in an integrated and optimal approach. Just as the name suggests, all this work is done in a concurrent approach. Producibility engineering is at the same time a root of the concurrent engineering philosophy and a subset of it, as a member of the concurrent engineering team. Therefore, in this chapter we explore producibility engineering as a prelude to and within the concurrent engineering concept. We will see how producibility engineering techniques are used in this unifying concept.

In this chapter we define the concurrent engineering role in modern manufacturing and then focus on the producibility engineering aspects of that role. We look at the techniques in which producibility engineering leads the concurrent engineering team, and also identify areas where producibility engineering assists other functions.

CONCURRENT ENGINEERING CONCEPTS

The definition of concurrent engineering is straightforward. It is:

> *The synergistic process of doing all the preproduction, production, and postproduction work in a manner such that efforts are scheduled and done in a parallel interactive fashion, rather than in a series singular manner.*

We all know the adage that teamwork is more important than individual achievement in reaching an organization's goal. Traditionally, we applied teamwork concepts to functions of a company and rarely to the company as a whole. The conventional wisdom was that there is no commonality between marketing, engineering, and manufacturing. They all do their thing independently, and there is certainly no similarity in management techniques. No commonality exists between managing Computer Numerical Control (CNC) machines and sales campaigns—or so it was thought 25 to 30 years ago. Hence teamwork existed, it was presumed, only within functions where all the activities were definitely adjacent in the spectrum of work. This meant the walls were up and work was done within a narrow universe. Marketing defined a need, engineering created a concept of a product to meet that need, and manufacturing took the concept and made a real thing out of it—and they all did it mostly independently of each other. Perhaps thought of as a relay race. Once the baton was passed, that lap would succeed or fall based on the spend and cunning of the runner. In this case the ability of the function to perform once the task was passed on it.

By allowing this thought process to dominate our organizations, we incurred significant additional costs and time in the process of creating and delivering products to customers. Costs were not optimized. Why? Simply because we were creating waste. The left hand did not know what the right hand was doing. In many cases the instructions from marketing to engineering or from engineering to manufacturing were incomplete or just plain wrong. Why was this so? Mainly because of unfamiliarity with what happens when the baton is passed and the next step of the product creation process takes place. When people simply do not know, strange things happen, and the laws of probability state that the vast majority of outcomes will be considerably less than optimum.

Concurrent engineering unites an organization's functions by creating synergy between those functions. The synergism is established by setting up parallel efforts between the functions through cross-functional teams, and the output is definitely greater than the sum of the parts. Via concurrent engineering we actually establish an atmosphere of trust and cooperation that allows all functions to do their jobs with the active and real-time advice of the other functions. Thus, the concurrent engineering process allows companies to serve their customers better by shortening the product cycle time. The parallel effort greatly reduces the cycle, from conceptualization of the product through delivery to the customer.

Concurrent engineering is not a precise and bounded body of knowledge such as thermodynamics or strength of materials. It is a concept, or perhaps some would call it a theory, of how a manufacturing company ought to structure its process of delivering its product to the customer. Along with this structure are techniques that are applied in carrying out the process, which are compatible with the parallel effort. Some are internally directed techniques and some are external. The internal category comprises mainly design engineering- and manufacturing engineering-based techniques. External techniques concern primarily marketing principles.

We will explore primarily the internal techniques, first with aspects of manufacturing engineering, then with those of design engineering. These two disciplines of engineering are quite broad, so the focus is only on the content that is applicable to concurrent engineering, that which makes the parallel effort work. In a supportive role, we will explore the external techniques, for example, marketing, from the viewpoint of how its principles are applied in the parallel team effort and where the ties to producibility engineering are.

Techniques of Producibility Engineering Used in the Concurrent Engineering Process

Figure 6-1 illustrates the power of concurrent engineering. But what exactly is the technology involved? What is the interface between the functions when they are together as a team? What do they really do? What is the expected contribution of each member? What is design for producibility? These are but a few of the questions that need to be answered and understood before concurrent engineering can be applied properly. Let's investigate the answers to these questions by starting at the beginning, which is producibility engineering.

Producibility engineering is the discipline responsible for creating producible designs. Note that there is nothing in the definition about teamwork or parallel efforts. Producibility engineering was created as a checking mechanism. The idea was to make sure that the factory was not being asked to make products beyond its capabilities. The original concept was not concerned with fostering parallel efforts to meet customer needs (although this later became a valuable attribute of the discipline, as it was used as part of the concurrent engineering process). It had a very focused approach to minimizing manufacturing losses by imparting information to design engineering vis-à-vis the specific drawing and whether or not the factory would have difficulty in achieving the desired design goals.

Producibility engineering is the beginning of the understanding that there has to be cooperation between functions in order to optimize a manufacturing company's performance. It is a process by which manufacturing-based engineers are assigned as liaison personnel to design engineering. The purpose was primarily to impart "manufacturing know-how" to the designer as part of the design process, which is a big part of what later became the full concurrent engineering process. It is important that we understand these procedures.

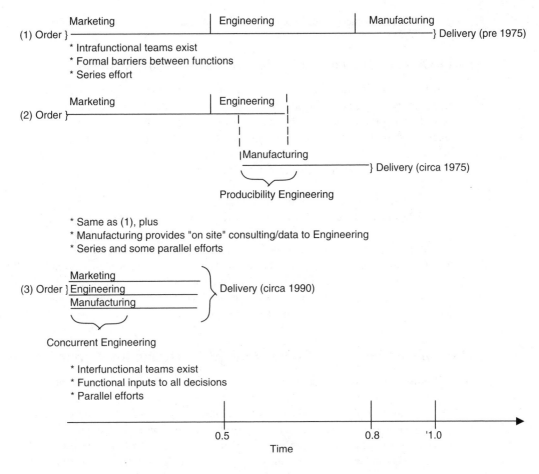

Figure 6-1. Evolution of Concurrent Engineering

One way to think about producibility engineering's place in the concurrent engineering continuum is to think of producibility engineering as the manufacturing technical contribution to the concurrent engineering team. Likewise, the designer would be the design engineering technical contributor to the team. We must understand this manufacturing engineering technical role thoroughly in making concurrent engineering a successful process.

The Producibility Engineering Process

The producibility engineer is expected to tell the design engineer how to make the design producible, which is commonly known as design for manufacturability. This means that he or she is expected to critique the design from the viewpoint of how to optimize manufacturing process costs. This is a very demanding task. Process costs comprise cycle time and material utilization, and keep in mind that the product still has to perform to the specifications required. Therefore, for the producibility engineer to do the job right, he or she must understand the theory of the design. As the factory's representative, the producibility engineer needs to understand the intent of the design, such as why certain dimensions, tolerances, and materials were selected, before he or she can make credible comments to the designer. Also, as the factory representative, it is absolutely essential that the producibility engineer understands the factory's limitations. So, for the producibility engineer to assure that designs are capable of being produced in the factory, two tasks must be mastered:

1. Understand the intent of the concept design.
2. Understand the factory's capabilities relative to the design.

If this is done successfully, the producibility engineer can offer cogent criticism to the designer.

When to Accept or Reject a Design for Producibility Criteria

Suppose the producibility engineer found that the intent of the design simply did not match the capability of the factory. Suppose also that the producibility engineer was certain that the intent must be met for the product to be successful. What are the choices? Remember, the producibility engineer's task is to prevent designs from being accepted by manufacturing if they cannot be produced economically. Furthermore, the producibility engineer is morally committed to the success of the cooperative venture between manufacturing and engineering. There are two basic choices:

1. Reject the design for production.
2. Recommend major changes in the factory to accommodate the design.

At first glance, the first choice appears to be a stonewalling strategy. But really it's not. Informing the design team that they must reject what they're proposing because it would be folly for the manufacturing group to attempt is a sufficient answer to give design engineering. This is a hard choice to make but a correct one. The last thing the company could tolerate would be manufacturing trying to do something outside its realm of competence.

An Example of the Logic of When to Reject a Design for Production

Dilemmas such as this typically occur when companies try to redefine what type of company they are. Let me give an example.

Company A is trying to redefine itself as a musical instrument maker instead of a piano maker. This could lead to serious mismatches. There may be problems in its desire to serve a market versus its capability to do so. When this happens, engineering is apt to respond in a positive manner much more quickly and readily than manufacturing. For example, as a musical instrument maker, instead of only a piano maker, company A might be tempted to offer a line of electronic organs. After all, there appears to be synergy: they are both keyboard instruments and both require fine furniture fabrication. But there the similarity ends, and this might not be a good choice.

Suppose company A were to embark on such a venture. Engineering could acquire the technological expertise by going to the market and hiring bright electronics engineers with organ design backgrounds. They would then produce the design for building organs. But what does manufacturing do? They could contract out the electronic components, to the point of purchasing completed subassemblies. Then they would assemble, test, package, and ship. How confident would manufacturing be in its ability to produce organs at the same quality levels as pianos? Not very. Piano making is 90+% mechanical in design, as the piano is a product dominated by mechanical engineering. Organs, on the other hand, are more than 50% electronic. So there is no synergy. The factory would have to go through a long and arduous learning cycle. It would be difficult to predict realistically a successful business outcome. This being the fact, the producibility engineer would make the proper choice in recommending rejection of the design even though he or she agrees with the intent of the design.

When It Is Proper to Accept a Design for Manufacturing Even Though Capability to Do So Does Not Exist

The second choice of action puts the onus on manufacturing. In this case the producibility engineer agrees with the intent of the design but does not reject it because manufacturing does not have the capability to produce it. Instead, the producibility engineer says that manufacturing has no choice but to acquire the capability to produce, or the company will suffer the consequence. The last phrase is key. If the producibility engineer is convinced that the design is proper and within the range of products the company must offer, then he or she should recommend that manufacturing take on the challenge of tooling up to make it.

What is the line of demarcation? How does the producibility engineer determine whether the company should be urged to invest capital and human resources to be able to make the product depicted in the design? Sometimes the choice is obvious. For example, in the silicon chip business, the ability to crowd more and more electronic circuits on a silicon wafer is the difference between staying in business or permanent demise. So, if a new chip design requires improvement in photo-etching capability, the producibility engineer will probably recommend that his or her manufacturing colleagues develop the necessary equipment. In this case, the need to meet the design requirement is obvious to all.

Sometimes the choice is not so obvious. Remember, we are dealing with cases in which the producibility engineer agrees with the intent of the design but knows the factory cannot make it at satisfactory cost levels. The choice of accepting the design commits the company to spending funds to gain the capability to produce. There is no rule book on how to make the choice. I can only offer some guidelines.

1. If compatibility exists between the proposed design and previous designs, then accept.
2. If tooling requirements to produce are within the realm of state-of-the-art equipment, then accept.
3. If employee experiences are similar to those required for successful manufacture of the proposed design, then accept.
4. If training requirements are extensions of current skills, then accept.

Let me expand a bit on these guidelines.

1. *Compatibility exists.* Compatibility existing probably means that the new design, while not within the capability of the factory, requires processes similar to existing processes. This is the chip maker example. Getting more transistors on the chip will require new machinery, but in truth it is just more of the same thing. It is somewhat like driving a car and having to go from zero to 60 miles per hour in 7.0 seconds. The process for doing it is the same; however, some automobiles will achieve the goal and others will not. If the goal is critical and you are unfortunate in having a car unable to accelerate fast enough, there is only one choice: trade up to a faster automobile.

2. *Tooling within the current state of the art.* The second condition, tooling within the current state of the art, means that no new technology needs to be developed for the factory to comply. This means that new invention is not required. For example, if the design calls for an electric-powered automobile with a 100-mile radius, this technology exists. The consequence is that the factory can be tooled up to make the product with state-of-the-art technology. On the other hand, if the design requires a range of 1,000 miles, the design theory might be satisfactory but the edge of practicality as we presently know it may be breached, and there may be reason to reject.

3 and 4. *Employee and training issues.* The third and fourth items are people issues: current capabilities and the ability to learn new technologies. These are the hardest to evaluate. We are dealing with many levels of experience and capability, and the producibility engineer will be hard pressed to come to any conclusions independently. Here the proper course of action is to involve human resources as well as the various levels of manufacturing management. The goal is to determine the median experience level and compare that with the tasks that will have to be accomplished to produce the intended design. If it is felt that the differences are not too great, the design can be accepted. Likewise, whether training can be successful depends on how close the new skills level is to the current level.

A Guideline for the Producibility Engineer's Job Performance

Summarizing responsibilities and how the producibility engineer should perform them is a complex matter. I have explained the various factors, but they are not simple. Over the years I have

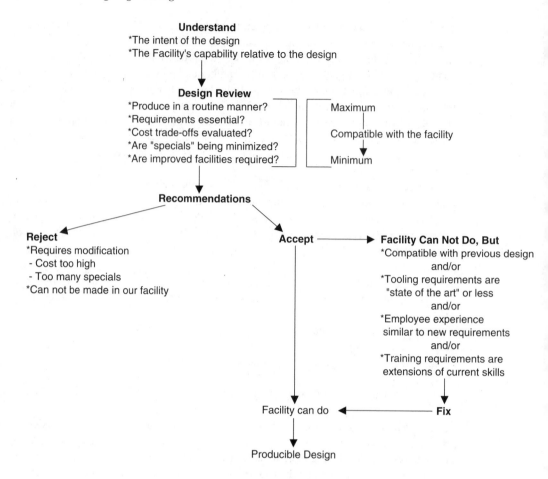

Figure 6-2. Producibility Engineer's Decision Tree

developed what I call the "producibility engineer decision tree," which is shown as Fig. 6-2. I trust it will be a useful aid in understanding the role of the producibility engineer. This is a very real and valuable role that needs to be performed within the continuum of concurrent engineering.

So far I have discussed what the producibility engineer's responsibilities are. But how are they accomplished? What methodology is used to do the job? Let's take a look.

THE PRODUCIBILITY DESIGN REVIEW PROCESS

Producibility engineers perform a liaison role, primarily via the design review. Prior to concurrent engineering this was done at the completion of significant milestones of the design task. With concurrent engineering teams, it is a continuous process. Nevertheless, the process remains the same. The process is not the normal design review for functionality and meeting of the customer specifications. It is a review to discover whether or not a producible design exists.

A producible design is defined as:

A design that can be manufactured correctly and economically in the factory for which it was intended.

The producibility design review starts with a look at the basics. Can the design be produced while maintaining the status quo in the factory? If this is so, the design is producible. It gets more difficult from here on. Referring to Fig. 6-2, we see that the degree of difficulty increases as the requirements of the design become less compatible with the factory's capability. The trick is to determine how much the factory's ability can be stretched to meet the design requirements. The practitioners of producibility engineering do this by looking at all five items listed under the design review segment of the decision tree. The secret is to list the reasons for incompatibility and then see what can be done to overcome these deficiencies. At the same time as we are looking at the factory side of the equation, the design requirements must be looked at with equal intensity. This ensures that unnecessary costs are not incurred. Creating a producible design is a systems integration activity and one that optimizes the process for the entire company.

To fully understand the producibility design review, let's look at an example. We will use a simple product so that the essence of the technique is not lost in the technology being applied.

Producibility Design Review Example

The scenario is as follows. We have a typical job shop supplied with normal CNC metalworking and woodworking machines capable of doing sawing, milling drilling, turning, and shaping. The current products are simple hand tools, specifically screwdrivers and chisels. The basic manufacturing process in use is to cut machine tool grade steel to length and then turn it to the desired diameter. Sometimes the process requires drilling holes in the shanks to mount the handles. The final metalworking operation is to mill the flats, either for screwdriver blades or chisel blades. The shop also does some simple wood machining, such as sawing to length, lathe turning, and drilling and shaping to make handles for screwdrivers and chisels. The final operation is an assembly task. Here handles are glued onto shafts. The more deluxe models have a rivet-like pin pressed through the handle and shaft before gluing.

Because of the good quality of work, business has been good and the company is contemplating expanding the hand tool line. In order to do so, a concurrent engineering team has been assembled to bring new products to market. They determined that the market is ripe for a high-quality, simple design needle-nose pliers. The designer has prepared a preliminary product scope to meet the need and is requesting that the producibility engineer review it for practicality of manufacturing in the factory. In addition, the concurrent engineering team is looking at material procurement cost, sales promotions, production run quantities, selling prices, capitalization, and other related and equally important issues for introducing a new product. However, for this example we will make the simplifying assumption that the go/no go decision will be solely a producibility issue. Let's now go through the steps of the producibility design review.

The questions to be answered during the producibility design review are:

1. Produce in a routine manner?
2. Requirements essential?
3. Cost trade-offs evaluated?
4. Are "specials" being minimized?
5. Are improved facilities required?

The needle-nose plier design shown in Fig. 6-3 is the item that will be reviewed for producibility.

1. *Produce in a routine manner?* The current production requires straight metal rods turned to a specific diameter. The rod is then milled flat at one end into either a screwdriver or chisel point, and a portion of the production of metal rods is drilled through to accept a pinned handle. The pliers will be made of two rods bent approximately one-third of the way along the shaft to a 60° interior angle. This will require milling an undercut near the pivot joint. Also, one side of the short end of the rod beyond the pivot will require tapered milling and knurling to create the needle-nose shape. Finally the two pliers halves will have to be drilled at the pivot point to accept the pivot pin.

In the example factory the only thing currently not done is making metal parts with angle shafts. This is different and will be so noted by the producibility engineer.

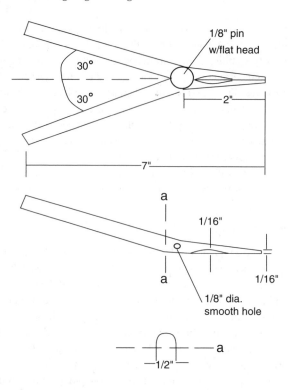

Figure 6-3. Sketch: Simple Needle-nose Pliers

2. *Requirements essential?* The designer is not requiring anything but bare essentials for the finished product to perform as an adequate needle-nose pliers. The producibility engineer would agree that everything specified is required for functionality to be achieved.

3. *Cost trade-offs evaluated?* In this case the answer is yes, they have been. The material to be used is the same tool-grade steel used for the screwdrivers and chisels. Because the company is a quality maker of hand tools, marketing would insist that the same material, which is already perceived as a quality commodity, be continued.

4. *Are specials being minimized?* One special is being considered, that of the bending of the pliers half. No bending is currently being done, so this must be considered a special for the factory. Keep in mind that the evaluation is for a producible design, that is, whether the design can be produced in this specific factory. The question is not whether the product can be made in a generic factory. There is no doubt it can. The question to be answered is whether this special will pose an unacceptable burden for the current manufacturing situation.

5. *Are improved facilities required?* The producibility engineer now has to decide how the angle bend can be accomplished. There are many ways to do this. It can be cold bent around a mandril. The metal can be heated to reduce the yield strength and then bent much as a blacksmith does. The process of manufacturing can be changed to drop forge the pliers halves. The halves can be made out of much larger-diameter bar stock and "hogged out" via lathes and milling machines. There are many possibilities, some with very low initial costs and some with high costs. The answer to whether or not improved facilities are required in this case requires inputs from other members of the concurrent engineering team.

The volume of product being considered and what the target selling price will be must be determined. Once these are known, the team can judge how much capital can be utilized for this product. At this point the producibility engineer, the manufacturing engineer, and the designer need to consult with the marketing and finance team members. If the volume is to be very large, the blacksmith approach will not be viable. If keeping costs to rock-bottom minimum is paramount, the "hog it out" approach is not acceptable—it would have low initial equipment costs but very high unit costs. Similarly, the cold mandril technique can be related to medium- to low-volume choices. The team also has to evaluate the perception of quality. The marketing member must lead this discussion. Certainly, drop forging will be a favorite alternative because market conditioning through advertisements has led the buying public to relate this manufacturing process to high quality. However, it is an acceptable alternative only if the concurrent engineering team is going after a large share of the market, hence can afford the high initial capital costs. Even this solution may have to be tempered with the realities of equipment procurement availability and schedule.

In this example, we can conclude that the producibility engineer will accept the design for manufacturability. It will be adding a burden to the company to do so, but just what that will be requires a decision by the entire concurrent engineering team. The producibility design review is a vital part of the decision process, but it is not done in a vacuum. As shown above, the final decision depends on many issues. However, the strategic issues of competition come to a head faster via the concurrent engineering approach than the obsolete singular stand-alone chain-type method.

Producibility Design Review, Another Example

Sometimes the producibility design review requires a more extensive review during the second step, "Requirements essential?" When the complexity of the design requires it, the second step is expanded to include more specific questions such as:

1. What is the part for?
2. How does it work?
3. What engineering analysis is important to the design?
4. How can the part be made?
5. What can be done to make it cheaper to produce?

In essence, because the design is more complex, the producibility engineer is looking at the design very closely to understand the factors the design engineer considers important. Then the producibility engineer considers the manufacturing factors to see if normal procedures can be used to make the part or assembly. If not, then recommendations have to be made to the designer to modify the design for producibility concerns. Let's look at an example in which step two has to be expanded to handle a more complex design.

Figure 6-4 is a drawing of a stud to be used by a company with its product, a large centrifugal pump. The stud is designed to hold the pump base plate to the customer's foundation. The

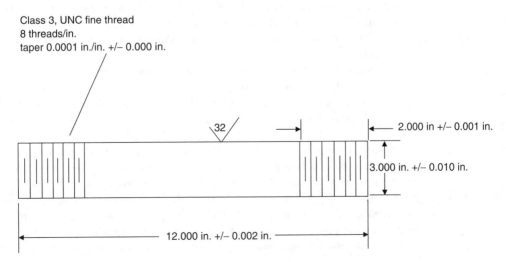

Figure 6-4. Drawing of a Stud to Be Used with a Large Centrifugal Pump

producibility engineer proceeds as follows:

1. What is the part for?
   a. Holding the base plate to the foundation. It will go through the base plate, then into a structure provided by the customer, secured at both ends by nuts.
2. How does it work?
   a. Puts plate and foundation in compression; therefore stud goes into tension.
   b. Taper in thread indicates that the designer intended to get a better distribution of load along the stud and nut threaded length.
   c. Prestress requirement shows that the designer intended to firmly anchor the pump, probably to guarantee known support conditions for rotor and bearing critical speed and vibration analysis as well as to overcome any possibility of joint loosening.
3. What engineering analysis is important to the design?
   a. Shear stress calculation for the body of the stud and for the threads.
   b. Bending stress calculation to ensure workability of the stud if an off-center load is applied due to misalignment during installation or pump axial thrust.
   c. Fatigue evaluation due to induced vibration when the pump is running in the customer's system.
4. How can the part be made?
   a. Bar stock cut to length by saw.
   b. Face mill on Horizontal Boring Mill (HBM) to desired precise length.
   c. Rough machine in lathe to a finish of 63 RMS.*
   d. Thread in screw machine; use go/no-go gauge, class 3, for check.
   e. Grind surface on cylindrical grinder to obtain a 32 RMS finish.
5. What can be done to make it cheaper to produce?
   a. The part can be made in the usual area or department for such parts; no savings opportunity exists here.
   b. The length is 12.000 ± 0.002 in.; is it necessary to produce such an exact length? *Recommended:* reduce length tolerance to ±0.015 in. ($\frac{1}{64}$ in.); this is within reason for a metal saw. *Result:* eliminate the face milling step on the HBM.
   c. Is the surface finish requirement of 32 RMS necessary? Obviously, the surface finish was specified to reduce stress risers on a relatively highly stressed part. *Recommended:* reduce finish requirements to 63 RMS; this can be produced easily on a lathe (lathes in good condition consistently produce finishes better than 63, approaching 32). *Result:* eliminate the necessity for surface grinding on a cylindrical grinder.
   d. Is the taper requirement of 0.001 in./in. over a 2-in. length realistic? Probably not for such a short length of thread. *Recommended:* eliminate the taper. Rely on class 3 thread to provide a tight thread mesh. *Result:* eliminate a more complicated setup on the screw machine.
   e. Will the part still do the job if producibility engineer's recommendations are accepted?
      (i) In this example, it obviously will.
      (ii) The design margin is somewhat reduced, but still safely within any reasonable expectations.

The above example results in less cost to produce the product. The number of steps is reduced from five to three. We have gone from a design to a producible design, from one that could do the job to one that still does the job but at a much lower cost. Notice that the producibility engineer must mentally design the part, then contemplate how it should be made, and then review the method to see whether it can be optimized. Cheaper and quicker ways to do it can be found if the tolerances or processes can be modified. By understanding what the design engineer intended to accomplish, and knowing how the part will be made, including an estimate of the cost, the

---

* RMS, root mean square. The value is a measure of surface smoothness.

producibility engineer can discuss his or her findings with the design engineer from a position of knowledge and can beneficially affect overall productivity.

How to Determine if a Factory Can Meet the Requirements of the Design

How do you evaluate whether or not the factory can meet the requirements of the design? There is no exact answer. The producibility design review example gives some insight into the complexity of determining whether a factory can handle a given design. It is evident that we cannot boil it down to a formula-driven exercise. A lot depends on an understanding of the specific factory. For example, is it capable of following deviations from norm through issuance of revised documentation? Or is it a factory in which change from the routine is abhorred and achievable only through brute-force methods, requiring constant supervision? Where the factory fits within this spectrum will determine the degree of change that can be tolerated. One of the strengths of the concurrent engineering process is that it gets this discussion going up front. If it appears that this will be an issue (more change needed than currently possible), the problem is spotlighted for solution much earlier in the process.

SETTING DESIGN TOLERANCES

One of the primary functions of producibility engineering is to define to design engineering the factory's ability to hold dimensional accuracy. With this information the concurrent engineering team can set tolerances that meet the design requirements and are compatible with the capabilities of the production facility. How do you determine the proper tolerance to apply? Mistakes in tolerance setting can cost significant amounts of money. The general rule is simply require what is necessary and nothing more. If design engineering does not know what the tolerance should be, they should work in consort with other members of the company team and learn together what will be required. Here are some basic rules that will help in setting tolerances.

1. *Understand the material you are working with.* To set realistic tolerances, it is absolutely vital to understand the physical and chemical properties of the material we are working with. For example, it makes no sense to require tolerances tighter than ±0.03 in. for hardwood because of its hygroscopic nature; it absorbs and gives off water vapor depending on the relative humidity of the immediate atmosphere. This means it is constantly undergoing small dimensional changes. So you must understand the limitations of the chosen materials and set tolerances accordingly. Set what is needed to maintain the intent of the design, but be aware of the need to be compatible with the requirements of the materials used.

2. *Use the statistics approach to setting tolerances.* Absolute dimensioning never occurs in the real world. We must understand that we cannot design for either the maximum or minimum side of the dimension. We must design to the given dimension and use statistics and probabilities to set the tolerances. This will give the most realistic values for tolerance ranges. Yes, there will be failures, but the statistics approach will yield fewer failures than any other method. So, when assembly tolerances are critical, use the statistics approach. This technique is demonstrated in the next section of this chapter.

3. *Use generally accepted methods of defining form, fit, and function.* Drawings are like any other language. They have to be understood, just as a paragraph in an instruction manual for how to operate a piece of equipment needs to be totally unambiguous. So too do drawings because they define what will be made. For this reason drawings should be made using generally accepted geometric dimensioning and tolerance rules. The American Society of Mechanical Engineers Y14.5 standard is the most accepted methodology for defining dimensions and tolerances and should be employed. I will discuss the Y14.5 standard in more detail later in this chapter.

4. *Avoid the need to set tolerances as much as possible by simplifying designs.* Whenever possible, use such techniques as chamfered holes and fastening devices, gaskets, and tapered pins in

assemblies to avoid precision assembly requirements. Consciously design for ease of assembly, thereby minimizing the need to set tight tolerances.

These four suggestions are all that is pertinent about tolerances. Set tolerances compatible with the materials being used. Use statistics methods for establishing tolerances instead of arithmetic methods. Follow accepted rules for defining tolerances on drawings. Finally, avoid the problem as much as possible by designing around the need for tolerances. Other concepts for setting tolerances are often used, such as employing a percentage of the dimension. However, these are not directly related to the primary need, that the part or assembly function as required. They are typically related to administrative procedures; therefore they are not discussed here.

STATISTICAL APPROACHES TO TOLERANCE SETTING

One of the important functions of producibility engineering is the development of techniques for design engineering that allows product reliability to be maintained while optimizing manufacturability. A significant body of work over the years has been done in the area of statistical tolerance setting. Statistics is used to negate or at least predict the effect of tolerance build-up during assembly operations. The purpose is to allow the widest possible latitude for manufacturing while still maintaining a very high probability that the parts will go together.

Tolerances are set to make sure that the finished part has the attributes required by design. Stress and strain analysis dictates the size and shape of a part. Tolerances are used to ensure that after manufacture, the part retains the proper size and shape. Unfortunately, the most popular method of setting tolerances is that of avoiding interferences. This method dictates that the largest (plus) tolerance must always be smaller than the part it mates to.

> **EXAMPLE**: Non-interference-fit keyway in a shaft; the plus tolerance of the key is always smaller than the minus tolerance of the keyway.

This is fine for a two-part assembly. But for a multipart assembly with many parts that are not directly related, we are concerned with tolerance build-up and we want to know what the probability of interference will be. Therefore, the normal scheme of tolerance setting is too severe, and many parts judged out of tolerance would fit with a high degree of probability. In order to not throw away perfectly usable parts, a statistical approach to tolerance setting is preferred by manufacturing, especially if the probability of failure (using a part that actually does create an interference) is small.

One such way of setting tolerances by statistical approaches is to look at load characteristics versus material strength characteristics. This can be done with a Warner diagram, which is a series of normal distributions on the same set of axes. Let us look at the Warner diagram in Fig. 6-5, using the joint stud of Fig. 6-4 for the necessary data.

The diagram shows an area where the two normal curves overlap. If these curves represent values pertaining to the same physical entity, we can deduce some vital information. From Fig. 6-5 we can see that the maximum load on the part is higher than the minimum strength of the material. That means if we are on the plus tolerance of load and the minus tolerance of strength, there will be failure. In fact, any combination of load and strength that falls into the failure probability zone will result in failure. Normally, when a machining drawing is produced, dimensions rather than stress levels are shown. Therefore, the Warner diagram must be translated into dimensions so that we can set tolerances.

The shear stress, T, was calculated by the design engineer to size a stud. In the example given, T is 30,000 pounds per square inch. If no bending occurs, $T = V/A$ or $T = V/\pi r^2$, and there is an axial stress effect only. If bending occurs, $T = 4V/3\pi r^2$, which is the maximum shear stress affecting the part. It is a simple matter to determine the value for r in the equations, hence the diameter. Assuming that bending can occur, referring back to the joint stud example, let $V = 159,000$ lb.

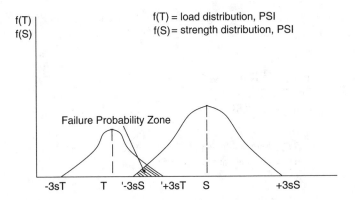

f(T) = load distribution, PSI
f(S) = strength distribution, PSI

f(T)
f(S)

Failure Probability Zone

-3sT   T   '-3sS   '+3sT   S   +3sS

Figure 6-5. Warner Diagram. Here T is the mean shear stress load on the bolt. $-3sT$ and $+3sT$ the lower and upper 3 sigma spread of the shear stress variation, respectivley; S the mean strength of the material used to make the bolt; and $-3s$ and $+3s$ the lower and upper 3 sigma spread of the material strength specification, respectively.

Then the diameter is

$$D = 2r = 2\left(\frac{4v}{3\pi T}\right)^{1/2} = 2\left(\frac{4 \times 159,000}{3\pi \times 30,000}\right)^{1/2} = 2.996 \approx 3.000 \text{ in.}$$

Now we have a mean diameter. In order to calculate a reasonable tolerance we must assume that the factor of safety for the material is such that slight variations in T would not ordinarily result in failure. The problem now becomes one of determining the variation in T caused by error in measurement (very likely to occur when prestressing a part). A normal measurement error would be 2½%. Then T (upper) is 30,750 lb/in.² and T (lower) is 29,250 lb/in.² Using the same formula for the diameter that we used above, we now have

$$D \text{ (upper)} = 2\left(\frac{4 \times 159,000}{3\pi \times 29,250}\right)^{1/2} = 3.038 \text{ in.}$$

$$D \text{ (lower)} = 2\left(\frac{4 \times 159,000}{3\pi \times 30,750}\right)^{1/2} = 2.963 \text{ in.}$$

Therefore the mean, upper bound, and lower bound are 3.000, 3.038, and 2.963, respectively, or a 3σ probabilistic normal curve distribution on the diameter of 3.000 ± 0.0375. Now these data are ready to be entered on the Warner diagram.

We now have data for the $f(T)$ distribution but lack data for the $f(S)$ distribution, that is, the material strength distribution. The maximum, minimum, and mean of this distribution will be found in material and test specification data sheets. For our example of the joint stud, let us assume that a steel with a 0.02% yield strength in shear of 34,000 lb/in.² was selected and that it has a strength range of ±4,000 lb/in.² Then

S(max) = 38,000 lb/in.²
S(mean) = 34,000 lb/in.²
S(min) = 30,000 lb/in.²

Now we can enter both load and strength values on the Warner diagram (see Fig. 6-6). Then, using the equation $D = 2(4V/3\pi T)^{1/2}$, where $T = 30.75$ KSI, 29.25 KSI, and 30.00 KSI (KSI = 1000 lb/in.²), we have

$D(\text{min}) = 2.963$ in. (30.75 KSI)
$D(\text{max}) = 3.038$ in. (29.25 KSI)
$D(\text{mean}) = 3.000$ in. (30.00 KSI)

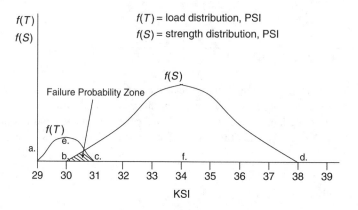

Figure 6-6. Warner Diagram with Data for Determining Diameter Tolerances for the Studs.
a.: $-3sT = 29.25$ KSI; b.: $-3sS = 30.00$ KSI; c.: $+3sT = 30.75$ KSI; d.: $+3sS = 38.00$ KSI; e.: T mean = 30.00 KSI; f.: S mean = 34.00 KSI (1 KSI = 1000 psi)

We can see there is a definite overlap into the failure zone and the probability of operating in that zone should be calculated. Before that is done, however, let us look at the ranges compared to that shown on the original proposal for the part. The proposal calls for a diameter $D = 3.000 \pm 0.010$ in., or a range of 3.010 to 2.990 in. A standard machine shop tolerance would be ± 0.015 in. (1/64 in.). The statistical analysis gives $D = 3.000 \pm 0.0375$ in., or a range of 3.038 to 2.963 in. This is a 1/27 in. range.

Without any further calculation, it is possible for the producibility engineer to recommend the normal machine shop tolerance of 1/64 in. for this part. This is correct since the statistically derived tolerance is broader than the normal good machine shop practice. If the outcome had been the reverse, with the original looser than good machine shop practice, the producibility engineer would still have chosen 1/64 in. If the statistical tolerance had been less than 1/64 in. but greater than the original selection, the producibility engineer would have selected the statistically derived version in order to prevent needless scrapping of parts with a high probability of being good. In this example, we can see that 1/27 in. is roughly twice the spread of 1/64 in.; therefore, noting the very small Warner diagram overlap, it is proper to conclude that the 1/64 in. tolerance is acceptable and end the problem here. For instructional purposes, however, let us continue and find the probability of failure based on the Warner diagram. The procedure involves finding the percentage of area

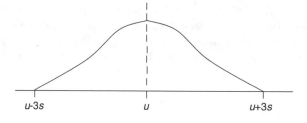

Figure 6-7. Normal Curve

TABLE 6-1. Areas Under Normal Curves

s	.00	.01	.02	.03	.04	.05	.06	.07	.08	.09
0.0	.0000	.0039	.0079	.0110	.0158	.0198	.0237	.0277	.0317	.0356
0.1	.0396	.0436	.0475	.0515	.0555	.0594	.0634	.0673	.0713	.0752
0.2	.0791	.0831	.0870	.0909	.0948	.0987	.1026	.1064	.1103	.1141
0.3	.1180	.1218	.1256	.1294	.1332	.1370	.1407	.1445	.1482	.1519
0.4	.1556	.1593	.1630	.1666	.1703	.1739	.1775	.1811	.1846	.1882
0.5	.1917	.1952	.1987	.2022	.2058	.2091	.2125	.2159	.2192	.2226
0.6	.2259	.2292	.2325	.2358	.2391	.2423	.2455	.2487	.2518	.2550
0.7	.2581	.2612	.2643	.2673	.2704	.2734	.2764	.2793	.2823	.2852
0.8	.2881	.2910	.2938	.2960	.2994	.3022	.3050	.3077	.3104	.3131
0.9	.3158	.3184	.3210	.3236	.3262	.3287	.3313	.3338	.3362	.3387
1.0	.3411	.3435	.3459	.3483	.3506	.3529	.3552	.3575	.3597	.3619
1.1	.3641	.3663	.3684	.3705	.3726	.3747	.3768	.3788	.3808	.3828
1.2	.3847	.3867	.3888	.3905	.3923	.3942	.3960	.3978	.3996	.4014
1.3	.4031	.4048	.4065	.4082	.4098	.4114	.4130	.4146	.4162	.4177
1.4	.4192	.4207	.4222	.4237	.4251	.4265	.4279	.4293	.4306	.4320
1.5	.4333	.4346	.4358	.4371	.4383	.4397	.4407	.4419	.4431	.4442
1.6	.4454	.4465	.4476	.4480	.4497	.4507	.4517	.4527	.4537	.4547
1.7	.4556	.4566	.4575	.4584	.4593	.4602	.4610	.4619	.4627	.4635
1.8	.4643	.4651	.4659	.4606	.4673	.4681	.4688	.4685	.4702	.4708
1.9	.4715	.4722	.4728	.4734	.4740	.4746	.4752	.4755	.4763	.4769
2.0	.4774	.4780	.4785	.4790	.4795	.4800	.4805	.4809	.4814	.4818
2.1	.4823	.4827	.4831	.4835	.4839	.4843	.4847	.4851	.4854	.4858
2.2	.4862	.4865	.4868	.4872	.4875	.4878	.4881	.4884	.4887	.4890
2.3	.4893	.4895	.4898	.4901	.4903	.4906	.4908	.4911	.4913	.4915
2.4	.4917	.4919	.4922	.4924	.4926	.4928	.4929	.4931	.4933	.4935
2.5	.4937	.4938	.4940	.4942	.4943	.4945	.4946	.4948	.4949	.4950
2.6	.4952	.4953	.4954	.4955	.4957	.4958	.4959	.4960	.4961	.4962
2.7	.4963	.4964	.4965	.4966	.4967	.4968	.4969	.4970	.4971	.4971
2.8	.4972	.4973	.4974	.4974	.4975	.4976	.4977	.4977	.4978	.4978
2.9	.4979	.4980	.4980	.4981	.4981	.4982	.4982	.4983	.4983	.4984
3.0	.4984	.4985	.4985	.4986	.4986	.4986	.4987	.4987	.4987	.4988
3.1	.4988	.4988	.4989	.4989	.4989	.4990	.4990	.4990	.4991	.4991
3.2	.4991	.4991	.4992	.4992	.4992	.4992	.4993	.4993	.4993	.4993
3.3	.4993	.4994	.4994	.4994	.4994	.4994	.4994	.4995	.4995	.4995
3.4	.4995	.4995	.4995	.4995	.4996	.4996	.4996	.4996	.4996	.4996
3.5	.4996									

under the normal probability curves. To do this we use the tabulated areas represented by the number of standard deviations calculated and measured along the abscissa of the normal-curve graph (see Fig. 6-7). Areas under normal curves are given in Table 6-1.

Table 6-1 is used to find the area proportions under the curve by entering with the number of standard deviations, Z:

$$Z = \frac{(x-u)}{s}$$

where x is the distance of the variable away from the mean, u is the mean, and s is the standard deviation. This is adequate for problems involving a single normal distribution, but a compatibility relationship must be developed to equate two normal distributions acting together as in the Warner diagram. Edward B. Haugen* has developed such a coupling formula:

$$Z = \frac{u_1 - u_2}{\left(s_1^2 + s_2^2\right)^{1/2}}$$

* E.B. Haugen, *Probabilistic Approaches to Design,* Wiley, New York, 1968.

where u_1 is the mean of the first distribution, u_2 the mean of the second distribution, s_1 the standard deviation of the first distribution, and s_2 the standard deviation of the second distribution. One other important fact must be kept in mind in order to use Haugen's equation; the distributions must be of like quantities, for example, both expressed in pounds per square inch. If this is the case, Z is calculated and the tables entered as described above.

We can now use the Haugen equation to find Z for our example.

Strength curve:

$$u_1 = 34,000 \text{ lb/in.}^2$$
$$s_1 = (38,000 - 34,000)/3 = 1333.3 \text{ lb/in.}^2**$$

Load curve:

$$u_2 = 30,000 \text{ lb/in.}^2$$
$$s_2 = (30,750 - 30,000)/3 = 250 \text{ lb/in.}^2$$

Then:

$$Z = \frac{u_1 - u_2}{\left(s_1^2 + s_2^2\right)^{1/2}} = 2.9487$$

and from the normal curve table:

$$\text{Area under curve} = 0.5 + 0.4984 = 0.9984$$

Therefore the probability of failure is

$$1 - 0.9984 = 0.0016 = 0.16\%$$

For this example the probability of failure, even using the entire range determined by the statistical tolerance method, is very small. This means that the tolerance for the diameter can be very liberal.

This example demonstrates the technique of statistical tolerance setting and implies that many normally condemned parts are indeed usable for most situations. The producibility engineer must be aware of this to prevent excessive manufacturing costs.

The method of statistical tolerance setting should be used where the costs of individual parts are high. It is scientifically accurate and will lead to lower scrappage rates. However, to use this technique for all tolerances would be a problem. Not all parts are designed with stress analysis as a part of the procedure. Therefore this technique, like so many others, cannot be applied universally. To try to force fit the data would detract from the producibility engineer's productivity. The statistical concept has been illustrated here to show that creative technical solutions are possible and can indeed lead to savings if properly applied. As computer numerical techniques become simpler to use, such techniques will probably become more commonplace.

When ever statistical tolerance setting can not be use, it is best to adhere to the ASME standard Y14.5. This standard will force the design engineer and producibility engineer to define importance of fits for the parts. In that manner only important dimensions will need close tolerances while the rest can be loosened to comply with standard machine shop practices. Let's now look at geometric dimensioning and tolerancing.

**The normal curve shows a 3 sigma spread; hence we divide by 3 to obtain the standard deviation.

GEOMETRIC DIMENSIONING AND TOLERANCING

Where are we? This is a question we certainly do not want to ask in reference to information on a drawing. We want the drawing to tell us exactly what the part or assembly is in a manner that we fully understand the intent of the draftsperson and or designer. We want to know what is critical for a successful part or an assembly to function as intended. However, if a drawing is simply a picture of an object the factory is required to produce and the only information is dimensions and tolerances, we may not know what is primarily important and what is of a secondary requirement.

Most good draftspersons will instinctively create reference notes that tell the reader of the drawing the relative degree of importance of each feature and what tolerances are critical and need to be complied with. If every draftsperson was left to his or her own devices to define criticality and levels of importance without any guidelines or procedures, to produce a drawing that the user could interpret, then every drawing would be an adventure in comparison to good detective fiction. We would be reduced to striving to find clues that would unlock the meaning of the drawing, where some drawings would be easy to interpret and others would be difficult. This would be an unfortunate situation, and is the situation that did exist before standardization of drawing presentation and layout. With standardization, a little familiarization training makes it possible for all readers of the drawing to interpret it in the same way and reach the same conclusions as to relative degrees and ranking of features importance. With this knowledge it is possible to manufacture parts tending to the most critical features as a top priority. Without standardization this is not possible unless the draftsperson is in close liaison with the manufacturing engineer.

Standardization, or at least the desire to do so, has been with us for quite a while. Certainly most companies have their preferred way of depicting features on a drawing and procedures for issuing explanatory notes. And for the most part, this works fine, as long as the drawing remains local; for example, not sent out for external manufacture (farm out). It works because the designers and the shop people are usually in the same building and face to face contact to resolve problems is possible and does happen (however, having to meet to determine meaning of a drawing is not conducive to high levels of efficiency and productivity improvement). When communications becomes difficult then the drawing has to be more explanatory. This is where universal standards come to the forefront. The universal standard creates a language of sorts that is interpreted the same way by all people who understand the meaning and intent of the standard. In fact universal standards for drawings is the vital cog in international and national commerce. Without it there would be no way to make parts in one factory and assemble in another that would be economically justifiable. ASME Y14.5 Geometric Dimensioning and Tolerancing standard is the Cadillac of the field for machine and assembly drawings. Therefore, all companies that intend to compete via providing drawings to others or work from drawings supplied by others need to be cognizant of this standard. There is no truly viable alternative position. The standard allows for people who speak different verbal languages to understand the meaning of their technical drawings simply by applying the standard. Perhaps I shouldn't say simple, because we do have to understand the standard, but it is doable in any factory.

Let's take a peek at the ASME Y14.5 standard. But keep in mind this is not a tutorial on how to apply the intricacies of the standard, but simply an explanation of how it is structured and a simple demonstration of its' use. For those who want to become truly competent in applying the standard, I recommend purchasing a copy of the standard from ASME and/or one of many excellent reference interpretation texts, several of which are listed in the bibliography, and perhaps taking an applications course for Geometric Dimensioning and Tolerancing (GD&T).

The ASME 14.5 standard is a language of symbols and rules that not only defines size and shape but also their relationships to each other. By doing so it allows the integration of size and shape and positions of various components to each other. This means it is possible, even desirable, to relate tolerances to shapes as well as sizes. We can define references for where to measure a feature from with respect to other features.

Definition: "A feature is specific component of a part such as holes, faces, slots, screw threads, etc."

We can determine tolerances based on synergistic relationships of features to each other. For example, a maximum material condition between a hole and a mating shaft can be easily defined.

Definition: A maximum material condition is defined as that condition of a part feature that contains the maximum amount of material within the desired limits of size: example minimum hole diameter versus maximum shaft diameter.

Definition: A least material condition is defined as that condition of a part feature that contains the least amount of material within the desired limits of size: example maximum hole diameter versus minimum shaft diameter.

We can identify virtual condition boundaries where accumulated tolerances based on several interrelated features maximum material conditions are defined.

Definition: Virtual condition is defined as the boundary generated by collective effects of maximum material conditions limits of size of any associated geometric tolerances.

Virtual condition is important because it defines whether associated parts will assemble. So we can see that GD&T is a powerful tool for defining and interpreting a drawing where the definition and interpretation would be compatible if not 100% identical. Let's use some examples to clarify the meaning of what I've stated above.

Figure 6-4 is an old style drawing that defines a stud to be manufactured for use in a company's centrifugal pump product. It would appear to have all the information necessary to successfully make it in the factory. That is absolutely correct where the draftsperson and the operator work in the same factory and have access to each other for consultation. By company policy and practice the operator knows which end of the stud is the primary datum reference from which to measure from. The operator also knows, by practice whether maximum material conditions apply as well as knowing how the thread form parallelism has to be in reference to the data. The operator knows all this because he has learned by experience and discussion with designer how it has to be to be a functional part. Now let's assume the factory is overwhelmed with work and needs to farm out the job of making the stud to another company. That company receives the drawing as shown as Fig. 6-4 with a purchase order to make a significant quantity (e.g., better price received through economies of scale). Perhaps they will make the part correctly because the thread form is standard. But perhaps not. Maybe the thread form is offset too much to be outside of the virtual envelop defined by the boundaries of the virtual condition, and the stud will not go through the hole, the threads are offset enough to create an interference where none can be. All this simply because the farm out shop does not have access to the thoughts of the designer and there is nothing to tell them where the reference datum plane is, or anything else about the relationships of features to each other. If the drawing was done as shown in Fig. 6-8, utilizing the standard format defined by ASME Y14.5 standard, it would have been perfectly clear what the relationships are and the ranking of importance to each other. It would of course required the farm out shop to understand GD&T.

Let's now explore what we see in the drawing shown in Fig. 6-8.

The basic characteristics employed in most drawings and symbols used to represent them is shown in Fig. 6-9, which is reproduced with permission of Quality Council of Indiana, from their instruction book, *CMI Primer.*

Notice the table has the previous edition of ASME Y14.5 symbols and the international version ISO 7083. This for reference. We will concentrate on the ASME Y14.5 1994 version. The table shows three kinds of features.

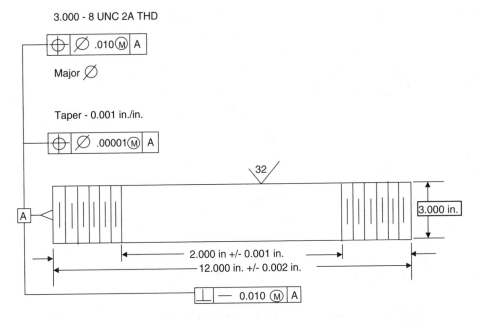

Figure 6-8. Drawing of a Stud To Be Used with a Large centrifugal Pump using Geometric Dimensioning and Tolerancing (GD&T) Format

a. Individual—referring to specifics about the part, in singularity
b. Individual or Related—referring to the specifics of the part, in plurality or singular
c. Related—referring to the specifics in relationship to its environment.

Also included to further explain information that is included in a GD&T formatted drawing are Figs. 6-10 and 6-11, again from the CMI Primer.

The basic feature employed in virtually every GD&T formatted drawing is the feature control frame. The feature control frame describes the feature it is attached to with all the information necessary to define its individual, individual and related, and related characteristics.

Definition: The feature control frame is a symbolic way in a rectangular box of showing a type of control- the geometic control symbol; the tolerance- form, run out, or location; and datum references and modifiers to the datum references.

As we can see from Fig. 6-11 not all feature control frames have the same number of box compartments. The practice is to only show those symbols that have meaning for the particular feature. They also rank in importance from left to right. Let's explain the meaning of the longer feature control frame in the left of the figure.

a. We start from the left of the feature frame and can see that the geometric characteristic for the feature is, by symbol, position. This says that the most important characteristic of that feature is its position with reference a datum.
b. The next box to the right within the frame tells us about the tolerance. Here we see a symbol that says it's a diameter and the geometric tolerance is 0.003 in. (or the appropriate metric measurement) followed by a modifier, in this case the symbol for maximum material condition. This means the rules for maximum material conditions would apply as to actual tolerance that can be allowed.
c. The next three boxes moving right show the datums in order of importance. Note the numbers 1, 2, & 3 next to primary, secondary, and tertiary. These represent the number of

Geometric Dimensioning and Tolerancing

Characterisitics & Symbols

Kind of Feature	Tolerance Type	Characteristic	ANSI Y14.5M 1982	ASME Y14.5M 1994	ISO 7083 1994
Individual	Form	Flatness	▱	▱	▱
		Straightness	—	—	—
		Circularity	○	○	○
		Cylindricity	⌭	⌭	⌭
Individual or Related	Profile	Profile of a line	⌒	⌒	⌒
		Profile of a Surface	⌓	⌓	⌓
Related	Orientation	Parallelism	//	//	//
		Angularity	∠	∠	∠
		Perpendicularity	⊥	⊥	⊥
	Runout	Circular runout	↗ *	↗ *	↗
		Total runout	↗↗*	↗↗*	↗↗
	Location	Position	⌖	⌖	⌖
		Symmetry	⌖	=	=
		Concentricity	◎	◎	◎

Table 5.15 Basic GD&T Symbols

* Arrowheads may or may not be filled

Figure 6-9. CMI Table 5.15 Basic GD&T Symbols
(From page V-19, *The Mechanical Inspector Primer, 4th edition*, Frank A. Bensley, Dale E. Clark, Bill L. Wortman, Wesley R. Richardson, Quality Council of Indiana, West Terre Haute, IN, 2004, reproduced with permission.)

Geometric Dimensioning and Tolerancing

Other Symbols and Terms

Symbol Description	ANSI Y14.5M 1982	ASME Y14.5M 1994	ISO 7083 1994
Maximum Material Condition	Ⓜ	Ⓜ	Ⓜ
Least Material Condition	Ⓛ	Ⓛ	Ⓛ
Regardless of Feature Size	Ⓢ	Ⓢ	Ⓢ
Projected Tolerance Zone	Ⓟ	Ⓟ	Ⓟ
Tangent Plane	Ⓣ	Ⓣ	None
Diameter	⌀	⌀	⌀
Basic Dimension	100	100	100
Reference Dimension	(100)	(100)	(100)
Datum Feature	-A-	A ▲	A ▲
Between	None	◄———►	None
Datum Target	⌀10/A1	⌀10/A1 or ⌀10/A1	⌀10/A1 or ⌀10/A1

Table 5.16 Other GD&T Symbols and Terms

Figure 6-10. CMI Table 5.16 Other GD&T Symbols and Terms
(From page V-19, *The Mechanical Inspector Primer, 4th edition*, Frank A. Bensley, Dale E. Clark, Bill L. Wortman, Wesley R. Richardson, Quality Council of Indiana, West Terre Haute, IN, 2004, reproduced with permission.)

contacts the part has to make with the datum to establish its position in space. Also note that the secondary datum location B has the maximum material condition modifier symbol to the left of it. This implies the MMC concepts need to be applied to datum.

With this feature control panel we know we're involved with defining the location of a diameter, with appropriate tolerances in reference to the maximum number of datum planes allowed, for example, 3. And we know that maximum material conditions apply.

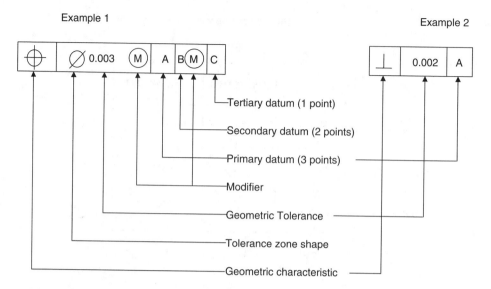

Figure 5.17 Example Feature Control Frames

Figure 6-11. CMI Figure 5.17 Example Feature Control Frame
(From page V-19, *The Mechanical Inspector Primer, 4th edition*, Frank A. Bensley, Dale E. Clark, Bill L. Wortman, Wesley R. Richardson, Quality Council of Indiana, West Terre Haute, IN, 2004, reproduced with permission.)

Definition: Datums are points, lines, planes, cylinders, axes, etc. by which the relationship of a feature to other features can be mathematically (geometry) defined.

Note that the feature control frame on the right side of Fig. 6-11 is considerably shorter in length. This doesn't mean information is missing. It just means that no more information is necessary to define the feature's relationship to the rest of the particular part or assembly. In this case we see that the feature is perpendicular within 0.002 in. to datum A. And that's all we need to know about this feature.

Perhaps we should discuss a little more about datums. Datums are simply orthogonal planes against which the part is located in reference to. By establishing datums we are then capable of truly locating a feature with respect to others, as we shall see in Fig. 6-8. We will be able to tell the relative perpendicularity and concentricity of features in a manner that there is no ambiguity. Hence the part made in a farm out location should assemble just fine when it arrives.

Now let's compare Figs. 6-4–6-8. Figure 6-4 is a conventional drawing made without reference to GD&T. It represents a bolting stud that is to be used as a connecting device for a large centrifugal pump. The drawing gives the dimensions and tolerances associated with dimensions. But it doesn't state what the relationships of the dimensions and tolerances to each other. Nor does it give us a datum to reference dimensions from. Now for a part made in a shop where the designer and operator are used to and understand the company's preferences how parts are to be made, this drawing is perfectly suitable and will get the job done. However, if we choose to make the part somewhere else, we're not quite sure if the information is suitable for another factory's common understanding of how to make the part so it does work in assembly. So it is preferable to use the ASME Y14.5 code to assure standardized interpretation of what the drawing tells us.

Figure 6.8 is the same bolting stud but using the GD&T format. The drawing has all the information that the conventional drawing had plus instead of relying on company shop practices for the "undocumented" information, it's now on the drawing.

The first thing we notice is that there are feature control frames associated with the important features of the bolting stud. Note in the upper left that we define the bolting stud's diameter and the thread form above the feature control frame. In the feature control frame we see the position symbol which is in reference to the diameter about which the thread form is to be cut. Then the position tolerance of 0.010 in. Then a circle M, indicating maximum material conditions apply and not as to the 180° opposite; circle L, least material condition. And all the way to the right of the feature control frame a capital A signifying the datum reference from where all dimensions are to be measured from. Under the feature control frame we see the word major followed by the diameter symbol. This tells us the thread form is to be cut at the major diameter, which would have been informally implied on the conventional drawing due to in-house shop practice/knowledge. But this may not be the case for a farm-out situation.

Going down the page we see another feature control frame essentially the same as above but this one is strictly for the taper. It also references the same datum. Even though any competent machinist knows the taper is associated with the thread form and has to be concentric it is shown here. In fact we could have added another location symbol, the concentricity symbol (a small circle within a larger circle), to the right of the position symbol. This would prevent any ambiguity at all, if there was any.

Continuing on down the drawing we see a capital A within a square with a line going horizontally to the right and terminating on a sideways isosceles triangle. That triangle being on the end surface of the bolting stud. This is the symbol used to indicate a datum surface. If there were secondary and tertiary datum surfaces (e.g., orthogonal axes), they would be symbolized with a capital B and a capital C in a square, and with the same triangle shape at the datum surface.

At the bottom of the drawing we see the same length dimensions as shown on the conventional drawing. But on the GD&T format drawing we also see the corresponding feature control frame. Here we see to the left the perpendicular symbol, followed by the straightness symbol telling us this is a linear as opposed to an angular dimension. This going right is followed with the tolerance for perpendicularity and the circle M once more saying that maximum material conditions apply. And all the way to the right the applicable datum surface.

The last GD&T symbol is to the right side of the bolting stud. Here we see a dimension with no tolerance bounded by a rectangle. This means this is the basic or theoretical dimension for that part of the bolting stud. We do not need tolerances because they are already accounted for with the information for the thread form and taper within their respective feature control frames.

We can see that the GD&T drawing contains more information than the conventional drawing. Essentially it adds relationships of the features to each other, which were only implied in the conventional drawing. You may now begin to wonder: "Wow, is all this necessary for such a simple part?" And it would not be a trivial question. Obviously the GD&T is overkill for a part made successfully in-house based on commonly understood shop practices. And perhaps not really necessary for a farm-out vendor who has done business with us before. But to be guaranteed success with far off vendors, especially if they speak different languages, I strongly urge that the GD&T approach be utilized. I doubt if the aerospace and automobile industries could be as successful as they are in global markets and manufacturing without GD&T.

Admittedly, GD&T to be applied successfully requires study and going through a learning curve. But we do that everyday with CNC and other high tech factory technologies. Why not do it for drafting practices too? After all, communicating what needs to be done to successfully manufacture and assemble our products is essential for assuring profitability. GD&T will eventually be as common place for drawings as simple feed and speed settings knowledge for machines are. Both represent the optimum performance paradigms for industry. As I stated at the start of this discussion, I couldn't possibly cover the entirety of the GD&T body of knowledge here. My purpose is simply to introduce the topic and show how it can be a useful tool for the producibility engineering continuum. I urge all engineers who deal with drawings to learn GD&T methodology. There are many courses and books available readily accessible via the internet which make it easy to do so.

ATTRIBUTES OF GOOD DESIGN USED IN THE CONCURRENT ENGINEERING PROCESS

In order to perform the producibility function it is necessary to understand design procedure. Concurrent engineering covers many aspects of product development, from conceptualization to delivery of the finished product, but central to this team-work process is the design. Design is absolutely critical to success. It is the plan for producing. If the plan is good, most likely it will succeed and be profitable. If the plan is bad, no amount of heroic effort is sufficient to make it succeed. This is why we need to consider attributes of design as part of any discussion about concurrent engineering and the role of the producibility engineer.

A Definition of a Design

A design is a combination of many things. First of all, it is a plan for achieving a goal. A goal can be anything. To give an extreme but totally proper example, successfully staging Mozart's opera *The Magic Flute* would be a goal. The producer and director would have to conceive how to do it to achieve the goal. This would be the design—how to meet their interpretation of the artistic merits they wish to portray.

When engineers think of design we narrow the definition. Design to an engineer usually means physical things, such as machines, vehicles, electronic devices, and virtually all the amenities of life. So, in reference to what factories do, a design is a plan that, if followed properly, will result in a physical thing that performs a function the designer intended it to do. In its most common form, the design is recorded in an electronic data base or, more traditionally, in a set of drawings and explanatory notes.

A design is also an expression of creativity. Designers are given a goal to meet and they use their experiences and education, with a good deal of intuitive creativity, to conceive an approach for reaching the goal. If we asked several people to move a 5-1b block from point A to point B, we would probably get as many unique solutions as there are people. This is a manifestation of the uniqueness designs produce. Within the imposed constraints, there will be many "correct" solutions to achieving the desired goal. The designer, using his or her skills and creative abilities, would offer a method for moving the block, and the method would be sufficient. A designer with a mechanical background might offer a solution based on principles of hydraulics. An electrical engineer might opt to move the block using an electric motor geared to a rack-and-pinion screw lift. Both would be appropriate solutions to the block problem. The thing all these solutions would have in common is their uniqueness, and because there are many "right" answers, discussing design is an abstract subject.

When we talk about the constraints, the practicality of the approach must be recognized. Here we begin to think of producibility. There are designs that are extolled for their simplicity and beauty. There are other approaches to design that we berate as "Rube Goldbergs" because they are too complicated and not pleasing to our eye or psyche. We usually praise simplicity because it is makable in the factory; it is producible. We hear terms such as functional, esthetic, and pleasing to the eye. These can be thought of as ways of saying the designer has created a scheme for achieving the goal that can easily be used by those responsible for doing it. In other words, manufacturing can make the product with only reasonable applications of their skills. This we judge to be a good design, a producible one.

The Set of Design Attributes

A good design has a set of attributes, and all good designs share the same attributes. Let's look at them, and then we'll investigate the various phases of the design process and their compatibility with the attributes. The five recognized attributes of design are:

1. Producibility
2. Simplicity

3. Lowest feasible cost
4. Esthetically pleasing
5. Meet quality requirements

1. *Producibility.* First, we must understand that a design, above all other requirements, has to be producible. Obviously, if the design cannot be produced, all other attributes have no meaning. Attributes 2 through 5 come to naught if the design is insufficient for the factory for which it was intended. The simple fact is that if the design is not producible, we have set up a no-win game.

2. *Simplicity.* Simplicity is the commonsense attribute. The desire to make a design producible virtually dictates that it be as uncomplicated as the physics of the situation allows. In the industrial environment this is affectionately referred to as the KISS principle. KISS means "keep it simple, stupid." This is an admonition to bright people not to demonstrate their technological mastery of their specific segment of engineering theory, but to deliver a design that serves the intended purpose and no more. We do not want to admire the clever nuances the designer employed to solve the design problem, unless they were truly needed to reach a successful conclusion. We do want straightforward understandable instructions that make use of the factory's capabilities. This is simplicity.

3. *Lowest feasible cost.* The cost attribute must be met. This means not only the targeted cost to produce but also the cost the company intends to charge the customer for the product. If the design results in a product that is beyond the means of the intended purchaser, it is a failure. Imagine if basic transportation, the "plain" automobile, had a price tag of $100,000.00. If this was the best that could be done, how many of these automobiles would be sold? Precious few indeed! We must understand that taking cost out of a product is not just a nice thing to do, but a necessary thing to do.

4. *Aesthetically pleasing.* Aesthetics are also important. We do not live by bread alone. This can be a difficult attribute for engineers to comprehend. For some products we can safely ignore the need to consider esthetics, but it is difficult to say where the dividing line is. However, we all know that some classes of products are virtually exempt from this requirement while others are not. For a screwdriver, aesthetics is a minor concern. As long as the screwdriver performs its function, the design is satisfactory. Grand pianos are another matter. Here we require not only supreme functionality but also demand beauty of appearance. Here esthetics are very important. There are no guidelines for determining when aesthetics are an important attribute, but failing to consider this can have significant negative consequences.

5. *Meet quality requirements.* This attribute sounds deceptively simple, but it is not. That the design must meet the quality requirements is an undisputed fact. However, what is the quality requirement? Is it that manufacturing meets the design specification? Or is it the pleasing appearance of the product? Or is it achieving the goal of the customer? It is all of these requirements—that the product is defect free, works correctly, meets the hopes of the user, and perhaps meets the expectations of admirers. Also, there may be government regulations that have to be complied with, even if the customer really does not care or is oblivious of the need. So we see that quality attributes vary depending on the viewpoints of various constituencies.

The Process of Creating a Design

Creating a design is very much a multifunctional task. This is something we are just coming to recognize. As recently as fifteen years ago this statement would have been vigorously disputed. Then we would have said that design is simply applying scientific principles to form and function, thus defining what should be produced. This is no longer a valid definition. There are three components of design: predesign, engineering phase, and postdesign. Predesign is the definition of the customer's needs and desires and is coordinated by marketing. The engineering phase is the traditional applying of scientific principles to achieve a workable plan for a product that meets marketing specifications. Postdesign is dominated by the customer service component of marketing and involves helping customers understand and properly use the product they have purchased.

The Engineering Phase of Creating a Design

The engineering phase consists of three components done in chronological order and concurrently iterating after the preliminary work is done:

1. Concept design
2. Producibility design
3. Manufacturing facilities design

Keep in mind that all phases of design, to be successful, must be compatible with the five attributes we discussed previously. Let's look at the engineering phase in more detail.

1. *Concept design.* The concept design is the rationalization of an idea in terms of science. Concepts are ideas for product offerings. These ideas can be new, such as when SONY introduced the commercially viable videocassette recorder. Or the idea can be an extension of or an improvement on an existing idea. The introduction of the automatic transmission for automobiles by Oldsmobile would be an example of the latter. In the concept design stage, we have to face the reality of scientific facts. This is the defining of the product in accordance with the laws of science. For example, when designing a ship, one must make sure that Archimedes' principle of buoyancy is complied with. We certainly cannot have a hull form that will not float. The task here is to make sure that the concept has been thoroughly vetted. It has to have been well conceived and thoroughly evaluated to make sure the goals of the design are achievable. The ideas employed have to be compatible in all aspects with the relevant science. Once a design is found to be based on science and is well established, we can begin to consider producibility.

2. *Producibility design.* The primary questions that have to be answered are where will the product be built and what are the capabilities of that source? The designer and the entire concurrent engineering team must make sure that whatever information is transmitted to the source factory is understandable and within the source factory's "normal" capabilities. If not, the probability of the design being executed faithfully is severely reduced. The producibility design phase is the process of "customizing" the design for the production source. Not doing this customization would more than likely lead to substandard production results. The process for achieving a producible design evolves about the design review process as explained previously.

The producibility design is not the end of the engineering phase of the design. Even though we have now tested the design for compatibility with the factory and we know that the requirements meet both the idea of the product and the practicality of building it, we are not done. Now we have to be concerned with fitting the product within the factory. This process is called the manufacturing facilities design.

3. *Manufacturing Facilities Design.* The manufacturing facilities design process is sometimes called methods engineering. You might ask, if the design is producible isn't this a redundant step? Absolutely not. Through the producible design development we have tested the concept design to see if it fits within the envelope of practicality. We have simply evaluated the design to see if it can be done in the factory it was intended for. Now we need to do the nitty-gritty work of tooling up the factory. Because the design is producible, we know we do not have to reinvent the factory. The tooling, although it could be very creative, does not have to be heroic to do the job. The manufacturing facilities design, then, is the specific designs for jigs, fixtures, and sometimes processes that are necessary to implement the producibility design.

SUMMARY

We have reviewed the role of the producibility engineer in the modern industrial organization. The working relationship between design engineering and manufacturing within the confines of concurrent engineering theory has been demonstrated. The role of the producibility engineer within the overall umbrella of concurrent engineering has been demonstrated to be that of the manufacturing engineering representative in the team process of conceiving and creating a product.

Creating producible designs for the factory is important to a manufacturing company's success. Without this activity, the probability of suboptimal production costs is greatly enhanced. Another important aspect of the producible design is that even though it is approached from the viewpoint of favoring optimum factory processes, it does not sacrifice the designer's basic intentions or increase the quality risk. On the contrary, producible designs tend to improve product quality and, because they require only needed processes, thus a simplifying technique, enhance the elegance of design.

Statistical setting is a method of monitoring and judging usability of assemblies. It evaluates risks objectively so that the engineer can make judgments based on facts. Setting of tolerances cannot be a game played between the designer and the factory to see who gets the greater level of insurance protection against risk of design failure. It has to be a win-win game for all parties; otherwise, the true loser is the entire organization. Geometric dimensioning and tolerancing is an ideal way of assuring design intent is fully understood. Therefore manufacturing engineers ought to embrace this drafting methodology to enhance the productivity of the factory.

The producibility design review is the focus of synergism within the concurrent engineering philosophy. Without it, costs of products could be considerably more than necessary, which would reduce the company's market and lower its profit potential. The producibility engineer is the catalyst for successful design reviews, creation of producible designs through practice of attributes for good designs, and the setting of realistic manufacturing requirements.

REVIEW QUESTIONS

1. Define concurrent engineering and the role of producibility engineering within the overall concurrent engineering concept.
2. Discuss the concept of producible design. Explain why a design engineer would often require the counsel of the producibility engineer to create a design that can be produced in the factory at reasonable costs.
3. Explain how tight tolerances increase costs. Devise a scenario in which tight tolerances are justified.
4. Discuss the reasons geometric dimensioning and tolerancing should or should not be used in a factory that does not normally farmout work.
5. Discuss the difference between least material conditions (LMC) and maximum material conditions (MMC) and what the reasons are to use one or the other in defining tolerances for an assembly.
6. Define datum as it relates to tolerancing and how it is used in making parts so they will assemble correctly.
7. A design for catalytic converters for automobiles is to be produced in a factory. Make a list of questions that the producibility engineer should ask before reaching any conclusion about whether the design is a workable one.
8. Explain what the statement "minimize the risk, then find a way to do it" means, and how it applies to introducing difficult-to-manufacture projects in the factory.
9. The drawing for a steam valve head joint bolt is shown below. Perform a producibility engineering analysis, answering the following questions. Is this drawing done in accordance with geometric dimensioning and tolerancing provisions? Why is proper to do the drawing as shown for producibility review?
   a. What is the part for?
   b. How does it work?
   c. What engineering analysis is important to the design?
   d. How can the part be made?
   e. What can be done with the design to make it cheaper to produce?
   f. Will the part still do the job if the producibility engineer's recommendations are accepted?

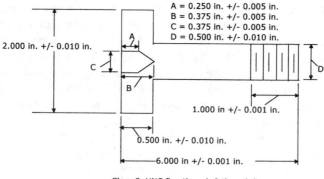

A = 0.250 in. +/- 0.005 in.
B = 0.375 in. +/- 0.005 in.
C = 0.375 in. +/- 0.005 in.
D = 0.500 in. +/- 0.010 in.

2.000 in. +/- 0.010 in.

1.000 in +/- 0.001 in.

0.500 in. +/- 0.010 in.

6.000 in +/- 0.001 in.

Class 3, UNC fine thread, 8 threads/in.
Taper 0.001 in./in. +/- 0.00001 in.

10. Perform a producibility engineering analysis of the following part. Use steps (a) through (f) of question 9.

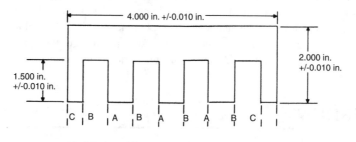

4.000 in. +/-0.010 in.

2.000 in. +/-0.010 in.

1.500 in. +/-0.010 in.

C B   A   B   A   B   A   B   C

Heating element for toaster oven
Required: 100 watts/in.
Thickness = 0.020 in., commercial grade
Material: silicone steel
A = B = 0.500 in. +/-0.001 in.; C = 0.250 in. +/-0.001 in.

11. Explain why dimensional tolerances must be converted to strength values in using statistical tolerance techniques.

12. A stud must carry a steady-state load of 25,000 lb/in.² with a cyclic load of ±1,500 lb/in.² Assuming that the basic stress mechanism is bending-induced shear, determine the diameter of the stud and the tolerance range. Also determine the probability of failure if the material used to make the stud has a strength of 30,000 lb/in.² with a range of 4,000 lb/in.² and the shear force is 50,000 lb.

13. Repeat question 12 if the steady-state load is 27,500 lb/in.² and the cyclic load is ±2,500 lb/in.²

14. Discuss limitations to the use of statistical tolerances. State reasons for not using statistical tolerances in all situations.

15. With reference to question 9, outline a procedure via a list of specific items to be covered for a design review to be held between producibility engineering and the concurrent engineering team.

16. Repeat question 15, using data from question 10.

17. Using the design data of question 9, define how each of the five steps of the attributes of good design is complied with.

18. Repeat question 17 using the data of question 10.

19. Referring to question 9, define what elements would be considered concept design, producibility design, and manufacturing facilities design. The design evaluation questions (a) through (f) of question 9 may apply to more than one of the three design subsets.

20. Repeat question 19. using the data of question 10.

Chapter *7*

METHODS, PLANNING, AND WORK MEASUREMENT

Methods, Planning, and Work Measurement (MP&WM) is the design of work performance on the factory floor to optimize output of the product. The function of methods engineering is to convert broad-based methods or procedures into detailed, easy to follow plans for the workstation operator. These plans must be detailed enough to specify the tools to be used, the materials to be used, the sequence of events, and even the time to complete each event.

The MP&WM unit consists of methods engineers, planning specialists, and time standard analysts. Methods engineers create the broad-based sequence for producing the part. The planning specialists then create the detailed instruction sheet from which the operator will do the work. The time standard analysts work with the method sheets to determine the time it should take to perform each operation. This involves study of the task to be performed at the workstations to determine what the optimum body movements, in accordance with the Principles of Motion Economy, should be to minimize the time and effort required. Finally, the MP&WM unit will have methods engineers who are totally involved in measuring the performance of the factory against objectively set standards. For an MP&WM unit to work effectively, smooth transitions must be made from phase to phase in this sequence from broad-based to particular. How these transitions are made and how the disciplines work with each other to produce manufacturing instructions for shop operations are the subjects of this chapter.

THE SCIENTIFIC METHOD APPLIED TO MANUFACTURING ENGINEERING

The purpose of the MP&WM unit is to produce a manufacturing plan, utilizing the existing facilities, to optimize production of the company's products. This is not a "by guess by golly" procedure. The philosophy of the activity is to use the scientific method:

1. Make observations.
2. Develop a hypothesis.
3. Test the hypothesis.
4. Make revisions to the hypothesis based on the test.
5. Test the revised hypothesis.
6. Reach a workable conclusion.

Let's demonstrate the scientific method in terms of a factory operation problem.

A new producible design is transmitted to manufacturing to produce. The methods engineer devises a sequence of events, called a method, to produce the product. This method is turned over to the planning specialist to create a detailed written plan for the shop to follow. The detailed plan will include times to complete each step, which will be obtained from the time standard equations or computer programs developed for the workstations by the time standard analyst. This series of activities constitutes steps 1 and 2 of the scientific method; observations are akin to studying the design, and developing a hypothesis is the work of creating the method and a written plan.

The written plan is given to shop operations, which will produce the part. Since it is the first time the part is being made in the factory, the methods engineer will be present at the workstation as a combination observer and consultant. At this point the method is being tested to make sure it is the best one possible and optimally produces the part. This corresponds to step 3 of the scientific method.

Having carefully observed the prototype production, the methods engineer now has practical experience to compare to the method and can make necessary adjustments (step 4 of the scientific method). Adjustments are necessary because, with prototypes, it is usually the case that several items are unknown and assumptions must be made.

> **EXAMPLE 1:** The method is derived on the basis of feeds and speeds being delivered precisely by the machine tool. Quite often, machine tool performance varies with the condition of the machine. Therefore modifications could be required in the method.

> **EXAMPLE 2:** The part is produced with a cutting tool from a different vendor, which was thought to be generically identical to a previous supplier's tool, but is not. A change in method is needed to accommodate the different results.

> **EXAMPLE 3:** The cutting tool is not properly sharpened by the operator. This could cause significant variation, especially in stationary or spade drilling activities, where the drill blade is held stationary and the part is rotated. If the methods engineer considered a mean time to perform the operation based on a particular level of drill blade sharpness and the actual mean time was different, a modification could be called for.

Once the revisions to the written planning document are made, the factory is ready to try again on a second work piece or series of work pieces. Again, the methods engineer will observe the manufacturing activity and offer consultation where required. This is step 5 of the scientific method. The methods engineer is testing the revised hypothesis, that is, the method, and will continue to revise the method until an acceptable manufacturing procedure is developed. This iterative process is continued as long as it is economically justifiable to do so. For instance, adjusting the method to bring it within 99% of optimal may not be economically justifiable if it is already at 95%. The termination of the experiments with the method is similar to step 6 of the scientific method.

The MP&WM staffs carries out the scientific method applied to manufacturing, continuously. It is done so routinely that its practitioners often forget that it is an application of the scientific method. It is the optimizing technique used by the methods engineer and the supporting staff of planning specialists and time standard analysts to find the best possible way to produce the product.

METHODS ENGINEERING

Methods engineering encompasses all the activities of the MP&WM unit: tool design, fixtures, setup optimization, time standards, feeds and speeds, detailed planning, operator training, and many other related activities. In this section we will discuss only the activities of the methods engineer as defined earlier. The work of the planning specialist and time standard analyst will be covered in later sections of this chapter.

Principles of Motion Economy

Principles of Motion Economy (PME), originally known as the "Laws of Motion," as developed circa 1920 by pioneering industrial engineers Frank and Lillian Gilbreth, are the foundation upon which all of methods engineering, scientific time standards, and even all business planning is based on. Including even those current reincarnations called Just In Time, Lean Manufacturing, and Six Sigma. It all boils down to conserving human energy, and utilization of energy in the most efficient manner in making products and delivering services. In its most fundamental definition:

> Principles of Motion Economy—an Industrial Engineering body of knowledge based on the physics of human body motion related to energy expenditure.

The thought behind using PME philosophy in setting up workstations is if we can ensure that people can work longer without fatigue setting in, we can make more products in a given period of time. Hence be more effective in production; for example, the unit cost goes down. For methods engineering work we refer to placement of tools and positions the operator must assume to do the required work. We think of this as ergonomics, on how to place equipment, people and the work piece to assure optimum output. In essence this is an exercise in the conservation of energy as defined by human movement. See Table 5.1 for a list of most common considerations.

While much of the early industrial engineering theory and observation work went into defining the PME for production lines, we have since found it is equally as important and meaningful for all types of office and clerical jobs. Extensive tests and observations of operations were done first in factory settings and later on in office/clerical activities to correlate the energy expenditure of people in doing work to find the least energy intensive methodologies. It was thought that people could be considered as machines utilizing energy. And since it is possible to relate machine design to kilowatts and horsepower to make shafts and levers and slides and gears, etc. moves based on how they are connected; then the same analysis could be used to define, based on human physiology, which movements used least to the most energy. Objectively this is true. In fact energy expenditures have been measured in doing simple body motions. Sort of defining how many calories are expended, running walking, standing, sitting, and so forth. The one overriding difference between machines and people is dependable repeatability. Since we have free will and independence of thought, what is entirely true for energy utilization by machines is only partially true for people. Machines are boringly repetitive and repeatable in their motions. People are not. However through PME we can learn what the most efficient human motions are and apply them to designing how a job is done. We can then train people to follow the best procedures and gain a win/win situation in performing work.

If we make a far out but objective assumption that the person in the work place is for all purposes a substitute for a machine, the knowing of how many calories are burned per motion is important for determining the energy cost for performing a task. For people, of course this is the fatigue factor, not how many kilowatts are needed. Fatigue has many physiological causes, but if we can simplify it down to expending energy by burning calories we can find those motions that consume the least calories, then design work places and required work motions accordingly. And this is exactly what is done. By doing so we have a much better chance of having work flow in a more even and predetermined pace. We can then use that pace to set the manufacturing output plan, and ultimately the entire business plan.

The Gilbreths formulated 19 laws of motion which we now know as the Principles of Motion Economy. General Electric Co. further collated the 19 laws into three major groups. I, in my then capacity as Manager, Industrial Engineering Consulting, and my colleagues at General Electric Co., used these three groups to formulate and teach "Motion Time Survey," a scientific time standards process (more about scientific time standards later in this chapter). In the 1970s General Electric Co. made this data available as public domain engineering information. These data are contained in a General Electric Co. publication titled "Motion Time Survey", the latest edition I'm aware of and I believe final revision, was published in 1977. This book is a compilation of General Electric Co. developed industrial engineering data (which I'm proud to say I made contributions to) and contemporary publicly available information from other sources. It is as valid today as it was then. The data I am presenting in this chapter related to PME and scientific time standards is from this source unless otherwise stated. The Gilbreths' 19 principles collated into three groups is shown below and paraphrased for brevity:

Group I—Use of Body Members

1. Movements are confined as much as possible to those requiring the least amount of time. Shown in order of least time:
   a. Finger
   b. Wrist
   c. Ankle
   d. Elbow
   e. Shoulder
   f. Knee
   g. Leg
   h. Back and hip
2. When using two hands; they should begin, as well as complete their movements at the same time.
3. The two hands should not be idle at the same time except for rest periods.
4. Motions of the arms should be made simultaneously in opposite directions over symmetrical paths. This means arms should move toward the body together and way from the body together. Adherence to this principle enables an operator to develop a smooth rhythm in performing an operation.
5. Rhythm is essential to the smooth and automatic performance of an operation.
6. Continuous curved movements are preferable to straight line movements involving sudden and sharp changes of direction.
7. Hesitation implies wasted time which can be caused by Search, Select, Plan, or Avoidable Delay occuring in an operation. These activities should be analyzed for cause and if possible eliminated.

Group II—Work Place Arrangement

8. Definition and fixed locations should be provided for all tools and equipment. Tools and materials should always be kept in the same location to avoid wasting time searching for them.
9. Tools and materials should be prepositioned where ever possible so they may be grasped in the position in which they are to be used.
10. Tools, materials and controls should be located in front of the operator within the normal working area.
11. Gravity feed bins or containers (tilted approximately 20–30°, depending on the size and type of material) should be used to deliver material as close to the point of assembly or use as possible.
12. Materials and tools should be located to permit the best sequence of motions. The work place should be arranged in such a manner that the time for the total required moves are kept at a minimum.
13. Drop disposal should be used where ever possible to avoid the need for stacking finished work.
14. Provision should be made for adequate visual perception of the work place, including but not limited to task lighting, magnifying glass, reflectors, electronic imaging, and other visual aids.
15. A chair of the type and height for good posture should be provided. The chair should be adjustable so as to permit free and easy movement by the operator in performing the operation.

Group III—Tool and Equipment Design

16. Weight should be reduced or used to assist motions where ever possible. Designing tools, jigs, etc. with ease of handling through weight reduction techniques, and using counter balance weights and springs in mind should be pursued.

17. Hands should be relieved of all work that can be performed equally as well by other body members. Foot pedals can be used to open and close fixtures, operate machines, etc., allowing hands to perform other productive work.
18. Two or more tools should be combined where ever possible. Often the combination of two or more tools into one will reduce the number of movements and the amount of time for an operation.
19. Levers, cross bars, hand wheels, and other operating devices should be located in such positions that the operator can manipulate them with the least change in body position and designed so they may be used with the greatest mechanical advantage.

These are the 19 principles of PME. If you review them carefully you will be hard pressed to find any type of physical work situation that is not covered by at least one of the 19 principles. To illustrate this point, let me give you a bit of engineering folklore. Perhaps more than folklore, because I have no reason to doubt it really happened. As you might be aware Frank and Lillian Gilbreth had a true personal need to be personally efficient. They were blessed with 12 children. Frank Gilbreth supported his family through his work as an industrial engineer. It is reported that when two of his children needed tonsillectomy surgery, he made a deal with the surgeon and the hospital, that in return for performing the medical work he would organize their operating room to be more efficient; for example, greater throughput of patients through more efficient use of the facility. He observed operations to find out how they could be done more effectively. In fact to get enough data we are told, he had the doctor take out the tonsils of all of his children, deciding that if two children needed tonsillectomies the rest soon would. What he found and later reorganized the operating room in accordance to was directly aligned with principle 8, 9, and 12. All dealing with tool placement location, in the proper position to be used, and to permit the best sequence of motions by the doctors and nurses. In fact when you see medical dramas on television or in the movie theatre, the doctor asking the nurse for the scalpel, clamps, etc.; the doctor putting his or her hand out without looking and receiving the tool in the proper orientation had its' origination in this famous work by Frank Gilbreth. The point to be made is that PME is the basis for effective work station design and operation, what ever the physical work is. Following the dictates will provide for higher profitability.

Obviously people are not machines, so we cannot guarantee that they will work according to the prescribed best motions. This has led to numerous, to say the least, motivational studies and practices. And they're all aimed at getting people to do work in the prescribed manner, which are all based on PME. So PME defines the best way to do work, at the least calorie burning rate. With this information in hand it is up to the leaders of the team to convince all the members of the team to embrace and use the least calorie expending method. You might say, for the non-creative portion of work, this is the sole goal for all management.

PME is fundament for all method and time standards work as we shall see. The basic of motions have been defined and tabulated. And in many cases the amount of time to do these various virtually micro motions have also been defined. With this catalogue of motions and associated times available it is possible to design, in this case we can say choreograph the motions, such that we can customize the design of work to be most efficient. Let's progress down that path to see how it is done and how the modern factory translates this theory into pragmatic practice.

The Methods Sheet

Methods engineers are experts in what can be accomplished at workstations, including machine tools and processes. They design the manufacturing technique. They do this by carefully reviewing the design information, ferreting out specific details that are critical for the functionality of the completed part, and specifying the individual steps to be taken to complete the manufacture. Figure 7-1 is an example of a broad-based methods sheet. This is considered a broad-based method for two reasons: it assumes that the operator is familiar with the workstation and the typical parts going across the workstation, and no time to perform or allowance is calculated for each step. Nevertheless, it includes such detailed instructions for the operator as what type of tools to use

---

Methods Sheet
Sealing Rings Setup and Metal Cutting Data

1 Change spindle if required, use wherever possible Setco no. 3615 spindle for increase rigidity.

2 Clean table of machine and take from cabinet (4) 3-in. square parallels. Wipe clean and position on table at 90 degrees.

3 Put in table T-slot nuts. Assemble studs to nuts and clamps.

4 Pick up sealing ring. Wipe ground surface, clean, and place on parallels.

5 Assemble 3-in. travel indicator to magnetic base. Mount on inside of machine splash pan, positioning indicator to contact sealing ring.

6 Rotate table by hand. Note dial reading on indicator. Tap sealing rings with no. 4 Compethane mallet as indicator reading increases to plus.

7 Repeat above 3-4 times until indicator is zeroed out.
7A Should sealing ring be distorted or out of round, center ring until readings at 180 degrees are the same.

8 Position clamps on sealing ring as per sketch and secure snug only.

9 Recheck runout. Adjust if required. If part did not move, secure tight.

10 Recheck runout.

11 Mount borozon wheel to spindle if required.

12 Move spindle to left side of sealing ring.

13 Set stops on vertical slide, top and bottom to clear portion being ground.

14 Set bottom positive stop and check stroke of vertical head.

15 Set rotation of table clockwise.

16 Move spindle toward left side of scaling ring so borozon wheel is within 0.020 of sealing ring.

17 Start table, set table speed at 20 RPM.

18 Turn cross feed wheel to feed borozon wheel into sealing ring slowly and continue to feed in until wheel starts to move. Note reading on handwheel, back off and repeat to verify touch off point and recheck number on feed dial. Then back off approximately 0.010.

19 Set vertical slide control valve to read A-2 which equals 5.25 in./M.

20 Set automatic feed at 0.0005.

21 Turn on coolant.

Figure 7-1. A Broad-based Methods Sheet

and even where to get the tools. Therefore it is broad-based only in comparison to the planning details and time standard documentation. Conversely, note that when work is to be performed, either setup or production, the operator is not told specifically how to do it. For example, in Fig. 7-1, step 11, "mount borazon wheel to spindle if required," the methods engineer is not specifying how that is to be done, but merely states when in the chronological order of events it should be done and assumes that the operator knows how to do it. In a detailed planning sheet the actual sequence of mounting the borazon wheel with the time allowed to accomplish it would be adequately described.

The Workplace Layout

The methods sheet is a key output of the methods engineer, but not the only output. In addition, a competent methods engineer also produces a workplace layout, as shown in Fig. 7-2. This layout differs from one produced by advanced manufacturing engineering in that it contains data pertinent to operating the workstation at an optimal productivity level. The workplace layout is a schematic of the workstation showing the primary and secondary equipment at the particular location.

Primary equipment is the main producing equipment. In Fig. 7-2 the primary equipment would be the rotary table, 15 × 15 ft iron floor, 10 × 10 ft iron floor pit level, tape control, ram, and head. Secondary equipment is the auxiliary equipment. In Fig. 7-2 it would be the vidmar cabinet, oil and water storage tanks, tables, electric cabinets, chip collector dumpsters, and tool cabinets. Auxiliaries are usually items not directly related to working on the part.

Other interesting items are found on the workplace layout, including (see Fig. 7-2) information such as, for example, how far the dispatch cage is from the workstation and the distance to go to sign up for a crane lift.

The workplace layout is oriented in the factory by designating its location with an address. The circled 19 and 20 in Fig. 7-2 indicate the location in J-Bay where this workplace is, for example, its address. This information is vital to the time standard analyst, who needs it to determine the optimal time to perform work at the workstation. The time it will take to move around the workstation and the time to obtain services are included in allowance factors, which are added to the actual performance times to obtain total optimal times.

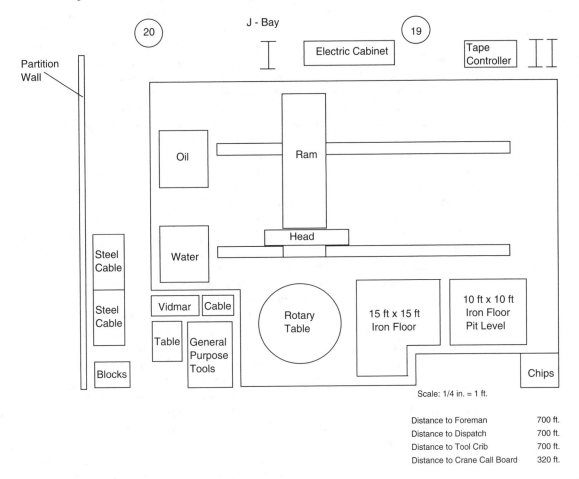

Figure 7-2. Workplace Layout

With the machine tool characteristics supplied by advanced manufacturing engineering, the methods sheet, and the workplace layout, the methods engineer has compiled a complete description of the workstation and its capacity or capability. From this broad-based but detailed data base, specific plans for producing the product and measuring the efficiency of the work in process are generated.

Productivity and Efficiency Measurements

Another task of the methods engineer is to evaluate the productivity level of the workplace. Since a large portion of manufacturing costs is due to the labor input required during production, it is important to know how well the labor force works in order to know how much to charge for the product.

The simplest measurement of how well the factory is working is output divided by input. The output is the number of product or component parts made over a finite period of time, and the input is the finite period of time. This is the basic productivity measurement:

$$\text{Productivity} = \frac{\text{output}}{\text{input}}$$

Usually the denominator of this equation is kept constant and is taken as 8 hours, the standard time for one shift of work in a factory. Therefore, if the output of the workstation is 8 units per shift, the rate of production is 1 unit per hour:

$$P = \frac{8\,\text{units}}{8\,\text{hr}} = 1\,\text{unit/hr}$$

This system is useful for measuring trends in factory performance, with results recorded on a productivity chart. The input value is almost always 8 hours; however, this may be adjusted to account for unusual circumstances. For example, if a machine breakdown caused the productive time available to be reduced to 6.5 hours, the denominator in the productivity equation would be 6.5. Or if a shift was extended or overtime applied, the denominator could be as high as 12 hours.

Methods engineers use productivity charts to judge whether the workstations are producing at acceptable rates and to evaluate the worth of overtime. In Fig. 7-3 the notation O/T for days 5 and 6 means that overtime was used to meet the daily piece count requirement. It is apparent that on these days the rate per hour decreased to the lowest level of the measurement period. For both O/T days the workstation produced $12 \times 3 = 36$ units per day. If the piece count requirement was less than 36 units per day, the overtime could be justified. Otherwise, it was an unjustified expenditure of labor hours. For comparison, on day 9 the workstation produced $8 \times 6 = 48$ units.

Productivity charts may also be used to investigate method adherence by the operator, if the rate was less than anticipated. In the case of the overtime-low rates in Fig. 7-3, the methods engineer would investigate adherence to the prescribed method and also whether the extended shift introduced an unfavorable fatigue factor.

Productivity rate measurements are simple and easy to apply if discrete work can be measured at the workstation. They are useful for relatively long production runs where a specific type of part will be produced repeatedly. For example, such measurements can be employed in the production of electric motor coils or stampings or in drop forging manufacturing. They would not work in a machine shop where there are daily or weekly variations in the size of the parts going across the workstations and the length of time it takes to perform the work. For this type of work the efficiency measurement is used.

Efficiency measurements depend on predetermination of the time it should take to do a specific segment of work. Therefore, rather than specifying how much product will be produced in a finite period of time, the efficiency method of measurement specifies how much time a specific segment of work should take and compares this with the actual time taken.

The mechanics of the efficiency measurement is as simple as that of the productivity rate measurement. The measurement is planned time divided by actual time, giving a numerical value that is almost always less than unity. The reason is that the planned time is the ideal time corresponding to everything being done optimally with no interferences or aberrations occurring during the performance of the work.

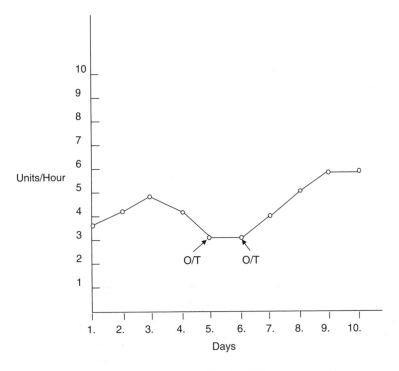

Figure 7-3. Productivity Chart: Workstation 932

$$\text{Efficiency} = \frac{\text{planned time}}{\text{actual time}}$$

Efficiency measurements can be either absolute or relative measurements, depending on how the planned time is arrived at. The planned time may be a simple estimate made just prior to the assignment of the work, or it may be the result of a scientific document taking into account the physics of human body movements and placement of the tools and fixtures at the workstation; e.g., application of principles of motion economy. The latter type of planned time using PME is called a scientific time standard and will be discussed later in this chapter.

There has been a debate within manufacturing management as to whether scientific time standards to determine planned times are worth the effort. The scientific time standard gives management an absolute value to measure performance against; the question is whether it is really worth knowing how far from the absolute optimum an operation is.

Figure 7-4 shows three efficiency lines. The top one is the planned time based on a scientific time standard. This is the absolute value. It represents the least possible time to do a job because it is the sum of the times it should take optimally to do subsets of the job, which are then added together to get the total time to do the job.

The second line is the estimated planned time to do the work at the workstation. This may be derived by calculating the feeds and speeds to determine how much time the actual machining will take, then estimating how long the setup and teardown times will be. By adding these together, the estimated planned time to do the job is obtained. This, unfortunately, is the most common form of time standard development used today. Primarily because it is perceived to be the easiest to do. It is generally okay, but leaves lots of profit dollars on the table as we shall see. This can be a significant problem if the company is in a very competitive situation.

The third line, the irregular one, is the actual time to do the work plotted on the chart. Notice that the efficiency value of the actual time is approaching that of the estimated planned time. Also notice that the efficiency value of the estimated planned time is significantly lower than that of the absolute planned time, which is the 100% efficiency value. The question then is whether management should

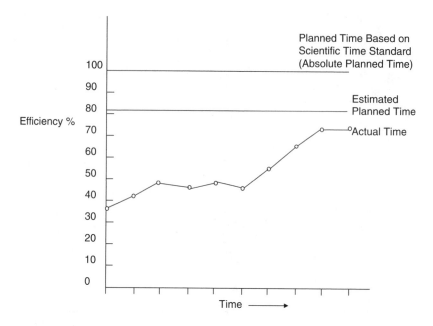

Figure 7-4. Efficiency Measurement

be satisfied with asymptotically approaching the relative measurement or should be appalled that the actual measurement is so far below the absolute planned time. The difference is really profit dollars not being realized by the company.

The answer to this question has to do with how much money is involved in minimizing the spread between the lines. The solution lies in the return on investment. If the estimated standards yield the targeted profit and market share, then lowering costs to gain more profit and a larger market share may not be desirable, or at least not politically desirable. Gaining a larger market share may be a poorer return on investment than other opportunities available to the company. On the other hand, lowering operating costs may be essential just to stay in business. This would be particularly true if significant pressure on sales is being felt from foreign competitors who start out with a lower labor cost base. Therefore, the question of whether to use scientific time standards must be evaluated in the same way as any proposal to invest funds to improve productivity. Another warning to consider when making this evaluation. Saying we will gain profit by implementing scientific time standards may be difficult to quantify in dollars. It is not a simple if then relationship. It is full of intangibles, difficult to quantify actions. It has more to do with showing what can be achieved if the people doing the job can be convinced that the scientific time standard will not effect their "working quality of life" in a negative fashion. Then getting the people to use the scientific standard as their goal to strive for. So adopting a scientific time standards program requires more than the expenditure of time for writing the standards. It also includes the effort, not necessarily in dollars, to convince the work place team that it is a good thing to do. Win/win for everybody.

Regardless of the choice eventually made, an absolute standard focuses attention on minimizing costs to an absolute value, while a relative standard focuses attention on minimizing costs to a relative goal. The relative goal is almost always easier and cheaper to achieve than the absolute one but is never worth as much.

Methods Analysis

Now that we have surveyed the activities of methods engineers, we will put the key parts together, using an example to demonstrate how methods engineers solve manufacturing problems. This example is a demonstration of the systematic technique called methods analysis, used to produce a manufactured component.

Let us assume we are asked to make a simple screwdriver in lot sizes of 1,000 pieces and the factory available for use has all the necessary equipment to produce hand-held tools. The methods engineer must:

1. Determine how to make the screwdriver.
2. Develop tooling and fixtures to optimize production.
3. Develop a useful measurement scheme.
4. Estimate the time it will take to produce the required amount of product.

As we will see, a great deal of the work can be classified as subjective, but it is subjective only in the sense that the approaches used by the methods engineer are based on previous similar experiences. Let us now take each step separately, even though they may develop concurrently.

Determine How to Make the Screwdriver. The methods engineer will consult the drawing for the design, which we will assume for this example to be a workable design, and extract all the physical factors such as dimensions, shapes, materials selected, assembly requirements, and any other special requirements specified by the designer. For this example we are producing steel-shanked screwdrivers with wooden handles. Therefore, the manufacture can be divided into three areas: making the shank, making the handle, and assembling the handle to the shank.

For the shank the methods engineers can envision the following sequence:

1. Cut steel rod, 3/8 in. diameter, to length on a saw.
2. Rough turn the entire length and thread opposite to the blade end on a lathe; class 2, UNC* coarse thread 8 threads/inch.
3. Form the blade on a grinder to a preset template.
4. Grind the entire shank to a commercial finish or, if specified, to the finish required on the drawing.

For the handle, the following sequence could be proposed:

1. Cut 1½-in. diameter wood dowels to drawing length.
2. Produce the exterior shape on a wood lathe.
3. Produce the internal thread for mounting the shank on a wood lathe; class 2, UNC coarse thread 8 threads/in.
4. Paint exterior surfaces in accordance with wood finish requirements specified on the drawing.

Next, it is necessary to create the assembly method. A suitable sequence could be:

1. Put glue on threaded portion of the shank.
2. Hold shank in a viselike device.
3. Thread handles onto shanks.
4. Let dry.
5. Package.

We have now created one acceptable way to produce screwdrivers in the desired 1,000-piece lot size. The methods engineer must now evaluate this method to see whether there is a better, that is, cheaper, way to do the job.

EXAMPLE: Could the shanks be drop forged instead of doing the grinding and lathe operations? For a lot size of only 1,000, the answer would probably be no. But the question illustrates the type of thinking the methods engineer must employ.

* UNC refers to a specific thread design standard found in most reputable machine design reference books.

Develop Tooling and Fixtures to Optimize Production. The method described above shows the basic steps to be taken to produce the screwdriver. Obviously, if left to their own devices, the workstation operators would, over time, develop fixtures and tools to help them do the job. Unfortunately, this would involve considerable trial and error and would be very inefficient. For this reason, the methods engineer must devise new or specify existing fixtures and tooling before the start of production. For the major components, the types of fixtures and tools that a methods engineer could possibly propose are outlined below.

For the shank:

1. A device that would limit the feed of the bar stock to the proper length, eliminating time to measure the length before cutting.
2. For the grinder, an angle fixture to hold the shank so that the proper blade angle is ground.
3. A CNC chucking lathe to rough turn, thread, and finish turn the shank. Once a CNC program is produced and debugged, all shanks will be produced identically.

For the handle:

1. A device for measuring the handle length for the saw cut, similar to the shank device.
2. Templates for the outside shape of the handle, and a plunge depth stop device to regulate the depth of internal thread cut in the handle. (Since CNC controls for wood lathes are not as common as for metal parts and more expensive, the idea of utilizing a computer-controlled machine to form the handle would probably be rejected.)

For assembly:

1. A special vise-like device to hold multiple shanks vertical, with blades down.
2. An overhead or suspended glue applicator with an easy to use trigger system to dispense a predetermined amount of glue on each shank threaded portion.
3. A device to hold the handles in a proper attitude to engulf the shanks and to spin the handles down onto the shanks.
4. A speed drying device similar to a hair dryer to shorten the glue setting time.
5. A fixture to hold shipping cartons so that finished screwdrivers can be packaged rapidly.

As we can see from the above description, manufacturing something as simple as a screwdriver requires considerable creative thought on the part of the methods engineer to optimize production. The methods engineer will always conceive of a total system, theoretically considering all possible uses of labor saving devices and items that improve cycle time. Whether the devices are actually designed and built depends on the economic justification evaluations. Just as a drop forge method would probably not be used, an automated spinning device would probably be eliminated, because of the lot size being too small. For a lot size of 1,000, complex, relatively expensive fixtures would not pay for themselves over the length of the contract. If they do not pay for themselves, then they are not economical and would cost the company more than the possible savings to be generated.

To summarize, the methods engineer initially considers all sorts of fixtures, then hones down the actual plan for implementation of tools and fixtures to those that are essential to the job plus those that yield good savings over the duration of the order.

Develop a Useful Measurement System. The methods engineer first determines whether the product lends itself to either the productivity rate measurement system or the efficiency measurement system. In either case, the technique for measuring must be one familiar to the personnel involved, or provisions must be made to train personnel. For the screwdriver example, it is probable that the productivity rate measurement system will apply. It is now a matter of determining how many shanks, handles, and assemblies should be produced per hour. This leads to the fourth and final task to be performed by the methods engineer.

Estimate Time It Will Take to Produce the Required Amount of Product. The only accurate way to know how long it should take to produce a finished product would be to do a scientific time standard analysis. Usually, for products that will be made over a long period of time such a study would be commissioned. For a long production run, saving 1 or 2 minutes per screwdriver by studying the method for optimization would result in significant cost savings in the direct labor and work in process inventory areas. Saving over $100,000 per year for a product like a screwdriver is a reasonable expectation.

For a production run of only 1000 screwdrivers before the model is changed, scientific time standard analysis is probably not justified. (However, I will use the screwdriver shank as an example of how to develop scientific time standards later in this chapter.) Therefore, the methods engineer will have to approximate. By knowing the feeds and speeds of the machine-dominated portions of the work, the engineer can determine how long it should take to complete the work. For the assembly portion, which is manual work, the engineer will observe the work being performed on similar types of products, time these operations, and then estimate how much time it should take to assemble the screwdriver. If the factory is one in which timing of workers is a sensitive issue, the methods engineer may have to duplicate the assembly format in a laboratory and time a colleague doing the work. Once an estimated elapsed time is calculated for each segment of the work, the values are summed and the time to complete the screwdriver is determined.

Knowing the direct labor time value for the entire order, the methods engineer can now give management the information needed to adequately price the product. Material is known, labor content is known, and tooling required is known. Therefore, price can be determined by simply adding on the appropriate overhead percentage and required profit margin.

It should be realized that the time to produce the product determined from the observations and machine sequence times does not represent the most efficient method of producing the product. The procedure described above does not judge the validity of the production method, and it is reasonable to assume that there could be a better method. Only a scientific time standard analysis could determine whether the method is optimal and, if not, how it could be optimized.

It should also be kept in mind that quotes usually must be prepared prior to the making of a prototype. Therefore, the observation method cannot be used. In its place, the methods engineer will estimate the amount of non-machine time in the production cycle on the basis of similar experiences. This leads to further inaccuracies. Again, the only way to eliminate such inaccuracies would be to commission a scientific time standard analysis. This would be done at the quote stage for a large enough project, where for marketing purposes the true costs must be known beforehand.

SCIENTIFIC TIME STANDARD ANALYSIS

Many consider this aspect of methods engineering to be the most abstract and theoretical. Of all the work performed within the MP&WM unit, it requires the greatest attention to detail. The results usually justify the significant work effort required.

Scientific time standard analysis, the product of a desire to use logic and reason to improve the output of manufacturing activities, began in the earliest phases of the industrial revolution. Such names as Frederick W. Taylor, Frank B. Gilbreth, and Henry L. Gantt are associated with its development. It has come a long way since then with many contributors. The version discussed here will be the last version I used at General Electric Co. compiled for training purposes. I will present the managerial techniques used in organizing for doing the work of a scientific time standard analysis. I will also summarize the specific details of methods developed for the actual writing of such standards and discuss the underlying philosophy of such techniques. If projects are organized correctly, all the data necessary to produce a good standard becomes available. A standard can then be written using the precise techniques taught in typical work measurement courses and documented in the many good texts on work measurements. The methodology I will demonstrate is based on the work I did as Manager of Industrial Engineering Consulting for General Electric Company in the 1980s and also previous assignments as the Industrial Engineering Manager and Manager of Manufacturing Engineering for turbine, generator, and motors product producing divisions of General Electric Co. The techniques as described are from my personal work notes and the manual titled "Motion Time Survey," General Electric Co., circa 1977, and as previously noted a public domain set of data. During my tenure in GE corporate IE consulting, I had the privilege and responsibility for teaching time standards applications and developing supporting software. The following discussions are based on those experiences.

Five tasks must be carried out to produce a scientific time standard, and it is management's responsibility to see that these tasks are accomplished in a useful way. The standards written by time standard analysts reflect the data presented to them, and the quality of these data is determined by how well the five tasks are performed. These tasks comprise:

1. Facility information
2. Workstation information—description of the workplace
3. Tool lists information
4. Methods instructions—description of the work to be performed
5. Application data—code used to translate environment and work to be done into time elements

The first four types of information listed above are derived and supplied by others, not the time standard analyst (however, they may be modified by the analyst). The data in item 5 are the analyst's input, derived by using the specific motion standards tables, or their computer program derivative, that tell the analyst how much time it takes to accomplish increments of the work. It is difficult to understand this last task without the background of the first four tasks. Therefore, it's important we review the tasks in order.

Facility Information

Tasks 1, 2, and 3 describe the physical entity called the workstation. This becomes the description of the environment within which the work will be performed. The time standard developed will only be applicable to this specific location with its unique geometry. That is, there will be variations among workstations in how much time it will take to do the entire job. Although feeds and speeds may be the same at different workstations, the supporting services are not and cannot be. For example, a facility 50 feet away from the inspection station is bound to receive quicker inspection service than a similar workstation 500 feet away. This is vital information for the time standard analyst, who now knows that the unavoidable delays incurred in waiting for inspection will be ten times greater for the distant station than the nearer station. Therefore, the time to accomplish work will almost certainly vary at the two workstations.

In addition to providing the geometric location of the facility within the factory, the methods engineer will obtain a detailed description of the primary components of the facility itself. This will include the specification data for the machine or process from the advanced manufacturing engineer—that is, how the machine or process works, the range of capabilities, and so forth. In essence, this is the type of information placed on the bidding specification when the machine was purchased or upgraded.

Workstation Information

This information usually consists of the workplace layout (see Fig. 7-2) enhanced with descriptions as required. Such descriptions would include comparisons with similar workplace layouts or peculiarities of services. For example, a peculiarity that does not show up on a workplace layout but could affect the time standard would be an overhead crane causing machine vibrations when it passes, so that finish fine tolerance cuts must be interrupted. A time standard that did not take such an interruption into account would be in error and would result in unrealistic data for shop loading calculations. If the interruption caused a loss of 3 minutes every 15 minutes, failure to include it would lead to an error of 20%.

The example above may be improbable, but it points out the need for the methods engineer to supply complete information to the time standard analyst.

Tool Lists Information

The time standard analyst will be interested in the performance characteristics of the tools available to the operator, since the time to complete a task can vary considerably with the choice of tooling. Also, the life cycle of tools will vary with the material used.

In addition, the analyst will want to know the physical characteristics of the tools. Such properties as size and weight will affect the time it takes to change tooling. A bulky and heavy tool may have to be moved with a jib crane and sling. Therefore, it would require a considerably longer tool change time than a relatively light tool that could be held in one hand.

If the tools were not adequately described, or if the analyst assumed what the tools would be but was wrong, the resulting time standard could be in serious error. The time allowed to change tooling could be totally inadequate or significantly too long. In both cases the time standard would not represent the true optimum time to perform the task. Again, this points out the need for the good communications between the methods engineer and the time standard analyst.

Methods Instructions

This information is the chronological sequence the methods engineer designed to produce the part at the workstation. If several products are to be produced at the workstation, a method sheet for each product will be produced and given to the analyst. Obviously, the more products going across the workstation, the lengthier and more complex the time standard will become.

Application Data

At this point the time standard analyst applies the reference library of motion times to the various segments of the chronological sequence. Although this is the least well understood part of the production of scientific time standards, it is quite simple in concept. The first four tasks outlined above present the time standard analyst with a dossier on the workstation for which the standard is to be developed. The analyst envisions what happens at the workstation, what the parts are like, where the operator stands, what tools are available for use, and where the tools are stored, as well as the requirements for quality and the services available to the operator in making the parts. In summary, with the inputs of information from tasks 1 through 4, the analyst is in the position of an operator being given a job to do and having to determine how it should be done. The analyst then mentally walks through the process, determining what kinds of physical movements must be made to accomplish the task. These movements are the keys to the method, because the analyst must determine how much time it should take to do them and whether they can be further reduced if physical changes can be made at the workstation, or if their sequence is altered. The analyst is choreographing movements in accordance with defined requirements and assessing which movements are superfluous and should be discarded.

Once the choreography is complete, the analyst can use the application data. These data come in many forms and many levels of detail, but essentially they comprise a catalogue of times for making various body movements. The information can be as simple and specific as how many milliseconds it takes to rotate an arm, or as complex and general as how long it takes to disassemble an automobile automatic transmission. Incidentally, most automobile repair standard hours are combinations of many body movement times summed together to give a time to perform a generic task.

We can see from the above descriptions that task 5 adds the human element to the definition of the job to be done. We start with hardware information on the facility, the workstation, and the tools; progress through a plan to produce the part, the methods sheet; and finally add the human element, the application data. In summary, the time standard analyst is required to complete a very thorough study of the method sequence and the workplace environment, then apply times to the individual incremental steps necessary to do the specific task. These times are obtained from the feeds and speeds of the machine tool, from human motion timetables, and from geometric locations for delay times. The times are tabulated as fixed or variable. Machine times are considered variable, as they depend on the size of the part, while installing a tool into a tool holder would be fixed for each specific tool.

If a scientific time standard is being developed for only one type of product to be produced at the workstation, the analyst will construct a table in which the only variable is the size of the part, hence the machine cycle time. If the time standard is for various parts, a variable table for different operator requirements as well as tables for the machine cycle times will be designed. The finished time standard then comprises tables or computer programs that allow the planner to extract 100% optimum times for a wide variety of applications at the workstation.

The scientific time standard differs from the estimated time standard in that all the non-machining or non-processing time is specifically accounted for. Hence it is more accurate than the estimated time standard.

> **EXAMPLE:** If the process requires an inspection, the estimated time standard could not single out this requirement, but rather would tally it along with all other nonperformance times. The scientific time standard would have calculated the time to call for the inspector and the time to perform the inspection. All this would be incorporated in the portal-to-portal time to do the work at the workstation.

A scientific time standard also acts as a check on the method and quite often leads to an improved method. By looking at the motions involved and placement of the tools, the analyst has a unique opportunity to make improvements in the method sequences and workplace layouts.

Clearly, a scientific time standard has beneficial results. But are the results worth the effort? As I've alluded to before, this is a difficult question to answer. If a job is to be repeated over and over again and taking out minutes is important for overall productivity, or if the job is very complex and must be meticulously designed and competitively priced for bidding purposes, then the effort is probably justified. On the other hand, if the job is simple or nonrepeatable, then an estimated time standard is probably sufficient. It is the grey areas between these two extremes that require careful consideration. Often a good scientific time standard can be used for many other products and workstations as an enhanced or upgraded estimated time standard. One may reason that the new area is not very different from one for which a scientific time standard has been developed. In this case, the estimated time standard may be enhanced by basing it on a scientific time standard, thus upgrading the overall planning activity. In such cases the benefit of the scientific time standard somewhat transcends the original purpose for its development.

The Specific Example of How to Create a Scientific Time Standard

Now that we have an overall view of scientific time standards development lets review the basic techniques as if we had to develop a specific standard. Let's consider the screw drive example again and let's assume for what ever purpose has come about this job is so critical to the company's future that management is requesting a scientific time standard to be developed to be used for quotation purposes. We'll use the five steps to develop the scientific time standard. To further clarify the process we will limit this discussion to portions of the screw driver shank only. Keep in mind that the remaining components of the screw driver would undergo a very similar analysis. As too would the assembly operation.

Facility information

This is a new work station to be set up to manufacture the screw driver shank. It would contain a saw to cut the shank, a lathe, and a grinder. Since the lot size is small, the equipment would be primarily manual with the exception of the lathe. A small CNC lathe similar to a Bridgeport lathe would be employed. This would be the choice because it is simpler to purchase a small CNC bench type lathe than to find a small manual lathe. Also, most manual lathes of this size are usually highly precise machine tools used in tool room applications, thus overkill tolerance wise for this application.

Workstation Information—Description of the Workplace

This is the layout the analyst would use to measure distances and configure motion paths to apply times to. Note the layout is annotated with information that the analyst uses to define the motions involved in performing the work at the workstation. You can see that the layout has been modified to show typical distances from the operator's station to the various pieces of equipment that will be used during manufacturing. Also notice that the flow path of the process from raw material to finished screw drive shank is super imposed on the layout. With this layout, as shown, the analyst can visualize how the work will be performed.

Tool Lists Information

This is a complete listing of the tools available to the operator for normal production at the work-station. From Fig. 7-5, going clockwise from the top and including primary and auxiliary equip necessary to do the job, the first cut at preparing a list is shown below. Keep in mind that a finalized tool list cannot be developed until the analyst thoroughly reviews the method to be employed in making the product.

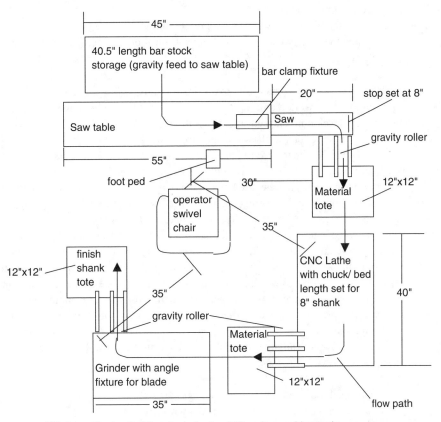

All totes on hydraulic lifters to make level 6" under machine tools

Figure 7-5. Workstation: Screw Driver Shank Machining

a. Wood rod to organize and control the bar stock on the saw table.
b. Vise grip to tighten the bar clamp fixture after insertion to the saw and grinder.
c. Steel rule to check length of bar stock coming off the saw.
d. Wire brush to clean cuttings off the saw and lathe bed.
e. Chuck key to seat the cut to length bar stock in the lathe.
f. Micrometer to check diameter of shank after machining.
g. Thread gauge to check thread form after machining on the lathe.
h. Portable vacuum (Dust Buster type) to clean grinder.
i. Socket set to make machine adjustments.
k. Needle nose pliers for general maintenance.
l. Screwdriver set for general maintenance.

Notice that the tool list is just that. It does not include the machine tools themselves or any permanent jigs or fixtures. It is only the devices that are in effect extension of the operator's hands to give the proper mechanical advantage and safety protection.

Methods Instructions—Description of the Work to Be Performed

The starting point is the methods sheet the analyst receives from the methods engineer. For our example this would be the broad based methods listed above in the Methods Analysis discussion and is repeated here.

1. Cut steel rod, 3/8 in. diameter, to length on a saw.
2. Rough turn the entire length and thread opposite to the blade end on a lathe; class 2, UNC coarse thread 8 threads/in.
3. Form the blade on a grinder to a preset template.
4. Grind the entire shank to a commercial finish or, if specified, to the finish required on the drawing.

This certainly is adequate for giving instructions to the operator, especially an experienced operator, to do the work required. But it is insufficient for applying the principles of motion economy, hence developing elapsed time to do the work. What I'm going to do now is to expand the methods steps so we can see how the operator really moves about the workstation in doing the job. To do this I would either have to observe the operator and note how he or she moves as the work is being done. Or mentally conceive how the operator should move to do the job. In this case since the work has never been done, the latter is the only choice I have. I will actually be doing the choreography of moves as I envision then, with the constraints being the geography of the work station layout itself (Fig. 7-5). The more flowing and graceful the moves are, the most optimum it is from the viewpoint of energy and time minimization. You probably never thought of principles of motion economy in relationship to ballet dancing, but it truly is. So let's create our dance movements by expanding the methods document. Before we proceed, I should mention that when observing an existing work station and watching the operator do the job, the analyst is doing the same choreography but from a critique viewpoint. She is watching for wasted motions that prolong the job and tire the operator faster. Hence will recommend changes in the method as part of the construction of the time standard.

Here's a typical example of how the analyst would expand the method document. I will only do step 1 for brevity purposes.

1. Cut steel rod, 3/8 in. diameter, to length on a saw.
   a. Sitting in chair, depress foot pedal with left foot.
   b. Hold petal depressed to release one 40.5 in. length 3/8 in. dia. steel rod.
   c. Pick up wood rod from rack on table with right hand. Extend right arm holding wood rod and push steel rod to front of table to enable rod to be grasped by left hand.

    d. Put wood rod back in rack on table with right hand.

    e. Pick up steel rod in left hand, place in saw bar clamp fixture. Simultaneously pickup preset vise grip from holder on saw table with right hand.

    f. Hold in place with left hand, secure bar clamp fixture with right hand using vise grip. Insure bar stock is inserted all the way to the preset stop.

    g. Replace vise grip to holder on saw table with right hand.

    h. Ensure saw guard is in place. Power up saw with left hand. Cut first piece.

    i. Feed bar to preset stop with left hand. Cut second piece.

    j. Continue step g. until all of the bar stock is utilized (four cuts).

    k. Turn off saw with left hand.

    l. Rotate chair $45°$, clockwise. Push five pieces to gravity roller with right hand. Observe pieces being deposited in the material tote.

Note that I've gone into considerable detail to describe what the operator was doing, particularly with hands and feet and specifying which particular hand or foot. I've also specified that the operator is sitting and will rotate the chair as part of the movement plan. With this sequenced set of steps I've described how we would like the operator to go about the business of cutting bar stock. I am specifying the coordinated hand and body movements where appropriate, and independent movements where that is the preferred way to proceed. Steps a. through c. are typical coordinated movements. Step k. is a typical example of an independent movement.

With this list of movement steps we can now use the relationships broadly defined by the principles of motion economy and specifically times found in the applications tables to put time values for each step a. through l. Putting the times together and doing the appropriate arithmetic is the next and final step in preparing the scientific time standard for the job.

Application Data—Code Used to Translate Environment and Work To Be Done Into Time Elements

The methods document now spells out the process in enough detail to assign times for each of the steps a. through l. In order to do that expeditiously we breakdown the movements into five categories:

    Transports
    Gets
    Places
    Precision
    Miscellaneous

We then record each motion of the method on a specialized flow chart in accordance with the five movements. Then we do the "look up" on the specific tables for each of the movements to find assigned times. We do this sometimes with modifiers related to the constraints of the method and record the incremental time to achieve the movement. First I will define each of these five movements. Then I'll apply the technique to the shank example.

Transports

Transports are the reach or travel of a body member or a device, either "empty" or "full," meaning either holding a part, tool, etc. used to do or claim a piece to do work with or to. For example; reaching for a screwdriver is an empty transport, while moving that screwdriver to the work piece to do assembly work with is a full transport. Positioning oneself to pick up a piece of raw material is another example of an empty transport. And moving the work piece to a machine is an example of a full transport.

Transports can be simple or complex. They may involve hand and wrist only and all be done in one motion. A more complex transport may involve a combination of moves such as reaching up over a wall type barrier for a clearance then lowering to a final destination. This would be a two-motion transport. Transports can also involve arms, back, hips, legs, as well as mechanical assists such as cranes.

Gets

A get is the action of gaining control of an object. A transport simply moves the device (e.g., hand or crane, etc.) to position for gaining control. Examples of gets are: grasping a part with a hand and fingers or attaching a crane hook. There are multiple ways for gaining control of the part or tool and fortunately most are defined in the look up tables. And like transports they can be simple or complex.

Places

Places are the opposites of gets. While a get grips a piece or tool, a place releases it at the place the operator needs to do so to perform work on it. Once a place action is completed it is said that control of the object is lost. This is true in the sense that the operator no longer has direct control. But does retain indirect control. For example, if the place relinquishes control by putting the part into the chuck of a lathe, the operator has tremendous power over the piece. She can start machining, do measurements, or simply leave it there and go on break. But nothing is by chance. The piece is not physically being directly controlled by the operator, but what happens to the object is virtually 100% at the discretion of the operator.

Precision

Precision is the fine tuning of a place operation. It is quite literally the physical moving of a part either linearly or rotationally usually in a bracketing set of micro-motions to put the object precisely where we want it to be. The amount of time it takes to do a precision depends on the degree of tolerance allowed for the final location. A placement of a part in a machine tool to be worked on is going to be more precise, hence take a longer time than putting the part in a location eventually be put into an assembly. We say the smaller the precision index the longer this motion will take to complete. Imagine placing round pegs into almost size and size fitted holes as compared to a child's toy of placing pegs into generously large tolerance holes. This will give you a feel for the precision index, much smaller for the size and size fit than the toy. Depending on the tolerance and whether one and or two hands are required, times for this motion have been tabulated in the lookup tables.

Miscellaneous

You might say this is everything else and be very close to the truth. This category starts off with just that. It is everything that doesn't fit the definitions of the first four items. However, since it can be so broad a category it is not a true motion from the standpoint that it is universal; e.g., categories we can call up out of a lookup table and use for any analysis without first validating it for the specific application. This category is used when the previous four can not adequately prescribe a time for a motion. Here are some examples:

a. Getup or sit down in a chair

There are many types of chairs in many different location that could have many different degrees of difficulty in exiting and entering. Image getting into a driver's seat chair in an automobile classified as a sedan vs. a sports car. The degree of difficult increases, hence time to enter or exit. A chair on a narrow platform with little freeboard around it will require more caution, than perhaps a desk chair.

b. *Inspect or select a part to get, then transport, place and precision*

There are hosts of specific uniqueness we can conceive here. What kind of inspection? Visual?, A measurement? The degree of difficulty in making the selection. Like at the fruit stand, picking out the desired apple. The number of apples present complicate the selection. The more of them the more difficult to make the "ideal" choice.

c. *Typical machine and assembly miscellaneous categories*

Here are some motions that are more specific than the universal four: get, transport, place, precision. They are relatively common, but still not universal:

- Operate a foot pedal—requires knowledge of the location of the pedal in relationship to where the operator is suppose to be.
- Cut a wire—depends on wire size
- Cut with scissors—depends on length of cut
- Hammer—depends on what is being struck, nail, part, etc. Also the degree of force necessary.
- Twist—is it with fingers, hand, or palm?

All of these miscellaneous motions require some individual observations to end up into tabulated tables. Many companies have done that or make use of universal services that have categorized these miscellaneous motions. The above have been categorized by General Electric Co. many decades ago and form part of the lookup tables in Motion Time Survey. As I mentioned previously the data are available to the public.

To summarize where we are now. We know that all motions can be classified in accordance with the five motion categories. The first four, transport, get, place, precision are considered universal in that they are applicable for virtually any job. The fifth, miscellaneous is more specific, usually for specific classifications of jobs and range from very specific to almost universal. The values for the miscellaneous are often developed by the user's company for specific needs. However, there are large databases available from many sources such as technical libraries where university studies sponsored by companies and trade groups that have since been made public domain. Applicable miscellaneous motion times are not difficult to obtain through literature search. So much work has been done over the decades that it would be unusual to not be able to find a catalogued motion suitable for a specific need.

We're almost ready to get out the lookup table and calculate a time standard for the cut step of the screwdriver shank. But before we do that, I need to bring up the topic of experience leveling. This is the so-called learning curve. We all know that within our coordinative abilities we can perform a repetitive task more efficiently if we continually practice the motion. This applies to virtually any physical and mental activity. We can do multiplications almost instinctively because we practiced intensively as children. For physical motions we can do the same thing, like practicing a golf swing makes us more proficient. In manufacturing, performing work motions becomes more effective both in time and accuracy through repetition. We call this experience leveling.

Experience leveling is the adjustment of the time allowed to perform a task based on the number of times the job is to be done. If a job is to be done only once it will be judged as if the operator has no practiced skill to do the job and the time will be longer. If the job is to be done thousands of times, it is expected that the job will take less time per individual cycle. The experience leveling section of the lookup tables take this into account in establishing the time allowed.

Okay, lets get to the application tables. Applications tables, also known as lookup tables, as I said are databases where after describing the motions, the engineer uses the data to assign a time to the specific motion subset. Figure 7-6 is an applications table. This particular table is based on General Electric Company's Motion Time Survey (MTS). Which is one of several tables, all of which are based on the principles of motion economy. MTS, which I'm intimately familiar with based on my previous management experience being its care taker, is historical data based on literally

Time Standards Applications Tables
(based on General Electric Motion Time Survey data)

Transports

four categories:
1. *Hand*
2. *Back and Hip*
3. *Turn*
4. *Walk*

Distances				
over	0 inch	6 inches	12 inches	18 inches
includes	6 inches	12 inches	18 inches	24 inches

Motions (times in minutes/100)				
1.Hand	0.34	0.42	0.50	0.58
2.Back and Hip	0.75	0.95	1.15	1.35

3.Turn	45 degrees	90 degrees	135 degrees	180 degrees
feet fixed	0.60	0.75	na	na
feet moving	0.79	0.86	0.94	1.01

4.Walk	
First Step or Pace	1.13
Each Additional Step or Pace	0.83

Gets

Class	A	B	C	D	E	F
Time (minutes/100)	0.09	0.18	0.54	0.90	1.08	1.58

Class A   Contact. Control is obtained by making contact with the part. Example: laying hand or finger over a part to gain control.

Class B   Grip. Control is obtained by gripping. Usually a singular piece independent of other pieces.

Class C   Sliding Grip. Control of a part is obtained by a sliding grip after it has been drawn away from the surface on which it lies or where it may disturb. Example: stacked parts on a surface, being lifted one at a time.

Class D   Down, Grip and Lift. Control of a part is obtained with a down, grip, and lift motion by moving the hand into a group or mass of parts gripping one.

Class E   Down, Grip, Lift and Separate. Same as Class D with the addition of movement required to separate parts that tend to slightly tangle or adhere.

Class F   Down, Grip, Lift, and Untangle or Shake. Same as Class D with the addition of movement required to separate parts that tend to definitely tangle or adhere.

Places

Class	1	2	3	4	5	6
Time (minutes/100)	0.07	0.14	0.44	0.63	0.74	0.93

Class 1   Drop Release. Control is relinquished by releasing pressure of the hand or when opening of the fingers is partially overlaped by the Transport.

Class 2   Full Release. Control is relinquished by opening the fingers after the transport has ended.

Class 3   Down and Release. A prepositioned part can be located in a predetermined position with a single movement following the transportand preceding the release by the fingers.

Class 4   Preposition, Down and Release. Same as Class 3 except that the part must be prepositioned before the single movement and release by the fingers can be made.

Class 5   Straighten, Down and Release. The act of placing a prepositioned part involves two moves before the release by the fingers.

Class 6   Preposition, Straighten, Down and Release. Same as Class 5 with the addition of a preposition before the two moves and the release by the fingers.

Precision

A modifyer of Places. Additional time to add to the place value depending on the degree of preciseness of the insertion of a part to a specific location. The tighter the tolerance of the part to the allowable area to be placed in, the longer the associated time to do so. Precision is most often used with one part being placed within another in assembly operations. Precision is only used if the dimension tolerances are less than 1/2 inch.

Precision								
under, inches	0.501	1/4	1/16	1/32	1/64	0.008	0.004	0.002
incudes, inches	1/4	1/8	1/32	1/64	0.008	0.004	0.002	0.000
One Hand								
Time minutes/100	0.16	0.32	0.48	0.64	0.95	1.11	1.27	1.43
Two Hands								
Time minutes/100	0.23	0.47	0.70	0.93	1.16	1.40	1.63	2.10

Figure 7-6. Scientific Time Standards Applications Tables
(From General Electric Co. Motion Time Survey.)

There are many unique motions beyond what can be categorized in Transports, Gets, Places, and Precisions. Perhaps making up 10% to 20% of motions involved in establishing a time standard. To establish time for a unique time standard there are three choices.

1). Observe. Do stop watch analysis. Minimum of 10 repetitions timed from start to finish. All repetitions in accordance with desired choreographed motions. Use average as the standard. This is the most accurate method.

2). Estimate a time for the unique motion based on experience of observing work patterns in similar situations. This is the least accurate method.

3). Research literature for exact match or close enough to motions where stop watch time standard have been determined. Depending on the closeness of the match to the required unique motion, this method can be as accurate as 1).

Times for common miscellaneous motions:

Motion	Time (minutes/100)
Cut wire with pliers	
light wire	0.6
heavy wire	0.95
Cut with scissors	
first snip	0.44
each additional snip	0.3
Get up or sit down in a chair	3.33
Hammer	
first blow	0.68
deliberate hammering	0.68
rapid hammering	0.3
Inspect or select- initial action	0.3
Operate foot pedal	
standing-lift foot from floor step on pedal	1.1
move foot down and up- foot on spring feed back pedal	0.9
sitting- step down on pedal	0.45
Palm and unpalm	
palm	0.28
unpalm	0.42
Twist with fingers	
first twist	0.21
additional twist	0.49
tightening twist	0.72
Twist with hand	
first twist	0.44
additional twist	0.88
tightening twist	1.23

Note: miscellaneous times include get instructions; allowance for interferences, fatigue and other unavoidable delays. Example: get up from chair includes securing work being done prior to action, removing interferences to and sliding chair back, standing up, securing chair in safe position.

Figure 7-6 (Continued)

thousands of stop watch time standards conducted for choreographing most efficient motions for over fifty years. It is very rich in validated experience and meticulously traces the motions measured and timed to the principles of motion economy. I have had the opportunity to compare MTS with other similar standards and can state that it is very comparable in the derived times. That is to say virtually all analysts of motion times adhering to the principles of motion economy will eventually end up with the same time values. We will use these applications tables along with experience leveling data to set the scientific time standard for the cut step of the screw driver shank example.

The analyst lists the detailed steps of the operation on a pad, in the example steps a. through l. She does the lookups from the applications tables, putting that time appropriately along side the specific step. You will note on the lookup tables that the time is listed: min./100. This means the number represents a time for 100 identical actions. This is just a convenience so the numbers are not microscopically small. Then she enters the experience leveling table (see Fig. 7.7) to see what correction factor if any needs to be applied to the standard. Finally, multiplying both values, the analyst comes up with a standard time for that particular motion element.

I should also point out that the times shown on the application tables have factors already built into them to account for all unavoidable delays. Sort of pro-rating such things as rest/break times, waiting for delivery of materials or tools, getting instructions from the supervisor, etc. This means that the times are inflated to contain the percentage of time allocated to the motion for these uncatalogued but true deflators to effectiveness. We know the typical work shift is 8 hours. We also know that due to intricacies of coordinated movements of people, materials, tools, inspections,

(Based on General Electric Motion Time Survey data.)

Quantity Produced	Time Adjustment Percentage	
	Get & Place	Precision
500	39	326
750	31	301
1,000	28	281
2,000	24	229
3,000	21	198
4,000	18	175
5,000	17	159
7,500	14	130
10,000	12	110
20,000	7	69
30,000	5	48
40,000	4	35
50,000	3	26
75,000	1	10
100,000	0	0
200,000	-2	-20
500	-4	-41
750,000	-5	-49
1,000,000	-6	-53
2,000,000	-7	-63

Linear interpolation allowed.

Figure 7-7. Time Standards Applications Tables Experience Leveling

etc., there are bound to be "holds" when productive work simply cannot be performed. These times have to be accounted for in establishing a standard. For example; referring to the applications tables Figure 7-6 under precision we see that accurately placing a part in its proper slot with a tolerance range of between 1/32″ and 1/64″ has a one-hand standard of 0.64 minutes/100 or 38.4 seconds/100. A 10% factor is added for unavoidable delays. This means for putting one pin in one hole the actual work time is: $0.384 \times 0.9 = 0.346$ seconds with 0.038 seconds added for the prorated unavoidable delay. However if an unusual situation exists, such as having to wait for an inspection before putting the pin into the hole, then that calculated delay time, based on observations, would also be inputted into the standard. The prorated time is a generic amount always in the standard to assure we are not dealing with ideal times for establishing the standard. Remember time standards are set up to determine how much work can be reliably done over the course of a shift, not setting individual one at a time speed records. So it is appropriate to have unavoidable delays built into the standard.

This work as you can surmise is meticulous and very detailed, therefore prone to clerical errors. We wish to minimize these errors. One way of doing so is to make sure all analysts do the work in the same way. Thus making it simpler to check work for mistakes. One of the ways that has been devised to do this is construct an applications form. A typical form is shown in Fig. 7-8. And with the cut step of the screw driver shank example data filled in Fig. 7-9.

Note that the format of the Time Standard Data Entry Form allows the analyst to list the action steps, or movements under both hands and all other body members in narrative form, then has columns for listing the values for motion categories. Note too that gets and places occupy the same column. This is done to conserve space and can be done since gets and places are the opposites of each other. The methodology of using this form is to first list the motions. Then fill in the information concerning quantity for which the standard will be issued for. With this information enter Experience leveling table, Fig. 7-7, and put the two numbers on the form. Henceforth use the get and place value or the precision value for the factor column. Whenever a motion is a combination of gets, places, or precision; my rule of thumb is to use the largest experience leveling factor. Which means using the precision values. This will give a higher total time, therefore a bit more conservative. Also

Scientific Time Standards Analysis Form

Job. No.:_____   Date:____
Job Description:_____ page no.___of___

Quantity Manufactured: _____   Experience Level Factor (percent) _____
   Analyst:_____

No.	Right Hand	Left hand	Other Body Members	transports	Gets & Places	Precision	Misc.	Factor	Total
						Minutes/100			
	Elements								
1									
2									
3									
4									
5									
6									
7									
8									
9									
10									
11									
12									
13									
14									
15									
16									
17									
18									
19									
20									

Notes:

Figure 7-8. Time Standard Data Entry Form

when a transport is associated with the motions affected by experience leveling, use that same factor with the transport.

Since time standard have such a large percentage of people movements vs. machine movements it is always prudent to err on the side of more time allowed than less. People are never 100%

Scientific Time Standards Analysis Form

Job. No _____1_____

Job Description: _____Cute steel rod 3/8 inch diameter, to length on sawpage no. 1 of 1_____

Date: _4/18/07_

Quantity Manufactured: _____1000_____   Experience Level Factor (percent _28% gets & places; 281 % precision_

Analyst: _A.B.Clark_

No.	Elements Right Hand	Left hand	Other Body Members	transports	Gets & Places	Precision	Misc.	Factor	Total
					Minutes/100				
1			Sitting, depress pedal with left foot.				0.45	1.00	0.45
2			Hold pedal, release one rod (foot)				0.90	1.00	0.90
3	Pick up wood stick from table rack, push rod to front of table	Grasp rod at front of table		0.50	0.18			1.28	0.87
					0.18			1.28	0.23
4	Put wood stick back in table rack			0.50	0.14			1.28	0.82
5	Pick up pre-set vise grip.	Move rod to saw clamp fixture.		0.58	0.81	0.32		2.81	4.81
		Precision place rod in saw		0.50	0.18			1.28	0.87
6	Tighten rod, by twisting vise gri Five total turns.	Hold rod in place					4.31	1.00	4.31
7	Replace pre-set vise grip to holder.			0.50	0.44			1.28	1.20
8		Turn on saw. Cut first piece	Insure saw guard is in place. eyes		0.09		0.01	1.28	0.13
							0.30	1.00	0.30
9		Feed rod to pre-set on saw. Cut second piece.		0.42	0.14			1.28	0.72
10		Repeat step 9. Three cuts.		1.26	0.42			1.28	2.15
11		Turn off saw.			0.09			1.28	0.12
12	Push five pieces to gravity roller.		Rotate chair 45 degrees, feet	0.79	0.70			1.28	1.91
			Observe pieces to tote. eyes				0.30	1.00	0.30
13									
14									
15									
16									
17									
18									
19									
20							Total:		20.08

Notes: ops. 1&2 foot pedal from table no factor adjust.   op.3 18" trans., class B get   op.4 18" trans. class 2 place

op.5 24" trans.rod, 18"trans.vise grip, Class B get-rod (0.18)+class4 place-rod(0.63) =0.81, precision(1/4"-1/8"), class B get-vise grip use precision over gets/places exp. factor for rod

op.6 hand twist from misc. tables(first =0.44, 3 additionals=3x 0.88,tightening twist =1.23).No additional time for right hand-no work done

op.7 18" trans., class 3 place   op.8 turn on saw is class A get., mach. time for saw cut =0.01, visual inspect saw guard is misc. inspect = 0.3.

op.9 12" trans., class 2 place   op.10 same as op.8 x3   op.11 turn off saw is class A get

op.12 turn 45 degrees feet moving-trans., visual inspect, class 2 place x 5

Figure 7-9.   Time Standard Data Entry Form with Data Entered (Data used based on General Electric Co. Motion Time Survey.)

repeatable in any activity. So by being more conservative we can have a higher degree of confidence that the standard would not be overly optimistic. The last thing we want to do is miss the promised schedule completion dates because of an overly optimistic standard. When this happens people lose faith in the entire time standards system.

Notice the form also has space for notes. This space is to be used by the analyst to explain what choices were made for data entry in the minutes/100 columns. I believe a note should be entered for each step. It is a fact that a few days later after doing work it will very difficult to recall why you did what in using the applications tables. This is especially true if the analyst is involve in many different job at the same time. The notes become the analyst's records of reasoning, his log book.

Figure 7-9 is the data entry form filled in for cutting the steel rod to length on the saw. This yields 5 shanks at the 8-in. length we desire to be further processed at the work station into finished screw driver blades.

During the creation of the standard I selected transport times. You will note on the applications tables that transport times are variable in accordance with distance. To select the appropriate one I needed to know these distances. They are derived from the layout sheet (Fig. 7-5). It is interesting to note some versions of data entry forms have the layout right on the form. This is okay, but limiting. I prefer the layout to be a separate drawing where more detail can be included. Also, by having the layout schematic on the data entry form, we inevitably create a cluttered, crowded document.

I also needed to determine the types of gets, places and precisions that were involved. This comes from experience. But I found sometimes even with people who have done considerable time standards analyst work one person's experience bias is not the same as the other. To try to get more consistency we introduced two decision trees. One for gets and one for places. These are shown in Figs. 7-10 and 7-11. Let's briefly review them with respect to the filled in data for cutting 3/8 in.-diameter rods to length.

Starting with the gets decision tree, let's look at operation 3. In the notes I have 18 in. transport and a Class B get. By definition a class B get is classified as a grip. It's fairly obvious that we are gripping the rod so this is the correct selection. Now, for illustrative purposes, lets suppose you're not quite sure what type of get we have. Then use the decision tree found in Fig. 7-10. It works by answering questions posed. For this example, starting from the top:

Question	Answer
Control required?	Yes
transport preceding get require slow down of hand before contact?	No
Do fingers close on object?	Yes
Is part or portion of part moved before control is obtained?	No
	USE "B" GET

You can use this logic branching network to arrive at the correct answer anytime you are unsure as to the type of get to use. Some time answering the questions may become some what subjective. For example, determining what percent of parts tangle. If there is no obvious choice the best thing to do would be to err on the side of the get that requires the most time to accomplish. In all cases common sense should apply. The places are slightly more complicated lets take a look.

For places we're doing more subtle things. With gets it's a matter of gaining control and linear or rotational measurement of how and where is not as important. With places we are relinquishing control but we need to do it in a way that control is being transferred for addition work or be readily available for additional actions some time in the future. If we look at operation 4 of the example (Fig. 7-9), we see we have "put stick back in table rack." This is not a place that has a significant location control factor. Going to the applications tables I chose a class 2 place, a full release whereby the fingers are opened. In this case at the rack where the stick is stored. Using the places decision tree we should find this to be a sufficient choice. Let's go through the same exercise as we did with gets but now using the companion logic chain for places. Gong to Fig. 7-11 we construct the following table.

Question	Answer
Must part be places in a definite spot?	Yes
*Is prepositioning required?	No
Number of moves other than release?	0
	USE "2"PLACE

*Note on the decision tree that there are two questions posed to help determine whether the answer is yes or no. In the example it makes no difference if the part is turned end to end for final placement.

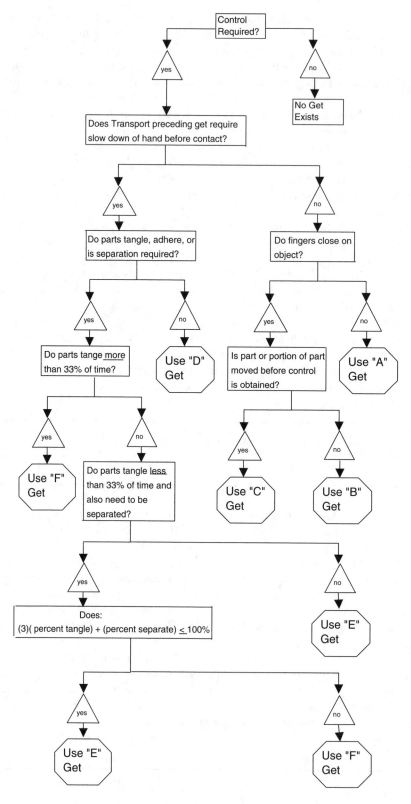

Figure 7-10. Gets Decision Tree (Based on General Electric Co. Motion Time Survey.)

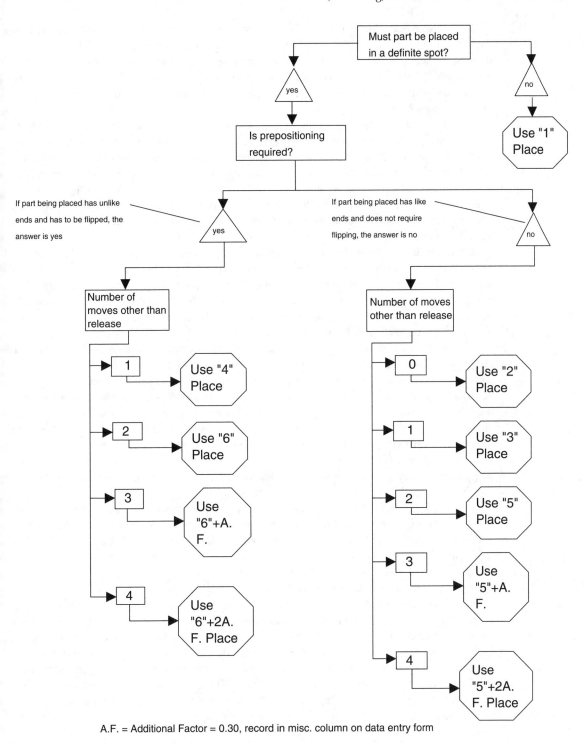

A.F. = Additional Factor = 0.30, record in misc. column on data entry form

Figure 7-11. Places Decision Tree (Based on General Electric Co. Motion Time Survey.)

Let's look at operation 7. This requires the operator to replace the vise grip in its holder. Here I've chosen to use a class 3 place. And I've done so because putting the tool back into the designated holder requires some amount of positioning moves to line it up before relinquishing control. So I've arbitrarily selected one. Is that Absolutely correct? I can't say so but experience tells me it's adequate. Again the human factor comes into play. We can never be absolute but we can be more right than wrong.

Another constraint using places must be dealt with. As I've mentioned above, places are more complex that gets because of how we chose to relinquish control. Control being the operative word. Notice we have something called an additional factor. I call it the "frustration factor." After making several attempts to place an object where we want it to be, I arbitrarily set it at two. There will be a tendency to slow down to make sure we get the part where it should be. We find this happening with class 5 and 6 places, which are the most complex places. Here if we can't place it correctly after two moves we add an additional time to the overall move. This is the additional factor. Fortunately most places are done using class 1 through 4 descriptions.

Places sometimes become precisions. There is a grey zone separating the two. How do you tell the difference between the two? Precisions deal with tolerances between the part and where it is put. Places do not. In the example shown in Fig. 7-9, we have one precision. The seating of the rod in the saw to assure it is correct to cut the rod to proper length. In this case the precision falls between less than ¼ in. to greater than $1/8$ in. This is a seating tolerance that makes sure we can get five pieces of suitable length to machine the screwdriver shanks.

The last item we should review is the miscellaneous items on the data entry form. As I've said, miscellaneous is a catch all that doesn't fit within the first four motion categories. They are unique, sometimes complex motions. But they happen and they make up ten to twenty percent of motions commonly found in manufacturing. Fortunately, most factories are not unique in the processes and procedures they use. Therefore a large collection of these motions have been studied over the years and are available through library sources for use with scientific time standards. The caution here to make sure what you pull out of a library resource is sufficiently compatible with what you do in your factory. I've listed some of the very common ones I've come across in my experiences and where necessary made sure they were included in the General Electric Co, Motion Time Survey. This is only a small listing for instructional purposes.

The way we use miscellaneous data is to literally take the numbers and place them into the data form. We tend to not use experience factors with them, except when we want to give the standard a trifle more elapsed time for conservative purposes. You can see in Fig. 7-9 that I only used an experience leveling factor for the saw cut time because I did not have any specific value for the machine cut time. Every other miscellaneous has no experience leveling factors. I think you will agree there is no learning curve for swiveling a chair $45°$ and other such similar items.

We've now concluded the explanation and demonstration setting scientific time standards. It is certainly more detailed than estimated standards. It is orders of magnitudes more accurate than estimated standards, but it requires more work. The question becomes is it worth the addition engineering time. I believe in today's very competitive world it probably is but we need to apply better methods of doing so in order to reduce the work in creating them. Doing manual scientific time standards work is fast becoming politically incorrect. This is mainly due to uneducated managers who don't see the true benefits. They can't see the forest for the trees. They think tactically rather than strategically. Remember , there are only two truly important things in running a competitive factory, know as "*The Basic Tenets of Manufacturing*":

1. KNOW HOW TO MAKE YOUR PRODUCT
2. KNOW HOW LONG IT SHOULD TAKE TO DO SO

Fortunately there has been a slow but growing rebirth in using scientific time standards through the use of the applying the massive data crunching capability of computers. This makes the engineering labor content of developing scientific time standards more palatable. In the next section we will explore this development.

Use of Computers in Creating Scientific Time Standards

As we've seen, producing time standards is a process involving meticulous attention to details. It is the accumulation of many pieces of data merged with the energy of the human body movements, expressed in time, to give repeatable calculations of how long it should take to accomplish a task. Computers are most helpful when applied to the task of keeping track of large quantities of data and being able to gather, or classify them, in accordance with a certain set of logic instructions. This definitely fits a need in scientific time standards development. Computer usage can be a great assistance to the time standards development. It is a great assistance to the time standards engineer to have a tool available to assist in this task, especially the drudgery of performing multiple simple calculations.

As with most engineering applications, there is definitely a role for computers in assisting in the development of scientific time standards. I and my colleagues recognized this in the late 1970s and early 1980s. We developed a program within General Electric called Computer Aided Time Standards (CASA) which was never commercialized, but through licensing lives on to this day and has found its way into many commercial offerings still available for purchase.

The CASA process was a way to eliminate the drudgery of doing lookups. It still required the analyst to define the work sequence in terms of transports, gets, place, precisions, and miscellaneous motions. But instead of entering a code and explanatory note onto a work sheet then doing a time lookup, the code was entered directly into the computer database and the computer did the look up internally. The big break through was that we were able to construct logic programs that mimicked the way the analyst analyzed the job and had the program do the mechanics, following the same rules the analyst used. This meant the analyst spent the majority of time being creative on how to do the job most efficiently and the machine (the computer) did the clerical and rules following work. The analyst time savings was in the order of 500%, with similar accuracy improvements.

The manual way to creating scientific time standards, as I've just demonstrated, is to analyze the method then determine the motions the operator has to go through to conform with the method. The analyst must decide whether the method can be modified to require less human motion or motions requiring less energy to accomplish. This is done by looking up time and efficiency contents of the motions the method requires, doing summations, and finally examining alternative scenarios to see if time could be minimized. The ultimate goal is to use the least amount of time, thus energy, to accomplish the prescribed method for both the operator and the machine. The computer allows many scenarios to be tried relatively fast, then pick the best alternative.

Let's look and see where computers can be of assistance in the above process.

Analyze the Method.   We can certainly expect that the method transcribed to the time standard analyst will be in a database format available for electronic interaction between functions. At the least, this will mean the analyst should always have access to the most current version of the method, reflecting any workstation or product changes. This communication enhancement alone ought to prevent wasted effort on the part of the analyst.

More important, we can use the time standard program to evaluate the method broadly to see if it is feasible, particularly in terms of what is or is not physically possible. We can code the steps per the five motions categories then make the data entries. For example, the program could easily see if we are asking an operator to do something that is beyond a person's physical capabilities to do. Usually this would involve lifting and placing of parts and calibration of settings beyond certain limitations—for example, lifting more than 20 lb and adjusting the position of a part to better than within ¼ in. without precise measurement tools. These may seem to be trivial examples. Unfortunately, they do arise quite often in the real world of manufacturing through simple oversight and end up being the cause of emotional labor-management grievances.

Determine Motions to Comply with the Method.   This is historically done by looking up tables of motions and movements and then compiling data on incremental motions with their associated times to make up the whole representation of the method. It involves meticulously looking up data, entering on tabulation sheets, and finally summing the data to get a time to accomplish for the total method. Obviously, this is a detailed library-research-like activity. The more complex the method, the more complex the task of defining the time elements will be. It is also quite apparent that the possibility for error is very high. In fact, recognizing this, most MP&WM units divide time standard work into analyst and checker activities. The analyst determines the motions to be employed and does the initial looking up to get the time values and create the summations. A category of analyst called a checker then reviews the analyst's work to minimize errors before the time standard is issued.

With available computer programs, the analyst enters the motions into the program and the program then does the lookups to get the incremental times involved in performing the method step (as briefly explained with the CASA development). This has been made possible by the input of the manual data books of incremental motions and times into a computer database. Also, in most cases, programmers working with expert time standard analysts have combined many incremental motions into commonly occurring macro motions, or sets of motions. This makes the lookups by the computer program even simpler. This is one of "tricks" we used to make CASA feasible.

Evaluate Accuracy and Alternative Methods.   Using these types of programs makes the task of developing time standards from methods inputs easier. In fact, the need for a checker function has virtually disappeared. Most advanced time standards programs also have built into them what we call impossibility factor eliminations warnings. The programs actively search for combinations of motions or sequences of motions that are either impossible or out of sequence. For example, if by error the analyst specified getting a tool and simultaneously fitting the work piece into the machine chuck, the program would signal error and perhaps even suggest a proper sequence. Similarly, a set of alternative combinations of motions can be suggested by the program just as word processing programs can check for grammar and spelling. Also, advanced programs of this genre are often asked to find the least-time combinations to do a certain task. Obviously, human evaluation is still required, but the clerical portion of providing alternative suggestions is simplified.

Use of these types of programs has caused one additional thing to occur. They have lowered the barrier to use of scientific time standards. No longer are scientific time standards relegated to products with high costs or large production runs. It is now feasible to use scientific time standards in limited-volume job shops as well as for marketing quotation purposes. This is all due to the fact that the engineering time and cost to produce accurate scientific time standards have been greatly reduced.

There is no doubt that computer programs in the field of scientific time standard development have given new life to this phase of methods engineering. Before these developments it was thought that using scientific time standards was too much of an academic exercise to have much practical use, other than for very large production runs. Now, with these programs available, expert analysts can apply these techniques economically to virtually all planning and quotation activities. We are at the point where not advocating their use would be virtually surrendering obtainable profits.

However, the word is spreading very slowly because many companies have changed their engineering organizational structure to eliminate industrial engineering activities. They have done this because they misguidedly think team activities and other nuances of current favorite productivity enhancers eliminate the need for such in house expertise. The results being management in many cases are "oblivious" to the profit potentials they are abandoning through their ignorance of time standards computer programs. The fact is those companies that have purchased computer based time standards programs are using them to employ the century-old industrial engineering concepts of effectiveness and efficiency for manufacture of state of the art products very effectively. They are gaining a competitive edge on those people who are ignorant of scientific time standards.

As you can now see, it's no big mystery why those using scientific time standards programs are beating their competition. It is simply understanding the principles of motion economy and using a computer program based on scientific time standards to find the best way to make the product. It takes some effort to learn the program and use it, which ever commercial one you choose. But this effort is significantly less than the manual effort I've described above. I believe any manager introduced to the concept of the "two knows" and buys into it (I can't understand why anyone wouldn't.) will eventually realize that true competitive edge comes through application of the scientific time standard programs. If you don't already have one, get one and use it!

PLANNING OF OPERATIONS

This phase of MP&WM operations constitutes the presentation of information to shop operations so that operators know the exact sequence of the work. The planning operation is the part of the routine manufacturing production sequence assigned to manufacturing engineering and is the only production sequence manufacturing engineering is involved in. Virtually all other manufacturing engineering work is out of the direct line of producing the product. The planning operation, similar to the making of engineering drawings by drafting, is a scheduled sequence that is monitored as part of the routine production control activities.

Planning is the most visible of all manufacturing engineering activities in that failures in planning are vividly demonstrated by lack of capability to produce the required products. Therefore, to ensure high levels of competence in the planning activity, MP&WM management has rigidly structured this activity. That is, the formats to be followed for planning paperwork or computer output are specifically set with little room for variance in procedures and technique. In this way, planning can be reviewed quickly and efficiently before it is issued, and errors of omission and other types of errors are easy to detect.

The planning specialist is responsible for producing complete packets of information for work to be done at the various workstations. This activity historically involved considerable amounts of paperwork and, more recently, large computer databases. Planning has four major responsibilities in information dissemination and storage:

1. Review drawings produced by design engineering.
2. Define procedures and processes to be followed by shop operations.
3. Produce operator instruction documentation.
4. Integrate functional operating systems.

These four responsibilities are discussed below.

Review Drawings Produced by Design Engineering

It would not be possible for producibility engineers to review all the drawings produced by design engineering prior to submittal to manufacturing. In fact, producibility engineers work with a higher level of drawing, the engineering layout. This layout, after being approved by the producibility engineer as a producible design, becomes the database from which a multitude of detailed drawings are produced. It is these detailed drawings that the planning specialists review for errors, adherence to producibility requirements, and materials usage and quantity. The planning specialist plans how the part described on the drawing will be produced, essentially by combining the methods instruction and time standard for the workstation with the drawing and the required materials. The result is a step-by-step detailed sequence to be followed by the operator.

Define Procedures and Processes To Be Followed by Shop Operations

Using the detailed drawing and the parts list, the planning specialist identifies the kind of material to be used in producing the part. In some MP&WM organizations this includes determining whether the material is stock material or must be purchased; in other organizations this decision is made by the materials function. Regardless of who specifies whether the material is stock or purchase, the planner specifies the size, thus determining how much work will be required to transform the raw material into a finished part.

Once the material size is selected, the planner describes the specific production sequence to be followed for a specific part at the workstation, based on the overall generic method prescribed by the methods engineer. Usually workstations are set up to produce what is called a family of parts—for instance, parts that are similar in shape but have different dimensions and perhaps different auxiliary components. The planning specialist modifies the generic method to produce a customized method for the specific part.

Finally, the planner matches the customized method with the time standards to obtain the item-by-item time for each detailed step. This is done by using the tables, graphs, computer programs, and so forth that constitute the scientific time standard. Knowing what workstation will do the work, the planner uses the matching time standard to calculate the time to complete each operation. This is different from the work of the time standard analyst in much the same way as the planner's work is different from that of the methods engineer. The time standard analyst produces a generic scientific time standard that contains algorithms which, when entered with specific parameter inputs, yield the time to perform unique operations. Determining and entering the specific parameters are the responsibility of the planning specialist.

To summarize, the planning consists of specifying the material, the size to be used, the detailed step-by-step procedure to be used, the tooling required, and stating how much time it should take to complete the work.

Produce Operator Instruction Documentation

The procedure just described ultimately leads to the package of instructions given to the operator. This package contains the detailed drawing, the instruction sequence sheet, the tool list, the materials list, and some form of document for recording progress or completion. No other information should be necessary for the operator to understand what has to be done and how to do it.

Figure 7-12 is an example of an instruction sequence sheet. It is similar to the methods sheet shown in Fig. 7-1; the major difference is that incremental times are shown for each step. The instruction sequence sheet is the key document contained in the package sent to the workstation, and it must be correct. If it is in error, operators and foremen will lose faith in the planning and productivity will deteriorate. Inaccurate planning in effect gives shop operations a license to set its own standards. The foreman no longer has a persuasive model for the operator to match performance against, and the tendency is to work at an unsatisfactory pace. The shop will then tend to require higher and higher levels of overtime to compensate. To prevent this from happening, the planning must be accurate, verifiable, and believable by all facets of the manufacturing organization.

Integrate Functional Operating Systems

Planning is not an independent system. The plans constitute the basic data used by the master scheduler to schedule the factory for extended periods of time, the basic data used by production control to calculate the in-process time at each workstation, and the data used by marketing in quotes to customers for similar products. Therefore, the plans become a very important part of the overall database.

Most manufacturing organizations strive to create computer databases in which the plans are a primary component and the basic production data are entered along with the design engineering data. The importance of such data will become clearer in the subsequent chapter on fundamentals of Computer Integrated Manufacturing (CIM). As the creators of this database, the planners are also the imputers into the database storage system.

Computer databases are very important to MP&WM because they allow standardization of planning, which improves its acceptance by shop operations. By being able to recall existing satisfactory plans and simply modifying them to suit slightly changed conditions, the productivity of the planners is significantly enhanced.

CONCEPTS OF WORK MEASUREMENT

Now that we have reviewed the basic responsibilities of the MP&WM unit, it is of interest to place these responsibilities within the framework of measurement. The methods engineer, time standard analyst, and planning specialist are all engaged in documenting work to be performed within the factory. They specify the way in which work is to be performed and how it is to be measured.

		Operation of Element Description	Bearing Caps Fabrication	
			Variable Hour / inch	Constant Hour / Each
1		SU400 set-up		0.14
2	405	layout, assemble, burn, grind, and tack weld		3.99
3	423	assemble and tack weld temporary braces (2 braces)		0.04
4	401	burn openings in wrapper	0.0036	0.021
5	402	burn crop ends of bearing wrapper	0.0111	0.02
6	420	get ring, deslag, check size, position to front vertical flange or thrust bearing, and tack weld		0.18
7	420	get cover flange, deslag, check size, position to wrapper, and tack weld		0.25
8	420	get pipe, check size, position to wrapper, and tack weld		0.05
9	420	get pad, check size, position to wrapper, and tack weld		0.07
10	420	get brace. check size, position to wrapper, and tack weld		0.06
11	420	get rib, deslag, check size, position to wrapper, and tack weld		0.3
12	420	get trunnion,check size, position to wrapper, and tack weld		0.06
13	420	get boss, check size, position to wrapper, and tack weld		0.06
14	405	remove clamps and position cap assembly for weld		0.49
15	418	lay out, assemble, burn, grind, and tack weld		4.03
16	424	assemble and tack weld back plate to wrapper and horizontal flanges		0.39
17	421	get front gib supports, deslag, check size, position to wrapper, and tack weld		0.4
18	421	get front gib block, deslag, check size, position to gib supports and front vertical supports, and tack weld		0.21
19	421	get boss, check size, position to wrapper, and tack weld		0.06

Figure 7-12. Instruction Sequence Sheet

However, how this function is carried out depends on how motivated the operators are to do their jobs effectively. The area known as concepts of work measurement deals with how the hourly work force is perceived and led. This perception will set the stage of how the MP&WM unit is to be managed.

There are two basic systems for motivating the hourly work force: an incentive system and a day work system. In the incentive system the operator is paid for each incremental amount of work performed. In the day work system the reward is continued employment, and it is based on a team approach to manufacturing where the team is able to successfully produce the company's products such that the desired level of profitability is maintained. Both systems have their merits and drawbacks and both affect the way in which MP&WM units approach their work.

The incentive system in many ways is self-regulating. In essence the operators are in business for themselves. They earn their living by producing parts and being paid for each one they produce. Therefore, the operators are management's allies in ensuring that support services are adequate, since lack of such services could cause lack of ability to produce the product. In an incentive system the main thrust of MP&WM work must be in the time standard and planning areas, which determine how much product an operator can produce during the work shift and ultimately how much the operator earns. Methods sheets are less important; the operator will tend to improvise if necessary in order to meet or be under the planned time. Incentive systems result in MP&WM units heavily staffed with planning specialists and time standard analysts, with less emphasis on methods engineering activities.

The day work system can be an authoritative system. The operators know they will be paid for hours worked at the workstation and that they must follow the methods sheet in producing the product. They realize that the only way they can lose their job is to constantly abandon the process they are told to follow; they have only a secondary interest in the number of parts they produce. In this system the methods engineer is most important. The methods engineer's creativity will lead to improved tooling, fixtures, and sequences in how the product is to be produced. The day work system puts a premium on creative engineering to produce optimal results and considers operators to be extensions of the workstation where they work. Day work systems typically lead to MP&WM units that emphasize intensive methods engineering activities at the expense of time standard analysis. These units stress the importance of proper layout of the workstation and making sure the operator follows the prescribed method. This is admittedly a drab picture of the philosophy of the work place and can become a drudgery of a work place scenario if it is allowed to happen. That is allowing an authoritarian philosophy to dominate. We have learned that team approaches whereby day work operators are solicited and courted for their ideas on how to make the work place more effective does yield better results. Replacing authoritarianism with a participative management approach is the current trend in day work system management.

Today there are few pure incentive or day work systems. Most organizations employ combinations of both but favor one. It is common in day work systems to use suggestion plans to elicit ideas from the operators for improving quality and productivity, and to pay the operators for all suggestions adopted. Similarly, incentive systems tend more and more to require good methods engineers to create uniformity of approach to product manufacturing.

The idea of using the experience of operators in developing methods has steadily gained favor and is now the preferred way to operate. Especially in a pure day work system, not using the accumulated practical experience of workers in optimizing workplace methods would be a waste of that experience. Because they live with the products and their workstations daily, operators come in closer contact with production problems than any other employees of the company. They have had to solve many practical, if small, problems in order to produce their required quotas. Therefore, to encourage operator participation in establishing workplace methods, many companies have established formal programs, which are classified under the general title of "participatory programs."

Participatory programs try to improve factory productivity by having the operators participate in setting goals, or establishing methods, or both. One of their purposes is to establish a sense of proprietorship for how the product will be made and how much will be made in a set period of time. In the day work system this raises the interest level of the operators so that error levels decrease and productivity improves. In the incentive system, the participatory approach leads to modified time standards, which should allow the operator to earn more by being able to produce more, again resulting in improved productivity.

Two common participatory programs are called team production and quality circles. Both of these programs have been applied with success in a wide variety of manufacturing operations. However, they do not replace the traditional responsibilities of MP&WM; they simply enhance the activities of that unit to gain further productivity improvements.

Team Production

In team production a group of hourly operators are assigned the task of producing a set number of units of a product, usually one requiring major assembly operations. The team is given a specification (similar to an acceptance standard given to a vendor asked to produce a given product), told the maximum time allowable to produce the product, and then left to plan how to do it. Of course, the team members are familiar with the product and can draw on methods and planning documentation already created by the MP&WM unit. This is essentially a group incentive system that replaces incentive systems designed for piece part production. Team production allows short-time-interval assembly operations to be put on a meaningful reward pay system.

The team has full access to all the support functions that would normally be available in a factory, including maintenance, quality assurance, methods, and so forth. While MP&WM does not develop detailed methods instructions, it still has the responsibility of developing general instructions pointing out the optimal ways of achieving specific tasks. However, this is an advisory role. The teams have full control of how they do their tasks and are accountable for meeting minimum production goals within cost constraints.

Many conservative manufacturing managers are uneasy with such an approach to factory management. They are concerned that day-to-day decisions on how the factory is run are being made by less qualified or unqualified people and that the results will be unsatisfactory. They say that they lack control to make required changes, which could be a valid criticism. Team production is not for all factories; it requires a skilled work force, people who are more like technicians than production line workers. The team members must be very knowledgeable about their product, its design, and how it is made. Therefore, this approach is usually found in job shops making complex mechanical, electrical, or electronic assemblies, where the operators are indeed experts. It is also often found in service and repair shops.

Quality Circles

A much more frequently used participatory program is the quality circle, sometimes also known as the productivity circle or methods improvement team. Like the team production concept, its purpose is to elicit ideas from the people directly involved in making the product and to give them a sense of proprietorship. There are many ways of organizing quality circles. One frequently used approach is described below.

A group of workers from a common manufacturing area are invited to join an improvement committee. The purpose of the committee is to suggest ways to improve their work situation and the quality of their product, and to reduce the time and cost to make the product. This group consists of the actual machine/ process or assembly workers, support people such as inspectors and maintenance personnel, material movers, technical support people such as manufacturing engineers, designers, and management representatives, usually the foreman. Office clerical staff may also be assigned. The management representative is usually the facilitator, that is, the discussion leader. The leader's purpose is to get the circle to put forth ideas and to discuss frankly whether their implementation will improve working conditions, improve quality, or lower costs. All three of these categories will ultimately result in improved productivity. This is not a brainstorming session, although some concepts of that technique may be used. It is a working session where the circle members draft recommendations to management. These recommendations are recorded in the minutes of the circle's meeting and actions are assigned to various members to be achieved, if possible, before the next meeting. These actions are usually in response to requests for information such as clarifications of design requirements, company operating policies, and business outlooks.

The circle's recommendations are reviewed by the manager of manufacturing for possible implementation. What happens at this step determines whether or not the quality circle activity will be successful. As with its antecedent, the suggestion system, a suitable number of the quality circle's ideas must be accepted or the system will collapse. By not accepting the ideas, management is saying that only it can be creative and innovative. If this happens, the purpose of the participatory program has been aborted. To avoid this, management must accept a good proportion of the recommendations. In this way employee interest in the circle is maintained and chances are enhanced that a truly significant idea will emerge.

Even with full and active management support of quality circles, they are rarely maintained for over a year. Once they become routine, their freshness and member enthusiasm begin to diminish, until they are no longer useful. It is important for management to recognize this and plan for a replacement activity. This can take the form of reshuffling people into new circles, creating competition between circles for the most successful ideas, with corresponding rewards, or any other idea that stimulates operator participation in productivity improvement.

Quality circles do improve work methods, often through a summation of small items that the methods engineer was unaware of but that make it easier for the operator to do the job. The ideas usually cost relatively little to implement and improve morale if they are implemented quickly. They create an awareness on the part of the operators that may lead to significant ideas and paybacks, and this is what makes them attractive to management.

MP&WM should never consider participatory programs as substitutes for or threats to the unit and its ongoing responsibilities. No sensible management team would consider an idea-generating scheme such as a quality circle to be a replacement for MP&WM. The MP&WM unit is a professional engineering-based organization with highly trained people dedicated to manufacturing optimization. Participatory programs are supplements to the normal creative and innovative tasks required of a progressive engineering unit, and serve an allied but the different purpose of creating enthusiastic interest in the success of the company. Therefore, they are tools to be used by MP&WM to improve their ability to achieve their stated goals.

SUMMARY

This chapter has described the activities and responsibilities of the MP&WM unit. The methods engineer, time standards analyst, and planning specialist are intimately involved in the daily workings of the factory, and MP&WM is the direct technical contact with shop operations. Whenever it must be decided how a part will be made, whether the part is being processed correctly, or whether the output level is sufficient, manufacturing engineering via the MP&WM unit will be a key contributor.

The difference between a methods sheet and a planning sheet has been explained; that is, the methods sheet precedes the application of the time standard, which, when completed and merged with the methods sheet, becomes the planning sheet. Scientific and estimated time standards have been defined and their uses illustrated.

Productivity measurements and the concept of work measurements, including participatory programs, have also been discussed. Both piece count and efficiency measurements have their place; the overriding concern is being able to determine the success or failure of an activity. This led us to the concepts of work measurements, the differences in how an MP&WM unit will function depending on the type of manufacturing operation it is supporting, that is, incentive or day work or a combination with some type of participatory program.

The concept of the scientific method was introduced early in this chapter, although it could equally well have been discussed with any of the manufacturing engineering units. The scientific method is a very important concept for success in any engineering organization and is one of the basic techniques applied in manufacturing engineering.

REVIEW QUESTIONS

1. In general, MP&WM work consists of developing specific instructions for the factory from broad directives. Describe how an engineering drawing can be considered a broad directive in reference to instruction details required by the factory.

2. Use the six-step scientific method to explain how a paper clip manufacturing plan could be devised.

3. Referring to the methods sheet (Fig. 7-1), how many steps provide information only and require no action by the operator?

4. The following are items found at a workstation. Which are associated with the primary equipment and which with the secondary equipment? (a) Tool changer, (b) operator platform, (c) compressed air in reserve tank, (d) hand tool rack, (e) tool spindle motor, (f) inspection gauges.

5. Given the following data, which workstation would have a shorter time assigned to it to produce a given product?

	A	B	C
Distance to inspection station, ft	500	250	35
Distance to foreman's station, ft	30	120	100
Surface ft per 85-minute capability	42	30	

6. What type of manufactured product would make satisfactory use of a productivity rate measurement and what types would not?

7. Describe the differences between the productivity rate measurement and the efficiency measurement.

8. Discuss the advantages and disadvantages of adopting an absolute planned time system versus an estimated planned time system.

9. A company has a contract to produce 16,000 steel claw hammers with steel and taped handles. As the methods engineer in a factory equipped to make such hand tools, outline how the following problems will be solved.
   (a) Determine how to make the hammer.
   (b) Develop tooling and fixtures to optimize production.
   (c) Develop a useful measurement scheme.
   (d) Estimate the time it will take to produce the required amount of product.

10. A scientific time standard is needed for the steel claw hammer of question 9. Using the five tasks discussed in the text, explain how the work of the methods engineer will be used by the time standard analyst to produce a scientific time standard.

11. Using the methods instructions in the text for producing a screw driver blade, expand step 2, rough turn and thread, to create an enhanced methods document for creating a scientific time standard.

12. Use the results of the enhanced methods document of question 11 to create an applications sequence of transports, gets, places, precisions, and miscellaneous for the work to be done to complete step 2. Enter into a data sheet (recommend format of Fig. 7-8) and calculate the total time required to do step 2.

13. Figure 7-12 represents an instruction sequence sheet produced by a planning specialist. Discuss the difference between this and the methods sheet.

14. Discuss the differences that could exist in instruction sequence sheets for a day work factory and a piecework factory. What factory would require the more detailed planning and why?

15. Explain how scientific time standard computer programs can be used to enhance the accuracy of an instruction sequence sheet and how shop operations would effectively employ this information.

16. Using the data in Fig. 7-12, explain how a scientific time standard computer program would review the method sequence to determine human feasibility to accomplish. List the basic conditions you think the program would search for and compare within its programmed logic steps.

17. Using the data in Fig. 7-12, give your opinion as to what steps, or combination of steps, could form the basis for scientific time standards computer program motion macros. Include your reasons.

Chapter 8

JOB EVALUATIONS, PAY PLANS, AND ACCEPTANCE

This chapter deals with establishment of job worth for pay purposes and also how to gain acceptance of the results by the work force. It is very evident that having the work force believing the pay plan is based on merit is extremely important for the success of the company. Nothing is as large a productivity deflator than the work force, both salaried and hourly, believing they are not justly compensated. Industrial engineering is intimately involved in this task because constructing jobs and placing them in their proper hierarchy with respect to each other is closely related to the methods, planning, time standards, and work measurement work we elaborated on in the previous chapter.

Where does it all start? As with job defining work for methods, planning, time standards, and work measurements, it has its basis with the principles of motion economy and the "Two Knows." The principles of motion economy because of physical effort required and the "Two Knows" for mental efforts associated with the job.

The "Two Knows" require that we know how to do the job and how long it should take to do so. This is the pre-requisite for all manufacturing success. But it's one thing to be knowledgeable on these fundamental issues and another to accomplish the task. Doing the work is primarily a people team thing. While the dream of a fully automated factory is out there, being realistic we know that only people operating equipment and having the necessary skills and knowledge will accomplish the goal of making the company's products. And in doing so in a manner that they are attractive to customers who will buy them and allow the company to make a profit. So all the planning, methods definitions, time standards, layouts for effective flow, using the most apropos equipment to do the job are all insufficient if we do not have the properly skilled and motivated employees to work the entire system. We must convince them that the reward, pay they receive is commensurate with the effort they put forth.

Getting the proper employees to operate the system is the task of Human Resources (HR). However, designing the jobs and creating order out of chaos as to how to rank the jobs so that morale is maintained and proper skills are mated with each specific job is the responsibility of Manufacturing Engineering through its Industrial Engineering subset. Just as we define how work is to be performed we must define what the skill sets are required for doing the planned work. When we design jobs we are well aware of the complexity of each task, and are also aware of the degree of difficulty each job entails. With this knowledge in hand we can design jobs and rate the degree of difficulty it contains to achieve the goal of the job, much like an Olympic diving competition stating the degree of difficulty of each dive. In the diving competition we award higher ideal scores for more difficult dives than for simple ones. In the work place we define jobs in a similar manner. The more difficult ones are rated higher than the simpler ones. And then we pay more for doing the more difficult jobs than the less difficult assignments. This is fair and proper. The task is how to design jobs so we can understand as objectively as possible the degree of difficulty each possesses, then make up pay plans to match these ratings. If we are successful everyone will agree that Harry, working a less complex job than Joe, ought to be compensated less than Joe.

In this chapter we will examine how industrial engineers evaluate jobs, structure them in accordance with overall worth and construct pay plans. We will also see how these pay plans are communicated to the work force and how to keep them competitive with other societal factors. This is not a simple task, but one that lends itself to engineering logic. Unfortunately it also lends itself to emotionalism and company politics. In this chapter we will examine the structure of job evaluations hence pay plans not in a vacuum but always considering the subjective nature of human opinion and emotionalism.

FUNDAMENTALS OF JOB EVALUATION

(Some of the content of this section has portions from my book *The Engineer Entrepreneur,* Chapter 9, "The Business Plan for the Entrepreneur," pages 253–258; published by ASME Press, New York, 2003.)

How do we go about the task of evaluating a job? We do it with as objective as possible a rating system. So before we can evaluate a job, lets learn about job ratings. A job evaluation is done by defining what the job is supposed to accomplish. We create a job statement for the job. We do this by defining the end result of the job if specifically possible, such as:

Egg Packer:
 This job is for packaging eggs into a crate without breakage for shipment to customers.

Then by defining the major steps in doing the job, such as:

Egg Packer Job Description:
1. Pick eggs from a tray.
2. Scale them as per categories, small, medium, large, extra large using supplied gauges.
3. Put eggs into corrugated packing trays, per scale classification.
4. Put protective covers over trays.
5. Place trays in specially designed egg crates, making sure only same scale size eggs are in each crate.
6. Seal crates
7. Mark date and egg scale size on crate.
8. Move crate to shipping zone.

Look familiar? It should. It's a method sheet. We now know what the job is all about. From methods and time standards work we know the degree of difficulty and how long each step should take to accomplish. We know what the incumbent will have to do. In addition we will add associated skills and education levels we believe are necessary for the incumbent to successfully complete the job task. Such as recognizing different size eggs and being able to read and write (perhaps I can summarized by saying having an education level through high school). Knowing all this information, it's possible to place a score, a rating, of some type against these requirements and compare this score with other jobs. In effect creating a relative worth comparison.

Sometimes jobs are more generically based and we can not define precisely what the incumbent is to do to complete the job task. When this happens we have to rely on other means to define the job content. Lots of times skilled trades fall into this category. For example, how do we define what a plumber does? We can certainly list the general areas of expertise we expect the plumber to have, such as the ability to thread pipe. But we would be hard pressed to say that threading pipe is the extent of the plumbing skill set necessary for the job. As an aside, if that's all of the plumbing skills a job required then we wouldn't hire a plumber. Instead we'd train a person to do that small subset of plumbing in the course of defining the job and would be listed as one of the skills necessary to do the job. The point is for a general job such as a plumber all we can say from a pragmatic viewpoint is that the company needs plumbing skills and we say perhaps: current license required. Then the job statement would look something like this:

Plumber Job Description
 This job requires plumbing skills.
The basic skills needed are:
1. Thread pipe.
2. Make piping and tubing connections via solder or mechanical means.
3. Run pipe in accordance to a drawing.
4. Read mechanical drawings.
5. Diagnose leak problems.

6. Install valves.

7. Install pumps.

We see that this type of job has no specific end goal as we had with packing eggs. There is no specific work leading directly to creation of the company's products that are stated in the job description. It's obviously possible to do plumbing work and have nothing directly to do with sorting, packing and shipping of produce to customers. In that manner it is more abstract. But we still have to evaluate the job content and create a score for it. Sometimes to help us do that we rely on industry standards. For example if we want a licensed plumber, there is a degree of competency that a license implies. We will make sure that the scoring system reflects that by sort of backing into it. First by first knowing what the going $ pay rate per hour for a plumber is. Then by the assuring the license has enough points to equate to that pay rate in the scale we will construct.

That gets us to the next step, developing pay scales. If we have a scoring system and rate jobs and then develop a certain "pecking order" of importance. What do we do with that. Well, quite simply we assign salaries to each job in accordance with its points. We establish the highest rate we're willing to pay (usually based on current trends and wanting to be competitive with other firms trying to hire similarly skilled people), that goes to highest scored job and so forth down to the lowest rate job. Fortunately the governments have been kind to us and prescribed a minimum wage, so we know what the lowest bottom job rate can be. This is the bare bones of job evaluation. Simple in concept, not so simple to do because of its subjective nature, thus making it difficult to be fair.

Let's take a look at a typical job evaluation system and after that, its associated pay plan to see how it's structured and how the differentiation between pay for different jobs is achieved.

The purpose is to establish a rational and defensible method of differentiating pay levels between various jobs. We do this by objectivizing an essentially subjective system. We know instinctively that some jobs are worth more than others for the success of the company. For example, a licensed electrician is going to be worth more than an electrician's assistant. That's simple and there would be few if any arguments to the contrary position. However, is the licensed electrician worth more to the company's success than the licensed plumber? This is hard to tell and in truth it would probably change from time to time depending on the scope of jobs a company undertakes, and the laws of supply and demand for plumbers vs. electricians.

How we do the differentiation can be complex or simple, depending on how objective we wish to paint the rational. The most common method is to create an evaluation point system to measure each job against, and then create pay grades based on the points. And to do this based on methods evaluations with time standards factored in. Let's take a look at the mechanics of creating the pay system.

A pay system for hourly personnel has been developed to achieve equitable pay for the various positions based on a job evaluation schedule, as shown below.

ABC Company

Grade	Point Range	Start	2 weeks	6 months	12 months	anniversary–24 mos. (after start date) review
A	52–100	$6.50	$7.00	$7.00	$7.00	inflation adjust.
B	101–150	$6.50	$7.00	$7.50	$8.00	inflation adjust.
C	151–200	$7.00	$7.25	$8.00	$9.00	inflation adjust.
D	201–250	$7.50	$7.50	$8.50	$10.00	inflation adjust.
E	251–300	$8.00	$8.00	$9.75	$11.50	inflation adjust.
F	301–350	$9.00	$9.00	$11.00	$13.00	inflation adjust.
G	351–400	$10.00	$10.00	$12.50	$15.00	inflation adjust.

Pay Grades: Pay Progression $/hr (Max. Allowed per Evaluation Period) — Job Rate Pay — Level

Other allowable pay adjustments

1. $0.25/hr. for each cross-trained[a] skill above pay grade, or out of dept. same pay grade. Max. 3 allowed.

Skill adjustments:	+1	+2	+3
$amount	$0.25	$0.50	$0.75

2. $0.25/hr. for acting as a designated leader.
3. $0.50–$1.50 range for being designated a dept. leader (acting as an assistant supervisor)
4. Progression may start at any point: management decision based on starting skills of individual.
6. Progression shown are for satisfactory performers as per table below. Lesser performance, poor attitudes, attendance problems will stretch out the time to maximum job rate.

Evaluation of Performance and Attitude Overall Rating	Pay Adjustment Factor
0–6.0	0
6.1–7.0	60%
7.1–8.0	75%
8.1–9.0	90%
9.1–10	100%

[a]*Cross training*: Supervisors nominate candidates to the Shop Operations Manager to receive cross training. Shop Operations Manager approves and sets up a schedule for–cross training for the individual. An employee is certified as cross trained when he/she achieves the accumulated time for training designated for the specific job and demonstrates proficiency.

TABLE 8-1. Hourly Job Rating System

	Relative Importance (Used to Develop Degree Pts.)	Lowest	Low	Degrees Average	High	Highest
Job Skills required	25%	13	25	50	75	100
Education required	10%	5	10	20	38	40
Responsibilities	15%	8	15	30	45	60
Effort required	25%	13	25	50	75	100
Working conditions	20%	10	20	40	60	80
Compete (ease of finding a similar job)	5%	3	5	10	19	20
TOTAL POINTS	100%	52	100	200	312	400

How to apply table:

1. Rate each job for the degree of importance for each factor. Select a point factor where it fits (lowest to highest) best. Assign the points for that rating from the table.
   a. *Job Skills Required*: (1) decide if skilled, semiskilled, or unskilled. (2) Depending on the level of skills involved, select the appropriate degree and its point value from the table.
   b. *Education Required*: Determine what level of academic education or practical training is required. High school graduate (or equivalent) is the midpoint degree level. Judge more or less education needed to do the job accordingly. Keep in mind the need for administrative capabilities and English speaking.
   Training period to become proficient in the job: The lowest skilled/knowledge jobs take the shortest time to become proficient. For classification purposes; lowest =1–2 wks, low =3–5 wks, ave.= 6–10 wks, high =11–15, highest = over 15 wks. For jobs that require both formal education and practical training , score each separately and then average the results for the amount of points awarded.

    c. *Responsibilities*: This applies to all aspects of the job. The lowest point value is assigned if it only includes the individual and not others, while the higher point value goes to jobs where responsibility is for the entire group doing the work. The larger the group, or the consequence of inadequate work are, the higher the point value assigned.

    d. *Effort Required*: The degree of energy expended to do a job is considered. For jobs requiring no physical effort the lowest point value is assigned. Jobs requiring the most physical labor manifested in requiring more than two rest periods in a 4 hour continuous time frame should receive the highest point score. Use principles of motion economy as the objective arbiter of work effort required. Also rate in accordance with the proper method as would be defined for establishing a time standard.

    e. *Working Conditions*: This is the physical comfort factor. The least comfortable the work environment is the higher the degree point value selected. Working in an un-air conditioned space would receive a point score in the high range. Working in perilous conditions such as on roofs, scaffolding, and ladders would rate a highest point score. Conversely working in an air conditioned office would rate the lowest point score.

    f. *Compete*: This factor relates to the ability to keep personnel in the job. If the skill is in high demand, Therefore the incumbent can be lured away, give a higher point score commensurate with the risk. If the skill is in low demand then the point score assigned should be at the low end of the scale.

2. List reasons for all selections made in step 1. This will help in making comparison for future job point evaluations.
3. Tabulate scores and record.

It is necessary to rate each existing job and any newly created hourly job with this point evaluation system to assure all jobs are compensated fairly. The staff uses the following form (see Fig. 8-1) to evaluate jobs under their jurisdiction to rate jobs for pay purposes. Job titles for all departments are drawn from the list in the evaluation form. The titles are for illustrative purposes only. Each department can change out and add in their respective job titles.

    A similar job evaluation system could also be set up for staff/management positions to assure that the company is competitive in keeping and attracting competent management Personnel. In both cases job descriptions have to be developed so proper evaluations can be made.

CREATION OF JOB DESCRIPTIONS

Writing fair job descriptions that can be evaluated for compensation levels is critical for maintaining a competent set of managers and staff. If people feel they are not being compensated adequately and fairly they will leave. And not surprising more will leave if they feel the compensation system doesn't fairly reflect the value of the work they do.

    Creating job descriptions for a company, then, has two purpose; to assure that all the steps of the "Seven Steps of the Manufacturing System" are assigned to individuals to accomplish, and to communicate the responsibilities a person has in carrying out his or her assigned job. When reviewing the entire set of job descriptions it should be very evident how the company goes about doing its' business. It is also a way of determining if all seven steps of the classical business system have been accounted for in the company's strategy, or are some being left to an ad hoc by chance planning level of accomplishment. If any of the steps are in the ad hoc camp, then the efficiency of the firm will be less than optimum. Most likely the efficiency will be significantly less than optimum. While the process is similar for staff and hourly positions (exempt and non-exempt per labor law) we will concentrate on the hourly jobs where people are directly involved with transforming raw material into finished products. The identical system with broad approach to the seven steps of the manufacturing instead of the specific method can be used for exempt and indirect labor jobs.

Job Title	Job Skills	Education	Responsibility	Effort Req.	Work Con.	Compete	Total	Job No.	
				Points Assigned					
Cleaner									
Electrician									
Class I									
Class II									
Licensed									
Plumber									
Class I									
Class II									
Licensed									
Welder									
Class I									
Class II									
Painter									
Class I									
Machine Operator									
Class I									
Class II									
CNC Machine Operator									
Class I									
Class II									

Figure 8-1. Job evaluation Form

The process of creating job descriptions for direct hourly employees, is as follows.

Strategic Evaluation

A. List all the job titles you have in your company on the left hand side of the page.
B. Draw lines from the job titles to the appropriate step of the method for creating the product. One job title may cover a multiple of steps. This is especially true for smaller companies.
C. For the company to be performing in a planned mode rather than ad hoc, every step of the method should be linked to at least one job.
D. If there are blanks, method steps without corresponding job titles, then create new job titles or expand existing ones to cover the blanks.
E. Create job descriptions as explained next. Good job descriptions contain topics I through V. Section VI shows the general format the document will entail.

Tactical Evaluation

I. Purpose of a job description:
  a. Define the scope of work to be performed.
  b. Create measurements for evaluating whether the scope of work is being accomplished.
  c. Establish the job within the hierarchy of the organization.
  d. Define skills required for the job.
II. Scope of work to be performed:
  a. Classification
    i. Management

        ii. Professional

        iii. Clerical

        iv. Value-added/non-value-added labor

    b. Specific tasks

        i. Mental

        ii. Physical

III. Measurements of Accomplishment:

    a. Generic measurements for the hierarchy level

        i. Management

        ii. Professional

        iii. Clerical

        iv. Value-added/non-value-added labor

    b. Specific methods related measurements

IV. Organizational Placement:

    a. Immediate superior position the job reports to.

    b. Responsibility for lower-level jobs reporting to this job.

    c. Evaluation responsibilities for subordinate jobs reporting to this one.

V. Required Skills:

    a. Education level—external to the job.

        i. General education achieved through high school.

        ii. Specific professional education, such as a welder certification

    b. Training—specifically related to the job.

        i. Example: Specific tradesman qualification usually in-house training such as a machine operator rating

    c. Experience performing immediately subordinate jobs.

        i. Example: To qualify as a heating, ventilation, and air conditioning (HVAC) technician, require 2 years experience in electrical and plumbing work.

VI. General format of a Job Description:

    a. *Job Statement.* A general explanation of the job explaining what the incumbent does and how he or she goes about doing the task. Includes the "Classification of the scope of the work to be performed." The last part of the job statement is an Organizational Placement sentence showing where the job fits within the company's structure

    b. *Specific Requirements.* A listing of all the tasks to be performed included in this job. Can be narrative or outline format. Recommended order is:

        i. Mental tasks—list specifics: for example, maintain job logs

        ii. Physical tasks—a listing of the job content: for example, prep surface for paint, mix paint, spray paint walls, etc.

            1. list biological requirements: for example, lift x pounds, etc.

            2. list environmental considerations: for example, work outdoors on ladders, etc.

    c. Evaluation Measurements

        i. A statement on company policy for measuring and evaluating job incumbents

            1. how often

            2. how done

        ii. Specific measurements for the job

            1. list all measurements used to make out evaluations that are specific to the job; for example, any specific quality measurements particular to the job; perhaps number of complaints received from customers, etc.

            2. Generic measurements: attendance, care of company property, etc.

    d. Education/Training Levels Requirements

        i. Generic education—school grade achievement required, through college (grade 16).

        ii. Specific skills gained through job training, or trade school

        iii. Experience levels obtained through lower level jobs to qualify for this job.

Job descriptions are never done by only one person. The draft is done by the initiator, then it is carefully reviewed and compared to other job descriptions for compatibility. The review is usually done by a small committee familiar with the job and similar jobs. To chance acceptance by company personnel this committee must be as broad based a representation of the company's various constituencies as possible. This means having hourly workers as well as staff and management on the committee. In fact it is quite common to have labor union representation on this committee in the case where the work force is represented by a duly recognized labor union. In all cases it is the responsibility of management (usually through manufacturing engineering) to ensure that all committee members are trained in doing job evaluations and job description writing. Also, the review committee is familiar with company policy concerning what is thought to be the prime strengths, therefore prime jobs the company has to carry out its' business. This is necessary for proper ranking of job importance with respect to each other.

Job descriptions spell out how a company performs. It is essential that all companies have some sort of definition as to what the responsibilities are for each of its incumbents. Even small companies need to have an understanding of what the work content is for each of its employees are. In the latter case the job description may be nothing more than a bullet listing of responsibilities and be quite informal. This is so because with a very small company each person has unique skills and does just that. What ever the case, defining job responsibilities is one task every company has to take on and master.

Let's develop a job description for the manufacturing cell operator for the machining of screw driver shanks as shown in Fig. 7.5. In order to describe the job we need to refer to the method prescribed for the job. The method is repeated below from Chapter 7.

1. Cut steel rod, 3/8 in. diameter, to length on a saw.
2. Rough turn the entire length and thread opposite to the blade end on a lathe; class 2, UNC coarse thread 8 threads/in.
3. Form the blade on a grinder to a preset template.
4. Grind the entire shank to a commercial finish or, if specified, to the finish required on the drawing.

The job title can be anything we chose that fits the vocabulary used by the company and has instantaneous recognition by those who are cognizant of the work and the product. So let's title the job " Shank Work Center Operator." The first step in creating the job description is to write a few general explanatory sentences about the job. This is called the Job Statement. For the shank work center operator job this would be an acceptable job statement:

> Operate the combination of machines and auxiliary equipment in making metal shanks for screw drivers. Do normal maintenance on the equipment. Keep records as defined by supervision for orderly management of production. Maintain quality as required by engineering designs and methods instructions. This job reports to the Supervisor of Screwdriver Production.

The statement is concise and in a macro view tells the person assigned to that job what the responsibilities entail and where in the organization the incumbent reports to.

The next step in job description preparation is the specific requirements. This can be either narrative or outline form. I prefer outline form for shop floor jobs. This way there is no ambiguity about what the specific task responsibilities are. Continuing on, here is a typical specific requirements listing that would be generated for the shank work center job.

1. Mental Tasks
    i. Review order quantities.
    ii. Review design requirements.
    iii. Review Quality requirements.
    iv. Record quantity and quality data on process sheets as required by instruction.
    v. Report material deviations and out of tolerance machining and grinding to supervision.

2. Physical Tasks
    i. Cut steel rod to length on a saw.
    ii. Rough turn and thread on a lathe.
    iii. Grind to a preset template.
    iv. Measure using micrometers, scales, callipers, and other gauges as required.
    v. Sweep and wipe clean work station equipment at end of each shift.
    vi. Lift, move totes, maximum of 25 lb.

With the job specific task laid out we now define the evaluation measurements. We do this first with an overall company policy enunciation pertaining to measurements, then with specific measurements for this job.

1. Measurement Policy.
    a. Job performance is measured for quantity and quality on an ongoing basis commensurate with the specifics of the particular production.
    b. Job performance is primarily operator generated, via operator certification.
    c. Employees are expected to work in accordance with the employee handbook.
2. Specific measurements
    a. Quantity of screw driver shanks made per shift
    b. Productivity rate per shift based on designated time standard
    c. Reasons for deviation to defined standards of production
    d. Quality defects on occurrence.
    e. Attendance in accordance with employee handbook.
    f. Performance in accordance to maintaining the work station and other company equipment on a semiannual basis performance review or more frequently as circumstances require.

The last section of the job description is the education/training requirements necessary to perform the job successfully. A nominal example is:

1. Job minimum education/training requirements.
    a. High school diploma or equivalent.
    b. Certified CNC machine operator.
    c. Minimum 6 mos. experience doing grinding to a template.
    d. Ability to read blueprints and engineering drawings.
    e. Ability to enter data to statistical process control forms.

The above is all the factors for the job description for the shank work center operator. See the compiled version in Fig. 8-2.

DEFINING KEY JOBS

In order to rate jobs fairly using the job rating system for hourly or salaried employees it is necessary to have a base to start from, for comparison purposes. We create this base by designating certain jobs as key jobs, usually two or three, that we can very easily compare to each other. For example we know that the CNC machine operator needs less skill than a tool maker working in die shop making precision dies. Therefore the tool maker should earn a higher hourly pay rate than the CNC machine operator. We also know that the CNC machine operator requires more skill than say a machine oiler, therefore compensated higher. Knowing these facts we can create comparison point values from the job rating system that justify the salary levels for each of the jobs. If these three jobs are designated as the company's key jobs, all other jobs individual worth should be equitably

Job Description:
Shank Work Center Operator

Job Statement:

"Operate the combination of machines and auxiliaryequipment in making metal shanks for screw drivers. Do normal maintenance on the equipment. Keep records as defined by supervision for orderly management of production. Maintain quality as required by engineering designs and methods instructions. This job reports to the Supervisor of Screw Drive r Production"

Job Specific Requirements:

Mental Tasks

Review order quantities.

Review design requirements.

Review Quality requirements.

Record quantity and quality data on process sheets as required by instruction.

Report material deviations and out of tolerance machining and grinding to supervision.

Physical Tasks

Cut steel rod to length on a saw.

Rough turn and thread on a lathe.

Grind to a preset template.

Measure using micrometers, scales, callipers, and other gauges as required.

Sweep and wipe clean work station equipment at end of each shift.

Lift, move totes, maximum of 25 lbs.

Job Measurements:

Measurement Policy.

Job performance is measured for quantity and quality on an ongoing basis commensurate with the specifics of the particular production.

Job performance is primarily operator generated, via operator certification.

Employees are expected to work in accordance with the employee handbook.

Specific measurements

Quantity of screw driver shanks made per shift

Productivity rate per shift based on designated time standard

Reasons for deviation to defined standards of production

Quality defects on occurrence.

Attendance in accordance with employee handbook.

Performance in accordance to maintaining the work station and other company equipment on a semi -annual basis performance review or more frequently as circumstances require.

Education

Job minimum education/training requirements.

High School diploma or equivalent.

Certified CNC machine operator.

Minimum 6 mos. Experience doing grinding to a template.

Ability to read blueprints and engineering drawings.

Ability to enter data to statistical process control forms.

Figure 8-2. Job Description, Shank Work Center Operator

compared to them, hence the pay for each is rationalized to be proper. In this manner points for all jobs are not skewed up or down with respect to known job values. The task then becomes that of defining what are the true key jobs for the company, selecting a proper few that have easily definable value differentiation, and what their compensation rate should be.

All companies have key jobs. The easiest way to determine what the key jobs are is to ask the question, "Can the company stay in business without performing this function?" And the answer cannot be: " It's a very necessary job but it can be farmed out." If the nature of company changes because the task is farmed out, then the job is a key job. By changing the nature of the company means no longer doing the core value tasks the company was founded to do. For example, if the company was founded to do miscellaneous metal lathe and milling work, and decides not to do milling work any longer, the nature of the company has changed. So when we ask the question, "What happens if I eliminate this job?" And the answer is, "The nature of company will change." Then the job is a key job.

Normally for manufacturing companies the key jobs are those that are core to performing the method of manufacturing. But don't be surprised if some jobs end up being expansions of key jobs that make doing the key jobs more effectively. For example, in the screwdriver shank example we see that we have the operator cutting the blade to length as well as machining, threading, and grinding. It is feasible to conceive that if volume increases we may want to split the job up three or more ways. One of the splits could be separating the machining and threading from the grinding of the shanks. Now if we ask the question what happens if we eliminate the grinding job in reference to determining if it is a key job, the answer would be it would have to revert back to the screwdriver shank workstation job. It is not a key job, therefore not used to create the job rating point system.

USING THE JOB RATING SYSTEM TO DETERMINE POINT VALUE FOR A JOB AND ASSIGNING A COMPENSATION RATE

Let's now find out how much we should compensate the shank work station operator. We will do this first by using the job description (Fig. 8-2) to create a job evaluation for that job. To do so we'll assume that the point scale for hourly work shown in Table 8-2 is that for the screwdriver manufacturing company. I say this because it is critical that we use a point system that is valid for the specific company and not a generic scale. Looking at the percent value for each factor and points assigned it is recognizable that it is not universal for all jobs anywhere. For example, a company using similar skills as for making screwdrivers, but making prosthetic devices would probably want to rate responsibilities as a higher fraction of total worth than 15%. An error in tolerances for a screw driver blade is going to be less severe than an error in bearing tolerance for an artificial leg. So the percentage may be increased to say 20%. This means other relative importance values would have to be adjusted downward.

After completing the job evaluation we will then enter Table 8-1 to find the compensation range for the job. And we will see that the pay rate is a range and not a fixed value. The reason for this will be explained later.

The data for the job evaluation is filled out as shown in Fig. 8-3. Let's review the data.

Using the hourly rating system (Table 8-2) we compare the points awarding schema with the job description (Fig. 8-2) to subjectively award points for each category. I use the term *subjectively* because for the most part that's all it is with some guidelines. That's why we always have more than one person do a job evaluation. Then we ask all the evaluators to discuss there assessments and come up with a consensus evaluation. It is like judging a competitive sport such as figure skating or diving—99% of the time the judges arrive at fair and just conclusions. Same for job evaluation. If we strive to be honest, the conclusions will be just. The total point score for the shank work center operator I arrived at was 218. This is a summation of points for all categories. I should point out that the evaluaor can only award points as shown and not arbitrarily

Job Title	Job Skills	Education	Responsibility	Effort Req.	Work Con.	Compete	Total	Job No.	
Shank W.C. Operator									
					Points Assigned				
	75	29	30	25	40	19	218	CNC 123	

Figure 8-3. Job Evaluation, Shank Work Center Operator

select points,. For example; if we rate a job as being totally unskilled—someone who only has to carry boxes to a truck, then we would award that job as lowest skill and give 13 points. If the job required carrying boxes to a truck and then stacking them in a manner to conserve space, then we could award a low skills rating and award 25 points in accordance with the rating system. We would not be able to give 20 points because we felt that while some minimal skill was required it wasn't much. The system requires set amounts of points to be given in order to maintain the relative importance between categories we established for the company. There is one exception allowed, for education, which I'll describe later. Let's go over my selections and I'll explain how I arrived at the point values.

Job skills. I selected 75 for this category. Since the operator has to know how to use three different types of equipment, one being a CNC machine, I felt that the degree of skill required is greater than average, but not the highest for all categories of machinists.

Education. This is a combination of education and training required. The evaluation requires a point value for education and one for training, then average the two. The reason we do this combination is because for this type of work we have different types of education to consider; practical and theoretical. One is suited to classroom learning, hence considered formal, and the other is on the job, hence apprentice like in nature requiring time on the job to master.

Using the job description, I assigned 20 points for being a high school graduate. I evaluated the training to be high, judging on the job training for the three machines to be about 3 months. The job description requires CNC certification and 6 months', experience doing template grinding. For these reasons I assigned the high point score for this portion of education, 38 points. The average of the two education scores is 29, and that's the points awarded.

Responsibility. The points awarded are 30. Using the Responsibility criteria, I chose the average value because while the operator is not responsible for work of others, the incumbent's performance certainly does effect others in the factory. If the shanks are not made correctly or in the quantity required it definitely would affect output. In this case the incumbent is the only one making shanks. However, that's simply because one operator on one shift is sufficient for demand. On the other hand, the job is midlevel in complexity therefore it ought not be a difficult set of responsibilities to maintain. So my judgment call came down to average, 30 points.

Working conditions. Here the job rating system takes into account the degree of discomfort or exposure to hazards the incumbent must protect herself against. Please keep in mind that if a job is considered hazardous, the company is still required to take all precautions so the operator is guaranteed he or she is safe doing the work. This category simply recognizes what the incumbent faces if he or she is not working in a safe manner. In this case there is no significant hazards that warrant more than an average rating. The comfort level of the job is certainly well within the norm for a factory floor employment. It may or may not be air conditioned in the summer months, and probably doesn't have the normal decorative amenities of an office, so it rates a higher point score than an office clerical job. But I cannot see where anything but average is justified. Therefore I awarded the average point score of 40 points.

Compete, ease of finding similar jobs in the geographic locale. This a tough category because it can change over time as companies come and go due to economic factors. In fact this is the economic factor of job evaluations. It in essence says we are awarding points, hence dollars, to keep the incumbent on the job. We are evaluating the odds of the incumbent being attracted to another company and what we have to do to keep him. Thus if the incumbent will stay we will not be burdened with the cost of training another individual to do the job and lose out in productivity during the "learning curve" time period.

Notice in this job rating system we are only awarding a 5% relative importance to this category. This implies we do not think losing people to other companies in our geographic area is much of a threat. If a company finds itself in an area of significant skilled labor scarcity, then this % would go up.

For this job I rated the compete category to be high. I did this because the incumbent is rather highly trained and to replace the person would temporarily lose productivity perhaps for up to 3 months. Therefore the point score awarded as per the matrix is 19 points.

The total point score for this job is 219. With this value we enter the pay progression Table 8-1 to determine the pay rate for the shank work center operator job. In this case we see that for the point value the job falls within pay grade D with a starting salary of $7.50 per hour and topping out at $10.00 per hour without any other adjustments.

Let's look at the pay rate matrix and trace it's development lineage.

DEVELOPING A PAY RATE MATRIX

We've now reached the point where it is feasible to determine how much we will pay employees for working in our company. Remember pay has to be considered fair in amount and in differentiation between jobs to have a content workforce. This is equally true for hourly and salary employees. If employees feel that compensation isn't enough, or the distinctions between jobs is not fair they will leave as soon as they can. Pay is the prime driver for employee morale, bar none. Obviously, if two companies have equal pay and one has better working conditions (which can be a very abstract factor), then the one with a better working conditions will draw the better employees. It is also true that companies with very good working conditions can pay less for equivalent jobs. But the amount of difference can never exceed 10% for any significant period of time. In fact 5% is all a management team can bank on if they think there strategy should be to pay less but have superior working conditions. The need for funds to live comfortable for an employee and his or her family is always the overriding condition. With this in mind we need to develop a company matrix that makes the company competitive.

To do so we first have to set some parameters. Unless we feel we need to steal employees from other firms (which could be the case if our company is starting to make a new product and can not afford to have poor quality.) the company will aim to be at the mid point for compensation for its positions. To determine what that midpoint is we need to compare our jobs with key jobs used by the U.S. Government's Bureau of Labor Statistics. The BLS prepares documents showing pay rates for key jobs by geographical areas. Using this as a benchmark, the company sets the rate for its key jobs.

With the dollar value for the key jobs set we go up the scale and down the scale for all other jobs. The job evaluation system gives us this range. It is not by mere coincidence that the point level for the pay matrix goes from 52 to 400. This needs to be the same as the job evaluation system. Note from Table 8-2 that the lowest job would be evaluated at 52 points while the highest job is at 400 total points. All jobs fall within this point matrix. Thus pay needs to be prorated somewhat proportionally to the point values as much as practical. We try to do this as much as we can; keeping in mind seniority, federal wage laws, and past practices (such as company–labor union contracts). For a company starting from scratch the proportionality is easier to maintain than for company in existence for a while. We would also want to make sure that the higher-paying jobs are skewed higher since there are fewer of them, thus less impact on the budget; and the fact these are the people we want to retain. So in summary, the starting point is BLS pay levels for key jobs. Then use a proportionality factoring for all other jobs based on the overall job evaluation point range. A bottom set by minimum wage laws and tradition in the area for how far above the minimum the company's products category and industry fall into (e.g., fast food emporiums set their bottom at the minimum legal wage). And finally, adjustments for what we can call political and competitive reasons.

Now, a company with over 20 jobs does not want to have 20 job rates. It becomes too cumbersome to administer. We also want to create comparable pay jobs to keep people from wanting to change jobs just to make a few more cents an hour. A desire to change jobs internally should be due to an employee truly wanting to improve herself. The pay plan should recognize jobs that have the same comparable worth to the company but in different fields. This helps the fairness doctrine. It says a welder is just as important to the company as an accounts payable clerk because the pay rate is the same. To do this we need to restrict the number of pay grade. Hence on the pay rate matrix, (Table 8-1) we see grades A through G with roughly 50-point spreads for job evaluations at each pay grade. This is done to enhance the credibility of the importance of fairness between dissimilar jobs, and also to make administration of compensation manageable. 7 grades is easier to administer than 20 or more. If more jobs are added to the company's roster the job evaluation system is used to define their place in the hierarchy of importance, and the points are used to define what the pay grade ought to be.

Look at Table 8-1 once again. We see a progress of increases for each grade from start through 12 months. This is the learning curve. Obviously a person who has more experience we believe will work more accurately and faster doing the same work as would a rookie. Therefore that person should be compensated greater. This is what the progressions are for. At the end of each time specified on the progression, the supervisor makes an evaluation as to performance progress. If the incumbent passes a specifically defined hurdle her or his compensation is adjusted. The last column-anniverary—24 months—tells the employee that an inflation adjustment will be made automatically to the pay rate. This is to keep the buying power of the earnings on an equal playing field. It also tells the incumbent job holder that continuous improvement is required for the individual to make more money. Notice in the notes appended to the pay rate matrix the operator is shown how he or she can gain more income in the same job. It also spells out that the incumbent needs to perform satisfactory to get the adjustments shown in the progression.

I should also like to point out that the pay rate table has fixed and variable components. The fixed components are the grades, point ranges, and progressions. All other components are variables and subject to change. Change occurs for inflation adjustments and supply and demand factors for the skills. Although I should point out, it is rare indeed for a job compensation rate to be devalued. Usually if a job is no longer needed, the incumbent is taken off the job.

When the rate table is adjusted we say the card (usually rates on the table are published on cards for employee information) received a raise. This causes a cascading of adjustments to salaries across the board. Companies handle this in many ways, from across the board adjusts for all employees, to corrections to salaries at review anniversary periods. For hourly employees these adjustments tend to be across the board, especially for union represented people. For salaried personnel, they tend to be individualized at the employee's review time. Yes, the exact same methodology is used to set salaried exempt and senior management pay ranges as hourly employees, but usually more secretive.

THOUGHTS ON ALTERNATE WAYS OF COMPENSATING EMPLOYEES, AND FAIRNESS INVOLVED

For a company involved in producing manufactured products, its too easy to create cast systems whereby the factory work force is considered different than the so called administrative and professional employees. This can create a problem in people's sense of worth and importance to the company. Manufacturing engineering through its industrial engineering unit has responsibility for the "scientific" development of pay plans, standards, methods, etc. It also has responsibility for being sure proper psychology is used to get the best productivity rates possible for effective manufacturing. This means team building and creating a sense of fairness for treatment of all employees. Thus any type of compensation system used throughout the factory ought to be equally administered amongst all categories of employees.

What I have described so far is the current popular way of compensating hourly employees. It wasn't always. At first companies literally paid employees a set fee for a day as a hired hand on a farm. There was no differentiation between skills other than between individuals and supervisors, normally called foremen. Factory management emulated that of well run farms. However, as the industrial revolution took shape, and instead of every factory employee being a jack of all trades (apprentice, journeyman, master) but being assigned to one specific type of job, the laws of supply and demand really began to take hold. For those jobs that were either harder to do, or there were fewer people willing to do them, incumbents demanded and got higher wages. Finally, with the work of the pioneer industrial engineers, jobs became codified and were paid as per "scientific" determinations. But this manifested itself into piecework. Where employees were paid for the amount of product they produced. There was no pay grades, etc. The factory owner decided how much to pay for an item, and the employee got paid for as many of those items they produced. This is the so-called piecework system, which until after World War II was the most prevalent way to pay employees. Let's compare this system with the current day work pay system I presented above.

Piecework Compensation System

The historical piecework system depends on the management team defining all the parts to be made and using methods and standards work to determine how to do it and how long it should take. (This is still the two knows, which was valid then and still is today.) Based on supply and demand a financial rate , usually in cents/per piece or cents/100 pieces depending on how long it takes to do the specific operation, is set. For the early piecework systems, late nineteenth to early twentieth century. management defined what had to be made that day in what quantity and posted the jobs. Individuals bid on the jobs, were selected, and went to work; typically based on favoritism or a seniority scenario. At the end of the day a tally was done and employees were paid for what they did, usually then and there and in cash. The remnants of this system are still with us today. Think of day laborers hired to do work on a temporary basis, usually construction or agricultural. Or perhaps as made more famous through literature and movies, the longshoreman hiring halls for offloading a newly arrived ship. The pertinent fact with the historical piecework system is that management set the wage rate based on pieces completed and no matter how well a person did the job there was no chance for advancement to a higher rate for the work done. Compensation was solely based on output.

The modern piecework system has evolved whereby employees are permanently hired and report for work every day. They do their jobs as assigned, and the assignments are virtually permanent as per the day work system. However, their rate of compensation is still per piece. Management has the right to assign different parts to the work station, and the compensation for doing those parts may differ from previous parts. So if an employee is told to make 10 part a components and 20 part b pieces, his compensations would be the mathematical computation of ten part a pieces times its rate plus twenty part b pieces times its rate, and the rates are probably different depending on the time standard. Again the employee only receives pay for what he or she produces. It is also common for the quality clause to be imposed which states that pay is only received for good parts. In fact many companies "punish" workers by making them correct quality problems without receiving payment for the rework.

The system is still ripe for favoritism. Operators are always scrambling for the next job and prefer to get the job that pays the best per piece, as they perceive it. In piecework systems operators are the de facto schedule maker and expeditors, with the supervisor's benign blessing. This is so since they only get paid if they make parts. So they are constantly hustling parts into their work area as best they can with the foreman's blessing and agreement that they can have the work.

Since there could be many people assigned the same job description, competition for incoming work can become tense. The foremen is critical as to who gets work because it is his responsibility to get parts made to proper count and quality. In critical situations the foreman will definitely want the fastest and most competent worker to get the job. If the shop is full and lots of work is available, schedules being the way they are, everyone works and are happy. If not, those employees out of favor for what ever reason may not get work. As can be seen with a disreputable foreman the system is ripe for corruption to exist. This usually manifests itself with operators giving bribes to supervisors to get work.

Contrast this with the day work system. Potential for favoritism is purposely taken out of the equation. Operators are paid for being available to do work on what ever comes across their workstation, and get paid the same no matter what the part is. It is management's responsibility to keep the work going because operators are not responsible for loading the workstations, only to follow prescribed methods with their associated time standards. In day work systems operators are not in competition with each other to find work and foremen have much less liberty of assigning specific work to favored operators. As we will see in later chapters, efficient management depends on effective predictable assignment of jobs to work stations under centralized control. This ensures that synergism takes place, hence optimum times for production prevails. Dispatch of work is more objective and less subjective.

Incentive Systems

Incentive systems are much like piecework but have none of the favoritism, and potential corruption associated with it. Incentives were placed on top of day work pay systems to get workers to be more interested in getting work done. Since the philosophy of day work is to provide instruction and materials to the properly equipped work station, the flow of work becomes an observational phenomena. The operator may be interested in seeing how work flows and know how to improve it, however there is no financial incentive to do so. To overcome this, many companies have an incentive system whereby operators are rewarded for meeting "stretch" goals of more output than the time standard requires. Since every time standard is conservative, based on averages, there is definitely an upside potential.

Initially incentives were offered to individuals. This worked fine for moving parts from one work station to another. But bottlenecks occurred out of control of the operators and no real benefits accrued to the company. To solve this problem, companies began offering group incentives whereby all operators shared in 'bonus" pay if the overall speedup accrued benefits to the company.

Of course for people to really do better than the time standard requires improvements to the methods and better equipment. This has led to quality circles, suggestion systems, etc. as discussed in a previous chapter.

Incentives have proved to be useful for productivity improvement. But my caveat is they have to be fair. We have to be very careful that incentive opportunity exists for all employees. They don't have to be the same for each employee, they just have to be fair and considered equitable. For example, salaried employees shouldn't be offered ways of making year end bonuses unless there are similar systems available for hourly employees. The amount of compensation percentage also needs to be comparable.

To give an illustration, let's say we have a group incentive system that allows the hourly work force to earn about 10% more pay over the course of a year, and this is computed based on output of finished products produced as compared to time standard computed output. It is computed and paid monthly. Let's also say that the foreman can earn an annual bonus based on the performance of his production department. At the end of the year the foreman can receive a bonus equal to 15% of his salary. Is this fair and comparable? Perhaps. The hourly employees are receiving an incentive pay based on monthly results. So if they have good results for the first half of the year, they get an incentive payout for each of 6 months. Then let's say there is a supplier strike and materials are not delivered on time for the next 5 months. The employees can't get their incentive pay for that period. Finally the last month of the year the employees achieve the monthly incentive goal and get their incentive pay. So for 7 out of 12 months they receive an incentive. They haven't received the entire 10%. Mathematically they probably received 5.8%. But what about the foreman? Well, this poor fellow got no incentive pay. His department could not meet the stretch output goal because of output problems. Therefore he received his salary and no more. He had greater risk for greater reward. So we can say in this case there was fairness between the salaried and hourly employees.

The message is fairness has to be perceived to be there for any separate but equivalent incentive system. This has led to many problems when the perception is that there is no fairness, management is easily perceived to be favoring salaried and making "big" money on the backs of the common laborer. When this happens companies are in trouble. Production usually suffers. Companies cannot allow this type of situation to develop. To avoid this, companies have to tie overall company performance to a prorated share of benefits for doing good work. This is simply good common sense because the goal is to produce more hence make more corporate profit. Unfortunately many companies forget this and err in giving bonuses to senior managers and executives regardless if the profit goals have been achieved. This altogether too common current compensation practices for executives and senior management to receive bonuses even when their companies suffer loses is the anti-thesis of the fairness doctrine, and should be abandoned.

My view is the most efficient and successful companies conduct business as if all employees are team members. They work together for common goals and should all share in an equitable manner if they achieve those goal. And suffer in an equitable manner if they do not achieve those goals. To that end some more enlightened companies have achieved significant success, financially and in morale and reputation by having company wide incentives whereby each employee enjoys a bonus if the company meets its goals. The amount of bonus being determined by combination of factors usually based on levels of responsibility and some degree of seniority. The point is, it is possible to have very effective incentive systems where there are no "separate but equal" programs that can muddy the water.

Off-Assignment Payments

In some instances companies have policies where employees can earn extra money by doing jobs that are outside the scope of their normal job description. There are two classifications of such programs; ones that directly increase company profits, and others that increase profit potentials through creating good will. Let look at both starting with good will programs.

Sometimes companies use these off-line incentives to encourage employees to partake in activities that are good for company's perception within the community, thereby creating good will. Such activities as getting employees to volunteer for big brother or big sister outreach programs, to be school tutors, to participate in charity programs, etc. are examples. You might wonder why a company is willing to pay for such things. Isn't doing volunteer or charity work an individual thing? And shouldn't people be willing to do so? The answer is obviously yes. But people only do what their attracted to do and if a company wants to have a significant impact for certain activities it needs to focus on those activities. There is no doubt that if left to their own choices people working for the company will direct their charitable and volunteer time in haphazard and very multidirectional causes all over the spectrum. The result being that a company can get very little good will credit if at all. It would be highly unlikely for the company to be recognized as a significant factor. Left to unguided choices the company would never be perceived to be actively involved in charities through its employees. So it chooses to guide its employees to certain charitable and volunteer activities it has selected for what ever good will strategy it is embarked on.

To do this, companies solicit volunteers for their selected programs. About half of the time there might be no compensation at all, except to make company resources available to help. Sometimes companies will pay travel and other expenses to defray out of pocket costs the employee may incur. Sometimes the company will either allow employees time off from work with pay to participate in such programs, or pay to make up for lost internal opportunities. In cases where employees may miss out on opportunities to receive other group bonuses because they are not there, such as missing out on overtime pay on a weekend, they will actually pay the employee the equivalent for overtime even though he or she is doing volunteer work. The ingenuity as to how to pay is boundless. The goal being to simply make sure the employee is kept whole even though he or she is doing this good will type work instead of his or her normal job.

Programs that directly increase company profits are much more straightforward. An employee does something that directly impacts company profit outside of her normal job, and is rewarded. The most common form of such a program is the suggestion system, as discussed previously. Here the off-assignment work is only a little off-assignment and the main reason for such programs is to generate good morale and teamwork. The second most common program is to pay for sales leads. Factory workers are not expected to find new customers for the company's products. So when they do, it is an added bonus to the company that they would probably not have received otherwise. Let's discuss this program a bit.

First, the company must have a policy in place that says it will pay for sales that come about through a lead given to the sales department from people who are not in sales. It also has to have rules about what constitutes a sale brought about because of the employee. For example, perhaps an operator knows of a machine shop that needs hand tools, such as screwdrivers—the product she is involved in manufacturing. She gives this lead to the sales department which investigates. It turns out the machine shop doesn't need screwdrivers but did need a new industrial vacuum, and it just so happens the company has another division that makes industrial vacuums. That division then has the opportunity to sell and does sell a new vacuum to the machine shop. Does the employee still get a fee for the lead? Would the machine shop have gone to the company independently to solicit a quote for the new vacuum? Was the employee a factor in getting the sale? As you can see there need to be rules as to what qualifies. This is especially important since the company is usually not actively asking employees to help with sales.

Second, the company has to have a clearcut policy on how the amount of payment is determined. Some companies give a flat finders fee of perhaps $200.00 and that's it. Others use elaborate formulas based on worth of the sale and how much the sales costs were reduced because of the lead and so forth. Some companies even pay a merit award of something like $25.00 for any legitimate lead. However it is done, it must be succinctly defined and publicized, and above all it must be construed to be fair and worth the effort for the employee to find legitimate sales leads.

Sometimes these incentives for obtaining additional sales can become quite aggressive and even involve vendors and other firms, such as consultants, that have reasons for wanting to help the company gain sales. Usually because it could conceivably increase their own sales as suppliers of materials or technical support for the additional business their client company receives.

As sales incentive programs become ever more inclusive of more people from a variety of organizations within or without the company, the need for a clearly defined policy becomes even more necessary. Often industrial engineering is asked to participate in developing such policies because of their prime role in setting up pay plans and evaluations for achieving the fairness/objectivity goals. The process of setting up such a policy is the same as that for incentive pay as described above. The key is to balance risk and reward. Employees who need to take greater risks should have opportunities for greater award. For example, people who create situations for gaining sales and take the initiative to go do it ought to get a greater percentage of the bonus pool if successful, than those who simply keep score or facilitate implementation. An example of a typical all-inclusive policy which provides incentives for sales leads for internal and external personnel is shown in Appendix B.

THE EMPLOYEE HANDBOOK FOR SELLING THE FAIRNESS DOCTRINE OF JOB RATINGS AND PAY LEVELS

The employee handbook is an information document given to all employees. It is management's best way to proactively make sure each employee is exposed to all company rules pertaining to conduct of employment, How people are paid, benefits, time of work shift, work rules, disciplinary policies, and other items deemed important for the specific company. Another unstated reason for the employee handbook is to put across the notion that the company treats all employees the same, thus that the company is fair. Since how people are paid is one of the most important consideration in treating people fairly, I've decided to discuss the employee handbook in this chapter. And since industrial engineering precepts are major factors for work rules and pay it is appropriate subject matter for this book. An example of a full employee handbook is contained in Appendix A. It is a specific identity cleansed copy of a handbook I wrote on a consulting assignment for a combined small service/manufacturing firm. While some of the items may not apply to all manufacturing companies, I believe it is a good example of typical content and presentation format.

New employees cannot just be given the handbook and sent off to their work stations. The content and philosophy has to be explained to them. The handbook has several sections, and is relatively complex, requiring an explanation to assure employees understand the company's policies and reasons for them. To accomplish this, most companies have orientation sessions for new employees to assure they are exposed to all the work rules and benefits pertaining to the company. The orientation, usually done by the human resources department, has as its goal to get the employee familiar with the company's way of doing business such that the employee can focus his time on learning his job and not be overwhelmed with learning a new social organization. Realizing that just lecturing to people will not gain them instant knowledge, the handbook idea came to be for the expressed reason of giving people a reference for all of these administrative necessities required for a smooth running of the company. To ensure that the employee has been given a copy of the handbook, and has attended an orientation session; a receipt tear-out section is attached. The employee is asked to sign and it give to human resource stating she has been given a copy of the book and attended an orientation session. This form is then put in the employees service file. This is done to assure that every employee has been treated fairly in getting information pertaining to employment.

Handbooks come in every size and content coverage. The one in Appendix A is typical. At the very minimum they should contain information about pay policies, work rules and conduct, and benefits. I believe a good handbook contains the following sections.

Administration
Hours of work and pay
Benefits
Transfers, termination, and leaves of absence
Work policies and regulations
Conflicts of interest
Safety
Transportation and travel expenses
Suggestions

An example of each section is demonstrated in Appendix A. Lets briefly describe what's contained in each section and pertinent philosophies with regards to each section.

Administration

As we would expect this section sets the stage for the rest of the handbook. It contains definitions of terms that may be new to the incoming employees, particularly definitions about federal wage and hours laws. It will also contain a paragraph explaining orientation and what the employee's responsibilities are for a successful orientation.

This section must also contain an "out" paragraph. This essentially sets up a trial period whereby the company can discharge a new employee within a certain time period without any due process steps except a simple evaluation for fitness to work for the company. Every company needs such a clause in the hiring process to allow for corrections to made for mistakes in judgment when hiring a new employee. Unfortunately it's not unusual for managers to misjudge a person's capabilities during the hiring phase and bring someone into the company who just doesn't fit the prevailing culture.

So a means has to be available to remove the person without violating legal rights. Fortunately courts have upheld the right for companies to have trial periods of employment for new employees where they can be dismissed for cause without recourse. But—and this is key—the new employee has to be informed of this right the company has at the very beginning of her employment. Putting it in the employee handbook and discussing it during orientation preserves that right.

The rest of the administration section deals with employee records that will be kept on file and making sure that the employee knows what his continuous service date is. This date will be used to compute various eligibility requirements whereby the employee qualifies based on minimum employment longevity. For example, most insurance benefits will traditionally not commence until after 90 days of continuous employment. Vacation eligibility, and in some companies payment for holidays are contingent on certain accumulated time of service. It's important that the employee understands the aspects of seniority that are employed by the company for various qualifications, and that her continuous service date is crucial for defining her eligibilities.

Hours of work and pay

This section explains shift times and breaks. It defines overtime and the company's policy with regards to the subject. Some companies virtually require overtime to meet production requirements, so they "demand" overtime work from their employees, while others shun it totally. Working overtime can be a legitimate condition of employment, as long as the company complies with the federal and state laws pertaining to lengths of periods worked between days off, and maximum hours of work allowed per day. In most jurisdictions it's 13 days and 12 hours, respectively. The section also defines when employees will be paid.

In addition, all the administrative rules about attendance are covered. Usually the employee handbook spells out requirements for attendance, particularly lateness and calling in if not available for work. It makes veiled, and sometime not so veiled threats about disciplinary action for poor attendance and lateness performance. It also explains time clock responsibilities, for the employee's own time and prohibiting clocking in or out for others. Emergency away from work considerations and work break rules are also usually grouped in with or adjacent to attendance paragraphs. What is mentioned and is not varies from company to company. Those companies that operate under union contract are very specific and all inclusive within their employee handbooks. Smaller companies with no union contracts tend to be less specific, but do cover the items that affect them the most for their specific situations. If attendance is not a problem, then perhaps a statement saying good attendance is expected as a responsible employee is all that would written. On the other hand if the company is in an area where the culture is not one of work, and this may be the first opportunity for gainful employment for many of the employees, then the handbook becomes a tutorial that not only spells out the requirements for attendance but also include an explanation as to why it is important.

The remainder of section 2 deals with pay. It starts with a statement about job descriptions and pay associated with the job, and the fact that the employee will get a copy of the pay rate table, the "card." The fact that the employee gets a copy of the "card" says the company's pay policy is aboveboard. It says the pay plan is fair and we, the company, do not keep any aspects hidden. It is based on proper job rate evaluations. It implies that pay improvements are left to the ambition of the employee, for example, learn more—earn more. In fact any good employee handbook will always contain a statement defining how changes in pay occur. In the example in Appendix A this statement is shown as Change(s) in Pay.

The rest of the section on pay defines garnishments, errors, deductions, who can receive the employee's paycheck, and how an employee's final paycheck is handled. Many of these paragraphs are the dark side of pay items, but must be covered. There must be no ambiguities concerning pay policies if the fairness doctrine perception we are striving to achieve will be successful.

Benefits

This is the largest section of the employee handbook. First the company obviously wants to brag about its non-dollar payments to employees. Second, the company wants to define as clearly as possible, and of course with a positive spin, what the benefits are and the specific qualification to obtain them are.

This is the section where vacation eligibility and holiday payment eligibility and rules for payment and scheduling are explained. Sometimes salaried and office vacation eligibility are different. Usually the salaried vacation is slightly better than hourly. This may be because the laws of supply and demand may require the company to be more generous to salaried than hourly employees. And usually the differences may only be within the first 2 years of service. Where the salaried employee gets 2 weeks vacation after 1 year, while the hourly employee has a step function of 1 week after 1 year and 2 weeks after 2 years. Some companies try to justify this in the employee handbook. My advice is not to do this. Salaried employees have higher percentages of their population starting after 4 years of college. The hourly employee has a significant proportion of the total starting right out of high school. The perk for salaried employees is sort of a recognition of hard work and expense in gaining an education that is very valuable to their new employer. They bring more to the company right off than the hourly employee. So the extra week of vacation is theoretically a part of their pay package necessary to attract them to the company. These and similar reasons are justifications for salaried employees getting a better deal on vacation. This is still well within the fairness doctrine. However, it is not that simple to explain in the employee handbook. Therefore my advice is don't put it in writing in the handbook, just explain it if asked during orientation or later. The most important factor is that the policy eligibilities be stated.

Paid holidays are shown for all employees. Most companies pay all their employees for the paid holidays. Some still have some service time requirements that have to be complied with. If that's the case it will be stated. But virtually every company requires employees to work the scheduled day before and the day after a holiday to qualify for payment. This is usually not a problem, If an employee wants to take a vacation during a holiday week. Scheduling a vacation day before or after a holiday is the equivalent of working those days for eligibility purposes. The requirement is a form of discipline for those employees having attendance difficulties, and an inducement to show up for work.

This section also has the policies on life, health, and dental insurances the company can provide. What is in this section varies considerably as to what the specific company feels it can afford. Also, being affected by the insurances the company decides it must have to be competitive. The trend is toward more contribution on the part of the employee to partially pay for health and dental insurance. Lots of companies are getting out of life insurance provisions entirely, as an area to economize therefore have more funds available for the health insurances. The fact is current insurance packages are much less than historically, and unless something drastic happens this trend will continue.

Some companies have sick days and personal days as well as vacation days that employees earn through time in service. These have become de facto vacation days. Employees take sick days or personnel days for what ever reason. Many companies just don't want to be bothered with the intricacies of managing these days off, so they simply lump them with vacation days and only have the one category. Depending on when and how and if they ever existed determines if these types of days away from work are added to the vacation day total or simply ignored. The advice for writers of employee handbooks is don't mention these categories unless the company still maintains them in their benefits plan. If so, spell out the rules in a manner similar to vacation policies.

Companies are also getting away from defined pension plans. Since employment in any one company is not likely to be for a lifetime career as it was 30 to 40 years ago, companies no longer feel obligated to get involved with defined pensions contributions by them and vesting, and all the record keeping involved. Instead, companies offer 401K type investment plans for employees where there may or may not be company participation in funding. A 401K plan belongs to the employee and can be transferred from employer to employer as the employee changes jobs. This is the current trend, recognizing that lifetime employment with a company, while not totally out of the question is less likely. The employee handbook must spell out the company's policies for this type of retirement savings, informing employees how they may participate.

All other types of benefits a company offers should be mentioned in the employee handbook. Where they can be succinctly explained, they need to be. If the provisions of the benefits are complicated, then it is sufficient to summarize what the benefit is and where to go within the company to get specific help concerning the particular benefit. Benefits like medical leave of absence and military leave would fall under this category. Mention of worker compensation and unemployment compensation are also topics briefly explained in this section of the employee handbook.

Transfer, termination and leave of absence

Except for transfers, which can be for promotional purposes, this section of the employee handbook is all about unusual happenings and discipline.

The transfer paragraph is usually a short paragraph stating that the company can transfer an employee to another job at its discretion. That's it. Any other set of explanations just clouds the issue. If the transfer is due to a promotion it's good and all the happy events ensuing from company policies outlined in the employee handbook are adhered to. If the transfer is for other reasons, a lateral move for production purposes or perhaps for personnel reasons, then again all the provision of the required policies come into play.

Of necessity for legal compliance issues, the termination section is one of the biggest subsections in the employee handbook. This section typically states it wishes all employees well in their association with the company, but there are times when things do not work out as we would have wished, therefore rules for termination need to defined and understood. This section points out that termination may be due to voluntary resignation, reduction in forces (layoffs), or firing for disciplinary reasons. A resignation is initiated by the individual. The company for economic reasons initiates a reduction in forces. And firing is initiated by the company because the employee violated his terms of employment; for example, the employee did not live up to the provisions of the company's policies, procedures, or work rules. The vast majority of the bad things that make up not living up to terms of engagement are listed in this section. They're listed so an employee cannot make an argument that she simply didn't know she couldn't do such things.

If you review a typical listing of "don't and do's" as shown in Appendix A, you see they are items that contradict following proper methods for doing any job. Especially where team work is essential for business success. However, a single violation of a work rule is never sufficient to discharge an employee. With arbitration and court rulings as precedent, it is necessary for companies to try remedial action first before the "ultimate" penalty of dismissal. For that reason the employee handbook spells out the disciplinary steps for doing "bad" things. It is a cascading set of disciplinary "punishments" based on how many times the offending employee has done the same thing. The idea in theory is to help the individual correct his performance and become a good and productive employee. Then if that doesn't happen the company is free to discharge him. It should be pointed out that some wrongdoings can result in immediate discharge. Any item that can be construed to be violation of a local, state, or federal law, can and probably would lead to immediate firing. Also wrong doings such as giving competitors company proprietary information can lead to immediate dismissal.

The employee handbook's job in this entire termination topic is to spell out the code of conduct the company expects of its employees. We cannot expect to have an ethical work place unless we point out that ethical behavior is expected, and then to remove ambiguities as to what is meant by ethical behavior.

Leave of absence usually occurs at the initiation of the employee. Many times the employee wants to remain with the company but for some compelling reason she cannot come to work on a regular basis. Most companies will allow employees to take leaves of absence. However, most companies will not guarantee the employee will be reinstated to her former job when she decides to return. Rather, the only commitment the company has with the employee is to give her preferential treatment when she returns. There are some leaves of absence where an employee does have rights to an equivalent job upon return, such as pregnancy and military call-up. This section spells outs the terms and conditions for obtaining a leave of absence and what each parties rights are.

Work policies and regulations

The previous section states what employers must not do. This sections states what the employer must do. Of course there is an implied threat that not doing what's required could lead to disciplinary actions. Here we see discussion about care of company equipment and tools, and types of clothing that has to be worn such as uniforms and safety shoes. We also see discussions about energy conservation, security, company moonlighting policy, soliciting for donations (most companies do not allow this), and where current policy information is available—usually a designated bulletin board or set of them.

The biggest subset is about harassment. This has been become a legal issue for companies and virtually every company I am aware of has extensive anti-harrassment policies, particularly sexual harrassment. The impetus for this has been court rulings that states companies are responsible for criminal acts done by their people, particularly but limited to supervisors to subordinates. Therefore what constitutes harassment is defined along with how to report incidents. Every employee handbook needs to include this subsection. I also recommend that every company initiate sensitivity training for all of its employees. The employee handbook orientation can be the start of that activity.

Conflicts of interest

We continue on with pointing out behavior that is not in the best interest of the company. Engaging in such behavior can result in disciplinary actions against the employee. This section must point out what a conflict of interest is and at least imply how it can adversely affect the company. The details must be clear enough so any employee knows what is a conflict of interest. This detail will vary from company to company and industry to industry.

Ethical conduct is part of the interpretation of conflicts of interest. You don't lie, cheat, or steal is the basics of ethical behavior, and that is important for writing this section of the employee handbook. We cannot then say it is okay to imply a lie to vendors by not telling the entire truth, only the part that's beneficial to our company. In writing this section we have to be very careful to make sure the precepts put forth are in actuality the way we run our business.

Also included in this section are the prohibitions of bribes, kickbacks, and other quas-legal ways of making payments to get business. The company also has to be careful to define proper conduct for business entertainment. Again this will depend on the types of customers and vendors negotiating with and the norm for the particular industry. This ranges from absolutely no bought lunches or other gratuities for federal government employees to virtually the sky's the limit for dealings within the entertainment industry. Perhaps here the only sound advice is to do what your conscience believes to be ethical.

Safety

This a growing section of the employee handbook. As we codify more safety practices under Occupational Safety and Health Administration (OSHA) and Environmental Protection Agency (EPA) regulations, more items need to be included in this section. Creating a safe work environment is the best way I know of to create a highly productive work force. I'm in favor of adding as much as possible into the work rules on how to work safely. In fact I believe the employee handbook needs to be the first basic safety primer the individual is exposed to as an employee of the company. All the common sense dictates should be here. If they are, the employee is being warned to be careful. And if she comes across a situation that is different than what she's experienced before, the best course of action is to stop and check it out with her supervisor.

Safety is divided into two aspects; OSHA type safety, which is equipment related, and EPA type safety which is materials and process related. The former is within the normal reality experiences of people. We know not to touch down electric wires, or get too close to a spinning saw blade. For the latter we instinctively know it should be guarded. The EPA items are a world of a different color. The company receives a bag of white powder. Is it as harmless as a sack of flour? Or is it as dangerous as anthrax? Our worldly experiences simply cannot help us. We have to be aware of this dual approach for writing this section of the employee handbook. We can have effective do's and don't, for OSHA type safety. For EPA safety we are a little more abstract in identifying the dangers. So we need to be more explanatory as to how to handle unknowns. Referring to material data safety sheets does this and making sure they're available with every raw material shipment. Also we need to have explicit instruction on how to deal with unknowns written into to employee handbook. In fact the employee handbook on the subject of general EPA type safety needs to be identical to proper proceedings required by the EPA.

Transportation and travel expenses

Every company deals with is issue. For transportation, every time we direct an employee to go somewhere to do a job other than his primary based location, we are incurring additional expenses and also exposing the company to liability in case of an accident occurrence. For this reason we need to promulgate policies that minimize risks and cost.

For transportation the company must spell out the rules for using company vehicles, including who can use them and even ride in them. I should point out that having company vehicles always increases the risk. The example in Appendix A is a good reference for crafting a policy for a specific company.

Travel expenses is almost codified. The federal government periodically publishes vehicle operating expenses per mile. Right now it is approximately $0.40/mile. This includes vehicle wear and tear and fuel costs. I suggest that companies codify the federal standard into their policies, hence the employee handbook. Concerning air or train travel it is customary to reimburse employees for out of pocket cost, usually coach class, or have the company purchase the tickets directly. It is also customary to pay for employee meals and lodging while traveling away from their home base of operation. It is hard to put exact amounts into the employee handbook for this reimbursement. It is probably best to refer to current finance department policies for this and all travel expenses.

Employee suggestions

This section faithfully implies that employees are encouraged to make suggestions on how the company can perform better. Some companies include elaborate suggestion program rules and policies here. I don't think that's a good idea for the employee handbook. Unfortunately suggestion programs, along with quality circles, and other team building programs ebb and flow in popularity. Therefore to try to codify a policy that varies considerably from time to time becomes very difficult to enforce. We chance the problem of having an employee handbook out of synch with what's really being pursued. For that reason I prefer a simple statement as shown in Appendix A, which essentially says if you have a good idea please share it with us and if we use it you'll be compensated for it.

This concludes my discussion of the employee handbook. It is an essential tool for getting employees to work the way we need them to work to create a winning team. Please feel free to use the whole or parts of Appendix A in setting up your employee handbook for your company. But be forewarned, Appendix A was done for a specific client taking into account his specific needs for his industry. Your employee handbook will undoubtedly be different and need to consider you specific needs. What Appendix A shows is generically correct by not necessary specifically correct.

SUMMARY

The doctrine of fairness for pay and non-pay issues for all employees is the key issue for establishing pay plans and opportunities for advancement. How we pay employees must be based on a rational that those affected by the policies deem to be fair. It was demonstrated that job evaluation is based on a point system that takes into account each job's worth with respect to skills, education, responsibility level, physical effort required, working conditions, and the laws of supply and demand for availability for persons to fill the positions. And that the point score for each job sets the pay rate. We further saw that every job has a learning curve whereby entry-level persons earn less but have opportunity for increases in wages earned through learning and gaining experience doing the job. Also, to further ensure the point system is properly calibrated, key jobs are selected and used as benchmarks for ensuring that the relative worth of a job with respect to others makes sense.

The topic of incentive pay was also discussed. Here we saw that incentives can be different for hourly and salaried personnel as long as the doctrine of fairness was adhered to. There is also a relationship between risk taken and the amount of incentive that is received. This too has to be reasonable and fairness maintained.

The final part of the chapter dealt with the employee handbook. Which is a codification of company policies for fair treatment of employees. The employee handbook also defines an expected ethical behavior standard for all employees. It also includes a set of actions an enlightened company would feel duty bound to take against those who violate those standards. The employee handbook is the focus of the fairness doctrine, which puts in words and defined policies how the company intends to treat all employees fair and equal. It is the company's best documented argument that pay and non-pay policies are fair.

REVIEW QUESTIONS

1. Explain how job evaluation is based on: a) the principles of motion economy, and (b) the Two Knows. Does (a) or (b) take precedent in establishing the worth of a job?
2. Define why industrial engineering has the primary responsibility for establishing systems for job evaluation.
3. What is the difference between a specific and a generically based job as it relates to our ability to do a job evaluation and worth of these jobs to the company.
4. Why is developing a set of point values for a job rating system considered to be objectivising a subjective task? Is it possible to create an objective point rating system?
5. For the hourly rating system shown in Table 8-2, discuss why the percentages for each rating category are why they are as shown. Can they be changed? If so, why?
6. Why are job descriptions prepared by a committee representing different constituencies? What are the pluses and minuses of operating this way?
7. Write a job description for the egg crate packer position for which the method was defined. Use the format as presented for writing job descriptions.
8. Conduct a job rate point evaluation for the job description in question 7. Use the Hourly Job Rating System shown in Table 8-2.
9. Assuming the incumbent for the job description has been on the job for six months and is performing adequately, what should his rate of pay be? Use the Pay Progression Table, (Table 8-1) shown for the ABC Company.
10. Is the egg crate packer job a key job? Explain the reason for your answer.
11. Explain why pay grades have a spread of pay amounts.
12. Discuss how the fairness doctrine gains acceptance of the employees for the job evaluation and pay plan systems.
13. What are the benefits and drawbacks of both modern piecework and day work pay systems? Compare the two of them to each other.
14. Why are incentive pay systems considered to be enhancements to day work productivity?
15. Explain why the fairness doctrine needs to be an important factor in setting up an incentive/bonus system.
16. Are there any justifications for paying performance incentive/bonuses when goals are not achieved? If so, why?
17. Discuss why good will programs are important to company profitability, and the ethical precepts for paying employees to participate in them.
18. Why is it important to consider finder fee payments to nonemployees?
19. What is the purpose of a new employee orientation session, and what use is made of the employee handbook during the session?
20. Explain why an employee handbook has to contain job evaluation, pay plan, and benefits philosophies as well as work rules and legal remedies for violation of work rules.

Chapter 9

EMPLOYEE APPRAISALS AND EVALUATION

Employee appraisals and evaluations, while theoretically considered to be independent from the pay plan, really are not. Ostensibly appraisals are for critiquing an employee's performance and offering suggestions for improvement, often bordering on directives, and not for setting pay. However as we've seen every pay system for progression purposes is linked to how effectively an employee is doing in learning the job and performing all aspects of the job. Therefore to go along with the illusion that evaluations are independent of pay systems is simply wrong. Perhaps in an ideal world appraisals are set up simply for the purpose of critiquing performance, with the goal being to bootstrap all performance, hence productivity, to the next level. In the world we inhabit this is only a partial reason for the appraisal system. Couple this with the desire to rate people for pay increase decisions and we have the pragmatic world.

Manufacturing engineering, once more through its industrial engineering responsibilities is very much involved in the design of the appraisal system. The system is based on establishment of proper methodology of doing any job and the measurements we use to evaluate if the methods are being done correctly. Once again we are interested in proper design of jobs with specific measurable requirements. From this job description an appraisal system is derived. While human resources (HR) are usually assigned the task of managing the personnel evaluation system, they do their work based on the system designed in accordance with good industrial engineering practices. This chapter is about the development of the appraisal system we find in most industries, which has been successfully applied to both hourly and salaried personnel.

PURPOSE OF APPRAISALS AND EVALUATIONS

As I mentioned, the appraisal is designed to point out good and bad performance of the appraised person. The performance criteria are evaluated against a guideline for performance as defined by the job description and a raw score is contrived. The score is then used to evaluate against a value system to determine a generic performance level.

The number of performance differentiation levels is most often five, although sometimes performance evaluation is further differentiated. The levels traditionally are: outstanding, very good, satisfactory, meets minimum standards, and unsatisfactory. The next most popular ranking system has seven differentiations. Excellent is inserted after outstanding and needs improvement follows meets minimum standards. How many differentiations there are is immaterial as long as its possible to easily define each level.

By having these differentiations we can rate people in accordance to how they achieve of the goals of the job description. And if each differentiation has a point value, then we can give an individual a score and judge his overall performance accordingly. For the five-step differentiation system, the point values could be 4, 3, 2, 1, and 0. So we can have an overall score and subscores for each job description requirement. With such a system we can appraise performance for any job description requirement, and also compare performances between individuals (This person to person comparison is only done by the manager for personnel actions criteria and never discussed directly with an employee.). With weaknesses known, the supervisor can suggest ideas for improvement and the supervisor/subordinate team can establish improvement action programs.

The supervisor also uses the point score to determine if the employee has progressed sufficiently to have earned the next stepped pay increase defined for the job. That means if the employee hasn't achieved enough appraisal points, he won't get the specified rate increase, and will have to wait for the next appraisal, or at least until he has progressed satisfactorily through the improvement program he has agreed to with his supervisor.

For exempt employees, the supervisor does much of the same to determine how much of an increase in salary is warranted, again done by criteria established for the pay grade. Note, exempt positions are placed in pay grades as is hourly. But exempt employees have a percent range of salary, usually quoted in annualized dollar amounts: For example, grade 5; $55,000 – $66,000, not a set progression from entry to mature pay rate ranges. The exempt salary increase typically depends on point totals from the appraisal. A 4.0 rating equating to outstand would get 100% of the increase planned for the exempt salary plan per individual. So if the budget had a 4% increase, this person would get a 4% increase as calculated from his current base salary. If a 4% increase would put the incumbent above the salary range, it is capped at the high point of the salary range. And henceforth if the incumbent is not promoted to the next higher pay grade, he would only be eligible to receive increases the salary card receives. This is usually only inflation adjustments.

The range shown on the exempt salary card is the amount of spread the company says it's willing to pay for that pay grade. This means there is discretion in hiring as to how much the then prospective employee was offered. Many times when a salary offer is made too high as compared to current incumbents in that pay grade we end up with morale issues. Because two employees in same grade, doing equally good work are not compensated the same. Each may receive a 3.5 rating for their appraisal. In that case they would each receive (3.5/4.0) × (current plan % increase) × (base salary) to determine their respective new base salary. In this case the person starting out with the higher initial salary has an advantage.

If a company feels it has inequities, it can try to solve them by rationalizing that the person in grade with a higher than normal initial salary should be expected to do a better job, therefore judged to a more exacting interpretation of the standard required for the job description. Similarly the person who through no fault of his own was hired at a lower than typical base salary, may be made whole by having the supervisor judge him against a less exacting standard, thus having a higher point score. This is done too often and it is wrong! Such decisions can never be kept confidential, and it smacks of favoritism. The employee being judged to a too exacting standard will feel he is being unfairly treated. At best his work performance would deteriorate. At worst a discrimination law suite could be initiated and even if the company prevails the damage is done. The company is perceived to be unfair. The best solution to underpayment is to simply bit the bullet and give the low salaried employee an adjustment to get him to within the norm for his performance and pay grade. Of course this is a subjective call as to what the adjustment should be and could open up similar morale problems if not handled properly.

So we can see the purpose of appraisals. But we also see how not doing appraisal fairly and within the rules of the evaluation plan can lead to trouble. And in many cases create morale problems even though there was no malicious intention or favoritism shown by management. Now with this introduction to purpose of appraisals and the salary equity pitfalls that need to be successfully navigated around, lets look at how the appraisal system is constructed, and how it ought to be fairly applied.

THE MAIN COMPONENTS OF ANY APPRAISAL

There are three components of an appraisal regardless of its structure. They are:

- Accomplishments as measured against expectations
- Significant factors impacting performance
- People relationships

Keep in mind the purpose of an appraisal is to both differentiate performance levels between employees for allocating proper awards and to improve performance. By linking awards and performance criteria we are using the "carrot and stick" approach to foster striving for optimum productivity amongst the workforce. The three components of the appraisal are designed to encourage good behavior and discourage below standard performance. Good behavior is rewarded, most prominently through salary increases, bonuses, and consideration for promotion. While bad behavior can result in dismissal, and obviously no rewards. As far as using appraisals to improve performance; i.e., improve productivity, it is not an immediate type of benefit. What it does is allow management to abstractly compare an employee's past actions with achievement of goals. And if there are major shortfalls in goals achievement, it is possible to analyze performance to see if there are any correlations between how the employee carried out his assigned tasks and achievement of goals. If there are, then the manager can use the appraisal as a mechanism for re-directing how the employee is to perform in the future. We'll take a look at the three components and see how they lay the ground work for post appraisal activities designed to make the company more productive.

Accomplishments as measured against expectations

This is perhaps the easiest portion of the appraisal for the person preparing the appraisal to do. It is a list of what the employee did, versus what he or she was suppose to have done during the appraisal period.

For production line employees this is simply the quantity of items made vs. how many were suppose to be made and coupled to the quality factor. In this case the number of bad quality parts or quanta of work attributed to the appraised individual, sort of like an earned run average for a pitcher.

For exempt employees it is a little more complex. Exempt employees are often task or project oriented. Thus the list of what they did is a list of tasks or project steps completed, and what the results of these completions were. I fact in order for this portion of the appraisal to be significant, the accomplishments are normally measured against a previously agreed to work plan. Some exempt employees have very little or no project/task work assigned. For example a production control specialist may be assigned the job of inputting schedule data to the MRP program on a daily basis. If there are no projects/tasks to list for deciding if the person has done well or not, then the exempt employee is rated much like an hourly employee based on the specific job description.

Significant factors impacting performance

Significant factors are the modifiers to the accomplishment part of the appraisal. They are the nonproductive items that the employee had to do in order to set the stage for doing the productive work. Here we have quite simple events such as how well an operator keeps her work station "ship-shape" to do work (housekeeping) ranging up to very abstract activities such as how the employee overcame unforeseen obstacles to still achieve the stated goal.

This is the place where the appraiser can compliment the employee for initiative. For thinking outside of the box. For being a productive employee by striving to get around obstacles. It can also be a place in the appraisal for taking the employee to task for no reaction, or improper reaction to unanticipated events. More often than we would like to think goals assigned to the employee get changed. Here is the location in the appraisal where comments can be made as to how the employee handled change. Did she resist too much so as to be ineffective, or did she get on with the change in plans by refocusing effectively on the new goal?

Some appraisal forms lists items on the form requiring the appraiser to comment on these aspects of performance, especially for the simpler items such as housekeeping and attendance. They may also have some of the more abstract items as standard required review items, such as: "How well does the employee overcome obstacles in performing the job?" This is the part of the form where some HR organizations try to put in reviewer requirements for psychological type evaluations, ostensibly for determining employee suitability for specific team membership or promotion. This is fine if trained people do it properly, and if appraisers are adequately trained to handle such areas of inquirry. If these conditions do not exist it should never be attempted. Potential for morale damage is too high with ill-conceived questions and amateur analysis.

People relationships

This is the last part of the appraisal and becomes a catch-all for the appraiser to give her opinion of the employee. Here we have all of the very soft issues; those most devoid of supporting facts to back them up. Such things as attitude, initiative, leadership, potential for advancement and to what position are covered. In many cases it becomes a totally subjective portion of the appraisal and really measures the appraiser's and employee's personal relationship. Most good appraisal forms down play this portion of the appraisal, because it is so subjective. If it became the prime factor in appraisals it would be ripe for favoritism accusals. And this is something we want to eliminate.

In fact the trend is to eliminate this part of the appraisal completely. And instead cover this portion of evaluation only as a judgment between the appraiser and HR and senior management when looking at staffing plans, not part of an appraisal. Simply put, relationships between an appraiser and employee can and do get strained if there is "bad blood" between the two of them. If this is the case any subjective type evaluation is bound to be skewed. Even if the appraiser and employee are well matched and enjoy each other's professional relationship, this type of subjective evaluation will be skewed, obviously to the other end of the spectrum. Therefore the best way to get as objective an evaluation as possible for these "soft" issues is to have as many independent informal evaluations made by as many people as possible. Thereby the mean of the comments is more likely to be correct.

THE DESIGN OF THE APPRAISAL FORM

Appraisal forms come in many shapes sizes and contents. They may be a one page form or a multi-page document. The common attributes for all appraisal forms are that they contain the three elements discussed above: Accomplishments as measured against expectations, Significant factors impacting performance, and People relationships. They would have places for the appraiser to rate each particular point within the three topic headings, as well as places to list action plans as a result of the appraisal agreed to by the appraiser and employee. Most appraisal forms have a place for the employee to write his comments pertaining to the appraisal, if he chooses to do so. And that's it. These are the requirements. They can be as elaborate or as simple as the company desires. The important thing is that the appraisal is done correctly and consistently the same way through out the company.

Figure 9.1 is an example of a simple appraisal form format. This particular one is for an hourly compensated production work. However, by simply changing the topics rated, it can just as easily be used for professional and managerial employees.

Notice how the three portions of the appraisal topics are distinct and there is no chance for misinterpretation. This is important. Information based on facts, supposition, and opinion have to be distinct. Because the types of future actions will depend on the basis of either facts or opinions about performance and it is better that they not be co–mingled for acceptance of the need to change by the employee.

Notice also that instructions on how to fill out the form are right on the form, therefore both the person doing the appraisal and the person being evaluated know precisely how the scores and opinions are arrived at. We also have a space for action plans to be noted based on the current evaluation of the particular rating point. This is important so that the supervisor and employee can define what the employee should do to improve future performance, and also as a point of reference for the next appraisal. To show you what I mean, let's look at the egg packer job once more. The items the incumbent is expected to do in the course of her daily work are repeated below from Chapter 8. Let's make up a fictitious evaluation for the quality section of the standard appraisal form. Let us suppose that Janet, an employee on the job 3 months, is coming up to her first progression pay increase time and we need to appraise her for her performance. In this case her progression on the learning curve. As with most companies there is no specific go no go performance accomplishments along the way to total competency to do the entire job. It is simply a subjective impression made by the supervisor over the performance period. So let's look at what we think Janet has accomplished for the period.

ABC Company Performance Assessment

Employee Name	Job/Position Title
Supervisor/Manager	Period Covered by this Assessment
Preparation Date	

Rating Criteria
Mark the appropriate box with an "x" and provide comments to justify ratings. To achieve a specific rating, the following percentage of the specific requirements of the category must be met: Outstanding-95%; Very Good-85%; Satisfactory-75%; Meets Minimum Standards-65%; Unsatisfactory- below 65%

Scoring Criteria
Each rating criteria has specific points assigned. Average the numeric values for each of the three evaluation areas. Average the three evaluation areas for a total evaluation score. Outstanding=4, Very Good=3, Satisfactory=2, Meets Minimum Requirements=1 Unsatisfactory=0

Accomplishments as measured against expectations.

QUALITY:

Expectations	Rating	Action Plan

- Is attentive to detail and accuracy
- Is committed to the quality standards
- Makes continuous improvements
- Monitors quality levels in accordance
  with the quality plan(s)
- Takes initiative to correct quality
  problems
- Produces error free work
- Puts quality first
	Outstanding (4) [ ]
	Very Good (3) [ ]
	Satisfactory (2) [ ]
	Meets Min. Req. (1) [ ]
	Unsatisfactory [ ]
	Comments	

Employee Name

PRODUCTIVITY

Expectations	Rating	Action Plan
- Maintains a fair work load - Manages priorities - Follows defined methods - Manages time well - Handles information flow properly - Maintains pace set by standards - Puts forth effort required to meet production schedules	Outstanding (4) [ ] Very Good (3) [ ] Satisfactory (2) [ ] Meets Min. Req. (1) [ ] Unsatisfactory [ ] Comments	

JOB DESCRIPTION SPECIFIC ITEM, :(_____)

Expectations	Rating	Action Plan
	Outstanding (4) [ ] Very Good (3) [ ] Satisfactory (2) [ ] Meets Min. Req. (1) [ ] Unsatisfactory [ ] Comments	

Point score: Accomplishments as measured against expectations. [   ]

Figure 9-1. Simple Appraisal Form

JOB DESCRIPTION SPECIFIC ITEM, :(_____)

Expectations	Rating	Action Plan
	Outstanding (4) [ ] Very Good (3) [ ] Satisfactory (2) [ ] Meets Min. Req. (1) [ ] Unsatisfactory [ ] Comments	

JOB DESCRIPTION SPECIFIC ITEM, :(_____)

Expectations	Rating	Action Plan
	Outstanding (4) [ ] Very Good (3) [ ] Satisfactory (2) [ ] Meets Min. Req. (1) [ ] Unsatisfactory [ ] Comments	

Significant factors impacting performance

PROBLEM SOLVING

Expectations	Rating	Action Plan

- Correctly identifies and defines
  problems
- Identifies root cause
- Generates alternative solutions
- Seeks long term solutions
- Implements and tests proposed
  solutions
- Iterates to best solution
	Outstanding (4) [ ]
	Very Good (3) [ ]
	Satisfactory (2) [ ]
	Meets Min. Req. (1) [ ]
	Unsatisfactory [ ]
	Comments	

INITIATIVE

Expectations	Rating	Action Plan

- Takes independent action
- Recognizes opportunities and takes
  appropriate action(s)
- Seeks out new responsibilities
- Anticipates and prevents problems
- Increases skills through on the job
  and formal training
	Outstanding (4) [ ]
	Very Good (3) [ ]
	Satisfactory (2) [ ]
	Meets Min. Req. (1) [ ]
	Unsatisfactory [ ]
	Comments	

Figure 9-1   (Continued)

INNOVATION

Expectations	Rating	Action Plan
- Challenges status quo - Takes calculated risks - Considers ideas of others - Consistently looks for better methods - Successfully implements new ideas - encourages others to be innovative	Outstanding (4) [ ] Very Good (3) [ ] Satisfactory (2) [ ] Meets Min. Req. (1) [ ] Unsatisfactory [ ] Comments	

ADAPTABILITY / VERSATILITY

Expectations	Rating	Action Plan
- Accepts authorized changes readily - Give new ideas fair hearing - Willing to accept new responsibilities - Handles pressures and uncertainly without adverse effects on performance - adjusts work plans to meet evolving organizational needs - works in other job descriptions as needed - Can work succesfully on multiple job assignments, as required - All around team player	Outstanding (4) [ ] Very Good (3) [ ] Satisfactory (2) [ ] Meets Min. Req. (1) [ ] Unsatisfactory [ ] Comments	

Point score: Significant factors impacting performance [ ]

People relationships

INTERPERSONAL SKILLS / TEAMWORK

Expectations	Rating	Action Plan
- Uses good listening skills - Accepts constructive criticism - Builds good working relationships - Flexible and receptive to different ways of achieving team goals - Willing to exchange ideas and opinions on how to achieve team goals - Cooperative with others - Sensitive to feelings of others - Strives to prevent conflicts and will help in resolving differences - Puts forth team goals over own	Outstanding (4) [ ] Very Good (3) [ ] Satisfactory (2) [ ] Meets Min. Req. (1) [ ] Unsatisfactory [ ] Comments	

DEPENDABILITY

Expectations	Rating	Action Plan
- Meets attendance requirements - Works independently with little or no supervision - Meets commitments - Accepts responsibility for doing job correctly - Reports deviations to job methods and procedures promptly - Accepts and abides by company policies	Outstanding (4) [ ] Very Good (3) [ ] Satisfactory (2) [ ] Meets Min. Req. (1) [ ] Unsatisfactory [ ] Comments	

Point score: People Relationships [   ]

Figure 9-1   (Continued)

Employee Name

Point score: Total [ ]

General Comments and Recommendations

Employee Comments (optional)

Appraiser Signature:_____ date:_____ Employee Signature_____ date:___

Reviewing Manager Signature:_____ date:_____

NEXT SCHEDULED PERFORMANCE REVIEW DATE:_____

Figure 9-1. Simple Appraisal Form

Egg Packer Job Description:

1. Pick eggs from a tray.
2. Scale them as per categories, small, medium, large, extra large using supplied gauges.
3. Put eggs into corrugated packing trays, per scale classification.
4. Put protective covers over trays.
5. Place trays in specially designed egg crates, making sure only same scale size eggs are in each crate.
6. Seal crates
7. Mark date and egg scale size on crate.
8. Move crate to shipping zone.

Of the nine items we believe Janet has done well, (e.g., no known mistakes) for all items except 2, 5, and 7. She still gets confused at times on how to use the size gauges and has had trouble packing same size eggs in crates. She also has some trouble putting all the information required on the packing crates. With this as a background, let's rate her for quality performance.

For quality

We look at each statement under expectation, and judge whether or not she meets them, then make an overall judgment as to her rating. "Is attentive to detail and accuracy," is the first statement. The evaluation for this measurement point is probably moderately yes. So we would rate her probably as a satisfactory. Remember she gets confused some of the time, not all of the time. We would go through each statement and make similar judgments as to rating. When we get to the statement: "Produces error free work," we know the answer is not always. So we would rate her down the scale depending on how frequently her mistakes occur. If they are very infrequent then the judgment rates a higher rating. If the opposite is true a lower rating. Also keep in mind that Janet is a new employee. So if her attitude is deemed positive, we should give her the benefit of the doubt and perhaps give her a meets minimum requirements rating.

Finally after going through each statement we would come up with an overall rating. Let's say we rate Janet as "meets minimum requirements". Now we certainly are not content to let her stay at that level for the entire time she is employed by our company and doing this job. Therefore we need to create an action plan to help Janet become a better egg packer. Her basic problem is mistakes using the gauges, and also mislabeling some crates, probably as a result of gauge mistakes. So an action plan filled out as part of the appraisal could be.

> "Work with supervisor to learn correct usage of egg size gauges. Practice and demonstrate correct usage to the supervisor. Ask for help in gauging when ever in doubt. Also consult quality plan on how to mark egg crates. Ask for help when in doubt as to how to mark crates. Keep notebook on how to gauge eggs and mark crates. Complete these actions within three weeks after our evaluation meeting."

This action plan states what to do and when it has to be completed. And if it is followed with due diligence, Janet should learn how to gauge correctly and how to mark crates correctly, thereby improving her performance.

The last page of the appraisal form is the summary page. The appraiser puts down his overall judgment of the employee's performance. This has to be in words with specific facts, not platitudes. For example it is not sufficient to say:

> "Joe is doing an overall good job."

It should also say how and why. For example:

> "Joe is doing an overall good job by constantly striving to meet production goals. His attention to detail is very good in that he tries to understand the important factors associated with performance of the job. He is willing to ask for help and give help to others in order to achieve the unit's goals."

This statement tells how he approaches doing a good job—by understanding the factors of the job. It also states why he does a good job—he wants to achieve the unit's goals.

Following these general comments is a place for the employee to put his own comments down for the record. Most employees do not do this. This is optional, but a good idea nonetheless. If the appraisal is less than what the employee expected, it is a place for him to explain why. It is way to vent frustrations and perhaps surface some issues between him and his supervisor. Thus affording an opportunity to address and resolve them. It is always a good idea to have a safety valve of some sort within the appraisal, so resentment is not suppressed letting unhappiness simmer leading to poor morale hence productivity.

Finally there is a place for all the requisite signatures. It is important that the reviewing manager signs off before the appraisal is discussed with the employee. This way the employee knows the appraisal's basic content has the support of the appraisers boss, hence agreement on the evaluation. This leads some degree of credence to the opinions of the appraiser and makes it at least a little easier to sell unwelcome news. It also makes it easier for the appraiser to get her point across as to what the employee needs to do to make improvements.

The last line on the appraisal is the next schedule review date. This sets the tone for the urgency of completing the action plans. If an employee sees a next appraisal set for a year from now and his performance was not in the "very good" category, then it sends the message that management doesn't really care too much about him, and this appraisal was simply a device to follow required company policy and perhaps to justify a low ball salary increase. It could also be taken as a message that the employee is not going anywhere in the company, promotion wise, and may want to consider looking elsewhere. It could also be a telltale admission that come needs for a layoff, the employee would certainly be on the high probability list to be let go.

As a rule of thumb, anyone with an appraisal rating less than "very good" should be appraised at no longer than six month intervals. Since we know our employees are the most important assets a company can have, all efforts need to be made to get those employees to perform at the highest level possible. This means working with employees to upgrade their skills and the performance appraisal is one very important tool to help in that task. The next section of this chapter deals with just that. How do we use the appraisal system as a tool in the quest to improve performance.

USING THE APPRAISAL AS A MANAGEMENT TOOL

Constructive Critique

We should consider an appraisal as an on-going dialogue of performance vs. the job description. On-going because we are interested in continuous improvement on the part of the company's employees. We should never consider an appraisal, as say, an exit interview with an employee either having left his job voluntarily at his initiative or as one we wish to have leave the company's employ. Unfortunately we all to often take this attitude. An appraisal is not an end of term school grade. It is instead an assessment of a work in progress, which we are striving to improve.

Start with the meaning of the word *critique*: to make an objective assessment of a set of actions or results of a plan. This is what we want an appraisal to do. We want to assess objectively, then decide where we go from there. For an appraisal to be effective it needs to be as close as possible to a truly objective assessment. Therefore the evaluation points we use need to be as objectively measurable as possible. So the hallmark of a good appraisal is to be based on facts pertaining to performance. To do this we need to start with agreed goals, which is work planning. We also have to get past any semantic divide, and understand what a goal is for assessment purposes. Fortunately it is the same as for any goal, for example, "a measurable statement of intent, with a finite completion date." Typical examples would be:

"Goal—make10,000 hammers in 2006 calendar year, at a rate of 193/week."
"Goal—hire a qualified production clerk by August 31, 2006"

In order to perform a critique we need facts pertinent to the employee's performance. These facts are obtainable from reports, data, or perhaps observations, and maybe inputs from other employees. But what ever the source they have to be verified. One way to do so is to ask the employee if they are correct. This is sort of a pre-appraisal discussion whereby the appraiser and the employee reach a consensus as to the accuracy of the facts that will be used to prepare the employee's evaluation. By having this pre-meeting the employee has a head start in thinking about his job performance such that during the appraisal review better more focused action plans can be arrived at.

Using the agreed to facts the appraiser writes the appraisal by comparing the facts to the job description requirements. The appraiser gives his assessment as to how the employee did his job, reviewing goals vs. accomplishment. Then during the performance review session with the supervisor and employee can have a discussion, devoid of much emotionalism because they both know the areas to be discussed beforehand. The supervisor gives his assessment and asks the employee for any extenuating circumstance, and the discussion, now very focused, goes on from there. Both participants discuss reasons for variance and look for ways to make improvements. Ideally this discussion should be akin to a gymnastics coach critiquing the performance of his athlete's floor exercise routine. The end point should be an agreed upon plan of action to improve.

Admittedly no performance review can be totally objective without emotional content. But we can do the best we possibly can to keep it from being an ego destroying activity. We do this by stressing the work planning goal of the appraisal. This is done by using techniques to minimize the perception of failure on the part of the employee. One effective way of doing that is to use "the sandwich technique." This is stating a bad performance aspect only after appraising on a good performance prior to and again right after discussing the bad performance, hence the title "the sandwich technique." Another way to minimize emotionalism is to divorce salary discussion completely from performance discussion, by at least one day.

Critique Leading to Work Planning as Part of the Appraisal

Try very hard to make the appraisal part of the ongoing work planning activity. Remember, the objective of the appraisal is to make the employee an effective member of the goals accomplishment team. Use the appraisal a way of reminding the employee that work planning is based on the company's Objectives and Goals which boils down to a scheduled list of "things" to be accomplished, some of which are the employee's responsibility to get done within a set of resources parameters and deadline dates. Also emphasize that work planning requires one on one reviews of status and the appraisal is just the most comprehensive of those reviews.

An appraisal can and should be an effective work planning document. The critique portion is a review of the work planning history and sets the current baseline. The appraiser discusses the specific goals and where they stand from his viewpoint. The employee comments from his viewpoint. And then the both of them elicit a course of action for improvement. Improvement whereby the employee is doing his job better, vis-à-vis better techniques, work habits, etc. results in company goals being achieve as per the requirements schedule.

Once the points of performance that needs improvement are agreed upon. The best way to get to the agreed to work planning action steps is through the use of participatory management techniques. This usually results in the employee being a willing participant in the actions to be taken by him, and not just doing so because he is being instructed to. In participatory management the appraiser must first make sure the goals to be achieved are defined. Then the appraiser must give his viewpoint on how that may be accomplished. He must then ask the employee how he would do it, or at least he elicit, suggestions on how to improve the proposed method. Then have a discussion with the employee on how "we" (although we really mean "you" with "my" supervision) will proceed. Finally agree on a plan that becomes the next basis of measurement for future reviews, the next work planning session, or next appraisal.

The formal appraisal may be slightly different than the normal work planning sessions in that they add personal improvement goals to the company goal. But this is okay. In order for the company to achieve its goals it needs strong, effective people. So it is proper to consider employee skills building as part of work planning. But we can't let people skills improvement become more important than achieving company goals. This happens when the appraisal system is not integrated with achieving company goals. It does no good to have skilled happy people not focused on company goals. It is critical for the appraisal to be part of work planning. That it be a complimentary one on one session. This way appraisals do not become an independent management sub-system. But they are an integral method of assessment for reaching the company's goal. After all, reaching the company's goals is primary. They are necessary for the survival of the company. The appraisal system is just one part of achieving people performance goals which must be complimentary and beneficial to achieving the primary goals of the company.

THE APPRAISAL CYCLE

We know the appraisal is part of the work planning system. However, there are some particular demands for personnel improvements that require a specific integration of activities over the course of the year. While the formal appraisal is definitely a specific work planning session, integrated with all other work planning sessions, there are additional actions we do to make the formal appraisal more effective for people skills growth. Let's look at this cycle as if it were an independent entity (which I must emphasize it is not).

The cycle starts with the end of the formal appraisal. Goals have been set for the next cycle, normally a year. The goals are a combination of:

- Personal improvement
  - Skills
  - Performance
- Company goals (work planning)
  - For specific areas; examples:
    - Production accomplishments
    - Sales accomplishments
    - Projects accomplishments

At the conclusion of the formal appraisal the supervisor must strive to end it on a positive note regardless of the rating assessment of the just completed evaluation. We want to set the stage for positive accomplishment in the next cycle.

The next step is to schedule informal appraisals. This may seem oxymoronic to schedule for informal appraisal but it's not. We're simply scheduling a time to get together to discuss how the employee is doing with regards to the goals of the formal appraisal. As a guideline a manager should have a performance review to discuss personal improvement goals at least monthly with each direct report. A manager should do this more so for problem employees.

The informal appraisal can be and often is the traditional work planning session. The only time it should not be is for an employee who needs considerable tutoring for development of personal skills. In that case two meetings need to be scheduled; the work planning session and the informal appraisal review. For a successful review, the supervisor must insist that the employee bring all pertinent data needed to support contentions and problems associated with his projects. This way the meeting does not degenerate down to discussions of generalities and impressions. These meeting should take top priority for the supervisor and the employee and should not be cancelled unless for extreme urgency. The meetings should last no more than 30 minutes, although the first one or two may stretch a bit longer. The format is the same as for work planning session. The first 15 minutes belong to the employee to present his data and advocate the current status. The last 15 minutes are for the supervisor to comment, coach, encourage, and for both to decide whether there needs to be schedule changes or other modifications.

The next type of meeting occurs quarterly. They are called interim reviews. This is where managers can gain a reputation for being a concerned people manager. Managing is coaching and directing people to achieve individual and group goals. Those who seriously take on the challenge to coach well are de facto good people managers. The interim review represents the core management task. The goal is to achieve the most output possible from the employee and create a true win-win situation.

The interim review requires preparation and follow through. The manager gives the employee a week or two notice of when the meeting will occur, what the manager would like to be reviewed, and an opportunity for the employee to select additional topics. The manager then must prepare himself by getting current on what the employee is doing with respect to his assigned goals. The manager also needs to review those goals to see if there are any revisions that need to be made that the employee may not be aware of. The manager should also review notes from recent one on one sessions and any other pertinent information on current performance. Keep in mind that the employee will be presenting status and expect some form of coaching and critique. So the manager needs to be prepared to offer positive help and direction.

During the review, which is more like an interview, the manager wants to review each performance goal assigned to the employee. He wants to exchange views on progress or lack of. If there is a difference of opinion on either side, there must be facts to back up each position. This is particularly important for the manager. To not have facts would give a bad impression that the manager is forming opinions on intuition and rumors, both detrimental to achieving a good coach/ employee relationship. In no case should a manager make subjective conclusions.

The manager should be prepared to discuss measurements that can lead to objective conclusions. She should use her independently arrived at measurements as well as those obtained by the employee. Based on these facts the manager has an obligation to identify good and bad performance and discuss them for future actions with the employee. Remember to use the "sandwich technique." We want the employee to leave the meeting feeling charged to do a good job. The results of the meeting should be agreement to continue carrying on the formal appraisal's documented action plans, or modifications as the situation demands. Finally the manager should tell the employee what his performance rating would have been if this was a formal appraisal, emphasizing that it is not a formal appraisal and the information is for the employee's use only.

The manager should use interim reviews as opportunities to keep employees focused on company and personal goals. Interim reviews should also be use by managers to get and maintain buy-in for company goals, and give and get feedback on policies and procedures and to demonstrate an interest in the employee's progress.

At the end of the cycle the manager and employee have the opportunity to conduct a formal appraisal, as discussed previously. It is very important for the appraiser to suppress subjectiveness as much as possible. One way to help is for the appraiser to ask himself the following questions during the preparation of the appraisal.

- Am I being fair?
- Do I have all the information I need?
- Am I giving the employee enough time to respond with information?
- Am I asking for the right information?

During the preparation of the appraisal start by reviewing the last one and how the employee has performed during the monthly one on ones and interim reviews. Also the appraiser should ask himself what prevented him from giving the employee the next higher rating the last time he was formally reviewed. Key in on this to see if the employee has made sufficient progress in those hold back areas to rate an improved performance evaluation. Also be careful to mitigate the halo effect, for example, the tendency to let one favorable or unfavorable incident unjustly sway the overall appraisal. Be careful too, to not judge late cycle events to have greater weight than early cycle events, Also be on guard against favoritism, prejudices of any kind, giving a person high or low scores based on a group performance, and stereotyping based on preconceived notions based on race, sex, color, religion, age, or national origin.

Writing appraisals is easy to do if you don't care. But it is a difficult task if you do care. Good managers do care and work hard to produce fair and objective appraisals that improve an employee's performance, hence that of the company. Always be a good manager.

SUMMARY

The appraisal system is not an independent system. It is fully integrated in the overall system of achieving the company's goals. In fact it can be thought of as the personnel improvement part of the work planning cycle.

Appraisals are means for developing an employees work skills. They should not be primarily tied to the pay system, but at least equally tied to the achievement of company goals. Good appraisals are used for constructive critique and to build rapport between a supervisor and employee. The employee should look upon the supervisor as a mentor and coach on how to do the job better. Not as a task master and only an assigner of jobs. Good appraisals assist in teaching employees how to become better employs.

REVIEW QUESTIONS

1. What is the relationship between appraisals and salary adjustments?
2. How are job descriptions used for formal appraisal purposes?
3. Is it proper to compare performances of several individuals for appraisal purposes? If so why? If not why not?
4. How does a supervisor use appraisals to induce improved performance by an employee?
5. How are accomplishments measured for appraisal purposes?
6. Why are appraisals divided into three parts: accomplishments measured against expectations, significant factors impacting performance, and people relationships? What are the relationships between the three?
7. What part of the appraisal form is coincidental with work planning and how is it used to measure performance?
8. Evaluate the egg packer employee for the productivity accomplishment section in Fig. 9.1. Base your evaluation on the narrative example shown for the quality accomplishment section.
9. Why do employees have a space on the appraisal form (Fig.9.1) to make comments? Give reasons why employees may or may not want to take advantage of this opportunity.
10. What would be the reasons for writing an appraisal at less than 1 year intervals?
11. Why is an appraisal considered an assessment of a work in progress?
12. Where does a supervisor obtain facts to write an appraisal? How are facts checked for validity?
13. What are the reasons for writing appraisals?
14. Describe what the "sandwich technique" is and how it is used.
15. What is the difference between informal appraisal reviews and work planning sessions?
16. What is the relationship between coaching and good appraisal system management?

Chapter 10

PROCESS CONTROL ENGINEERING AND QUALITY CONTROL IN JOB SHOPS

Job shops are the most numerous manufacturing firms as compared to continuous flow (production/assembly line) manufacturing. Roughly 75% of all manufacturing occurs in batch sizes of less than 10,000 pieces, and probably two-thirds of that population have batch (lot) sizes of less than 1,000. The basic premise of quality control including statistical quality control applies to all types of factories. However, as lot sizes decrease it becomes ever more difficult to use statistical process control. In many factories, since every item in extremis may be distinctly different from the one before or the one following it, typical quality control statistical techniques often fail to achieve the goal of ensuring that the products meet the design objectives. Therefore, individualized process control is essential for success. In job shop process control, analytical evaluations are made for adequacy of design, producibility, shop performance, minimization of costs, and so forth. A fully implemented process control system affects almost all phases of engineering and manufacturing operations and is a vital link in the design engineering/manufacturing interface.

This chapter explains how a successful job shop process control system works. Since job shop process control includes most continuous process quality control procedures, those will also be included. We will consider the dynamic nature of the system, how manufacturing losses are controlled, how goals are measured, how statistical concepts can be utilized, how Total Quality Management (TQM) and continuous improvement concepts such as six sigma are utilized, and, most important, why it works.

SEVEN STEPS OF THE MANUFACTURING SYSTEM

(This section is based on my article "Making it in the competitive world," published in *Mechanical Engineering,* May 1998, ASME International.)

The seven steps of the manufacturing system have already been mentioned several time in this text. The seven steps, as I describe in detail below, are the premise for making products of required quality and in accordance with the scheduled needs of the customer. And process control is the glue that holds it all together. Because it is the system by which we ensure each step is done in accordance with the total product plan for supplying the finished goods to the customer.

It all starts with the "Two Knows"; know how to make the product, know how long it should take to do so. These are the Basic Tenets of manufacturing. The seven steps are an expansion of the "Two Knows" in that we need to know and place in logical sequence, where orders come from, how to receive those orders, how we acquire the appropriate material, and where we ship the product to after it is made. This is required information that greatly affects the process of manufacturing and the derived cycle times. Also, during this process it is vital that we know how we will assure we've done the job correctly, the quality step.

This leads to formulation of the seven steps, which is a logical manner for ideas to become reality in the manufacturing universe. There exists a series of events that all companies who are producing goods and services traverse through for each and every product they produce. The odd thing about it is that only the very best are consciously aware that they are doing it. And it should come as no surprise that those that are aware, are coincidentally the companies that are the most efficient and profitable. They are "making it" in the competitive world. This series of events is the manufacturing system and they can be described in seven discrete but interlinked steps. This is not a computer system. It is not software, although it forms the basis for developing integrated business software. It is a logical approach to solving the problem of business; for example, how to deliver goods and services on time and at a profit; and is as old as the industrial revolution itself. The seven steps are:

1. Obtain product specification.
2. Design a method for producing the product, including design and purchase of equipment and or processes to produce, if required.
3. Schedule to produce.
4. Purchase raw materials in accordance with the schedule.
5. Produce in the factory.
6. Monitor results for technical compliance and cost control.
7. Ship the completed product to the customer.

Let me reiterate. Those companies that are aware that the seven steps exist and consciously try to make sure they do all of them to the best of their ability in sequential order, are the successful companies. Those companies that fail, do not try to be in compliance and more than likely don't even know the seven steps exist.

These steps follow a definite sequence and should except for rare circumstances be done in sequential order. We can see that there are definite linkages between all the steps and that it is "intuitively obvious" that the predecessor step must be done before the successor step. But is it really so obvious? How many manufacturing companies try to make a product before the design is finalized and then are baffled as to why costs are so prohibitively high? This leads to operations rule number one.

Rule 1.   Never attempt a successor step until the predecessor step is "really" complete.

To do otherwise is like taking off on your leg of a relay race without first receiving the baton.

The other "intuitively obvious" observation is that the content of what happens at each step needs to relate to same set of facts. Many times in industry we find that sales has not fully transferred customer requirements to design engineering and manufacturing so the company blithely goes about designing and building a product that the customer "almost" wants. This leads to operations rule number two.

Rule 2.   Make sure the data being acted on at each step of the manufacturing system are consistent and identical between predecessor and successor steps.

Confusion reigns supreme when this rule is violated. It is like sales saying the customer wants green tomatoes. In response, the engineering team concocts a way to make green fried tomatoes because its more elegant to do so. But it turns out that all the customer really wanted were tomatoes that weren't overripe.

What rules one and two show us is that we should do everything in business with well thought-out logic of process and that we should make sure that the entire team is aware of what the logic is. We use the process control system to require we are constantly checking to ensure that the well-thought-out process is being complied with. Now let's look at the particulars of each step of the manufacturing system and see what are the optimum approaches so your company can be World-class and truly be "making it."

Before we proceed, let me state that the seven steps applies to services as well as physical goods. Providing a service is just as much a deliverable as making hard goods. A company that contracts to mow your lawn and trim your shrubs has to go through the exact same logic as one that is contracting to build your patio. As we proceed with the description of each of the seven steps we'll see that world-class companies fulfil the intent of each step for everything they do regardless of what the deliverable is.

Obtain Product Specification

This is the sales and product design phase of producing a product. Sales people have to determine what the customer wants then transfer the information to the design team to conceptualize how to create it. This is a very iterative step whereby sales has to be very careful to fully understand the true needs of the customer. Sometimes it's easy because your business is a readiness to serve business and the customer comes to you. Sometimes this is difficult because customers do not know what they want until the new product is a reality. It is the latter where world-class companies differentiate themselves from the also-rans.

Often times we think we see products created then a market created for it. Or so it seems. In reality what is happening is that marketing and sales are discerning what the customer wants by focusing on future desires. For example, modern production technology has allowed us to mass-produce products thus lowering their costs. Which in turn, makes these products more readily available for the general population. Now people are enjoying products that weren't previously within their means and they want more. They want products to be individualized, at mass-produced costs. Knowing that this is the trend, astute marketers, teaming with design engineers and manufacturing engineers, are stretching the technological envelope. They're saying, "see, we can give you individualized designer jeans at the same price as the mass-produced variety were last year." They know they can do this because they've bought "smart" sewing machines utilizing the latest microchip technology. They are obtaining product specification in an abstract manner. They are using so-called soft science—psychology—to do so. This is obtaining product specification generically speaking. They tap the unconscious desires of the body public market place and discern what their company should produce (always in line with their company's capabilities) to obtain product specification. Does your company have the capability of understanding its markets and obtain product specifications compatible with your capabilities in this manner? If not, it's not world-class.

How do the best companies ensure that product specifications meet their factory's capabilities? At a minimum they employ some form of concurrent engineering team whereby the team is formed from all functions of the company to usher products from conception to distribution. The best companies do one more thing. In addition to simply discussing products with their customers, they expand it to a more definitive analysis by employing a technique called pragmatic Quality Functional Deployment (QFD). QFD allows them to discern customer needs and compare it with their firm's ability to meet those needs. It is a systematic approach of matching a client company's product specification requirements with the producing company's capability to do so, and creating a probability model for assessing success rates. Therefore the producing company knows the level of difficulty it is likely to encounter before a project is started. This way management can make a go no go decision based on the best available data. If your company doesn't have this capability it is not in the category of "making it." It is spinning the roulette wheel on every job it takes on.

Design a Method for Producing the Product, Including Design and Purchase of Equipment and/or Processes to Produce, If Required

This second step used to be considered the manufacturing engineering step. In actuality it is that plus a continuation of the design phase of step one. This may sound like a contradiction of terms, but it's not. Ever since the advent of producibility engineering in the early 1980s and its successor, concurrent engineering, became the integrated way to manufacture; we have truly recognized that design includes three phases and they are dependent on each other.

The first phase is the concept design phase and it relates the product specification to test for compatibility with the laws of science. Which it must do. The second phase is the producibility design phase where the design is tested to see if it is technically and economically feasible to produce in the intended factory. Yours. The third phase is the manufacturing facilities design phase where the jigs and fixtures and tooling are designed to be compatible with the proposed concept design. This is the portion of design where we figure out how to create a reality of the concept in a manner that satisfies the customer and all of our stakeholders.

Everybody must do the concept phase. Where World-class companies excel is that they put as much effort into the second and third stage as they do the first. And it pays off. Here is where the concept design is tested for its robustness. If it isn't, then there will be many failures in manufacturing which means low product yield and high manufacturing losses. All of which probably spells doom for the economic viability of the product. Does your company engage effectively in all three phases of design? If not, it is not "making it."

Examples of companies doing all three phases effectively abound. There is no reason not to. And in fact, not to is the easiest way to go bankrupt. In a very difficult manufacturing situation, semiconductor chip making, we see strict adherence to the practice of all three phases of design because they know that to do so is a prerequisite for success. They've learned that there is no choice but to do equally effective design work in all portions of design to survive. Perhaps the severity of their environment causes them to be very precise. But I feel that the world-class performers would have done it anyway because they are determined to keep their costs down and profits high. If your company doesn't have this attitude then it is not world-class.

Simply put we must have the very best designed factory procedures in order produce the very best concept design. Companies that put all their creative talent into concept design and then treat creating methods for producing as an afterthought, or worse yet, put lesser talent on that task, will fail, perhaps not tomorrow but relatively soon.

Schedule to Produce

This is the coordination step and if not done well will spell doom to the company. How many companies do you know of that have great designs, and touring their factories you see great equipment and facilities, but still can't deliver on time? How often does the lowest cost quote producer not meet his production due date to your needs? Probably more than you'd like to admit. Scheduling is as critical to world-class company performance as having the products customers want to buy.

The design step output is used to create a coordinated production schedule for all parts, sub-assemblies, and assemblies related to your company's products. We do this by creating a work station route for where work will be done and in what sequence, all derived from an engineering Bill Of Materials (BOM). The BOM is really part of the concept design which shows what the product will be made of and what the sequence order of fabrication has to be. With the route and BOM as a guide the company can construct a coordinated schedule to ensure that the proper parts are done on time to meet all the assembly needs.

I am often flabbergasted that companies that spend enormous sums in perfecting designs get themselves into company killing conflicts by not being able to schedule their factories. The results then being economic failure because they can't deliver. This doesn't happen to world-class companies because they've learned two very important things about competition. In order to sell, they need to know two things; "The Two Knows," know how to make their products, and know how long it should take to do so which, as I've mentioned before, are the basic tenets of manufacturing success. By knowing the method and how long it should take they can schedule their factory. By scheduling their factory they know what promise dates to give their customers and they have a very good prognosis of meeting the schedule. This makes them reliable vendors with the ability to generate additional orders, as long as their product meets customer needs.

There is only one way for companies to schedule effectively. They need to have a Manufacturing Resources Planning system (MRP II) integrated with a Computer Aided Process Planning system (CAPP) all driven by cycle times derived from a scientific time standards system. It is virtually impossible to schedule dynamically without these three systems.

These three items sometimes are marketed under different terminologies. A common one is Enterprises Resources Planning system. But they all have one thing in common. They're driven by a common integrated information data system. We call that system CIM for computer integrated manufacturing. Many times we hear this referred to as the factory of the future. I agree it has a very bright future, but it's here now and companies that will "make it" need to embrace it wholeheart-edly. There is no choice. To say no I'll pass, is tantamount to committing corporate suicide. CIM with its subsets, one of which is MRP II, is an absolute necessity for the world-class company. No other way of performing integrated schedules exists in a manner that you could say is even a reasonable alternative.

Some may say, what about Just In Time (JIT)? Isn't that an alternative to MRP II, and perhaps better? No. JIT is not a scheduling system. It is a philosophy borrowing heavily from traditional industrial engineering theory that stated simply: *eliminate waste*. In its most popularized form that's taken to mean the elimination of the waste of excessive inventory on hand. There is nothing wrong with this, but it is not a scheduling system what so ever and should not be confused with one. I heartily endorse the precepts of JIT to eliminate waste, and in fact we should all understand that world-class companies do this in an intuitive manner. We will cover both MRP II and JIT in depth later on in this book.

Before I leave the scheduling topic it would be prudent to address service businesses. Those of you engaged in this endeavor may ask, " I don't make anything. Why do I need a CIM based MRP II system?" Service businesses do need MRP II because you do make something in the broadest sense of the definition. You provide some sort of value added function otherwise you wouldn't be compensated for it. Most likely it is a time-dependent function in that the customer wants the service at a particular time. Therefore the need to schedule is there. Also, if you're successful you need to juggle demands from many customers at the same time. You need to satisfy your customers based on their time frames or risk losing them to competitors. This means you have to schedule and you need to comply with the two knows; know how to perform your service and know how long it should take. So virtually everything a hard goods manufacturer has to schedule for, you do too. Even workstations are similar. The service firm's workstations do not make chips or transform ma-terials, but they do create the basis of your service. It may be a design workstation or even a word processor, but the fact is it only does one job at a time. Therefore, the through-put of work has to be scheduled. World-class service firms need an MRP II type scheduling algorithm. They cannot compete effectively without one.

Purchase Raw Materials in Accordance with the Schedule

Again coordination is the key to success that companies "making it" employ. To purchase effec-tively we need to know precisely what it is that has to be purchased and when will the materials be required. World-class companies take advantage of the integrated nature of CIM to use the same database information from design and scheduling to create purchase orders for materials. In fact the modern MRP II system evolved from its progenitor Materials Requirements Planning (MRP). Where make or buy decisions were made for every item on the engineering BOM. Those that were buy were time phased via MRP and ordered accordingly.

By integrating make or buy via the MRP II system, world-class companies create seamless integration of internal make items with external purchased items. Materials arrive at workstation when needed in accordance with the routing instructions. The integrated nature of the scheduling algorithm makes it possible to purchase materials with enough lead time to ensure on time delivery vs. actual need. Also ran companies continuously struggle because vendor supplied items are not coordinated tightly with needs, hence their ability to deliver on time is jeopardized.

World-class companies do some additional things to gain competitive edges in purchasing. The main one is they create a supply chain with their vendors. They treat vendors as an extension of their own in house workstations. However, before they engage a vendor they make sure the vendor is qualified to meet their high quality standards, the same as required for internal operations. They go so far as to assist vendors in upgrading their quality and production management skills as required so they do not become weak links in the supply chain. As policy, they also strive to maintain long-term relationships with specific sets of vendors through long term contracts for services and supplies. By doing all of this, companies that are "making it" are steps ahead of their competitors in guaranteeing high-quality materials and services as well as setting fair and reasonable prices.

Produce in the Factory

This is the transformation phase of raw materials to finished product commonly called the value-added step. This is where work is accomplished that directly affects the customer's receivable. So we can see that it applies equally to service firms as well as goods producers.

World-class companies make no distinction between external and internal work stations for exercising management control. In fact the only differentiation they make is that internal work stations are considered to be "owned" while external are "rented." They integrate workstation activities based on their master schedule output from MRP II and make no distinctions as to where the workstation is located. They allow, and in fact insist, that vendors have the same scheduling information as their in house workstations receive.

Companies that are "making it" mitigate differential in labor costs by maintaining tight control on how work is handled at workstations. They employ "short interval scheduling" techniques that factor in workstation methods, time standards, maintenance criteria, suitability of materials, and operator training. They factor this in with accomplishment expectations over short periods, usually no longer than one half of a work shift. They also vigorously investigate failures of any kind for root cause and set immediate corrective actions. And with all this, they use the workstation operator as a member of their production team, not a cipher to be controlled as a mere nuisance to be put up with. This is much the same as viewing a test pilot is a member of the technical team developing a new aircraft and not just a human guinea pig in the cockpit. This philosophy assures the best performance on the shop floor and gives a company the highest probability of success. Does your company do this? If not it won't "make it" and will be constantly squandering resources by chasing low labor cost around the globe.

Monitor Results for Technical Compliance and Cost Control

World-class companies consider this quality assurance phase to be an ongoing process of constant vigilance and the seeking of continuous improvements. Quality assurance encompasses ensuring that the product or service is being provided in accordance with the plan. And the plan includes technical, schedule, and cost goals.

The best companies work on a philosophy of Total Quality Management whereby continuous improvement is ongoing, as exemplified symbolically by the TQM triangle. Let me explain the concept of the triangle. On the triangle, the customer is at the apex spinning off data about the validity of the work received. This information goes down the leg of the triangle to the right corner as data to be processed. The data goes along the base of the triangle becoming an improved process. Process improvements commence at the left corner and back up to the apex for customer judgment. Then the process starts over again.

There are many approaches on how to monitor and control processes falling under this step of the Manufacturing System. By far the most popular is the Statistical Process Control technique (SPC), especially that tied into the CAPP system for developing process monitoring steps as part of the methods plan. In fact many world-class companies consider SPC to be part of the integrated CIM approach whereby SPC action steps are included in the MRP II scheduling algorithm. Most recently we've seen world-class companies employ the six sigma variant of SPC to gain even further competitive advantage by approaching a very difficult goal of "zero defect." The message here

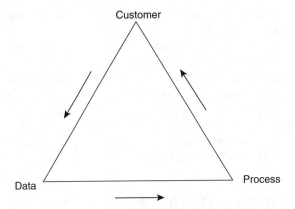

the relationship of customer, process, and data
in the pursuit of continuous improvement

Figure 10-1. The TQM Triangle

is that companies that are "making it" employ some form of SPC usually tied into their CIM based MRP II system. If your company doesn't, then it needs to.

You'll notice that I haven't mentioned ISO 9000 registration. That's on purpose. To register or not to register is an issue of whether or not your company needs to advertise that it is a world-class company. ISO 9000 by itself signifies little in the regards of adequacy of your quality system vis-à-vis the needs of your company. It only states that you have a control system that you follow faithfully and have records to prove it. It is employing TQM philosophy along with tools such as SPC tailored to your customers' needs that makes a viable effective quality system. This is the only way to comply with the spirit and intent of this sixth step of the Manufacturing System." We will discuss ramifications of ISO 9000 later on.

Ship the Completed Product to the Customer

The job is not done until the product purchased from your company is delivered to the customer on time, complete, and at the expected quality level. Just as your vendors are part of your supply chain, you are part of your customers' supply chains. World-class companies have as a goal to never be the weak link in any supply chain.

Shipping completed good to customers means just that. You are using a valid system to track materials through your operation to make sure the actions are performed correctly and that they are done on time to your customers needs. You in fact ship completed product to customers and do not continually ship incomplete jobs, that are perhaps only 90% complete. These are traits of world-class performers.

Also-ran companies do very well in being what I call 90% companies. They can get the first 90% of the customer's order out on time but the remainder is always backlogged. Now 90% is sufficient in school but not in the real world. Would you be satisfied with an automobile with the trim missing? Or how about the spare tire not being in the well? Of course not. World-class companies always find ways to ship 100% on time.

Companies that are "making it" use integrated scheduling and control systems to track each and every aspect of production, then kit goods systematically and only release for shipment when it is all there. They use MRP II as their tool and control mechanism to make it work. Their shipping and warehousing people are trained in distribution controls and are held responsible for inventory control. In fact their inventory control records are always 99% accurate as a minimum. They know that what ever happened beforehand, only the customer taking delivery of the product and agreeing that all is in order and as expected could complete the job. Can your company make these claims? You need to, to be a world-class performer.

As you can see the thread of information cascading from step one through step seven is very evident. World-class companies exude communications excellence. Not only do they consciously subscribe to the tenets of the Manufacturing System but they've integrated the information flow within an all encompassing CIM system. By doing so, they can respond to opportunities in a dynamic fashion. They can make changes to schedule, designs, SPC checking parameters, contents of shipments and virtually any other demand made by their customers many orders of magnitude faster than those companies not cognizant aware of the Seventh Step of the Manufacturing System. These are the companies that are "making it" and will continue to in the future.

THE QUALITY ORGANIZATION WITH RESPECT TO PROCESS CONTROL ENGINEERING

As we can see the seven steps of the manufacturing system embrace all aspects of activities necessary to produce a desirable product for a company's clientele. Every aspect of the process is linked for it to be in synchronous unity for excellent results. The quality organization has as its one and only purpose to monitor the activities of the entire company to assure that the linked activities stay in synchronous motion and are being adhered to in accordance with the overall business plan.

We see several bits of nomenclature being used: quality assurance, quality control, process control, inspection, and just plain quality. They all refer to the basic same premise, that is, make sure that the basic business plan is being adhered to and products and/or services are delivered in accordance with specifications. This aspect of management is the "official worrier" of the company and ombudsman of the clients. They measure how well the company does in making products to client specification as well as getting things "fixed" for clients when they aren't. Obviously it is much better when the necessity for interceding on client's behalf is a rarity.

Where does process control fit into this equation? Perhaps the best way to explain that is to describe the quality organization. The quality organization is a matrix, both formal and informal, set up to conduct step 6 of the seven steps of the quality system.

"6. Monitor results for technical compliance and cost control."

Some of it comes under the direction of manufacturing engineering, some of it does not. Under the overall heading of Quality Control we have broad policies, strategies so to speak, and tactics. Strategies are the domain of senior management. Thus the management and development of quality strategies falls under the control of the general manager, or manufacturing manager at the lowest level. We call this activity quality assurance. It deals with overall policies and procedures regarding how the company derives and carries out quality philosophies, for example, the quality of how a company would derive a fair quality warranty to be included in sales activities. These and other esoteric concerns related to the manner in which the company espouses quality to the rest of the world are worked on by quality assurance. Some people call this the "touchy-feely" part of quality. It is the perception part of quality, and quality assurance has the responsibility that the company does maintain and doesn't lose its reputation for fine quality, always.

Quality control is just that. Doing the tactical activities of putting in place controls and checks to assure that only proper procedures and materials are used to produce and/or services. Mostly quality control works outside of shop operations. It sets up methods of checking to ensure materials have the proper properties as called for by the design specifications. Also, quality control ensures that vendors are performing in accordance with contractual agreements. It coordinates in process checks with process control to ensure proper documentation is recorded for all activities necessary to certify correctness of design and manufacturing. Quality control is the "official" head of quality operations within manufacturing.

Process control is a manufacturing engineering unit that designs and implements methods for checking in process of manufacturing. It is much like methods engineering doing methods, planning, and time standards work to produce a product. But in this case the task is to design methods, planning, and to some extent time standards to measure the technical effectiveness of the methods for delivering the product in accordance with the design. For every methods step there needs to be a process control check to ensure adequacy in performing the step.

The terms *inspection, inspectors,* and *test* typically refer to those carrying out process control steps. Sometimes they report to the process control unit, and sometimes directly to quality control. It all depends on the philosophy of the company. But carrying out the process control steps is the essence of what the general public considers to be the responsibilities of the overall quality function.

Process control exists in both mass flow as well as job shop type companies. For mass flow companies the emphasis is on statistical quality control. For job shops it is more focused on methods validation, as we shall see. However, both are used in both forms of manufacturing; it's just a matter of emphasis.

HUMAN MOTIVATION AND QUALITY

Given the opportunity, the vast majority of people want to do well. They would rather do things right than wrong. If a person is trained properly he or she will do the work properly. The quality of a product depends on this truism. So management has to make it easier to do the right thing rather than do wrong. Here is where motivation comes into the equation. If we give people the ability to do right, encourage them to do, and reward them for proper actions, we are much more than likely to get the performance we desire.

Achieving satisfactory quality depends at least equally to motivation as it does to employing proper methodology and equipment. Proper quality planning takes advantage of this fact. We see modern planning using operator certification as a more potent force for obtaining good quality than inspection. The old saying, "you can't inspect in good quality, you have to make it that way" has gained acceptance from all world-class companies. This means the emphasis is on asking the operator to tell us if the quality is below standard. This is done through a technique called operator certification. Here the operator inspects her own work for compliance to design requirements and makes the judgement as to whether its good or bad.

This is at first glance a high risk strategy. But is it? Research says not. First, it is impossible to 100% inspect anything, and certainly not as an ongoing activity. Therefore there is no reason to expect the inspection process to be successful. Even if we could inspect every part and every operation required to make those parts, we would still fall far short of assuring 100% quality. The statement familiarity breed's contempt is certainly true here. If we had 100% inspection, just the act of doing so would ensure less than 100% good parts being accepted. Familiarity, in this case dull routine, causes carelessness. We daydream and don't pay attention to what we're doing. It becomes routine, and only a totally obvious bad product is caught. Any deviation that requires intense scrutiny to find has a good chance of not being discovered. In fact old quality procedures recognize this truism and set up statistical concepts that predict percent bad parts being accepted for this reason. And statistical process control gives us probabilities of good parts being made. All this remains true even if we set up simple go-no-go gauge inspections. Because we still depend on the inspector doing the correct procedure every time.

Second, when people are given responsibility they tend to accept it. So if an operator is told he is responsible for making the part right and no one else is going to check it, it is highly likely that the part is going to be right. This becomes even more likely if we ask the operator to sign a form for the work and thereby attesting to have done the operation in accordance with the prescribed method. Couple this with proper respect for the job and the operator, the probability is very high that the proper parts will be produced.

THE PROCESS CONTROL SYSTEM

A process control system is a series of coordinated events in a manufacturing cycle that allows achievement of stated quality and production goals.

Coordinated events can be thought of as process steps or sequenced manufacturing operations, or a chronological listing of things to be accomplished that combine the quality requirements with the shop operations requirements. Essentially, the process control system is a method of specifying who does what, when, and with what required quality measurements or checks. Note that it does not specify how a function is to be accomplished. However, by setting the quality measurement parameters with some forethought, it indirectly specifies, or at least strongly influences, the choice of tool or process.

> **EXAMPLE:** Assume that we are machining surface plates for heavy machine mountings. The design calls for a 2×2 ft. plate 2 in. thick made of carbon steel with the surface finished to a 32 RMS finish. This is normally considered a moderately fine finish, and in many cases a variation of 10% would mean little. The process control system can dictate how the surface is obtained. If it specifies that a simple comparison gauge be used to measure the operation, then a simple, single-point tool on a vertical boring mill may be used to perform the manufacturing function. However, if an elaborate electronic surface profilometer were to be employed, something better than a VBM would probably be needed; this would most likely be a surface grinder.

The example above shows how a choice of measuring tools can affect the ultimate quality and manufacturing choice without mentioning tolerances. In a real case, a tolerance would be specified on the engineering drawing and a measuring tool capable of differentiating between the tolerance band would be selected. But would the correct tool have been selected and the requirements stated on the engineering drawing have been met without the pressure of the process control system? This is always a question in job shop activities, where there is no production run of any consequence to fine-tune quality methods. It is a lesser problem with mass flow manufacturing. However a whole pilot production run could be in jeopardy if the proper method is not employed. It is evident that with short production runs and individualized products, mistakes cannot be easily absorbed. Therefore, process control in a job shop is vital. Discrepancies must be found and corrected very quickly.

The process control system represented by Fig. 10-2 is a closed-loop feedback system in which the results are reduced manufacturing losses, minimized production delays, and improved product quality. The interesting thing about the system is that it never becomes static. In manufacturing there is no reason why the three outputs should ever reach a plateau or maximum value, because the jobs are constantly changing. Thus more opportunities to improve on past performances continually become available.

The system allows extremely rapid corrective action and long-term solutions for similar individualized products. In a sense, since job shops do not have long-term identical product runs, the process control system has a memory loop that allows corrective action to be taken on future products made by similar techniques. Experience gained on one job can be related to others, and these data contribute to an accurate planning and estimating activity as an offshoot of the process control system. In a high-technology job shop, this planning ability often means the difference between success and failure.

The process control system serves another vital function. In many ways it acts as an avenue of communications between manufacturing and design engineering, between different manufacturing units, and between manufacturing and finance. The closed-loop feedback system virtually guarantees this communication with little or no further attention needed from management.

The six blocks shown in Fig. 10-2 constitute the process control system. Because they interact continuously, it is difficult to isolate one block of activity from the rest. Therefore, to explain the system, let us assume that we can freeze everything in place and examine one block at a time.

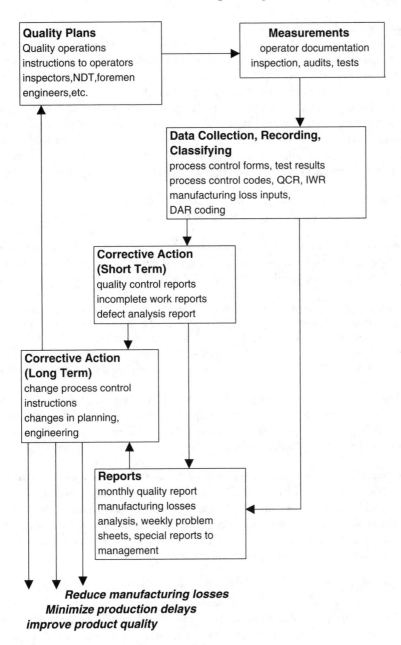

Figure 10-2. The Process Control System

QUALITY PLANS

Who does what, when, and with what quality measurements is spelled out in considerable detail in good quality plans. Quality plans can take many different forms; however, every good quality plan must spell out requirements, assign responsibilities, and be capable of having its specified actions measured. Because job shops handle a great variety of work, their quality plans must be concise, easily understood by all, and complete in themselves.

For both job shop and mass flow quality planning, the "cookbook approach," in which information is presented with very little room for interpretation or misunderstanding, is excellent. In this approach we spell out exactly when to measure something, who does it, with what facilities, and what he or she does with the information. This type of quality plan may even be completely obvious. The quality plan is made up of three sections: (1) purpose, (2) responsibility for quality, and (3) procedure. An examination of this format will lead to a thorough understanding of the components of the typical quality plan and demonstrate the philosophy of good process control.

Since job shops features mostly non-repetitive or short production runs, therefore require a more complex and numerous plannings; I will use it as an example of quality planning. The same philosophies and formats will hold true for mass flow manufacturing, but usually less complex. However, mass flow manufacturing may have more SPC requirements that we will discuss in a later section of this chapter. Let's proceed with an examination of a typical job shop quality plan

Purpose.   This section of the plan states what the quality objectives are and how they will be achieved. It can be thought of as a preamble or introduction to the quality plan.

Responsibility for quality.   This section states who is responsible for the quality of the product to be inspected, reviewed, and so forth. It also gives a general description of what has to be reported and how it is reported. For example, this section might state:

> "For each operation listed, the operator is responsible for the quality and completeness of the work. The operator is obligated to notify supervision and inspection of incomplete and defective work whenever it is discovered, whether generated at his work station or elsewhere.
>
> All requests for inspection are to be made by signing a sheet on the inspection call board located at the inspection station. Requests should give operator's name, part to be inspected, and location of the part."

There is no mention of the foreman or any member of supervision or process control management. It states simply that the operator is responsible for quality; therefore supervision must be automatically pledged to quality. Having the responsible managers review and approve the quality plan by signing it further reinforces this.

Procedure.   This section states where and when process control activities will be conducted. It also states who will make these checks, audits, and certifications, and to a considerable extent how they will be made. This section of the quality plan must be completely synchronized with the workstation planning. Operations are listed in chronological order with the quality requirements for those operations and how the requirements are to be satisfied. It is important that each operation be explicitly defined along with the process control requirements.

EXAMPLE (a):
3.6   *Operation:* Finish machine valves on vertical boring mill.
3.6.1   *Operator:* Complete form PC-210, listing actual dimensions, and return to the inspector at the J-15 inspection station. Hold piece on machine for process control release.
3.6.2   *Inspector:* Audit operator certification form PC-210 that actual dimensions are within drawing tolerances. Sign off PC-210 and send to A-8 inspection station. Notify operator for release of part from machine if dimensions are within tolerance. Otherwise, report all out-of-tolerance dimensions on quality control report and list QCR number on bottom of PC-210.

The operator and inspector are directed to perform particular process control actions. Other indirect work-performing personnel can also be instructed by the quality plan.

EXAMPLE (b):
3.14   *Operation:* X-ray welded connections and other casting areas. Refer to Process Instruction P6G-AL-1223.
3.14.1   *X-Ray Technician:* Completely fill out x-ray folder (TG-1238) noting x-ray numbers and types of defects, if any.

3.14.2 *Materials Engineer:* Review x-rays; determine whether or not x-rays meet standard. P6G-AL-1223. Disposition on face of x-ray folder (TG-1238).

3.14.3 *Foreman:* Have x-ray defect areas laid out on casting for removal. After defect removal sign x-ray folder (TG-1238).

The two examples show the brevity and conciseness of this type of quality plan. It is ideal for a job shop where the size, shape, and design of parts vary considerably but the basic processes vary comparatively little. Many different critical paths may be laid out with this type of quality plan, while only the path employed for the particular operation is implemented.

The numbering system has a distinct meaning. In example (a), 3 indicates that this is the procedure section of the quality plan; 3.6 signifies that this is the sixth sequenced operation of that particular manufacturing area dealing with valves listed under the procedure section; 3.6.1 references the operator's quality action requirements for the sixth sequenced operation; and 3.6.2 specifies the inspector's duties.

The control documents and information needed are also spelled out. In example (a), the operator is instructed to enter data on process control form 210. In example (b), the technician, foreman, and engineer are informed that the x-ray standard is contained in process instruction P6G-AL-1223. By referring to other documents for specific needs, it is possible to have the information on the quality plan and to keep the size of the quality plan manageable. Also note that the language is simple and direct; abbreviations are used sparingly and only when they are easily definable. The object is to avoid confusion and to tell only what needs telling. Obvious things are normally left out. In example (a), there is no reference to who receives the Quality Control Report (QCR), because it is known throughout the factory that QCRs are sent to design engineering for action and dispositions and that an overall QCR system is in place. In example (b), there is no doubt that the materials engineer makes dispositions regarding acceptance or rejection of the x-ray of the casting. Instructions are given for recording the engineer's actions, because it is not obvious what the disposition recording method should be.

The cookbook-type quality plan puts a great deal of information on paper, but leaves out the obvious for the sake of brevity and clarity. Therefore, some simple rules should be followed when deciding who should be listed for inputs under each operation and how much information should be given.

1. List everyone who has a routine input for the operation.
2. List all the information necessary to do the job; that is, process instructions, material specifications, process control forms, and so forth.
3. Do not list any information that would be classified as background material.
4. List all operations regardless of whether there is a corresponding process control requirement. For example, a queuing operation is vital to the process, but rarely to the quality of the product. By listing all operations, continuity is maintained.

Quality plans are essentially simple things. Unfortunately, many writers of quality plans lose track of what they are trying to do. To summarize, a quality plan should include the purpose for its existence, a designation of responsibility for quality, and step-by-step procedures that are easy to follow.

MEASUREMENTS

The type of measurements used, accuracy required, and redundancy depend primarily on the product being produced. In a job shop it is common to assign like work to like areas, thereby minimizing the need for quality plans and the types of measurements to be made in one area. An exception would be job shops that employ group technology layouts. Even here, commonality of parts being processed also minimizes requirements for different quality plans and measurements. For instance, in heavy fabrication and machining businesses, it is common to have all welding assigned to one area; therefore all the measurements, regardless of the products involved, are basically similar. This permits skill development by operators and inspectors and cuts down considerably on errors of omission. Since all the welding is done in one area, there is no need to achieve proficiency in weld auditing in different areas, and it is not necessary in this area to teach inspectors to audit machining techniques. This segregation technique facilitates the development of specific quality plans even though the product mix varies, sometimes almost daily. Some people would say that there is a basic contradiction applying this philosophy in Group Technology (GT) style layout where there are no weld departments, machining departments, paint departments, but instead many independent production lines for products (families of parts) that have similar manufacturing process flows. To some degree this is probably true, but for the most part is overcome with only a minimal additional proficiency required. For example, in a GT layout requiring welding, the welding technique used for the part will most likely be very limited in scope. Therefore, a proficiency just enough to satisfy the need for the particular family of parts can be readily acquired.

In a job shop where like work is dispatched to the same general area, the quality plan is generated to handle a similar product line. For example, heavy-duty induction motor frame welding and high-volume, low-velocity pump convolutes fabrication welding could take place in the same location. The parts are considerably different and may even differ among themselves (e.g., there may be many different styles of motors). However, they are all welded fabrications to be produced in the same shop, so that the weld measurements can be similar, if not identical. In this case, the primary quality plan would be written for welded fabrications, treating all of them in the same way as far as this is practical. Only when a specific requirement for one does not agree with that for the others will there be a need for a subsection of the quality plan. This subsection will be entered in the regular area quality plan and will be very clear about what it is and why it differs from normal. In this section, and nowhere else, specific measurements will be called for. The measurements called for in the general section apply everywhere, while the specific section is narrowly defined.

Measurements must be carefully planned for maximum effect. In job shop environments the planning is such that general techniques must be effective for many specific types of products. Therefore, it is desirable for as many measurements as possible to be made by the operator [see example (a)], inspections should be basically audits, and true inspector-conducted measurements should be of the tollgate type. Hence the basic body of documentation should be the operator's certification (measurements) backed up by in-process audits (sampling) conducted by inspection personnel. This documentation is then backed up by the tollgate inspections to ensure the needed degree of quality.

Tollgate inspections are the control points for process control. Through these inspections the process control unit can assure management that the product is being made to design specifications, that everything up to the point of the inspection is correct, and that the process control system is working adequately. If something goes wrong in the future, it is possible to go back to the tollgate status and either reconstructs the cause of the failure and corrects it, or scraps the procedure and start over.

Tollgate inspections should normally be conducted by an inspector and perhaps involve supervisory process control personnel. The procedure involves check-off sheets and verifications that required operations have been completed. Occasionally a physical measurement or test may be called for, but usually this is not required.

EXAMPLE: A tollgate may be required before steam turbine high-pressure shells are furnace heat-treated after welding. The tollgate inspection verifies that all required welding and all required non-destructive testing have been completed and, most important, that the requirements have been completed successfully, that is, have met design engineering requirements.

Another feature of a tollgate inspection is that the authorized performer must sign for the work. In the process control system, measurements are the basis for all long- and short-term corrective actions. It is necessary to make sure that measurements are made correctly, that the work is done correctly, and that incomplete and incorrect work is reported immediately. By making people accountable, there is a much higher probability that the work and the measurements will be done correctly. This is the Theory of Accountability.

The Theory of Accountability states that people will perform at a conscious level consistent with high quality work only when they can be held accountable for that work.

To put this theory into practice, everything a worker is asked to do must be measurable on a quality basis. The quality plan must state what kind of measurement is required, how it is to be made, and how the action is to be recorded. There must also be a way to record the worker's performance or percent acceptable, and a reward for doing the job correctly. The reward can be positive or neutral. For instance, a welder may be hired to do x-ray quality work. This is a basis for employment. The reward for doing the job to the quality standard is continuance of employment, hence a neutral reward.

We cannot leave this section on quality measurements without mentioning tests. Tests are everybody's idea of process control. There seems to be a feeling that testing is always good and that more testing is better than less. What is often not realized is that testing, unless it is designed for a specific purpose and well planned, can be useless and sometimes harmful. Testing in a manufacturing environment is not like testing in a laboratory. It is difficult to control the variables. Facilities for fine testing normally are not rugged enough to survive the factory environment, and properly trained technicians either are not available or are working in an unfamiliar and often hostile personnel environment. If a test is needed and can overcome these difficulties, it should be successful. But those requesting tests in a factory must plan the exercise to the smallest detail. Design for Experiments, which we will cover later in this chapter, is a body of knowledge of how to set up meaningful tests that focus on questions that need to be answered through obtaining valid test data. But in order to get good data we need to set up to do so. In its simplest form, Design for Experiments application rules have been developed from experience to satisfy this need. A set of rules for achieving good tests in a job shop have evolved:

1. Design the test to fit into the production schedule.
2. Make the test as simple and straightforward as possible.
3. Make the physical part of the test accomplishable by the factory personnel if it is a process requirement, continuous-production type. To do this, supply clear instructions with no ambiguities in writing, specify easily used equipment, and require little interpretation by the operators.
4. Where possible, make the interpretation outside the factory.
5. For one-of-a-kind or seldom used tests, have a knowledgeable person available at all times during the test.

DATA COLLECTION, RECORDING, AND CLASSIFICATION

The quality plan states what information is to be recorded and measured. Data collection, recording, and classification sorts the inputs, categorizes them, and eventually becomes the basis for reports and corrective actions. The quality plan does not specify what reports will be issued; however, when a quality plan is developed, some thought must be given to the types of reports desired. Otherwise, the wrong information may be collected and an erroneous picture of quality presented.

When analyzing data, it is necessary to know beforehand what to look for. The real world of manufacturing is far removed from the scientifically ideal one. We must predetermine the kind of information we wish to have, and then decide whether the available data are sufficient to tell us what we need to know. Another truth about factory-derived data is that they do not generally result in precise curves or fit nicely into a mathematical model. We are dealing with the ability of many people to produce data; they all have their own tolerance bands, so that dispersal of data is the norm.

In analyzing process control data, we must be willing to make judgments and take actions based on incomplete observations or short spans of data because of the compressed time scale involved in manufacturing. Every decision based on data collected must be backed up with a plan for what to do if it is "all wrong." The engineer involved in process control work must be extremely familiar with the manufacturing processes, know the products, and be willing to make decisions based on the data available at that time. This does not mean that the engineer stops analyzing the data because the decision is made. The decision may be modified at a later point. The important thing is that a decision is made and the manufacturing process continues.

I do not wish to give the impression that process control data collected by inspectors and taken by operators is suspect. In fact, it is usually of high quality and taken with diligent care. However, unlike a laboratory experiment, verification is usually not possible and the conditions required to take the readings are not always under tight control. The chance for error is naturally higher than under controlled laboratory conditions. Since this is the case, the process control engineer must know the expected scatter band for any process in order to correctly interpret the data. If there are enough data points, the engineer uses statistical methods. Often, however, especially in job shop process control, it is necessary to make assumptions because of a lack of statistically significant data. Therefore, in order to minimize the risk factor, the engineer is continually testing the assumptions against the incoming data. This leads us to how the data are recorded and analyzed.

The first and most important input is found on the quality control report (QCR), which is sometimes called a deviation report. This document is an exception report listing discrepancies from required values. It is most important from a control standpoint because the first order of business is to correct mistakes. Therefore, the process control engineer can monitor the temper of the manufacturing operation by keeping a close watch on QCRs by number and defect category. This may be done manually or with computer programs. It is normally possible to tell by the frequency of occurrence over a period of time whether the quality of the area is increasing, decreasing, or remaining stable. This is done simply by relating the number of QCRs to the number of man-hours worked. If the engineer sees an increasing QCR frequency, it is an indication to look for problems. The simplest way to do this is through a QCR analysis called a defect analysis.

The defect analysis is usually a computerized report with the QCR as its basic input. It is simple to code all QCRs by operation number or description, by responsibility for causing the discrepancy, and by part name involved. The QCRs may then be sorted by these three groupings—operation, responsibility, or part—and much meaningful information obtained. For instance, if a certain part consistently turns up on the report, one would further sort by responsibility to find the primary cause of the defect and by operation to determine where the defective work is occurring. With this information, the process control engineer can now make a detailed investigation and take corrective action. Another important feature of the defect analysis report is that it can be keyed to the costs of labor and material expended in correcting mistakes. If the costs of correcting mistakes (manufacturing losses) are keyed to the QCR, it will be possible through the use of the defect analysis report to associate true costs with individual problems, allowing the business to decide where it is best to expend its resources for corrective actions.

The other major source of data is the in-process data sheets known as Process Control (PC) forms, which contain the primary information obtained from the manufacturing operations as required by the quality plans. In many cases this type of data is used primarily for the specialized reports needed for short- or long-term managerial information and only secondarily as an input for corrective action. The PC form data are preplanned data that measure specific operations. They are often valuable in comparing similar products from time to time to obtain trend information. The process control engineer uses PC form information as a check on various assumptions and decisions that have been made.

With the various inputs—QCRs, the defect analysis report, and PC forms—the process control engineer is capable of monitoring the manufacturing operation, reacting to situations that require

Quality Control Report

QCR No._____

Department_____   Date_____

Part Description_____

Description of Deficiency   Operation No._____

_____
_____
_____
_____
_____
_____
_____
_____
_____
_____

Responsibility (list department)_____

Cause: (check appropriate box or boxes)

☐ operator error   ☐ defective material

☐ machine failure   ☐ process failure

☐ engineering error   ☐ transport error

☐ CNC program error   ☐ method instruction error

☐ Other -describe below
_____

Corrective Action (check box and describe as required)
       (use back of form as necessary)
☐ use as is ☐ rework ☐ scrap

_____
_____
_____
_____
_____
_____

Inspector   Engineer_____
                             date: _____
Corrective action complete
          Inspector_____   Cost, $_____
          date: _____   Finance _____
                                           date: _____

Figure 10-3. Typical Quality Control Report Form

corrective actions in a logical manner, and judging the results of the actions taken. Without the process control system inputs, the engineer would be using a scattergun approach to problem solving. The detailed information that can be obtained through the process control system enables manufacturing management to act in a precise way, eliminating the problems without sacrificing any of the good elements of the system.

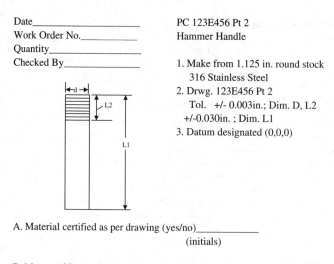

Date_____
Work Order No._____
Quantity_____
Checked By_____

PC 123E456 Pt 2
Hammer Handle

1. Make from 1.125 in. round stock
   316 Stainless Steel
2. Drwg. 123E456 Pt 2
   Tol.   +/- 0.003in.; Dim. D, L2
   +/-0.030in. ; Dim. L1
3. Datum designated (0,0,0)

A. Material certified as per drawing (yes/no)_____
(initials)

B. Measure 20 parts from lot selected randomly. Circle out of tolerance dimensions.

	D	L1	L2
1	__	__	__
2	__	__	__
3	__	__	__
4	__	__	__
5	__	__	__
6	__	__	__
7	__	__	__
8	__	__	__
8	__	__	__
10	__	__	__
11	__	__	__
12	__	__	__
13	__	__	__
14	__	__	__
15	__	__	__
16	__	__	__
17	__	__	__
18	__	__	__
19	__	__	__
20	__	__	__

from:"Fundamentals of Shop Operations Management: Work Station Dynamics"
fig.4.3, pg 70; Daniel T. Koenig, ASME Press, NY NY, 2000

Figure 10-4. Typical Process Control Form for component part

CORRECTIVE ACTION (SHORT TERM)

Immediate response to problems is essential for success in business. The process control system, through the use of QCRs and PC forms, offers the ability to respond to problems quickly and intelligently. This "fire fighting" capability is the hallmark of the successful process control organization.

In process control work, one always plans to eliminate or at least minimize the impact of "fire fighting," and the fact that such problems occur should not always be construed as a failure of the process control system. If the problems are due to poor planning or anything within the effective domain of the process control organization, then the system has a flaw and any corrective action must have a dual purpose: to correct the problem and to correct the system. However, if the problem is a result of outside influences, for instance, deterioration of vendor quality, then corrective action can be directed to the deficiency and concern about the system eliminated. Fire fighting is detective work coupled with good intuition based on an understanding of the manufacturing processes involved. How we go about "fire fighting" is almost an art form. However there are some rules that make it easier to learn the art and make the practitioner more effective. Rules for "fire fighting" can be stated as follows:

1. Define the problem as quickly as possible and in as much detail as possible.
2. Determine whether it is a process control system problem as well as a quality problem.
3. Take action to solve the problem, first addressing the immediate quality problem and then, if required, the process control system problem.
4. Make an intensive effort to solve the quality problem and gain control of the situation. Corrections to the process control system can be made at an appropriately controlled pace.

When engaged in "fire fighting" activities, the process control unit is asked to take a leadership role in the problem solution. This means effectively coordinating the interactions between manufacturing, design engineering, marketing, and occasionally finance. To do this successfully, the process control unit must command the technical and managerial respect of the other functions involved, or its leadership role will fall vacant. Cultivating technical and managerial leadership is an independent personal obligation for all members of the process control unit. Since process control is often the art of gentle and not so gentle persuasion, the people involved must be technically proficient beyond reproach and possess good common sense. The successful process control engineer must be both an idealist concerning the quality of the system and a pragmatist about what is achievable with respect to cost in time and money and the prospective payouts.

With this attitude, process control can often temper too much conservation on the part of design engineering and too much radicalism on the part of production control. This balance is essential in successful "fire fighting," for without it, erroneous decisions can be made. Process control often finds itself playing the role of advocate for the missing party when conversing with either design engineering or shop operations and occasionally marketing or finance. Besides maintaining the balance, process control must decide what the correct course of action is and have it agreed to by all parties. This ability to sift through the data and arguments presented by all parties takes considerable technical engineering skills and rapport with those involved. By virtue of dissecting and presenting the facts about the problem and knowing where it fits into the manufacturing time frame, the process control engineer is in the best position to solve these problems.

The main indicator of trouble requiring short-term corrective action is the quality control report, which details the nature of a specific problem. While all QCRs indicate a problem of some type, not all require short-term corrective action. Most QCRs are routine and the procedure for QCR disposition should be sufficient for solution. It is the QCRs that involve a high cost in time and money that require a high-level, immediate response.

It is the process control unit's responsibility to determine whether a QCR needs short-term corrective action or whether the standard routines should be used. This judgment can be made more objectively by establishing basic working guidelines.

EXAMPLE: A QCR involving an error that costs approximately $100 to correct would not normally justify short-term corrective action. But if the same deficiency was discovered on the day the product was to be shipped, the time frame would dictate that it be handled as a short-term corrective action problem. In some cases a shipping problem, regardless of cost, can require immediate correction.

The question of how to handle a QCR leads to the subject of documentation. In job shops, since every job may be different from its predecessor, the problem, its definition, and its solution constitute meaningful information in planning for similar or even dissimilar jobs. For instance, certain contour welds that failed on one job might be inadvertently tried on a future job if proper documentation on the original problem did not exist. In effect, the organization must keep records so that it does not have to "reinvent the wheel."

With the proper documentation, the next step in the short-term corrective action format can take place. This is establishing a temporary preventive maintenance routine, which is akin to an exploratory solution. Major changes in the quality plan to permanently resolve the problem should not be made until the temporary solution has been properly evaluated. The preventive maintenance routine may be a specially designed PC form to be filled out by the operators, an audit to be conducted periodically by the inspectors, special engineering drawings to correct a specific item, or some other extra effort. Whatever it is, it must be implemented with the complete understanding of all those involved. The temporary preventive maintenance routine is more than a patch to stop a leak; it is a patch that may or may not have to be redesigned to make it permanent. Once the temporary routine has proved effective, it is incorporated into the permanent process control system, usually through the quality plan but just as correctly, although less frequently, through the appropriate engineering drawing.

In summary, short-term corrective action is the part of process control work aimed at correcting important quality problems that require more attention than the normal discrepant quality routines. It involves using the data collection and analysis phase to identify problems, mainly through QCRs. An immediate, vigorous response action may need to be taken. After the immediate problem is resolved, a temporary preventive maintenance routine is established, and finally a long-term solution is put into effect. Long-term corrective action will be discussed in detail in a later section of this chapter.

REPORTS

Process control system reports display the current status of the manufacturing operation. Their main function is to give an accurate picture of the manufacturing operation at a particular point in time. The reports must be constructed so that they do not give a false sense of complacency to the reviewer or be overly alarming. They must be factual and clearly state how conclusions are reached. They must not contain any form of hidden editorial.

A secondary function of the reports is to aid in trend tracking. Therefore, wherever possible, they should show comparison results. For example, a report stating that there were 42 operator-responsible machining errors in March does not mean very much. For the reported 42 discrepancies to be meaningful, they must be compared with like data. The usual way to do this is through comparisons with similar previous time periods, taking care that the work was similar and of the same relative frequency. The goal in trend tracking is to have the value reported on be the only variable with everything else held constant. This gives the reviewer the ability to judge accurately whether the situation is remaining the same or changing.

In job shop operations, it is not easy to find commonality of data, so that before a measurement is developed and reported it is essential that it be reduced to the lowest common denominator.

EXAMPLE: For welding measurements it is essential that measurements be for like processes with like equipment. Comparing automated welding mixed with manual welding will always give a false picture. Each type of welding should be reported separately.

This leads to another requirement for reporting data. The method by which the data are obtained and what the data mean should be readily identifiable by the reviewer. There should be no secret formulas or codes; the report must be simple and understandable by its least familiar reader. I recommend a minimum of four report categories be forwarded to management:

1. Monthly quality reports
2. Manufacturing losses analysis reports
3. Weekly problem sheets
4. Special management reports

The first level of management would receive all four reports; the next higher level would receive reports 1, 2, and 4; and the highest level of management would receive only report 4.

Monthly quality reports explain the status of quality for that time frame. They are heavy on tollgate inspections and compliance with quality plans. One of their functions, in addition to showing the quality status of the products, is to show the status of the quality plan and whether it is functioning as it is supposed to.

Manufacturing losses analysis reports are detailed examinations and explanations of deficiency correction costs. Here manufacturing losses are broken down by part, responsibility, and operation in a format useful for planning corrective action. This is one of the basic reports used for long-term planned improvement programs. Since it is a report on money spent, it identifies the large-payoff items that should be given close attention.

Weekly problem sheets are the most basic of all reports, where the process control engineer lists the problems of the week. The engineer records specifically what the problem was, on what workstation, very briefly what cause the investigation showed, and what corrective action should be taken or already has been taken to correct the deficiency.

Special management reports are either condensations of the three types of reports mentioned above or reports dealing with a particularly difficult problem needing the direct attention of higher management. These reports are prepared quarterly or semiannually for higher management. They are more general and brief than the other reports, as the recipient must be aware of trends, not the specific details involved.

These are the primary process control reports; all other reports are variations on these types. The important thing in reports is to make sure that all vital items are accounted for in a very readable manner, and that the data are presented against a measurable base for comparison. Good reports lead to effective long- and short-term corrective action programs.

CORRECTIVE ACTION (LONG-TERM)

This is the area that makes or breaks a good process control system. The system's ability to evolve and improve through the use of the various feedbacks and its ability to function competently in a dynamic situation are the prerequisites for its success.

The method by which the process control system corrects itself is found within the corrective action (long-term) block in Fig. 10-2. Included here are improvements to quality plans, changes in methods and planning, and future problem forecasting. Also included would be the objectives and goals with supporting projects, as discussed in Chapter 2. Long-term corrective actions are taken only when their worth and cost have been carefully calculated and evaluated by the responsible process control personnel. The short-term corrective action is reviewed and steps are taken to make it permanent by issuing revised quality plans, or the short-term action is modified and subjected to more field trials.

In this phase, the results of temporary preventive maintenance routines, short-term corrective actions, and trends shown on reports are evaluated and tested to determine the need for permanent system improvement. While it is recognized that a short-term action would not have been taken without some justification, it must be determined whether the original set of conditions will still prevail over a longer period of time.

> **EXAMPLE:** If a temporary preventive maintenance routine was set up to control a particular weld variable, and the welded fabrication was later replaced by an easy-to-make forging, there would be no need to incorporate the temporary preventive maintenance routine. A record of the routine and how it worked would be in a reference file. Therefore, the short-term corrective action would be terminated, and no corrective action on a long-term basis would be taken.

Now let us assume that in the above example a new size assembly is required and it cannot be produced by the forging vendor at a reasonable price. Hence, the old welded fabrication design must be employed. Going through the experience file on this operation turns up the record of the temporary preventive maintenance routine used previously. At this point, it would be wise to incorporate this routine in the quality plan to prevent the problem from recurring. Thus the temporary preventive maintenance routine becomes a permanent part of the quality plan.

The objective of the process control organization is to use all the feedbacks available to develop evolutionary changes in the process control system. The organization should employ the trends in anticipating future needs so that it can supply management with information to be used in decision making. Solving small problems revealed by trends before they become big problems is the most cost-effective way to function.

Process control uses trend information to guide manufacturing engineering actions concerning producibility and equipment operation. It informs shop operations about potential operator performance and manufacturing problems. It gives design engineering the quality results related to their calculations. Process control tends to control the manufacturing process by being a vast information storehouse and by having the responsibility for quality plan initiation. Therefore, any long-term changes to the plans require considerable evaluation before they are made. But changes must be made, and the "red tape" procedures common in many well-meaning organizations cannot be allowed to prevent change. Therefore, another factor in long-term corrective action is that indecision and procrastination must not be allowed to gain a foothold. Process control has to be a pragmatic action center or it will fail.

The process control unit must always be aware that it is through long-term corrective action that the three continuously generated benefits of the process control system are gained. These benefits are reduced manufacturing losses, minimized production delays, and improved quality.

QUALITY COSTS

Management of costs, what ever the category, are essential for business success. We would therefore expect managing quality costs to be a vital part of process control and it is. When we speak of quality costs we are essentially referring to three integrated subsets. They are:

- Categories of quality costs
- Quality costs relationships
- Quality costs reporting

These are all tracked, analysed, and reported on via the company's cost accounting systems. We will look at each subset individually. But keep in mind, that as an integrated whole they actually react with each other continuously and without definition borders. Let's start with categories of quality costs.

Categories of Quality Costs

There are four defined categories, all tracked via the cost accounting procedures of the company. It is process control's responsibility to obtain these tracking data from finance, then analyze them for trends and recommend corrective actions to minimize these expenses. Sometimes it is difficult to extract these categories of expenses and finance can be obtuse in providing the data. But keep in mind it is to everyone's vital interest that they be monitored closely. To not do so can lead to severe competitive and profitability problems. Therefore it behoves process control to be insistent with finance functionaries to supply the data on a timely basis, which I believe to be weekly. From experience, I must say that I have come across bureaucratic financial management types who were more interested in form and function than providing services they are obligated to do. In that case you must urgently and voraciously insist they do their job so you can do yours. The four categories are: prevention costs, appraisal costs, internal failure costs, and external failure costs.

Prevention costs are those associated with identifying and minimizing activities that are detriments to good quality. They are work that has to be done to ensure that the company can produce its products to the specifications required by its customers. They include the costs to do:

- Audits
- Design reviews
- Process control monitoring
- Quality planning
- Quality reporting
- Training
- Vendor certification

These are considered operating expenses and are in the same costing categories as rent, telephone, computer charges, etc.

Appraisal costs are those associated with doing the actual evaluating processes to detect whether specification are being met or failures are occurring. They include:

- Incoming materials inspection
- In-process inspection
- Nondestructive testing (NDT)
- Final inspection
- Operator certification of process compliance.

As with prevention costs appraisal costs are considered operating expenses. I should also point out that the last category under appraisal costs, operator certification of process compliance, is too often buried under direct labor costs (which is not an operating expense, but part of the cost of goods sold) because of difficulties of separating out for time keeping purposes actual value added work versus all other activities. This is almost always due to laziness of supervisors in properly instructing and enforcing good time keeping amongst operators and must not be tolerated. A smart company needs to understand its appraisal costs in totality in order to manage properly and charge fair prices for its products. Also, by knowing each category of cost by its specifics it is possible to take effective cost reduction actions. In the operator certification area, that may mean development of faster and more accurate procedures and/or tools for doing this evaluation.

Internal failure costs are those necessary to fix bad products or doing them over again from the start operation. They include:

- Scrap
- Rework
- Identification of failure cause

Scrap and rework are considered costs incurred to produce products; materials and direct labor. These are costs of goods sold costs for accounting purposes. Identification of failure costs are considered to be operating expenses.

External failure costs are associated with defective products being delivered to customers in error. These are the worst type of quality costs to have because they give cause to customers and potential customers to question the quality commitment of the company and can quickly destroy a reputation for caring for customers and as a quality producer. They include:

- Repairs
- Replacements
- Recalls
- Price concessions in lieu of repairs
- Cost of returning goods to the factory

These are all listed under the warranty cost category except for the cost of returning goods to the company, which is an operating expense. Warranty costs are considered part of cost of goods sold.

Quality Costs Relationships

Quality costs are associated with productivity improvement measurements. I will discuss the overall topic of productivity improvement as the last subject of this book in Chapter 19, entitled "The Integrated Productivity Improvement Program." So I will limit this discussion to an introduction to quality cost measurements.

All companies productivity accomplishments can be categorized productivity improvement due to direct labor programs, productivity improvements due to all other programs, and quality costs reduction. The first two are plus type improvements, for example, the results are usually equated to sales divided by labor, either direct, indirect, or staff. Quality costs improvements are virtually always a reduction of a negative. Quality costs as the previous section shows are add on costs. They are either costs associated with "policing" the system; for example, preventive. Or they are the costs of mistakes that have to be overcome. These are the scrap, repair, and warranty costs. So a productivity improvement for quality costs is the reduction of mistake costs primarily. And occasionally the reduction in "policing" costs due to better "policing" methods.

We can see that quality costs are philosophically productivity and profit deflators where the "policing" costs are a necessary cost of business to minimize the mistakes costs. The quality cost equations are mitigating factors to apply against the purity of the positive productivity improvement factors. We express this in scrap, rework, and warranty services cost. We also create a measurement called the quality ratio to give an overall trend indicator as to how the company is progressing towards becoming a World-class performer. The following expression is typical for a quality ratio used in manufacturing and service industries:

$$\text{Quality ratio} = [\text{mistake costs} + \text{policing costs}]/[\text{total operations costs}]$$
$$\text{Where: total operations costs} = \text{cost of goods sold} + \text{operating expenses}$$

All functions of a company contribute to quality costs. This is another way of saying everyone makes mistakes and we need to have procedures in place to minimize and eventually eliminate these mistakes. The opportunity to minimize mistakes is boundless. The process control unit is charges with being the "champion" of all programs to do so. Therefore process control or its sometimes existing allied organization quality assurance is charged with developing, monitoring and creating measurements for programs to reduce quality costs. Here are some examples of opportunities for reducing quality costs throughout a typical product or service producing organization.

Marketing: poor contract negotiations, requiring rush manufacturing, not enough time to adhere to proper evaluation of processes during fabrication

Human resources: Lack of qualified/trained operators due to poor hiring tactics resulting in bumping rights in layoffs without regards to adequate skills levels

Design Engineering: Designs not matched with the capability of the factory, resulting in scrap and/or rework to fix and cull out poorer product output

Finance: Inability to finance proper equipment, hence lower yield in manufacturing

Quality Control: Failure to provide adequate inspection and quality planning for operator certification, resulting in defective product being sent to customers and incurring warranty costs

Shop operations: Failure to train operators, resulting in higher scrap and rework during the learning curve ramp up

Manufacturing Engineering: Poor fixture design, resulting in out of tolerance parts which are mostly scrap

As it is so evident, the quality costs relationships are definitely cross functional and all pervasive. It is a very important task to stay on top of this situation in a manner that objectively digs out these causes and effects, and programs are put in place minimize the negative effects.

Quality Costs Reporting

It is very important to log quality costs in the correct account so corrective action can be planned and implemented. This is so because of the exceedingly high cost of recovery associated with mistakes. Various studies have shown that the cost of correcting defects averages seven times the initial cost for each operation that has to be repeated. For example, if a defect is found at the third manufacturing operation, the cost of repair typically will have been $3 \times 7 = 21$ times the original cost of doing the work at these three work stations. No wonder astute managers go after quality defects elimination hammer and tongs!

Quality cost reports are highlight reports of the normal cost reporting system, not double counting. Prevention and appraisal costs are part of the monthly operating expense reports. The monthly reports can be considered to have two components:

Standard costs—examples: salaries, assessments, etc.
Special costs—examples: training costs, non-destructive tests, etc.

Standard costs are logged as per budget procedures. For example, if five staff people are assigned to a unit, their salaries for the month are automatically booked in that unit's account and can be predicted to the nearest penny, month by month. Special costs are booked against the unit to be charged for the expense as they occur, and are not predictable except as trends based on expected happenings compared to past experiences.

Failure costs (internal) are as a result of Quality Control Reports (QCR) and placed in the respective cost of goods sold (COGS) accounts. To give an idea as to how costs grow (a reason for the factor of 7). Let's look at the Quality Control Reports procedure. Here's the chronological listing of steps taken.

- Deviation identified
- Investigation for cause and extend instigated
- Disposition based on deviation extent determined
  - Scrap
  - Rework
  - Use as is
- Work order issued for scrap and rework, charged to the department at fault
- Rework, remake done in accordance with the disposition and charged to the responsible department
- Cause of deviation logged
- Corrective actions initiated to prevent (at least minimize) recurring problems

It's easy to see how cost can grow rapidly while taking these actions. It is very necessary to highlight these costs to give them visibility so all employees recognize how devastating they can be for the economic well being of the company. One way of doing this is to have monthly employee meetings to show these costs by category, usually in a pareto format. Then explain what actions have been taken and to also ask for inputs on how to solve these failure problems.

The other failure cost is the external account. This is the warranty costs and is initiated as a result of the returned goods authorization (RGA) process. This process recognizes that a mistake has been allowed to get to a customer and must be rectified. Once an RGA is authorized the method of accounting for and fixing a warranty problem is identical to the internal failure process. External failures need to be illuminated to all personnel just as internal failures, and in the same manner. In fact it is customary to combine internal and external failure presentations for this purpose. I believe external failures are more repugnant than internal because they show the company's failures to customers. This in turn makes it more difficult to generate future sales. So corrective actions for external failures need to be aggressively pursued, even more so than internal failures.

QUALITY AUDITS

Doing quality audits is one of the major activities of the process control unit. While verifying that every job is being and has been done correctly is the direct responsibility of operators, supervisors and line managers; ensuring that the quality procedures are being carried out properly is the responsibility of process control. The way that is done is through a well-designed audit program, including certain types of inspection. The inspection procedures will be discussed in the next section. For now, even though they are very much linked to each other, let's divorce audits from inspection and investigate audit procedures.

Audits are done to evaluate, measure so to speak, the organization's ability to carry out the quality requirements of producing goods and services. It does so by looking at the quality system as defined by the quality plan to see if the system is in synchronization with the plan. It also looks at the system's capability to deliver products and services to the specifications set by the customers.

Various people in a variety of job positions do audits. The idea is to get information on how the system is functioning from as many perspectives as possible. Therefore, audits are scheduled and become part of the job responsibilities of various personnel as part of their daily work schedule. For example; supervisors will be required to audit their own area of assigned responsibility on a daily basis, and do an audit of another area on a weekly basis. In both cases, the auditor is expected to follow an assigned regimen, which is designed by process control engineering. The purpose is to get consistency in data by requiring everyone to answer the same questions. In addition, a good audit process will have ample opportunity for the auditor to note specific one of a kind or unusual situations.

Auditors are given training in how conduct a successful audit. The training goes through the reasons for audits and how to do them, both in generalities and for the specific product or service profile being examined. The training program goes through the typical audit process, which if followed, stands the individual well for what ever area the auditor will be asked to observe and make comments on.

The audit process has six definitive steps:

Understand how the system is supposed to work. Before doing the audit the auditor must "learn" the area well enough, and know the chronological flow path, such that he or she can trace the process and comment on how well it is being complied with. For example; if a supervisor is asked to audit a weld process, even though she is not very familiar with welding techniques, she should be able to tell if proper procedure is being followed. If the quality plan requires a part preheat of 450° and recording that it has reached the required temperature on the appropriate PC form before welding, the auditor can certainly confirm whether or not that is happening. Simply reviewing the PC form to see if preheat is entered does this. If it hasn't been and the operator is welding on the part, then a quality system violation has occurred. This is so even if the auditor asks the welder if the part is up to pre-heat and the welder can demonstrate that it is via hand held infrared imaging. The fact that it is not written on the PC form is a violation. The need to document is part of the process. By doing so it makes it rather difficult to start welding before the conditions are correct. And it is possible to audit for this even though the auditor is not a qualified welder, or may not even be familiar with welding.

Compare actual system workings to the system plan. This is nothing more than saying "you're plan" requires you to pre-heat. Are you doing it? If so fine. If not it is a systems violation.

Determine if the system plan can ensure customer specification compliance. During an audit the auditor is constantly asking herself whether the system delivers the goods the way the customer desires them. Keep in mind that the customer is the next operation along the line. In the example, the customer for the welded product will expect the weld to be defect free and dimensionally correct. This is easily checked by looking at the quality log for QCRs pertaining to these categories and see if there are any incidents, how often, and when the last one occurred. If the process as audited appears correct and there are still QCR's being generated for these categories, then the system is deficient and needs to be written up as a violation.

Observe actual system workings to see if customer specifications are being met. This is the same as for the previous step except it focuses on internal step processes and not end of the line.

Document deficiencies for all four points, above. Deficiencies found should be described in as much detail as possible so readers of the audit report have full understanding as to what was discovered.

Recommend corrective actions. No audit is complete unless it summarizes what the audited function has to do to remedy found deficiencies. Completed audits go to process control who then record results in a tally document, then created corrective action activities, both long and short term.

As we can see auditing is a dynamic, ongoing activity designed to assure the process of creating products is being conducted in accordance with plan. It also encourages improvements to the plan as deficiencies in the plan are uncovered.

INSPECTION SYSTEMS AND PLANNING

Inspection is defined as:

"The evaluation of products or services to determine if it complies with the customer's specification."

Inspection is a form of auditing, and the most common application of auditing. Inspectors are trained in the audit procedure and use it for their daily work. They are quite often the front line of quality in the factory working directly for process control. Inspectors do not create good quality products and services. The axiom "You can't inspect in good quality to any product or service" always holds true. What inspection accomplishes, however, is an ever present reminder that the company is dedicated to producing quality products and that critical steps in any process must and will be corroborated by an inspector. As always the person responsible for quality is the operator. The inspector's job is to be an umpire calling balls and strikes as the operator performs the value added work. When an operator believes he has done the job in accordance with the method, and that everything is correct; he will submit the work to the inspector if it is a tollgate operation for review and corroboration. The inspector reviews the work for compliance to the quality plan. If it is correct, he passes it through the tollgate to the next process area. If not he writes a QCR defining the problem for process control to design a corrective action. The inspector also collects all completed PC forms at workstations, audits them for completeness, then files with the respective process engineers.

Methods used to do the inspection job which is, as mentioned before, the audit process but with added emphasis for observing the work very closely for compliance with the prescribed method and its companion quality plan. Unlike other audits the inspector is given authority to measure other people's work to determine if specifications have been accomplished using the same types of tools for measurement as used by the operator and other more sophisticated equipment such as comparators, surface plates and other mechanical, optical, and electronic tools, not normally available to the operator at the work station. We can consider the inspector to be the specialist with regards to measurements and be called upon to do critical measurements for the product at various stages of completion. This ensures that very precise information will be provided to operators and supervisors so that processes can be further refined to maintain process capability within the three sigma plus/minus spread for acceptable results for the required specifications. Inspectors also use testing devices usually designed specifically for the product or process in the course of their work. These may be destructive or nondestructive depending on the nature of the product to be evaluated.

Many times we may consider the work of the inspector to be duplication of that of the operator. And perhaps that is true. But it is necessary. It provides a third-person verification of the operator's findings, which is often necessary as a contractual requirement with customers. We only have this duplication at tollgate inspection because we want to be doubly sure that the product in its current state of development is correct in accordance to specification at this point. All other inspections are done solely by the operator as part of her daily work assignments. Decisions are made as to correctness at all work stations by the operation, and at the tollgate workstation by the operator and inspector.

There is a great range of leeway granted to the operator and inspector as the sole arbiter as to whether or not a part or process is in compliance; e.g., meets specifications. Operators and inspectors are trained to understand the specification requirements as defined by the methods plan and its companion quality plan such that they can determine whether or not a part is okay or if an out-of-compliance part can be fixed or must be scrapped, And for internal components, whether the deviation can be accepted and the final assembly still meet the specification. To summarize, inspection determines if a part is in compliance. If the cognizant operator and/or inspector deem it to be okay, it is by definition okay. Also, process control depends heavily the inspection report's recommendations for fix or scrap decisions.

Inspection planning is integrated into the planning document. The planning document's method sheet calls for inspection as a specific step in the process. Planners reference the overall quality plan for when inspection is to take place, who does it, and how. Inspection planning is a subset of overall planning. To write an inspection plan the planner needs to be aware of

- How the product or service is to function.
- Critical factors of the design.
- Robustness of the design, for example, the deviation from the mean that can be tolerated.
- What kind of evaluation the product or service can tolerate.
- Amount of time the evaluation process will take, for scheduling purposes.
- External documentations requirements the company will need to provide to the customer.

All of this information is provided to planning by process engineering. This information makes it possible for the inspection to planned, thus integrated into the production schedule. If additional methodology needs to be provided to the operator or inspector as to how to do the inspection, then process control will write a detailed methods instruction.

A very important part of inspection planning is to decide when to inspect and how often. The question has to be answered, do we need to inspect every part or is a sample audit sufficient? Inspection costs money and is considered a non-value-added cost. Therefore inspection frequency is a trade-off between reasonable ability to detect nonconforming products or services and the cost to do so.

Types of Inspections Versus Costs

Attributes: judgement characteristics, good or bad, go-no go, pass-fail. Easy to judge, relatively low cost to do so.

Typical terminology: percent defectives, number of defects per thousand units.

Variables: judgement for suitability of use inspection, based on magnitudes of several characteristics. Requires careful analysis of multiple measurements to make a judgement, relatively high cost to do so.

Typical terminology: tolerance range, average of mean values, within 3 sigma plus/minus spread.

The type of inspection technique used will depend on the type of products being made. We find that on average three times as many parts are checked via the attribute technique vs. using variables techniques. Obviously for parts made in the hundreds and thousands over the course of work week, we would want to emphasize attribute style inspection, especially if the cost to scrap and remake is less than or equal to fixing. On the other hand, high-value complex parts need to be handled via the variables route more often than not. So what it comes down to is that inspection costs are very heavily swayed by the type of inspection required, as dictated by the part to be inspected. But the frequency of inspection depends on the degree of risk a company is willing to accept. This too is modified by the type of inspection the company needs to employ. Each company needs to make its own decision on how much inspection cost it is willing to absorb. Types of inspection strategies will also affect costs. Here are the three inspection strategies going from lowest cost to highest cost.

- Sporadic inspection (audit only)
  - Used for low-value products or services
  - Intuitively obvious if products or services are done satisfactorily.
- Sampling
  - Medium- to high-value products or services
  - Statistically based plan primarily with variable based measurements
- 100% inspection
  - High-value products or services
  - Attribute-type measurements are dominant by necessity of ease to do.
  - Inspection never yields 100% certainty that only good parts are accepted. Best rate to be expected is 75% to 85%.
  - Needs to be automated to reduce error potential.
  - This method is gradually being replaced by "six sigma zero defect plans" combined with sampling.

Selecting an inspection plan is never a simple task. The best advice I can give is to err on the side of caution. No company ever lost market share by over inspecting. Many have and even gone out of business because of under inspecting. The worst thing any company can do is achieve a reputation for poor quality. That is the "kiss of death" for the company that is tagged with that reputation. Under inspecting presents grave risks that must be avoided.

RELATIONS WITH OTHER MANUFACTURING ENGINEERING UNITS

Process control is the newest addition to the manufacturing engineering sub function, and it is not derived from the traditional base of manufacturing engineering, that is, industrial engineering, tool and dye engineering, methods engineering, and design engineering. It was split off from quality control when quality control became a non-manufacturing entity called quality assurance. Process control is an amalgam of applied engineering that derives its strength from statistical concepts considerably modified by pragmatic approaches to manufacturing and design engineering. In some ways it is like producibility engineering, except that it deals in after-the-fact rather than before-the-fact evaluations of products. Similarly, it resembles methods engineering, but again in after-the-fact aspects of production, not before. Unlike any other manufacturing engineering functions, process control must react to and correct immediate production problems, while the other manufacturing engineering functions can deliberate for considerable periods of time in solving problems. Process control represents manufacturing engineering to shop operations for virtually all product-related technical questions.* Thus it is the counsellor, confidant, and sometimes disciplinarian of shop operations. However, in order to do its job effectively, it must draw on all the resources of the other manufacturing engineering units as well as support those other units. Let us now look at the relations between process control and the other manufacturing engineering units.

It is the responsibility of the manager of manufacturing engineering to ensure that all of the manufacturing engineering units function as a coordinated team. Since process control is a little

*In contrast, methods engineering is the shop operations contact in all technical matters pertaining to the operation of the machine tools and process equipment.

different from the traditional manufacturing engineering units, it is important to understand what it does and what its relations with the other units should be. I will start with advanced manufacturing engineering.

The advanced manufacturing engineering (AME) unit deals with the mechanical systems that the factory needs to produce its products. Thus it needs to know whether the equipment is working adequately to produce the parts in accordance with engineering drawings. This leads to a natural communication between the area planner and the process control engineer. The area planner is interested in knowing whether the plans for the manufacturing layout and the equipment are working well. The process control engineer cannot answer directly for the performance of the equipment, but can account for the human–machine interface and the results of putting it to work. The process control engineer supplies performance information related to failures and why they occur. In contrast, the performance information supplied by the methods engineer is related to the efficiency of operating the equipment. The process control engineer's information deals with the reasons for failure; whether it is operator-caused or machine-caused. In both cases the area planner will be interested in learning whether or how the equipment contributed to the failure, and in discussing possible corrective actions with the process control engineer. Knowing the nature of the failures, the process control engineer is in a position to review the equipment and offer avenues of exploration for correction to the area planner.

Producibility engineering is responsible for ensuring that only producible designs are released for manufacturing. Process control engineering is at the other end of the spectrum and evaluates the results of the work of producibility engineering by determining how producible the design really was. Its measurements of percentages of failure and costs of failure show how well producibility engineering did its job. A good deal of the success of a producible design on the factory floor must be credited to the methods engineer, but without a producible design to start with, the efforts of the methods engineer would be futile. Hence the failure rate of a producible design is basically a measure of the effectiveness of the producibility engineering function. This helps us to understand the relationship that exists between the process control engineering and producibility engineering units. The producibility engineer can ask the process control engineer about the probable effects of allowing specific designs to reach the factory floor. The process control engineer can provide needed information for producibility engineering to represent manufacturing in design evaluation activities.

The Methods, Planning, and Work Measurement (MP&WM) activities are close to the evaluation activities of process control. Both are interested in measuring the output of the work stations—process control for product conformance, and MP&WM for efficient production of a conforming product. To be efficient, both types of measurements should complement each other and, whenever possible, be derived from the same data. Methods engineering measures the rate of productivity, while process control measures the quality of productivity. Both must be optimized for a manufacturing operation to be competitive, and one cannot be optimized at the expense of the other, hence the need for cooperation. This cooperation usually takes the form of designing the quality plan to conform with the prescribed method for producing the product, so that measurements required by process control will also be useful in measuring productivity. A typical example would be a process control measurement requiring the number of items produced over a shift to be recorded along with the necessary quality characteristics. This gives a measurement of productivity rate as well as a measurement of quality.

Finally, process control provides a statistical service to maintenance engineering by categorizing and quantifying machine-caused quality problems. This is the type of information maintenance needs for planning corrective actions. Defective products may or may not be due to a complete machine failure but they can show equipment that is not functioning as it was specified to. Therefore, the process control data point out symptoms of equipment in the process of failing and enable maintenance engineering to take action while the deficiencies are still relatively inexpensive to fix. For this reason, the maintenance engineering unit must have a close relationship with the process control unit. It is not unusual for maintenance engineering to request specific statistical comparisons of equipment from one time period to another on machine-caused quality failures.

The ability to detect trends of machine performance and plan future maintenance activities helps the maintenance engineering unit to maximize its utilization of repair funds. Thus process control data represent not only quality trend data but also, in this case, machine deterioration data.

The examples presented above show that process control engineering is one of the technical support functions of manufacturing and explain why it is now considered part of manufacturing engineering.

AN INTRODUCTION TO STATISTICS AND PROBABILITY THEORY FOR STATISTICAL PROCESS CONTROL*

With the modern approach to quality control we want the operator at the work station to be the inspector as much as possible. Therefore we need to equip her or him with the tools to do the job and to do it in such a way that they do not reduce the rate of production. By far this is most successful when the operator can carry out inspections internal to the process cycle time. One of the most successful ways of doing this is via statistical process control. This allows the operator to monitor whether the process being employed is capable of delivering good parts. Thus the process serves as a data collection function as well as providing the operators with information to do their job properly.

In the past operators were never trained in SPC techniques because it was thought to be a quality function, not a production function. We now know the fallacy of that reasoning. We know that the operator is a full and equal member of the production team and as such is closest to the value added activities of making the product. Therefore it makes sense for operators to have a significant role in measuring the results of the process. This is so because it is the operator who will have the point responsibility in reporting and fixing any problems that comes along.

SPC deals with normal distribution theory and measures of central tendencies. These are all aimed at determining whether a process can deliver product within design specifications and to tell if the process is capable of continuing to do so with reasonable and ordinary diligence. Let's look at the terminology used and then see how it is applied.

Specifications

Desired attributes established by the customer and/or designer normally found on engineering drawings, and data sheets. They may incorporate accepted standards reached by some consensus or governing body for that particular class of product. Specifications, which are defined by dimensions and their tolerance range, serve as goals for establishing the mean and standard deviations.

Process

Any combination of equipment and systems with corresponding instructions for use, that if followed will result in a successful outcome to achieve the specifications in accordance with a plan.

Common Causes

Variations in processes, which randomly affect the process output. These are the nuances that are there all the time to which little can be done, without massive efforts, to control. Some example are variations in ambient temperatures and pressures; general condition of the equipment such as looseness or tightness of the assemblies, ball screws, slides, etc., and cleanliness of the atmosphere. These common causes are like the background static noise of the system. When we say that a process can produce within ±0.0005 in. we mean we can't achieve ±0.000000 without some "major"

*Parts of this section are excerpts from my book, *Fundamentals of Shop Operations Management, Work Station Dynamics*, Chapter 4, published by ASME Press &Society of Manufacturing Engineering, New York 2000.

change in how the product is made. Thus all the random events would result in all produced product falling within the ±0.0005 in. range. Since we really can't control the random events to any extent, this is the best the process can do.

Special Causes

Variations in processes caused by unusual, nonrandom occurrences in the process. We can assign a cause and effect reasons to them. They are usually easily detected by SPC methods and they must be corrected (unlike Common Causes) for the process to produce product to specification within the selected tolerance range.

Statistics

Numerical facts or data assembled, classified, and tabulated to present significant information about a subject we are interested in.

Statistical Process Control (or Sometimes Known as Statistical Quality Control)

Quality Control accomplished via statistical methods.

Parts per Million

This is a term used to signify virtually perfect quality once thought to be unattainable. This is in contrast to the old-style quality philosophy, which worked on establishing acceptable levels of defects between customer and vendor; and never considered perfection as a goal worth striving for. For six sigma theory, parts per million are set to correspond with the naturally occurring standard deviation of the system.

Control Chart

The control chart is a tool to first determine the process limitations by detecting and recording the variations limits. Then monitoring the process to determine when it is nonrandom. for example, something changing in the process that is controllable. It uses the mathematics of statistics so small samples can be taken which then can represent the characteristics of the whole.

Many companies are reluctant to introduce control charts at the workstation because they feel that the mathematics will overwhelm the operator. This is not so. Most people working in factories today have at least a high school education. This ought to mean they are conversant with simple algebra, and that is the extent of mathematics background required for understanding statistical process control. The following explanation is sufficient for the operator to successfully apply SPC in controlling the process he or she has been put in charge of. Make no mistake about it. In the modern theory of quality control, the operator needs to have the responsibility and authority to manage her or his work station. This many times means making go no go decisions on process output adequacy based on SPC data.

Let's look at the mathematics of statistical process control. We will want to apply statistical process control to control charts, which are a pragmatic use of normal curve distribution theory. We know that a naturally occurring process will have a random distribution, or variation of measurements, based on a ±3 sigma (S) spread. Where sigma is the standard deviation. S, standard deviation is a measurement of dispersion of data from an arithmetic mean. We are very interested in controlling dimensions or process outcomes, hence it is natural that a mathematical technique that tells us something about the likelihood of a process straying from the mean would be useful, especially if we know how wide or tight that spread is from the mean. We are interested in controlling central tendency, for example, to have the actual dimension be as close to the desired dimension as possible. Normal curves enclose a ±3S spread. This is simply a natural mathematical truth. This means that 99.7% of all naturally random occurring events will fall somewhere within

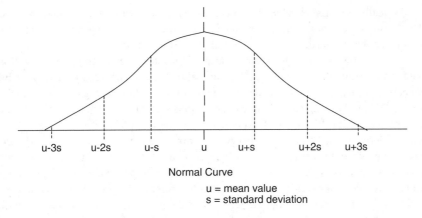

Normal Curve

u = mean value
s = standard deviation

Figure 10-5. Normal Distribution Curve

the parameters of the curve. We can use this fact to set dimensions and tolerance ranges and then measure to see if we are within the boundaries.

Useful formulas relating to normal distributions are:

U = Mean = sum(Xi)/N ;
where: sum(Xi) = arithmetic total of all the like measurements
N = no. of measurements taken
S = standard deviation = {sum(Xi – U)^2/N}^0.5

For SPC purposes, by definition we try to set the process whereby the value of U is the drawing dimension or process parameter and the tolerances fall within the ±3S spread. We then measure randomly a desired minimum of 20 events and calculate U and S. We plot these on our curve to see where they land. If they are random and fall within the limits of the curve, then the process is under control and as long as there are no Special Causes occurring to the process we will continue to make a good product.

In an attempt to determine whether a process has been set up to ensure that the plus or minus acceptable tolerance will fall within the ±3S spread, we measure central tendency. Normal curve theory is useful for determining the natural range of random events. And we strive to match a ±3S spread (99.7% of the randomly occurring events) with the ± tolerances of the design of the process. This means we are deliberately setting the spread we will accept such that the tolerance range and all normally occurring events are coincidental for product acceptability. But just saying we do so doesn't mean it will happen. Many times it takes tinkering with the process, making adjustments, to make it happen. Actually we are trying to make the process even more robust than the ±3S spread thus getting a better probability of success. It is common in many circumstances to set the tolerance range to a ±2S spread. This means having the true spread of occurring events more closely grouped about the mean, U, and having the narrowest spread possible. To measure whether or not we've been successful (e.g., having the tolerance band being equal to or greater than the process capability ±3S spread), we measure central tendency. This is the CP and Cpk measurements, which quite often is a measurement/calculation requirement on customer specifications. These formulas are:

CP = capability of a process = actual tolerance spread/process variation;
where: actual tolerance spread = measured range between lowest and highest values of the acceptable process measurements(e.g., drwg tol.), and process variation = 6S

Cpk = capability of process centering = [(±)Specifications limit – Process mean]/3S
where: (±) Specifications limit is either the plus or minus value of the acceptable value for the process.

Typically a minimum value of 1.33 is considered satisfactory for CP and Cpk to be acceptable. These values indicate that the central tendency of the process will generate acceptable results and the process has been successfully set to fall within the ±3S spread.

Let's use the formulas to create a mean and standard deviation. Then we will use the same data to determine the Cp and Cpk. The normal distribution curve is illustrated in Fig. 10-5. For process control applications we identify the mean, U, as the drawing dimension and either a two-sigma, $|2S|$, or three-sigma, $|3S|$, spread as the plus and minus tolerances. If we have a $|2S|$ spread we are stating that 95.5% of all parts made will be acceptable if the process is under control. This is derived from probability theory and corresponds to the area under the normal distribution curve, where the area from $-2S$ to $+2S$ includes 95.5% of the total area. This means that in a process under control 95.5% of all parts produced will be acceptable, that is, within the tolerance band—provided, of course, that the selected tolerances fall within the capability of the human-machine system. Selection of tolerances to fit a $|2S|$ spread is an iterative process based on design requirements and machine tool capabilities. If a $|3S|$ spread is used, the probability of good parts being made increases to 99.7%.

EXAMPLE: Suppose we are making roller bearings on a cylindrical grinder. The drawing calls for a diameter of 0.500 ±0.002 in. Therefore we wish to set the mean $U = 0.500$ in. and set the three standard deviations to be $|3S| = \pm0.002$ in., or $-3S = -0.002$ in., and $+3S = +0.002$ in. This means that if the process is under control, 99.7% of the sample population would fall within the 0.500 ± 0.002 in. tolerance band. Now we collect data to see whether we are performing satisfactorily. We take 10 samples of roller bearings at random as they are produced, measure the diameters, and calculate the mean:

Sample	Diameter	
1	0.499	
2	0.500	
3	0.501	N = number of samples = 10
4	0.501	
5	0.502	U = mean = $(\Sigma X_i)/(N) = (5.000)/(10) = 0.500$
6	0.500	
7	0.498	
8	0.500	
9	0.500	
10	0.499	
$\Sigma X_i = 5.000$		

and the standard deviation:

Sample	Diameter	Difference from Mean	Square of Difference from Mean
1	0.499	0.001	0.000001
2	0.500	0	0
3	0.501	0.001	0.000001
4	0.501	0.001	0.000001
5	0.502	0.002	0.000004
6	0.500	0	0
7	0.498	0.001	0.000004
8	0.500	0	0
9	0.500	0	0
10	0.499	0.001	0.000001
			$\Sigma(X_i - U)_2 = 0.000012 = Q$

*S = [Q/(N−1)]^0.5 = [(0.000012)/(10−1)]^0.5 = 0.001155

*Note: to find the standard deviation it is good practice to use N − 1 instead of N to eliminate the possibility of bias for samples less than 30. Over 30 the standard deviation calculation reverts back to using N.

Therefore, the standard deviation of this sample would be defining a process that is unacceptable, because the $|3S|$ spread is ±0.003465 in. We are well beyond the selected desired tolerance band of $|3S| = \pm0.002$ in. so we can expect the process to deliver out of tolerance parts.

We can also verify that this process is unacceptable by calculating the Cp and Cpk indices.

CP = capability of a process = [measured range between lowest and highest values of the acceptable process measurements (e.g., drwg tol.)]/process variation
= 0.502 −0.498)/6S = (0.004)/(6)(0.001155) = 0.577

and

Cpk = capability of process centering = [(±) Specifications limit − Process mean]/3S
where: (±) Specifications limit is either the plus or minus value of the acceptable value for the process. = (0.502 − 0.500)/(3S) = (0.002)/(3)(0.001155) = 0.577
both values are under the acceptable ratio of 1.33.

Now we would want to know the percentage of within tolerance parts that will be produced. We can find this value by using another commonly used statistical/probability tool we first introduced in Chapter 6. We will find the percentage of area under the normal curve that equates to good products produced (those within the tolerance range). We will need to find the number of standard deviations contained there in. The number of standard deviations is noted as: Z. And the equation for Z for a single normal curve is:

$$Z = (x − u)/S, \quad \text{where: } x = \text{data value, } U = \text{mean, } S = \text{standard deviation}$$

Figure 10-6 shows the tolerance range, upper and lower imposed on the abscissa of the normal curve. So we know the x distances but have no way of calculating the area under the curve contained between the upper and lower tolerance values. Fortunately many mathematicians have worked over the years to create values of area based on the linear measurements along the x-axis for a normal curve. This data is represented in Table 6-1, Areas Under a Normal Curve.

To find the probability of parts being made with 0.502 in. diameter to 0.498 in. diameter we enter values in the Z formula:

Using the positive half from the mean to the + tolerance value first.

$$Z = (0.502 − 0.500)/(0.001155) = 1.7316$$

We then enter Table 6-1. On the vertical axis we find 1.7, go across until under the column 0.03, this gives a value for 1.73 = 0.4584. But since the value we want is for 1.7316, we need a little more and we obtain this by interpolating between values for 1.73 and 1.74. In this case the value is 0.4585. That means the area value from the mean to the upper tolerance limit is 45.85% of the total area under the curve. Since the lower tolerance value of 0.498 yields an identical Z

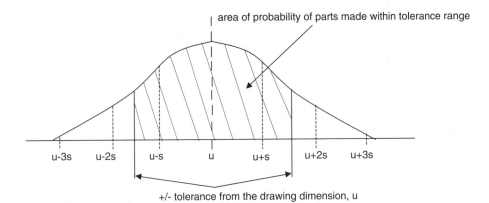

Normal Distribution Curve with Tolerance Range Shown
u = mean value
s = standard deviation

Figure 10-6. Normal Distribution Curve with Tolerance Range Shown

value of standard deviations, only minus, we can see that the total area under the curve between the tolerance range is 2 times the value we just calculated, or 91.7 %. This means we have a probability of 91.7% that the process will make a roller bearing within the desired diameter tolerance range.

USE OF CONTROL CHARTS*

Control charts are sometimes used in job shop process control work (more frequently for mass-flow production). They represent a statistical concept that can be very effective in monitoring machine tool quality. Control charts are based on the concept of the normal distribution, in which the values for like things, such as values for the diameter of roller bearings, cluster about an average measurement. This average measurement, called the mean, would be the diameter specified on the engineering drawing, and the plus and minus tolerances would be the allowable deviation from the mean.

We can see that trying to plot on a normal curve would be quite tedious and perhaps misleading since we are taking measurements over a period of time. For this reason we use a time dependent chart, the statistical control chart (Fig. 10-7). It is important for operators to be able to take data and plot the data on the control chart. They then need to interpret what the data is telling them so they can take appropriate corrective actions, if necessary. The control chart is a time dependent chart that records the measured attributes of the data. Let's describe it in detail.

First, the x-axis is the time dependent axis. It could be delineated in any increment of time that suites the process needs. The y-axis is the attribute axis. From top to bottom we see the Upper Control Limit (UCL), Process Average, and Lower Control Limit (LCL). If we were making a part, the roller bearings for example; the UCL would refer to the plus diameter tolerance (drwg. dim. + tolerance), the Process Average would coincide with the required drawing diameter, and the LCL would be the minus tolerance (drwg. dim. − tolerance). If we were measuring welding process parameters; say, we wanted to control voltage to ensure the spot weld of panels to a frame was proper, then instead of dimensions we could be recording actual real time voltages. The UCL and LCL would represent the plus and minus range of acceptable voltages and the specified voltage would be the Process Average.

If the time scale selected is days, then we are recording daily sampling averages. In Fig. 10-7 we can see that the mean steadily climbs toward the upper tolerance limit and on day 9 exceeds it.

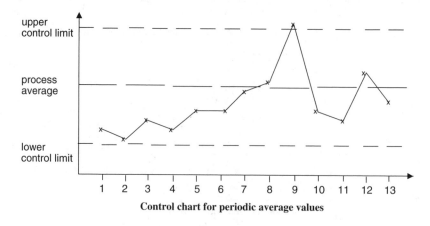

Control chart for periodic average values

Figure 10-7.   Control Chart for Periodic Average Values

*Parts of this section are excerpts from my book, *Fundamentals of Shop Operations Management, Work Station Dynamics,* chapter 4, published by ASME Press &Society of Manufacturing Engineering, New York, 2000.

This would indicate that the process needs to be brought back under control and would lead management to take action to investigate for cause and then take the necessary steps to regain control. In practice, this action would probably have been started as soon as a trend was discerned, possibly after day 3 but certainly after day 4. Days 10 through 13 show the effect of corrective action taken to regain control.*

The technique is for the operator to take a series of measurements, calculate the mean and standard deviation and plot them on the control chart at a point on the x-axis corresponding to the time period of the data collection. Quite often only the mean is recorded on the chart (if the 3S variation is consistently less than 10% of the ±tolerance spread allowed for the process). If, however, we do need to plot the ±3S values then it is often done as a vertical line at the designated time value, where the length of the line is the 6S spread, with U shown as a point or x or some other defining symbol on the vertical line. Figure 10-7 shows a typical set of data taken over time. Each point shown is a calculation of U at a specific time, with the data points of Us connected.

The control chart technique is useful, but only if we can relate actual tolerances to the basic ability to produce. If the tolerance band is too tight for the human-machine process, then no amount of statistical manoeuvring will be of any use. Therefore, two things are necessary precursors for successful use of control charts: (1) tolerances must be chosen within the capabilities of the processes, and (2) the design and manufacturing process must be correctly matched. We need to ensure that the control chart represents the true process, if it doesn't then the chart needs to be modified. That means each control chart numerical values for process average, UCL, and LCL have to be established to match the process we wish to be controlled. Control charts of this type are called $\overline{X}$-R charts (pronounced x bar and R charts). These are used for variables data. $\overline{X}$ is the average value for a group of sample data. R is the range of the data sample, largest minus the smallest. There are many other types of statistical control charts for variables data such as run charts, median charts, and other combinations. But for most uses the $\overline{X}$-R charts are the preferred selection. Setting up an $\overline{X}$-R chart is done by doing the following steps in sequential order.

1. *Select a sample size.* The exact number depends on how easy it is to collect data. But a minimum of 3 is required.
2. *Collect a series of groups of measurements.* Try for a minimum of 20 groups of at least three measurements per group.
3. *Calculate the average for each group of measurements.* This average is the $\overline{X}$ for each group.
4. *By observation, determine the range for each group of measurements.* These are the R values for each group of measurements.
5. *Calculate the average of all the $\overline{X}$s.* This becomes the centerline for the $\overline{X}$ chart and is known symbolically as $\overline{\overline{X}}$ (pronounced X double bar)
6. *Calculate the average of all the Rs.* This is the centerline of the R chart. The centerline of the R chart is known as $\overline{R}$ (pronounced R bar).
7. *Calculate the control limits.* For the $\overline{X}$ chart, UCL and LCL are the ±3S values discussed previously. But statisticians have developed tables of constants to be used based on the number of samples in a group for use in calculating UCL and LCL for both $\overline{X}$ and R charts.

    a. *For $\overline{X}$ charts.* $\quad \text{UCL}\overline{x} = \overline{\overline{X}} + A_2\overline{R}$
    $\qquad\qquad\qquad\qquad \text{LCL}\overline{x} = \overline{\overline{X}} - A_2\overline{R}$

    b. *For R charts.* $\quad\ \ \text{UCL}_R = D_4\overline{R}$
    $\qquad\qquad\qquad\ \ \text{LCL}_R = D_3\overline{R}$

8. *Plot the respective $\overline{X}$ and R data.* Interpret as described below.

---

*Figure 10-7 shows a nonrandom distribution. If the distribution were random, a trend line would not be discernible. The significance of a nonrandom distribution is that the process is changing, and it will continue to change until a new steady-state level is reached. At that point randomness will reoccur.

Table 10-1. Control Chart Constants Table

n (sample size)	A_2	D_3	D_4	d_2
2	1.88	0.00	3.27	1.13
3	1.02	0.00	2.57	1.69
4	0.73	0.00	2.28	2.06
5	0.58	0.00	2.11	2.33
6	0.48	0.00	2.00	2.53
7	0.42	0.08	1.92	2.70
8	0.37	0.14	1.86	2.85
9	0.34	0.18	1.82	2.97
10	0.31	0.22	1.78	3.08
11	0.29	0.26	1.74	3.17
12	0.27	0.28	1.72	3.26
13	0.25	0.31	1.69	3.34
14	0.24	0.33	1.67	3.41
15	0.22	0.35	1.65	3.47

Control chart values for $\overline{X}$-R charts are found in mathematical tables books having statistical data. The following is such a table (Table 10-1) I used as Manager of Industrial Engineering Consulting for General Electric Co.

Sometimes we want to do a quick check to see if data we are getting are falling within the proposed UCL and LCL. We do this to see if there is a match between the process we want to use and the drawing specifications. Statisticians have come up with a fast way to estimate the standard deviation, again through compiling lots of experimental data and creating a factor for easy use. This is the d_2 factor found in the Control Chart Constants Table (Table 10-1). To find the estimated standard deviation we use the following expression:

$$\text{Sest.} = \overline{R}/d2$$

In our example for the roller bearings we found S = 0.001155, and the range R is 0.502 – 0.498 = 0.004 in. The sample size is 10, the number of measurements we took. We can also see that R = $\overline{R}$ because we only took on group of measurements. Let's see how close our estimate of S will be to the actual computed by solving the above table. We enter the table for n = 10 and obtain a d_2 value of 3.08

$$\text{Sest.} = \overline{R}/d2 = (0.004)/(3.08) = 0.001299.$$

This is a 12.5% error from the actual of 0.001155. Considering we only had one set of data to get an average range, $\overline{R}$, this is very close and certainly would give us a useful comparison for matching the process to the desired results.

Let's look at an example of how control charts are developed. We will use the eight steps outlined above. Most control charts are initially constructed this way then become what is called run charts, whereby additional information is added on periodic scheduled basis. The quality plan will dictate when samples should be taken, for example, measurements obtained to then be plotted on the run chart.

EXAMPLE: We have a need to construct an $\overline{X}$-R chart for daily data from the manufacture of roller bearings because we are concerned with the potential for out of tolerance bearings being used in critical rotating devices for elevators. Using the data from 20 groups of samples determine the UCL, process mean, and LCL for the $\overline{X}$-R chart to be used to monitor daily results. The data is as shown in fig. 10-8. Where the upper and lower values for the control chart are as shown and the process mean is $\overline{\overline{X}}$.

The next step is to plot the data on $\overline{X}$-R chart control paper. This can be purchase from many stationery supply companies or you can construct your own. Figure 10-9 illustrates the chart with the roller bearing example data.

Data for $\overline{X}$ - R Chart Roller Bearing Diameter Measurements (inches)						
sample	measurements				average; $\overline{X}$	Range; R
number	1	2	3	4		
1	0.502	0.501	0.497	0.499	0.4998	0.005
2	0.498	0.500	0.499	0.501	0.4995	0.003
3	0.500	0.499	0.501	0.500	0.5000	0.002
4	0.501	0.501	0.502	0.499	0.5008	0.002
5	0.497	0.500	0.498	0.500	0.4988	0.003
6	0.502	0.501	0.499	0.500	0.5005	0.003
7	0.502	0.501	0.500	0.501	0.5010	0.002
8	0.499	0.498	0.500	0.502	0.4998	0.004
9	0.501	0.501	0.502	0.501	0.5013	0.001
10	0.500	0.500	0.499	0.500	0.4998	0.001
11	0.498	0.499	0.501	0.501	0.4998	0.003
12	0.497	0.501	0.501	0.500	0.4998	0.004
13	0.500	0.502	0.501	0.499	0.5005	0.003
14	0.498	0.499	0.498	0.501	0.4990	0.003
15	0.499	0.498	0.501	0.500	0.4995	0.003
16	0.499	0.501	0.502	0.501	0.5008	0.003
17	0.498	0.498	0.497	0.499	0.4980	0.002
18	0.500	0.500	0.498	0.499	0.4993	0.002
19	0.501	0.499	0.499	0.501	0.5000	0.002
20	0.500	0.500	0.498	0.500	0.4995	0.002
			Totals		9.9970	0.0530
			$\overline{\overline{X}}$		0.49985	
			$\overline{R}$			0.00265

$UCL\overline{x} = \overline{\overline{X}} + A2\,(\overline{R}\,) = (0.49985) + (0.73)(0.00265) = 0.501785$

round to 0.502

$LCL\overline{x} = \overline{\overline{X}} - A2\,(\overline{R}\,) = (0.49985) - (0.73)(0.00265) = 0.0.497916$

round to 0.498

$UCL_R = (D_4)\overline{R} = (2.28)(0.00265) = 0.006042$

round to 0.006

$LCL_R = (D_3)\overline{R} = (0)(0.00265) = 0$

Figure 10-8. Example of Calculations for Establishing a Control Chart

We can see that the process as described by the same data is under control and should deliver good roller bearings. Right now you have to take my word for it that the process is under control. How do you make that judgment? Let's investigate.

Interpretations of control charts are based on normal curve theory. We know that a process can be thought of as a series of random events grouped for observational purposes. Those random events will have a dispersion from a central tendency based on the controlling factors for that series of events. For machining, the controlling factors could be the speeds and feeds of the machine and how well the machine is capable of maintaining accuracy. This takes into consideration the bearings, the machine sliding surfaces, the motors voltage controls, and the sharpness of the tools, to name a few variables. There are many variables. The normal curve statistical theory says that all these many variables will create a central tendency measurement with a predetermined plus/minus range. Therefore, if everything remains the same the process will exhibit a spread of ±3S for 99.7% of all events. If we can match our needs for ± tolerance band within, the random events that happen anyway, then we can use the theory to observe the process.

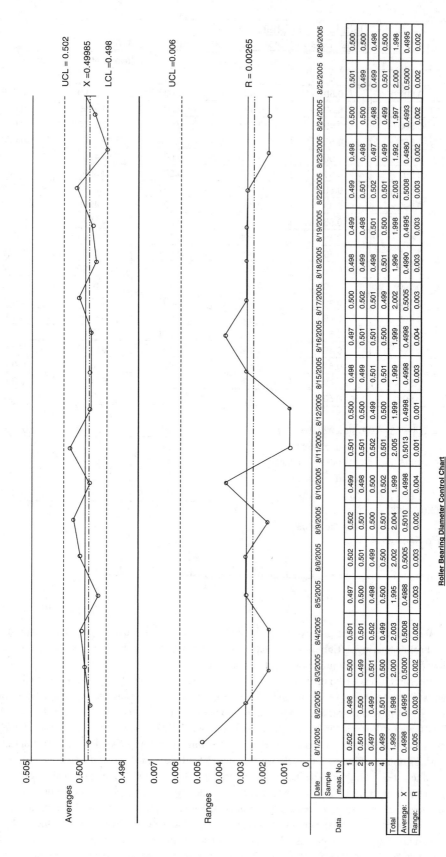

Roller Bearing Diameter Control Chart

Figure 10-9. Sample of an X̄R Control Chart

It is important to understand what the data is telling about the process. Let's quickly review specification limits versus process limits.

Referring to Fig. 10-10, for normal curve distribution Specification Limits vs. Process Limits, we see that first a random event has to be occurring, if not the theory is not applicable. If random events are happening, then that means common causes are occurring not special causes. A restriction to the process, such as a severe restriction on the rotational speed of the cutting tool perhaps outside the range of recommended settings would be a special cause. If this happens the data is no longer random for normal events but is random for this special event.

From Fig. 10-10 we see that the normal curve represents the spread of the tolerance. It is what it is unless we change the process. We also see from the figure that the specification limits are superimposed on the normal curve. If the specification limits (tolerances) fall within the process limits then 99.7% of the output will be acceptable. If they do not then there will be an inevitable percent of failure depending on how much of a misalignment there is between the specification limits and the process limits.

As I've said, the normal curve is not a particularly user friendly diagram to be used for interpreting data. We use control charts for this. There are some very reliable methods of interpretation we can make from the trends displayed on a control chart, based on normal curve theory. Figure 10-11 demonstrates these rules of interpretation.

First we look for randomness about the mean, U (or for the chart calculated and denoted as $\bar{\bar{X}}$). There should be about as many point above the mean (perhaps the drawing dimension, or a pressure nominal, etc.) as there are below the mean over time. If this is not the case the process and the specification are not in synch. This means, unless there is a very small standard deviation, there will probably be failures occurring from time to time at either the upper or lower control limit depending in which direction the mean of the process and the mean of the specification deviate. The corrective action would be to correct the process so that it is centered about the mean of the specification. For a pressure or temperature setting that is relatively easy to do. For machining operations it may mean changes in tooling and set up; and quite often this is a trial and error process.

Next we look to see if there are any unusual readings that appear to be different than the normally occurring data. This quite often indicates a non-random event has occurred. An event such as a broken drill bit or something of an equally severe catastrophic happening. The fix here is to look for any unusual occurrences, perhaps a faulty circuit breaker or the fixture used to load the part was used incorrectly. Quite often, this type of fault responds well to cause and effect analysis and for the preponderance of instances, there is a sole direct cause for the problem, which can be corrected quickly once unmasked.

The third item is to look for trends. For a process behaving in accordance with normal curve theory there ought not be any trends. The rule for trends is that there should never be more than seven reading in a row that we can discern either an upward or downward slope. Statistically there is only a 1 in 128 chance of such an event happening randomly, less than 1%. This type of fault indicates that there is an erosion of the common cause base. Quite frequently it is something as simple as a tool wear problem. As the tool becomes duller the result change. If the readings are several hours apart for a day or two this trend would become very obvious. If the readings were taken every few minutes we would probably miss it entirely. This is a good reason to compare historical data to see if any long-term trends are developing. The fix is to find the variation in the common cause category and correct it.

The fourth item is a little more subtle. It is a randomness settling about a set point that is not the mean, but within the tolerance band of the process and it happens to be within the required specification. This means that the output is acceptable, for now. The problem is that the situation is out of control and could go sour very quickly. The room for error is reduced. We are measuring randomness about a mean, but unfortunately the mean, U, in this case doesn't correspond with drawing dimension or process variable set point. The results being we have lots of room at one set point, but the degree of error allowed is reduced significantly at the other end of the tolerance band. We want the process to be centered about the specification mean to give use the greatest chance for success. Here we see the use of Cp and Cpk measurements of central tendency and why many companies

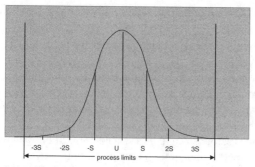

If only random variations are occurring in the process
* The distribution will be a "normal" bell-shaped curve.
* We can calculate and predict the process limits - which will include 99.7% of
the output.

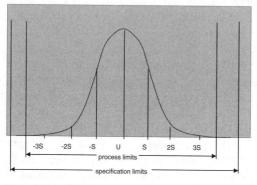

* Process limits should fit within specification limits.
_ if not, process unit produces rejects

*But process limits don't always match specification limits, and nonconforming
material will be produced.

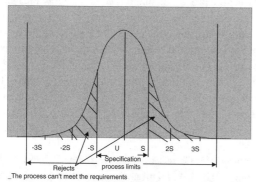

_The process can't meet the requirements

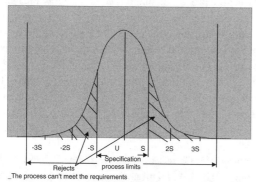

_The process is shifted to the right (or left) of the specification
note: process in control may produce product that does not meet specification

Figure 10-10.   Specification Limits Versus Process Limits
(From instructor notes prepared by Daniel T. Koenig for internal General Electric
Co. course, November 1985, first appeared in Daniel T. Koenig, *Fundamentals of
Shop Operations Management: Work Station Dynamics,* New York, 2000.)

* Learn what to look for

_Process in control

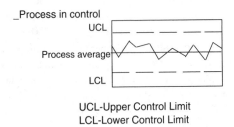

UCL-Upper Control Limit
LCL-Lower Control Limit

* Process _not_ in control

points are beyond control limits

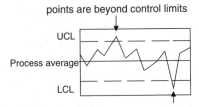

* What action do you take?
_Look for non-random variables
_Typically readjust process

* Process _not_ in control

Long run down could also be long run up

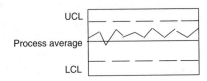

Long run above average could also be below average

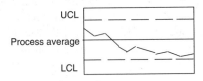

* Watch for a run of 7** points
_All running up or down
_All above or below average
* What action do you take?

_Look for process settings, material, or system change
_Make necessary adjustments

** Less than 1 chance in 128 that a random variation will produce these trends.

* Process _not_ in control

points too close to the control limits

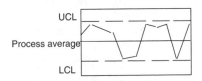

points too close to the average

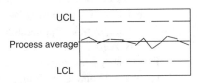

* Watch for nonrandom patterns
_2/3 of data points should be in middle 1/3 of chart limits
_Check, does data make sense?
* What action do you take?
_Look for process system change
_Recalculate control limits if pattern continues

If process is in control, then there will be
_No trends
_No out of control points
_No obvious random patterns

* Action: don't touch it ; you're likely to make it worse

When a point is out of control:
* Ask to see a record of what action was taken
_ Ideal method is to record on back of control chart

Figure 10-11. Interpreting Statistical Control Charts (From instructor notes prepared by Daniel T. Koenig for internal General Electric Co. course, November 1985. First appeared in Daniel T. Koenig, *Fundamentals of Shop Operations Management:* ASME Press, New York, 2000.)

now specify its use by vendors when manufacturing their products. It gives a better probability of higher effective yield. When this fourth type of error is detected the fix is to look for process system changes, determine cause and correct.

We can see that SPC can be a powerful tool for control process quality. However, we also see that it takes considerable attention to detail to make it work. Quite often companies set up SPC regimes and then promptly forget about using the data. They think by taking data their problems will diminish and perhaps disappear entirely. Well, that's asking an awful lot of the Hawthorn affect to cure. While the SPC system is new and everyone has a heightened awareness of it being employed, a certain amount of improvement will occur because we're giving it so much attention. However, like everything else, familiarity tends to build benign contempt and we find that the tools are in place but ignored. For SPC to work, it too has to become institutionalize. The SPC regime needs to be a vigorous part of the quality plan and must be reported on quite frequently. To the point where managers want to know the process SPC results second only to the numbers of product shipped versus promised. Only with this high degree of awareness will SPC be a useful tool for producing zero defect products.

Probabilities are used in process control activities in other ways, as in determining the percentages of defectives in a population of parts to see whether a satisfactory population is being maintained. In this case the satisfactory population includes a percentage defective agreed to by a customer and a vendor. There are also procedures for calculating the probability of accepting a bad lot of goods. These are not useful in job shops, which normally have small lot sizes where virtually every part must be acceptable. Also, the thought that even for mass flow production we are willing to risk shipping defective parts is the exact opposite of what we want to achieve: zero defects. Zero defects is the core philosophy for the six sigma approach which we will look at later. For job shops, the parts are usually complex and do not lend themselves to representation by large universes of simple but similar parts where acceptance criteria are simply measured. Within these constraints, this section shows the present practical, statistical techniques that can be used in job shops. The roller bearing example is not exactly a perfect match for job shop use, but it does illustrate the techniques.

These techniques can be used to monitor the performance of the human-machine system. For example, in job shops we may not have identical parts in great quantity, but we have similar parts. If we make shafts of 2.000 ±0.002 in. diameter for week 1 and switch to shafts of 2.350 ±0.002 in. diameter for week 2, and so forth, we can still calculate the mean and standard deviations of the sample and use control charts to monitor the process, provided that the process remains the same for the different-size shafts. However, our use of probability theory to determine the percentage of good shafts to expect would have to be revised constantly as the mix changes, so we would not use this tool very often. The results would be interesting but of lesser use for future manufacturing planning.

From our examples, we can see that the statistical concepts used for control charts are most useful for continuous manufacturing, but can be partially adapted for job shop use. We should also be aware that we are measuring an entire human-machine process, not just the human or the machine. This, of course, introduces subjective decision making into something that might at first glance be considered entirely objective. When a control chart indicates an out-of-tolerance situation, we must subjectively determine whether it is due to an operator deficiency, a machine malfunction, or a combination of both before we can determine how to correct the situation.

Areas of sampling for control charts must be carefully selected. The more complicated the part, the less likely it is that we can measure one or two key items that will accurately indicate overall quality, and the more difficult it will be to calibrate all the individuals who will be taking measurements. This implies correctly that there is a measurement accuracy problem in sampling work. Variations in taking measurements from individual to individual can lead to misleading interpretations. In many process control units, only a few specified individuals will make sampling measurements. Ultimately, if only one person takes measurements, the only concern is how that individual will perform from sample to sample. Obtaining consistency in sampling measurements is a difficult task.

In theory the use of control charts is appealing, but in practice it can be difficult, especially in a less than antiseptic factory. The control chart is a powerful tool, but the user must always be aware that it is only a tool and must be used correctly and interpreted wisely. I do not wish to discourage use of control charts with statistical theory, but simply to warn that due diligence has to be practiced. By all means I urge you to use statistical procedures, but make sure you are not doing so while viewing your factory through rose colored glasses. The opportunities are great but the dangers of misinterpretation can be even greater if you are not aware of the limitations your processes may impose.

DESIGN FOR EXPERIMENTS RELATIONSHIP TO STATISTICAL PROCESS CONTROL (SPC)

In Manufacturing activities, design of experiments (DOE) is a special use of statistics to fix out of control processes in very special circumstances. You might call it a last chance approach to fix a problem. In all honesty DOE is a complex process involving higher-level statistical mathematics which requires specific training to reach a competency level that the average engineer would feel comfortable using. It is relatively new, just now (within the last decade) emerging from a statistician's theoretical approach to real world "almost" common applications. While statistical process control is an observational approach to defining problems then coming up with solutions, DOE is a more proactive approach, purposely making changes based on mathematical assumptions that such changes are the controlling variables of a complex problem. So first we have observational versus proactive intervention. Second we have a more "go for broke" approach using DOE of finding all the likely controlling variables of the problem changing them in accordance with what the analysis of variables tells us to do and thereby getting a successful solution before (perhaps much faster than the SPC approach). This is great if it works. In my experiences, unless the leaders of such actions are both intimately familiar with the process needing fixing and intimately familiar and competent in advanced statistics DOE requires, we will have less than desired results. Being intimately familiar with the process and at the same time being intimately familiar (to say nothing about being very competent) with DOE is not common.

Therefore DOE, while offering some enticing benefits (mostly speed and less experimental costs) over the tried and true one variable changed at a time approach, in my opinion is not ready for prime time, yet. With the one at a time approach based on well over 100 years of development of the scientific method, we only allow ourselves to change one variable at a time and test its effectiveness on improving the process (e.g., fixing a problem). We maintain complete control of the process while experimenting, and it is easy to understand. DOE advocates say this is too slow and unnecessary. They say through mathematics we can determine which variable are key and which aren't, therefore change all the key ones at the same time and get to the solution faster. Theoretically this is true. But in the real world, an engineer working with a process will more than likely have an intuitive understanding of the variables and their effects, and select the most probable one to experiment with first. The 80/20 rule applies. 80% of all problems are caused by 20% of all possible variables. The DOE approach doesn't quite ignore this but is willing to deal with all the 20% at the same time. With the one at a time approach the experienced engineer familiar with the process probably will fix the problem just as fast as the DOE approach would and without the possibility of creating chaos because only a portion (probably a very small portion) of the 20% was selected.

Also, please note that most process control problems would not fall into the "complex" category. Perhaps 90%+ of the problems encountered that are processes out of control are readily solvable with nothing more than SPC, if that is needed at all. So while DOE is an intriguing, in my opinion almost utopian, methodology in virtually every case is supreme overkill for all but the most intractable problems. Before you become convinced that DOE is not worth learning based on the pragmatic realities I've put forth, let me state that you would be doing yourself an injustice to do so. DOE is a thought process that can help you understand the relationships of independent and dependent variables involved in manufacturing processes. The principles of what to do to set up a process for solving problem are valid. And in fact whether you choose to use the factorial process of considering multivariables at the same time or use the one variable at a time approach the process is the same. So let's look at the process, with at least the thought that the DOE process gives us a good way to approach process control problem solving.

DOE starts with the traditional industrial Engineering: "Plan, Do, Study, Act" (PDSA) process. (See my book *Fundamentals of Shop Operations Management: Work Station Dynamics,* ASME Press, New York, 2000, for a full discussion for PDSA.)

Planning the experiment. We must identify the end results we wish to see. In the case of process control it will be the reduction of a particular problem that is causing unacceptable results. Notice I don't necessarily say precisely what the problem is because if we knew we wouldn't be contemplating a DOE approach. We have to by some means identify variables that could be the cause of poor results. Simply asking ourselves what factors, such as temperature humidity, process compound composition can do this, etc. for a chemical process, for example, could affect results. We can do this through group discussion, brainstorming, theoretical cause and effect analysis, or any other method where we can define what we believe to be the causation variables. Now that we have these variables we design ways of testing them. In the one at a time approach, we would set up a multi-step plan hold all variables but one constant, make changes with one and observe results. For the DOE approach we would still define those variables and plan to do a factorial test. Here for example if we have 3 variables and 2 likely settings for each we would try to do a 3- factorial test, for example, $3 \times 2 \times 1 = 6$ virtually simultaneous tests than use a statistical approach to see if any of the variations are significant. If any are then we've found the culprit.

Do the experiment. Use either one at a time hold all but one variable constant, or the factorial approach. My advice is to always consider the one at a time approach before deciding on the factorial method. Remember the latter is much simpler and easier to come up with a certifiable yes/no decision. Very little statistics are involved, if any. Also the one at a time analysis can be just as fast if the engineer is really on top of his game in understanding the process. He may be able to suggest what the most probable, second most probable, etc cause of the problem is during the planning phase, based on experience. If he doesn't have a clue as to what the problem is then the factorial approach will save time in getting to the answer. In both cases, however, the answer can only be achieved if the correct variable is tested.

Study the results. This is the analysis step. Data has been collected. If it is the one variable at a time approach we compare the results for the modified variable with the process performance. Did it make an improvement? We do this for all the variables in turn, and tabulate the result. We look for patterns or trends that would indicate a change causes the process to do sometime different. If more than one variable gives improvement, then we have to consider both for corrective actions. For the factorial tests we will also end up with data to analyze. And again look for trends. But in this case we are looking at changes in process performance all mixed up with the inter-relationships of variables with one another. So we have to use statistics to determine to significance or lack of significance of these relationships. Those that have no significance we throw out. Those variables that do we keep and put them in our plans to correct. To do the analysis of the factorial tests the first thing we do is test for statistical significance, usually doing what is called the F test. There are several formulae developed by statisticians for this purpose and I will not go into that here. The important fact is that by using this approach, we can get a probability value that the relationships between two, three, etc. variables are significant. What that probability needs to be for us to consider the results depends on the nature of the problem. It is variable. Usually we say we want the probability to have 95% confidence level that what we're seeing is true. Then with this information we can come to conclusions as to what variable and combination of variables are important.

Act on the results. Now, knowing which variables are the culprit we put action plans in place to fix them.

As you can see this process will work for the factorial approach or the one variable at a time approach. The one variable at a time approach is used much more frequently. The factorial approach, which we normally think of as DOE, is used only for complex problems where we have not been able to solve through simpler means. There are lots of good books explaining the factorial approach under the general heading of DOE. What I believe to be the few best ones are listed in the bibliography and have a full explanation and demonstration of the mathematics involved.

I cannot leave the topic of DOE without expressing my concern that you understand that it should be reserved for use for very complex problems. It is best to remember the KISS principle when it comes to problem solving (Keep It Simple, Stupid!). Most often problems are solved by going up the hierarchy of causes. Look for the simplest cause first. If a computer doesn't work, first check to make sure it is plugged into a power supply before doing diagnostic analysis. To further illustrate this philosophy I want to relate an explanatory experience.

In my career, I have only come across one problem where I originally thought it was too complex to solve via simpler methods and required a DOE approach. As it turned out I was wrong. Let me tell you this story.

As part of my responsibilities of supervising manufacturing engineering in one plant, I had an aluminium anodizing line that was malfunctioning and shop operations asked for help to fix it. The problem was we could not get consistent depth of penetration of hardened surface with this four-dip tank process of acidic, wash, and electrolyzed anodizing solutions. There are at least 30 variables involved in anodizing; everything from acid concentration to voltage and amperage control. The process also had significant temperature controls and methodologies of adding chemicals to the bath. The problem was that the results were very inconsistent—sometimes very good, sometimes very bad with lots of scrapped parts. We tried SPC, to no avail. The data just didn't make sense. Finally I discussed this problem with some engineering professors at the local university. They suggested we try DOE. I agreed. I enrolled a few engineers in a 10 week DOE course and part of the course was to use the anodizing line as a real-life problem for the students to practice on. The professors saw this as an opportunity to go from theory to a practical application. Needless to say we trucked on with our high scrap rate, and my engineers learned about ANOVA (analysis of variations techniques), and key functions testing. We tried to define the range of variations and determine the ranges of probability that they were the problem. We never seemed to get the confidence level that we hit on the right variables, which greatly perplexed the professors and made setting up a factorial experiment very difficult. The problem seemed intractable. Results varied so enormously such that there was no degree of confidence that we had a normally occurring process at all! We couldn't define anything approaching a normally occurring process.

Finally I said we need to put our IE hats back on and retrace the process, which I thought had been done but really hadn't. We traced the process from one step to next, spending many hours in not so nice environments, and found that control was very sloppy. Primarily because the supervisors were reluctant to truly supervise the operators at some of the smelly, hot work stations, hence thing were being done "willy nilly." The operators themselves were not happy at being at their respective work stations, and did whatever they could to not be there. Recognizing this out of control situation I set about to fix it. First, by insisting that the methods were being followed explicitly. Second, by making sure the work environment was safe and as comfortable as could be (through erecting portable offices with air conditioning for the operators, and other amenities). By doing this, the problem got better. Within 3 months, it was no longer a problem. In our investigation we found that not only were key operators not following the method, but two of them couldn't read! So anything they did was purely by guess. And it just so happened that these guys were responsible for titrating samples for the SPC reports. To this day I don't really understand how they managed to create reports at all. The process was out of control simply because management let it get that way. As operators rotated out of these jobs they trained their replacements and management never checked to see if the training was adequate. We had a mess and as you can imagine heads rolled, but it was fixed.

The point of this story is that all the analysis in the world couldn't have fixed this problem. And the experts in DOE hadn't clue as to what was happening because the manufacturing process was completely alien to them. We were remiss in not identifying this out of control problem, simply because shop operations management and several engineers were reluctant to get their hands dirty. I was remiss because I didn't pay attention to the production reports and suspect the situation (not necessarily only the process) was out of control. The end results of this whole deplorable mess was that we learned that paying attention to details, both process and human, eliminates virtually all problems. And while it looked like DOE was necessary, it really wasn't. After we cleaned up this mess, retrained people in SPC procedures, we found we could control this multivariable process quite well. And any other glitches that came along were handled quite successfully using the tried and true scientific method. I have yet to see a situation where DOE is necessary to solve process problems.

Having expressed my opinion that DOE is a tool that isn't necessary to control a manufacturing situation. I should state it could have some use in making a good situation even better. If I wanted to make a step function improvement change in a process I might want to get there faster, than by using the tried and true method of only changing one variable at a time, and therefore try to identify the key factors, modify them simultaneously through a statistically controlled set of experiments and have a high degree of confidence that what I did was appropriate. That it would lead to the step function improvement in performance, faster than the one variable at time method is certainly plausible. This is the theoretical premise of DOE. The key words here are finding the key factors (e.g.; variables that control the process) and having a high degree of confidence in your identifications. With this knowledge, between the process engineer and the statistician it is possible to construct a test regime that simultaneously tests for what happens when the key variables are modified.

TOTAL QUALITY MANAGEMENT AND THE SIX SIGMA APPROACH

The topic of process control would not be complete without a discussion of Total Quality Management (TQM) and the most recent popular expression of that philosophy known as six sigma. Virtually everything said about TQM is identical for six sigma. I will discuss the relatively few deviations later in this section. In recent years TQM, and now six sigma has been touted as one of the few new management practices that will make companies competitive, particularly against the onslaught of Asian competition. The shorthand TQM has come to mean a powerful solution to all that ails modern North American industry. It implies that if we will only pay supreme attention to the needs and desires of our customers, and we deliver on those aspirations, we are bound to be successful. Obviously, no such nirvana will occur. There is no one magic cure. There is no doubt, however, that the philosophy and techniques attributed to TQM and six sigma will enhance a company's competitive position if they apply the principles properly. I should also state that these are not particularly new philosophies. They are in truth nothing more than a rebirth of the original industrial engineering theories espoused since the mid-to late-nineteenth century.

TQM and six sigma focuses on awareness techniques for making products to the best of the organization's abilities. Let's take a look at these principles and see how they lead to a set of objectives compatible with the process control philosophy.

TQM theory (therefore also six sigma) can be described via the TQM triangle, as depicted in Fig. 10-1. I discussed this figure previous, but it bears repeating. Let's look at it from the total synergistic viewpoint. We see at the points of this equilateral triangle the labels Customer, Process, and Data. The focus is on the customer's requirements at the apex of the triangle. This in turn generates a plan (the process) for achieving the requirements. The plan is implemented and data are generated, from which the effectiveness of the plan is evaluated. The results are then compared with the customer's requirements and the process is modified in order to get improved results. Notice that the statement is very direct and forceful. Modifications will be made, even if the initial results are satisfactory. This implies a very strong objective of forcing continuous improvement at all times. The modifications are implemented, and then the results (data) are analyzed again to see if they are in compliance with the customer's needs (see Fig. 10-1). Around and around the cycle goes, each iteration (hopefully) resulting in further improvement until eventually total compliance with customer need is achieved and we have reached a state of zero defects. This is the theory: continue improving until perfection is achieved. It is equally valid for non-manufacturing activities.

EXAMPLE: I think everyone would agree that by having a 100% efficient and effective payroll system is a good thing. And striving to have finance being 100% flexible in meeting the payment need for an employee, such as timely vacation, holiday, and special programs pay paid without bureaucratic delay is very beneficial for morale. A company doing well by its employees will benefit by those employees striving to do well for "their" company.

There are a few important concepts at work here: continuous improvement and the striving for zero defects. These are what I call awareness factors. No new techniques for achieving improvement are employed. In fact, the process control techniques described in previous sections of this chapter are probably optimal in attaining the TQM and six sigma objectives. We are simply focusing on using what we currently employ but not being satisfied with status quo results. We are focusing the entire organization's efforts on finding ways to improve the process so that the next time we will get even better results, and the cycle of improvement never ends until the system is perfect.

Perfection, of course, is like the mathematical concept of infinity—approachable but never achieved. This is also true in manufacturing via TQM and six sigma. However, we can see that by focusing on perfection we will easily surpass any strategy that assumes that some percentage of manufactured output will be substandard.

The example in the section on the use of control charts is an example of the old versus the new concept of management control of operations. In the example, we see that the process is capable of producing 91.7% acceptable roller bearings or, conversely, 8.3% defective. We then observe that this is a satisfactory population of percentage acceptable versus defective, probably agreed to by the customer and the producer. Agreeing to buy a percentage of parts that are undiscoverably defective is, unfortunately, still a common practice in majority of all companies, but it no longer need be. With TQM and using the six sigma approach toward achieving zero defects, the data of the example, would not be accepted as is. They would become the starting point for further pursuits of improvement along the road to the ultimate goal of 100% acceptable parts. With processes and procedures invented decades ago, we find that if we actively and resolutely pursue perfection the results are astounding. Now, the concept of Acceptable Quality Levels (AQLs) is no longer valid in commercial considerations between buyer and seller. Instead, the concept of parts per million defectives is the standard between buyer and seller and can be approached only via the concepts of continuous improvement and zero defects philosophies. Process control techniques approached as the *modi operandi* within the TQM frame will definitely generate superior results. TQM awareness concepts merged with process control techniques should be adopted as the overriding policy in applying quality control concepts within an industrial organization.

The question then becomes, how do we apply this awareness factor required by TQM? The answer is through identification of the customer on a generic and a specific level.

We all know the definition of a generic customer. Our organization performs its macro functions to satisfy the requirements placed on us by the ultimate customer, the buyer. This is the generic customer. The specific customer is an individual with a receiver-supplier relationship with another individual. Precisely, this means that the specific customer is the person who receives the immediate output of another person's work. For example, a machinist's immediate customer would be the individual who receives his or her work output. Therefore, the need to satisfy the customer is very focused. The machinist's part in his or her organization's desire to satisfy customer requirements boils down to satisfying the demands of the next workstation in the entire chain of producing the product. If the localized provider-customer relationship is satisfactory and all subsequent relationships are likewise satisfactory, then the probability of overall satisfactory results is high.

In essence, then, organizations practicing TQM espouse the following theme: Continuously strive to make your specific customer(s) 100% satisfied. This is done on the individual, mainly internal, level and also on the macro level as a provider to the client. This principle of customer identification is the key to successful process control practice in a TQM environment.

Now let's look specifically at six sigma. Everything mentioned above is also true for six sigma. And there are no additional technologies involved. Therefore six sigma boils down to a sales approach on how to get people to buy into the zero defect philosophy. General Electric Co. has been in the forefront of selling employees on the importance of this philosophy, thereby improving profits. They have stated that the philosophy transcends departments, job functions, and products. It is a universal method of doing business. The six sigma process per General Electric Co. Annual Report, 1997 consists of five basic activities:

1. Defining
2. Measuring
3. Analyzing

4. Improving
5. Controlling the process

My thoughts on the program the company has tried to, and mostly successfully, implemented are as follows. I have a sort of proprietary interest in its success because I was involved in some of the initial investigations and discussions the decade before that lead to it's promulgation within the company.

- It is a quality program to minimize the cost of quality through minimization of mistakes. Ultimate objective is to achieve zero defect for the targeted process (statistically this means 3.4 defects/1,000,000 occurrences). The targeted processes go beyond manufacturing and include all other aspects of work, such as finance and human resources. Anything that can be measured is fair game. And the company has taken huge steps in requiring all nonmanufacturing functions to develop measurable parameters.
- All of the above done in sequential order, with iteration as necessary.
- By defining means understanding what the goals of the task need to be to achieve zero defects or mistakes.
- Measuring means coming up with a scheme that allows for keeping score as to whether or not the goal is being achieved and presenting data that is specific enough to be interpreted. Probability and statistics are used where applicable.
- Analyzing means reviewing the data in a way that conclusions can be drawn, again using probability and statistics as an enhancer for understanding.
- Improvement takes place when analysis leads to conclusions that can be implemented for process improvements.
- Controlling the process is an application of the improvement scheme and monitoring results. Then iterating back to any step in the sequence necessary to make continuous improvement is done.

In actuality; six sigma is a packaged and perhaps "hyped" version of old-fashioned, IE techniques; that is, it involves using the scientific method. This involves using statistics and probability to evaluate data. And, of course, the TQM triangle analogy of Customer-Data-Process.

Six sigma has taken on a cultural aspect of its own through the green, brown, and black belt gamesmanship for defining, determining who is, and awarding those people who truly understand the intent of a zero defect program and have mastered the skills to do so. 99% of these skills are a thorough understanding of industrial engineering principles including practical application knowledge of probability and statistics. These are the topics I've covered in this chapter. Sometimes I think this "belt fetish" is a disguised attempt at getting nonengineers to be comfortable with basic industrial engineering techniques. I suppose if it does that and people practice good industrial engineering techniques, it's good. But let's not fool ourselves. The "belt fetish" is just a fad. It will fade away in time. Therefore it is important for manufacturing engineers to understand the principles of industrial engineering as a core competence of their profession and practice it competently whether or not they or other nonengineers are wearing belts of any color.

SUMMARY

This chapter has presented the process control system for use in job shops and by extension and explanation, also flow manufacturing. Most textbooks on quality control concentrate on continuous-flow manufacturing and emphasize statistical analysis. Continuous manufacture is used for commodity-type products, so that an individual failed part can be disposed of and be replaced with one of many more satisfactory parts. However, 70% of all manufacturing in North America involves discrete parts and is job shop-oriented. Hence there is a need for relevant information on job shop process control first as the ordinary status of manufacturing, then recognizing flow manufacturing as a special case.

Process control in job shops requires attention to detail for each part, whereas in continuous or mass production it would be impossible to check each part after each operation. Each part in a job shop usually carries an identification, that is, a part number and a shop order number. This means that each part has a distinct value for a specific project. In continuous-flow manufacturing, individual parts have no identity; therefore, techniques based on statistics and probability theory can be applied relatively easier. If a part fails in manufacturing in a job shop, an entire project may be delayed. This difference sets the tone for process control in job shops. The specific part is important and must be carefully evaluated throughout the manufacturing cycle. Statistical concepts can be used, but the importance of the individual part must be maintained. The checks and balances must be aimed at thorough control of processes to ensure that individual parts are indeed of adequate quality to achieve their design purpose. Finally, TQM and six sigma theory espousing continuous improvement leading to zero defects in manufacturing applies to job shop as well as continuous flow manufacturing process control.

REVIEW QUESTIONS

1. Refer to the methods sheet in Chapter 7, Fig. 7-1. For each step shown, list possible quality actions that the operator and/or inspector will perform.
2. Describe how the choice of a measuring tool can dictate the process by which a part will be made.
3. Discuss how process control engineering acts as the communications link between the manufacturing and engineering functions. Use Fig. 10-2 to illustrate two key points of the discussion.
4. Write a cookbook-type quality plan for the methods sheet in question 1.
5. Discuss why job shop quality plans are normally written around generic processes rather than parts.
6. Explain why operator certification is a better form of process control than inspector certification.
7. We wish to test frame support welds, by ultrasonic techniques, for railroad car undercarriages. The production schedule requires two undercarriages to be completed per 8-hour shift. No defects in the welds can be tolerated and an angle beam test ultrasonic equipment is available. Only trained ultrasonic technicians are allowed to use this equipment, of which there is one unit available per shift. It is estimated that one frame support weld test can take 15 to 30 minutes to perform. Using the five rules for testing in a factory, prepare a test plan outline for the frame support welds.
8. For question 7, discuss the type of data that can be gathered and how the data can be presented to give management an accurate evaluation of the quality of the work being performed.
9. Evaluate the quality trend of a shaft axial manufacturing area. The following data have been collected.

Week	1	2	3	4	5	6
Operator errors	6	8	13	5	9	12
Machine-caused errors	2	6	6	8	5	3
Defective material	2	1	2	3	2	4
Shaft axials produced	30	33	29	36	32	30

10. When is "fire fighting" a valid process control engineering technique? Discuss it in comparison to long-range corrective action programs.
11. With respect to process control techniques, state what a temporary preventive maintenance routine is.

12. For the data in question 9, discuss how the data and/or the interpretation would be reported to the following levels of management: (a) manager of process control engineering, (b) manager of manufacturing engineering, and (c) manager of manufacturing.

13. Explain how a temporary corrective action program becomes a permanent addition to the quality plan.

14. Using the data of question 9, construct a scenario showing how process control engineering would interact with each of the other manufacturing engineering units. Show how these data could affect the activities of the other units.

15. Valve cover joint studs of 3 in. diameter are produced on a CNC chucking lathe. The data below are from a sample audit taken to verify that the process is under control. The engineering drawing requires a stud of 3.000 ± 0.015 in. From the data, determine whether the process is under control by determining the mean and standard deviation.

Sample	Diameter
1	3.005
2	3.007
3	2.987
4	3.006
5	3.008
6	3.009
7	2.995
8	3.010
9	3.003
10	3.002

16. Sample means for the valve cover studs in question 15 are: 3.010, 3.009, 3.007, 3.005, 3.006, 3.008, 3.004, 3.001, 3.000, 2.998, 2.995, 2.992, 2.990, 2.991, and 2.989 for days 1 through 15, respectively. Construct a control chart for periodic average values and interpret the data.

17. Based on the data of question 15, use the probability technique of the standard deviation of a normal distribution to predict the percentage of valve cover studs expected to fall within this tolerance band of better than two standard deviations if the process remains the same in the future.

18. Three subgroups of shafts, five pieces each, had an $\bar{X} = 5.4$. Calculate the control limits if the average range was 1.2, and determine if the process is or is not under control.

19. Construct an $\bar{X}$-R chart for shaft pin manufacturing given the following 10 days sample data for the pin diameters

day 1	day 2	day 3	day 4	day 5
1.002	0.997	0.100	0.997	0.996
0.999	0.997	1.003	1.000	0.997
1.000	0.999	1.001	1.000	0.998
1.003	1.000	0.998	1.002	0.998
0.998	1.000	0.996	0.999	1.000
day 6	day 7	day 8	day 9	day 10
0.998	0.999	1.000	1.001	1.000
0.999	0.996	0.999	1.000	0.999
1.001	0.100	0.999	1.002	0.998
1.002	0.998	1.000	1.000	0.999
1.001	1.000	1.001	0.999	1.000

20. Discuss the issues related to converting a process control function's methods of operation in a company entering into contracts with suppliers based on AQLs to one requiring parts to be acceptable in the parts-per-million range.

21. Define the term "zero defects" in the context of a manufacturing organization: continuous flow and job shop. Is it possible to achieve zero-defect manufacturing? Explain your answer.

22. Discuss the philosophical difference of working with data for SPC purposes versus Design of Experiments.
23. What is the difference in approaches between "one at a time" and "factorial" with respect to experimentation?
24. Why do the process capabilities of job shops increase many times over when continuous-improvement awareness factors are introduced?
25. Discuss what is meant when we say we are appling the six sigma philosophy to design engineering output.

Chapter *11*

MAINTENANCE ENGINEERING

It is essential for a factory to maintain its equipment in order to produce a product within an acceptable cost and time period. If the equipment in the factory is erratic in performance, only an underoptimal result can be obtained. How erratic the equipment performance is, will determine the degree of under optimal operation. The mission of maintenance engineering is to severely depress the occurrences of erratic equipment performance, thereby minimizing a negative productivity factor.

The basic task of maintenance engineering is to develop preventive routines for equipment so that it is capable of running to specific design requirements. This means that maintenance engineering is interested in determining what checks, inspections, and replacement of parts should be performed during scheduled downtimes so that the machines run as designed and unscheduled downtimes are prevented. However, when breakdowns occur the unit must react quickly to return the affected equipment to service. A secondary objective of maintenance engineering is to keep the facilities and office spaces in good repair to foster excellent employee morale. With these objectives in mind, let us examine the techniques used by the maintenance engineering unit.

TASKS ASSIGNED TO MAINTENANCE ENGINEERING

Four basic tasks are assigned to maintenance engineering:

1. Preventive maintenance
2. Rapid emergency repair
3. Toolroom activities
4. Buildings and grounds maintenance

All four activities can require a wide variety of engineering disciplines to achieve successful results. When making repairs to machine tools, the skills and knowledge of electronic engineering may be required to diagnose a Computer Numerical Control (CNC) problem, and those of mechanical engineering may be necessary to determine the risks of bearing instability if a substitute lubricating oil is used.

In emergency repair work it is common to make replacement parts in the toolroom without having the vendor's detailed drawings, so that the maintenance engineer becomes a design engineer, producibility engineer, and methods engineer all at the same time. The maintenance engineer must have more than passing competence in these areas if the repair parts are to work and the equipment run again. We can see that maintenance engineering is complex and challenging. I often think of maintenance engineers as the land-based equivalent of shipboard marine engineers who can keep a ship afloat and operational in what appear to be impossible circumstances. The true maintenance engineer is of the same mold: creative, resourceful, and possessing a positive attitude toward keeping the equipment operational. Management of maintenance engineering then involves creating a system of controls and reporting within which these engineers can operate effectively. Before examining maintenance engineering management techniques, let us review the four basic tasks of the maintenance engineering unit.

Preventive Maintenance

The basic goal of maintenance engineering is to prevent unscheduled shutdown of equipment due to failure. Therefore, preventive maintenance is a primary activity of the unit.

Successful preventive maintenance requires a precise and thorough understanding of the equipment used in making the company's products. This includes understanding the design philosophy of the equipment, including what makes it work and what must be considered the weak links of the system, and understanding how the equipment is to be used in the factory. This dual view of design use versus actual use sets the tone for how preventive maintenance will be applied.

EXAMPLE (a): Suppose that a lathe designed to carry a load of up to 600 lb is to be used frequently at or slightly above this load. What would the preventive maintenance scheme be like? Obviously, operating above recommended load levels is not the best idea, but occasionally a firm considers the risk of overloading preferable to an expenditure of capital funds to purchase different equipment. Management in this case is deciding to cut into the equipment's factor of safety, recognizing that the load rate was set conservatively low by the vendor to protect against warranty failure. Therefore, the maintenance engineer would look for the weak links in the lathe design and schedule them for frequent examinations and possibly replacement of parts before a failure occurs. The design's weakest link would probably be the bearings of the machine, because they are being asked to carry a continuous overload. Therefore, after a certain period of operating time the bearings would be inspected, and after a longer period the next weakest link along with the bearings would be inspected, and so forth, until a time interval was reached where the bearings would be replaced before a failure occurred. The replacement time would be determined by calculating the expected bearing life with the sustained overload.

In the above example, the weakest link was identified and a preventive maintenance program was designed to fit the user's specific circumstances.

EXAMPLE (b): If the lathe was to be used commonly for far lighter pieces, then a preventive maintenance program might have been focused on other things—perhaps the lubrication systems to make sure oil lines are not plugged—and the bearings looked at only rarely. Here the usage dictates examining not load-carrying components but systems components.

These examples show that maintenance engineering cannot simply follow vendor-suggested preventive maintenance techniques and ignore the specific utilization of the equipment. Preventive maintenance techniques suggested by vendors are based on a generalized concept of the use of the equipment and a desire to protect themselves from negligence lawsuits. Therefore, maintenance engineering should use the vendor's techniques as guidelines but not as rigid requirements.

In these examples operational parts are sometimes replaced before they fail. This is done to control downtimes and not let failure dictate when machines are not usable. Many parts that appear to be usable have reached the end of their statistically usable lives and are in imminent danger of failing. Good preventive maintenance is based on knowing the useful life of components and replacing them prior to failure.

The techniques for determining the effective life of component parts, such as roller bearings, are well known and are covered extensively in textbooks on machine design. Maintenance engineers are competent in these techniques. However, the body of knowledge applied by maintenance engineers encompasses virtually the entire spectrum of engineering. What is necessary in a particular company depends on the type of product being manufactured and the type of equipment used. In an electronic assembly operation we would expect to find automated equipment and the equipment containing electronic circuits, especially for testing. Therefore, the basic skills employed by the maintenance engineer would be those of an electronics engineer. Similarly, a factory producing artificial fabrics would be basically a chemical plant, and the maintenance engineer would be skilled in chemical engineering techniques and perhaps the fluid flow techniques of mechanical engineering.

Rapid Emergency Repair

Most people think of emergency repair as the focus of maintenance engineering. However, like the emergency room of a hospital, it does not represent all the techniques and services represented, and is not the preferred method of treatment.

Rapid emergency repair is the ultimate fallback position of the maintenance engineering unit. If programs of preventive maintenance fail, or an accident occurs, or an operator error results in machine failure, the only practical recourse is repair in the most expeditious and economic manner. The only alternative to repair is to scrap the equipment and replace it. If this is the case, the maintenance engineering unit's involvement comes to an end and the project is assigned to advanced manufacturing engineering. Since the latter is a significantly rarer occurrence than making the required repair, let us look at the philosophy of emergency repair and see how it affects the actions of the maintenance engineering unit. Remember that this is emergency repair, not repairs made as part of a planned downtime where the activity has been extensively planned.

The first aspect of emergency repair is that there can be no specific preplanning. The nature of the repairs can vary widely, and the exact requirements are never known until the need for an emergency repair becomes evident. This means that the maintenance engineering unit must be competent to handle a large variety of repairs and be stocked in a general sense with a wide variety of repair materials. This broadness of application is often a problem in determining stocking levels for spare parts. How many sizes of pressure regulators or capacitors can the maintenance engineering unit afford to stock? Think of how often home repairs are thwarted because the jar of screws and nuts does not contain one of the proper length, diameter, or thread pitch. The same is often true in factory repair activities. To minimize the need to stock a variety of sizes, maintenance engineering makes sure that AME knows what components for all equipment should be standardized, down to the level of thread sizes and electrical hardware. One of the critical factors in being able to effectively cope with emergency repair is to have a well-stocked inventory of generalized and standardized spare parts.

The second aspect of successful emergency repair is the ability to diagnose the problem quickly and correctly. This has a large impact on the economics of the factory. A machine that is down costs the company productive capacity, hence lost profitability. Machines usually cost in the neighborhood of $150 per hour to run, so that every extended hour of lost production will cost $150 plus the cost of coping with the problem that one of a series of machines needed to produce the product is down. Therefore, there is a need for the maintenance engineer to respond quickly and make an accurate assessment of the damage and how to fix it.

The third requirement in successful emergency repair is that the equipment be fixed quickly after it is determined what has to be done. This quite often requires a balancing of resources by the manager of maintenance engineering. There are only a finite number of people available to be dispatched to any number of jobs, so it is necessary to set priorities. The manager of maintenance must be kept informed about the situational needs of production in order to correctly assign priorities to the repair jobs. The need to perform preventive maintenance economically must be balanced against the need to accomplish rapid emergency repairs. This may sound like a contradiction; however, not all breakdowns of equipment dictate a need to repair the equipment rapidly.

EXAMPLE: If a company has a half-dozen engine lathes engaged in making similar parts, it may not be necessary to fix one lathe when it breaks down. It may even be possible to schedule the repair at the convenience of the maintenance engineering unit. This would be the case when all six engine lathes are not needed to make the production quota at that time. If only four lathes are necessary, then two would be idle before the breakdown and work could be shifted to one of the idle machines without production output suffering. If the breakdown occurs at a critical machine, then maintenance engineering has no choice but to react rapidly to repair it.

The setting of priorities must have a rational basis. Critical machines must always be serviced first. Other equipment will receive immediate attention only if personnel are available. Preventive maintenance work is abandoned only to fix critical machines; otherwise it is carried out on schedule. Whether a machine is considered to be critical can change with production circumstances, and the manager of maintenance engineering must be aware of present conditions.

Preventive maintenance work can sometimes turn into rapid emergency repair. Suppose that a critical machine tool has been scheduled for a preventive maintenance check and minor overhaul. Since it is a critical piece of equipment, the maintenance engineering unit is given a specific time when it must be available for shop operations. If a serious problem is discovered during the work and the return to service by the due date is jeopardized, then a scheduled downtime must be treated as an emergency repair to meet the commitment. In this situation other critical machine breakdowns will not be allowed to supersede this preventive maintenance work.

Computer programs have been developed to help management decide how to set priorities for repairs and preventive maintenance. In most cases these programs are based on a hierarchical branching network decision-making algorithm. The answers to a series of questions are rated to indicate how critical an incident is to current production needs and therefore how high its priority should be. The problem is usually how to decide whether one product component manufacturing area is more vital than another. No matter what kind of algorithm is designed, this is still basically a subjective decision. All the programs can be expected to do is perform the much needed clerical function associated with job dispatching. To illustrate how transient priorities can be, suppose that the manager of manufacturing gets a call from an important customer who needs replacement bearings immediately. Since the customer is important, the manager will expedite the manufacture of the bearings, and any breakdown in equipment becomes a critical machine failure. Now suppose that the same customer calls later and needs electrical circuit boxes before the bearings. The machines that cannot be allowed to stay down now become the sheet metal forming and assembly equipment. The electrical circuit box machines replace the bearing machines at the top of the critical list. With such a dynamically changing situation, it is difficult to design computer programs that are flexible enough to set priorities for maintenance responses.

We have also seen the introduction of Artificial Intelligence (AI), particularly expert systems, computer systems for use in repair strategy diagnostics. AI programs are discussed in Chapter 13, dedicated to computer integrated manufacturing.

Toolroom Activities

The toolroom is a support function for preventive maintenance and repair activities. It is set up to supply "homemade" parts to the maintenance crews, and is also a primary source of manufacture of tools, jigs, fixtures, gauges, and other miscellaneous items for production needs specified by the methods engineer. In addition, the toolroom is a place where prototypes of new products can be built and evaluated.

Another important feature of toolrooms is that they are staffed by the most skilled machinists and craftsmen on the company's payroll. They must often make precise parts to meet urgent needs without definitive engineering drawings, commonly by using the failed parts as models. To accomplish this type of task, the toolmaker must possess not only a high level of competence in operating a variety of machine tools, but also an excellent knowledge of machine component parts in order to reconstruct the original from the failed reference part. The maintenance engineer aids in this process by explaining what the part does and how it works, but the major reconstruction task still falls to the toolmaker.

Toolrooms are characterized by having many precise but manually controlled machine tools. All are capable of very fine tolerances but usually have a limited output rate. Toolrooms are measured not by output volume but by the quality of the output and the ability to make single parts as required. Therefore, the key measurement associated with toolrooms is their readiness to serve. This is a highly valued attribute and most manufacturing companies are willing to fund activities on this basis in preference to most other measurement concerns.

Material stocks for toolrooms are usually kept separate from other company stocks. One reason for this is that the toolroom material stocks usually represent a cross section of materials that match the basic components of the firm's production equipment. Unless the company is a machine builder, it would be unusual for the tool steels and gear blanks needed for repairs to match the types of materials the company uses for its sales products. Another reason is that it is risky to keep the material in a common location with production stock. Its volume is always minuscule compared to that of production material, and it could be used in error by shop operations or could be lost. Finally, if the toolroom material is segregated it can be found and sent to the proper machine tool rapidly, which is often important.

Making repair parts for machines and equipment is a primary purpose of the toolroom and always receives top priority. However, the bulk of its activities revolves about the methods engineering task of producing jigs, fixtures, and gauges for production. Making production fixtures usually carries the major financial write-off and covers part of the cost for maintaining the readiness to serve capability. The toolroom will prepare quotes for making the jigs and fixtures, carrying the proper overhead rates of a designated portion of the costs for readiness to serve.

The maintenance engineering manager must be conscious of outside costs and must make sure the toolroom does not become a profit center at the expense of the shop operations units. This would contribute to raising the selling cost of the company's products and would ultimately lead to pressures to minimize the extent of toolroom activities. A proper balance must be struck in pricing toolroom work so that the toolroom costs are covered and costs to overall operations do not have a negative impact on competitiveness.

Maintenance engineers use the expertise of toolroom personnel in determining how repair parts can be made and estimating whether the repair part can be considered a permanent or a temporary fix. Often the materials stocked by the toolroom are insufficient for permanent repair but satisfactory for temporary repair. Since the toolroom does not have access to the vendor's design calculations, it is not surprising that a materials match is not made. Usually the toolmaker will test hardness to determine whether the available material is sufficiently strong to withstand the loads that caused the original part to fail. With this information the toolmaker can discuss the risks involved in using the homemade part with the maintenance engineer, who can then determine whether it will work and for how long. In evaluating the risks to be faced in getting the down machine on line again, the maintenance engineer benefits greatly from the master craftsmen expertise of the toolmaker. Therefore, one of the ongoing tasks of the maintenance engineering manager is to foster this relationship.

Buildings and Grounds Maintenance

The fourth and final responsibility of maintenance engineering is the "all other" category called buildings and grounds maintenance. This is a broad responsibility covering activities that range from good housekeeping, such as sweeping up metal chips in the factory aisles, to complex engineering activities such as managing energy conservation efforts. Virtually anything deemed to be technical but not specifically assigned to other manufacturing engineering units is assigned to the maintenance engineering unit via the buildings and grounds category.

No specialized organization is set up to manage buildings and grounds maintenance (i.e., nothing similar or parallel to a toolroom). Typically the work is dispatched in the same way as any preventive maintenance or repair activity. Buildings and grounds maintenance activities are usually assigned in an integrated dispatching system used for all maintenance work. For example, in making work assignments to the plumbers, no distinction would be made between serving the cooling unit of a machine tool or the hot water system in the plant cafeteria.

Minor exceptions to this commonality policy exist, but they always occur within very specialized areas of activities, such as the operation of the cleaning crews. Cleaning of the offices and workplaces is a specialized activity within the maintenance engineering unit because there is no other place to assign the responsibility.

Buildings and grounds maintenance covers so much that it is difficult to adequately describe the breadth and scope of the activity. The ability to respond creatively is of paramount importance. The maintenance engineering unit may consider this activity to be secondary to keeping the production machinery operational. However, the success of this activity usually has a positive effect on morale. Keeping the workplace environment pleasant tends to improve morale, hence indirectly helps to improve productivity.

One buildings and grounds maintenance activity, that of energy management, does have a direct influence on the company's profit and loss statement. Since the oil embargo of the early to middle 1970s and the subsequent rise in the price of energy, management of energy usage has become a vital aspect of business control. The maintenance engineering unit has the responsibility of monitoring energy costs and planning for energy cost savings. This is done through a multifaceted program. First, energy-producing and delivery equipment is monitored for efficient operations.

EXAMPLE: Steam traps should be inspected at frequent intervals to ensure that they are functioning properly. Steam traps that allow steam to pass directly to the condensate lines obviously cost the company money because of the inefficient use of energy.

Second, the maintenance engineering unit is involved in conducting surveys and programs to reduce energy costs.

EXAMPLE (a): A program to replace broken window frames, sashes, and glass with high impact plastic and insulated substructures.

EXAMPLE (b): Measurement of lighting levels throughout the offices and factory area. If lighting levels are found to be significantly above those recommended for particular types of work, energy savings can be realized by removing the excess lighting fixtures.

The third facet of the energy cost reduction activities would be conservation through awareness programs and the establishment of policies aimed at conservation.

EXAMPLE: Appointing area energy monitors with responsibility for turning out lights at the end of the workday.

Most companies are making strong efforts to reduce energy costs, and the significant progress needed will depend on a sound engineering approach. Maintenance engineering is usually assigned that task.

THE THREE TYPES OF MAINTENANCE MANAGEMENT THEORY AND WHEN TO USE EACH MAINTENANCE ENGINEERING*

The main focus of maintenance is to prevent loss of capacity through unplanned for underperforming or down workstations. Therefore, as shown with the strategy employed, we normally emphasize preventive methods as the normal course of conducting the maintenance business. For this reason, generally speaking, the maintenance philosophy will always favor preventing downtime instead of being very proficient in reacting to unplanned outages. However, having capability to react to unplanned outages is still a necessary factor in planning maintenance's capabilities. The basic structure of the maintenance organization is one aimed at maintaining a vigorous PM program while at the same time maintaining a reserve capability to react to unplanned downtime. But the thrust is always an offensive one of prevention rather than a defence against potential breakdowns. However, to present a full picture, we must recognize that PM-type strategies aren't always selected. It is conceivable that a reactive approach is suitable especially in situations where the

*This is an expanded version of this topic contained in the second edition of this text. This is a segment of what first appeared in my book *Fundamentals of Shop Operations Management: Work Station Dynamics,* ASME Press, New York, 2000.

equipment is relatively straight forward and little PM is feasible. For example, where most work is done via nonmoving jigs and fixtures to assist highly skilled manual operation; for example, jewelry manufacturing.

The interactive strategies are as follows:

- Reactive
- Predictive
- Preventive

Lets take a look at these three strategies.

Reactive

Both preventive and predictive strive to eliminate or at least minimize down time with proactive actions on working machines and processes before there is any indication of problems. A *reactive strategy* is one of minimizing the effects of down time by fixing the problem as soon as it's discovered. Reactive is the oldest strategy. Some people call it the "if it ain't broke, don't fix it" strategy, which at first glance would appear to be just good common sense. But not necessarily. When a machines breaks is not predictable. Therefore, the ability to plan for an outage is very limited. In the reactive mode we are in essence nothing more than superb volunteer fire fighters. We may be able to put down the blaze, but not without unplanned for cost. And as we know, unplanned outages will cost more because we need to respond quickly. When we respond quickly it is usually a priority to do so and any ability to control costs becomes secondary to getting the job done. Pure Reactive strategies just are not optimum from the viewpoint of profitability, therefore are seldom a prime maintenance strategy. We use a reactive strategy as a secondary approach to respond to spontaneous outages that weren't precluded by proactive techniques.

A reactive strategy for a company having simple machines, jigs, and fixtures is really a hybrid. It will include simple PM tasks done at the beginning of shifts or internal to the manufacturing process (as long as it doesn't lengthen the time to accomplish the prescribed manufacturing work). These PM tasks will be accomplished almost exclusively by the workstation operators. This is so because most reactive strategies occur in smaller manufacturing activities where the maintenance staff is very small and in many cases also handles janitorial and facilities upkeep roles.

Predictive

The proactive strategies of Preventive and Predictive try to interdict prior to workstation deteriorations to the failure levels. The difference between the two is essentially where in the life cycle of the workstation will the interdiction take place. Predictive is based on calculations of mean times to failure. To say it in simpler terms: we calculate the number of cycles the equipment can go through until its weakest link fails. For example, how many on and offs for a light fixture before the bulb burns out. In other words, how many times is the equipment revved up from a total stop to full throttle before steady state running commences? Using the host of fatigue equations employed in machine design offices, we can with some accuracy predict the number of cycles the machine will undergo before we have a failure. With this information in mind the strategy says replace parts just before they actually fail. Of course this means replacing components that are still performing satisfactory, and this is the main draw back for this strategy. But its also its main strength. We rarely, if ever, have unplanned outages so we never suffer the high cost consequences of the Reactive methodology.

The process is simple in concept.

- Calculate the number of cycles to failure for the major components.
- Set up a procedure to record operation cycles (like a baseball pitcher's pitch count up to a maximum before being taken out of the game).
- Estimate the time the cycle count will be reached and schedule an outage to correspond shortly before the cycle count limit occurs.

- Remove the components that have closed to within the prescribed cycle counts, and replace with new parts.
- Dispose of the removed parts even though they appear usable.

By following this process we tend to sufficiently shorten maintenance down periods but the costs are sufficiently higher than the older Preventive strategy. The other draw back is the ability to actually calculate with a high degree of certainty the number of cycles to failure. This assumes we know what the weak link in the system is and we can actually do the fatigue cycle calculation. Since no competent engineer would stake a reputation on such accuracy, in practice significant safety factors are added in to ensure no unplanned outage occurs. What this does is raise the cost of maintenance by disposing perfectly usable components significantly before they should be.

Preventive

The Preventive strategy is the compromise between Predictive and Reactive and the most widely used. It differs from the Predictive method in that we do prescribed upgrades at specific times only if we discover the first signs of deterioration. Otherwise we leave things alone. The preventive strategy depends on doing investigative periodic audits of the condition of the equipment. We determine how often we ought to look for deterioration and we couple this with prescriptions of known "good things to do" to further stretch out deterioration causing unplanned failures. This is the common strategy of the prescribed downtime for examination and the prescribed preventive tasks in between down times.

Figure 11-1 illustrates a typical plan based on a Preventive Strategy.

Let's look at some of the key factors this strategy relies on as shown in Fig. 11-1. I developed this plan for a wood working factory but it is applicable for virtually any factory using machines and processes for producing its products.

The first thing to note is I divided the factory into its major producing categories and the facility itself. Often companies buy subcomponents and have significantly large assembly operations. Also, quite often the assembly operations consist of multiple sets of automated workstations. If that is the case, these companies would have a different grouping emphasis. Perhaps instead of a machine group they would have an automated assembly group. The important point is that the structure of the preventive maintenance (PM) plan matches the factory.

Under each category we list the activities we will do to investigate the state of deterioration, or a simple preventive activity designed to minimize and stretch out deterioration. Again, I should point out that deterioration for any device that has moving parts is inevitable. Our purpose is to slow down the rate of deterioration as much as possible to ensure the equipment can perform in accordance with its specification as long as possible. Note I divided the tasks into categories sorted by time intervals to perform. In the figure the tasks go from daily to five years. Each company needs to decide what the intervals should be to suit its particular needs. This depends on the rate of natural deterioration you would expect for the machine or process. The only rule of thumb being that the shortest inspection period for the intended investigation be less than the theoretical time it takes the machine or system to reach an unacceptable level of deterioration for proper performance. For example, we would inspect a machine's bearings annually because a set of roller or ball bearings would probably be satisfactory for at least 1 year, possibly 18 months, for normal usage.

Notice these are general requirements for all the machines, processes, and buildings. A full plan would use the model to detail specific requirements tailored to the particular machine, process, and building system.

Continuing horizontally across the form, the third column lists the estimated time it should take to complete the task. Of course, for this generic figure the times are averages. When making up a specific PM plan for the factory's equipment these times can become more accurate for the specific tasks. In fact, often industrial engineers will do a time standard analysis on the PM times to get a more precise time to do the PM. This is then inputted to the MRP II scheduling system to allocate time for that purpose in the overall schedule.

GENERAL REQUIREMENTS item	MACHINES frequency	(As Applicable) est. time to do	done by	est. annual hrs. per machine
Clean (remove debris)	daily	10 min.	operator	39
Check gauges & controls for problems	daily/on-going	-	operator	-
Check oil levels, grease, mis. Lub.	daily	-	operator	-
Change oil, grease, lub. & filters	monthly	1 hour	oiler	12
Inspect brgs., gears, hydraulics, air	annual	16 hours	tech./maint.	16
Inspect motors & controls	annual	8 hours	tech./maint.	8
Inspect PC controller/logic circuits	annual	4 hours	tech./maint.	4
Inspect lead screws/structure/align	annual	16 hours	tech./maint.	16
Tumpane testing for accuracy	5 yr. intervals	24 hours	tech./maint.	5
Major overhaul/clean paint	5 yr. intervals	40 hours	tech./maint.	6
			painter/h-man	2
		Total Annual	S.O.	39
		Time/Machine	Maintenance	67

GENERAL REQUIREMENTS item	PROCESSES frequency	(As Applicable) est. time to do	done by	est. annual hrs. per machine
Clean (remove debris)	daily	10 min.	operator	39
Check gauges & controls for problems	daily/on-going	-	operator	-
Check moving parts for proper ops.	daily	-	operator	-
Lubricate	weekly	10 min.	operator	39
Inspect steam/air/hydraulics/etc.	monthly	1 hour	tech./maint.	12
Inspect motors and controls	annual	8 hours	tech./maint.	8
Major overhaul/clean/paint	annual	16 hours	tech./maint.	10
			painter/h-man	6
		Total Annual	S.O.	78
		Time/Machine	Maintenance	36

GENERAL REQUIREMENTS item	BLDG SYSTEMS frequency	(As Applicable) est. time to do	done by	est. annual hrs. per machine
Inspect steam system; traps, pipes valves, etc.	annual	320 hours	tech./maint.	320
Inspect dust collection system	annual	320 hours	tech./maint.	320
Inspect oil fired boilers	annual	80 hours	tech./maint.	80
Inspect wood fired boiler, silos, feed system	annual	80 hours	tech./maint.	80
Inspect air condition system	annual	40 hours	tech./maint.	40
Inspect and repair stretchers	annual	500 hours	tech./maint.	500
Paint bldg., internal	5 yr. cycle	continuous	painter	1856
		Total Annual	painter	1856
		Time/Machine	Maintenance	1340

SUMMARY	Number	Shop Opers	Tech./Maint.	Painter
Machines	100	3900	6700	
Processes	75	58,500	2700	
Building Systems			1340	1856
TOTAL HOURS			9750	1856
TOTAL MANPOWER/YEAR			5.23	1

Figure 11-1. The Generic PM System (based on Wood Furniture Co. Model)

The fourth column identifies the position assigned to do the task. In the example we see operators, technicians/maintenance specialist, oilers, handymen, and painters. Usually the shorter time and most frequently done items are assigned to the workstation operator. The only caveat being that the tasks are within the expected capability of the position assigned to it. This means that the operator has to be trained to do the PM tasks assigned to that workstation. Also, doing the PM segment has to be part of the official job description and time allocated to do it.

The fifth column sums the annual time it takes to do each PM. We use this information to determine the staffing requirements. This tells us how much time it will take out of the productive time to maintain the equipment and how many people we need to do this. Note, we show the staffing requirements by category of labor. In the example we have shop operations, the maintenance staff, and a painter. For convenience sake I grouped the handyman and oiler into the technician/maintenance category knowing that the latter is the culmination of the two junior positions as the incumbents gain experience. Knowing total hours and dividing by 2,000 hours/year we arrive at a staffing level.

The figure shows that shop operations will lose the equivalent of 5.23 people a year dedicated to PM. This is very necessary work but it does reduce capacity and has to be accounted for in the MRP II scheduling system. It's accounted for by scheduling PM as a job going through the workstation. The workstation load sheet will show all the jobs assigned to the particular workstation for the particular shift, including the planned for PM. By this method we know that maintenance is being done and accounted for. To further emphasize the need to do PM, shop supervision is measured on their ability to do all jobs assigned efficiently, not just value added jobs.

Similarly, the calculations for personnel needs are done for the technician/maintenance specialists. Only here, this become overhead expense and is not a deduction from time to do value added work. On the contrary, it is additional work that has to be done so value added work is capable of being done.

The PM program as outlined in Fig. 11-1 and its specific development for the particular factory are only the summation of the planning process for PM. We plan PM just as we do production jobs. We determine the method to be employed. We then estimate the time to do it, using whatever technique is appropriate, making choices as we do for production work. Knowing the sum total of all PM's necessary and the time to do it we can create a time phased plan (this is the Two Knows specialized for maintenance work) to achieve the desired outcome. When integrated with the company's planning and scheduling system, vis a vis the MRP II system, we create a synergistic system that occurs virtually automatically. And this is what we're striving to do. In the case of maintenance, it is working to have the longest up time at the lowest possible costs. We can see that the contribution required of maintenance is to have as much capacity available to make the company's products as it possible can at the lowest possible cost.

Measurements of Maintenance Effectiveness

Before we look at how we measure the effectiveness of a maintenance organization, let's recap the three approaches to extending the deterioration cycle. It is important to understand this to know what is vital to measure.

Reactive: Fix quickly after it breaks
Predictive: Remove and replace before it breaks
Preventive: Fix before it breaks

The major measurements we use to evaluate the effectiveness of the maintenance organization are:

1. Response time
2. Mean time between failures
3. % downtime
4. Budgetary control
5. Number of outages per time period

<u>Maintenance Measurement Criteria</u>

	A Response Time	B Mean Time Between Failures	C % Down Time	D Budgetary Control	E Number of Outages per Time Period
Reactive	1	2	3	2	3
Predictive	3	1	1	1	1
Preventive	2	1	2	2	1

1.= highest priority
2.= medium priority
3.= lowest priority

Figure 11-2.   Maintenance Strategy Measurements Criteria

Figure 11-2 shows how each of the three major management strategies approaches these measurements.

The figure shows the relative emphasis each strategy places on each of the measurements categories. The type of measurements an organization uses to evaluate its effectiveness needs to be able to judge objectively whether the selected strategy is succeeding or failing. This is true in general and particularly so for maintenance. In addition we have the daily operating status report, also known as the "down and limping report" (which will be discussed in detail later in this chapter) which is universal for all maintenance strategies. While we may think that all measurements may be equally important, they are not necessarily so and they vary in importance depending on the strategy employed. Let's evaluate this further by comparing the measurements with the three strategies.

Reactive Maintenance Measurements Criteria

For reactive maintenance, as expected, the most important measurement criteria would be response time. Since this strategy relies on "if it ain't broke don't fix it" the most important measurement is how fast we can fix a down or limping machine or process to get it back to normal performance levels.

The next level of importance would be the mean time between failures because this is an indication of how well the maintenance organization performed when they had to fix something. It indicates how well and thorough the repair was accomplished. Equally important is the cost associated with the repair. This is the only way that reactive strategies can be compared to the other two. If we can show that reactive strategies cost significantly less than predictive or preventive, then it is a viable alternative for management to consider.

There are two measurements that are least important; measurement of % down time, and the number of outages per period. We say % down time is a least important measurement because a reactive strategy doesn't try to prevent downtime, only minimize the extent through fast reaction. Of course if too many short down times occur; for example, too many work stations down at a time that they significantly effect capacity, then the reactive approach would be judged a failed strategy for that particular company. This leads to the fifth measurement, also rated low priority, the number of outages per period. Since this strategy reacts to outages, not trying to prevent them, this measurement is considered sufficiently nonrelated to gain a low priority level. In fact proponents of reactive maintenance say the only value of a number of outages per period measurement has is to tell us whether the machine or process is reaching its useful life. They say the data can then be used to justify capital expenditures for replacements, but have no meaning in evaluating the effectiveness of maintenance.

Predictive Maintenance Measurements Criteria

Predictive is the most scientific maintenance strategy in that it uses series of fatigue and probability equations to predict component parts useful life. Proponents base its success on the ability to significantly shorten downtime by replacing parts before they exhibit any indication of impending failure. The key to the strategy is using massive data on like parts to understand the number of cycles to failure. If we know this number we can productively change out the part before it fails. We simply have to count cycles and take out the part that will statistically fail at a cycle count of x before it reaches x. The trick of course is to have reliable data to do this.

Since predictive maintenance is such an enticing concept, a lot of research is being conducted to learn how to predict cycle times to failure with more accuracy. But even today, there is enough knowledge to accurately predict most bearing failures and virtually all lighting fixtures. One promising techniques is signature analysis, whereby machine shafts will exhibit certain acoustic signatures before journal bearing failures. Another signature analysis is temperature related. Here thermal sensor probes show characteristic rises in temperatures as bearings enter the end phase of their useful lives. There are also strain gage systems that allow us to predict materials getting close to their yield points, hence ready to crack and fail. These are just a few of the newer engineering techniques that make predictive maintenance possible. One of the drawbacks however is being able to predict with accuracy the time delay between a precursor being observed and the actual failure. Obviously there needs to be sufficient time to plan for an outage when the precursor is detected.

Predictive maintenance is still a risky business whereby it is possible to really be throwing away perfectly usable parts based on a theory of failure instead of observed fact. Since a large part of the theory is based on statistical inferences, the simplest being averages of past experiences, it is reasonable to expect the maintenance measurements criteria to favor investigation related to these calculations, and they are. Predictive maintenance, if done correctly, would maximize mean time between failure (these should be rarities) and minimize downtime (these should be planned in advance to change out parts on a schedule for replacement). So mean time between failures is a primary measurement. And so to would be the number of outages per time period.

In addition, budgetary control is equally important. The company is being asked to spend more money for new parts rather than less money to fix worn but repairable parts. These parts will have some additional life in them after repairs, but not as much life as new parts. Therefore the value gained for having virtually new machines and processes most of the time needs to achieve higher usable capacity than a less expensive strategy. Coupled with the budgetary control is the number of outages per period. This is an important measurement that needs to be low to justify the disposing of still functional parts.

The least important, and by a wide margin, is response time. With predictive theory there is no need to have a well-trained team to respond to outages. Since they will be so rare, according to proponents, that rapid response is a moot point.

Preventive Maintenance Measurements Criteria

Preventive theory is perhaps the middle ground. Although, in truth, it leans more toward the scientific of predictive theory rather than the rough and tumble of fix it if it breaks and do it fast philosophy.

Preventive has two tiers of measurements importance. The two in the higher tier of measurements emphasis are the mean time between failure and number of outages per period. The reasons are much of the same as for predictive. By shutting down machines for planned inspections and minor adjustments of parts (we do consider bearing oil, filters, and greases as parts) we should be catching all causes of failures before they happen. The measurements internal to the inspections themselves should assure the grander measurement scheme that mean time between failures is at least being kept in check and hopefully growing. Similarly, if shutdown adjustments are being done correctly, the number of outages per period should remain low. So the outage count is a strong direct measure of the efficacy of the theory.

The secondary measurements are useful but do not really pass judgment by themselves. Budgetary is a secondary measurement for the preventive strategy. It is not as significant as in predictive because the need for entirely new parts is less. We are fixing existing when required, and replacing at a less frequent rate, hence costs are lower.

Percent downtime is also a secondary measurement. It can be more than predictive but needs to be less than reactive. The cost of the strategy is less than predictive (from a planned purpose viewpoint) so the balance sheet will tolerate less productive capacity over the time period than for predictive. But the downtime has to be less than experienced with reactive because we are taking the work stations down for planned evaluations and preventive work based on those evaluations. By doing so, we are betting that there will be fewer unplanned down times.

The last secondary measurement is response time. The only strategy that emphasizes this criteria is reactive. But preventive maintenance, being sort of a middle of the road strategy has to perform reasonably well in this category. Predictive bases its existence on the fact that it can prevent unplanned downtime due to failures because it knows when to put in new replacement parts. Preventive maintenance doesn't make that claim. The strategy lies on the fact that periodic inspection can determine wear rate and even though a part shows wear it can perform well enough to do the job. We figuratively keep suspect parts under careful watch and perhaps at the next inspection period we will fix or replace it. So there is definitely a probability that we're waiting too long and there can be an unplanned outage. If that occurs the maintenance team needs to respond. So response time is a measurement a maintenance organization operating under this strategy has to be cognizant of.

How to Chose a Maintenance Strategy

The choice depends on the type of factory the company has. And this resolves down to the complexity of the equipment. By complexity I mean the ability to understand the nature of potential failures. If we understand the nature of potential failures and can predict them then one strategy would be employed, probably predictive. If we understand the nature of failure but cannot predict with any reasonable certainty when failures occur then a preventive strategy is called for. And finally, if the equipment is relatively simple thus potential repair choices minimized, then a reactive strategy would probably be sufficient. Figure 11-3 is useful in helping to select the best strategy for a particular company's set of circumstances. Let's do a brief tutorial based on the figure.

Figure 11-3 is based on the premise that various manufacturing scenarios can be predicted and thus matched with a preferred strategy, as mentioned above. The figure lists characteristics of factories, which the user employs to guide her or his selection of a probable best match scenario for a maintenance strategy. It is a matter of checking each statement that pertains to the particular factory, while skipping those that do not. Then the user adds up the various scores shown for the selected statements for each of the three strategies. One of the strategies will have the highest point score. This is the most probable match.

The word *probable* is a key factor here. No one can possibly conceive of all the various permutations and combinations a factory can have. So at best, the user of Fig. 11-3 and other similar mnemonic devises uses it as another data input factor. For example, consider availability of expert resources, for example, sufficiently trained people who are capable of doing the diagnostic job. This is equally as important as the strategy chosen. There is no sense employing a reactive strategy, even if the numbers add up that way, if there is no ability to react quickly and with assurance that the breakdown can be repaired properly. it would be much better to have a preventive or predictive strategy where still functioning parts are abandoned on a time schedule in order to take advantage of a probability theory strategy. Using these alternate strategies, there is a lesser chance that a breakdown will occur that will require an expert to fix.

Maintenance Strategy Selection Matrix

Criteria		Strategy Choices		
		Reactive	Preventive	Predictive
1	flow production	1	3	2
2	job shop	3	2	1
3	0–10 work stations	3	1	2
4	10–50 work stations	2	3	1
5	greater than 50 work stations	1	3	2
6	similar machines @ work stations	3	1	2
7	dissimilar machines @ work stations	2	3	1
8	mix of same/not same machines @ work stations	3	2	1
9	less than 25% computer control	3	2	1
10	more than 50% computer control	1	2	3
11	product mfg. cycle less than one week	3	2	1
12	premium for up-time critical	3	1	2
13	seasonal manufacturing	1	3	2
14	precision tolerances required	1	3	2
15	more than 2 axis movement	1	3	2
16	controlled environment	3	2	1
17	special training required	3	2	2
18	special diagnostic tools required	3	1	2

Instructions:
    [1] select criteria applicable to specific factory.
    [2] add point values of selected criteria for all three strategies.
    [3] highest point value indicates probable strategy choice.

Figure 11-3. Maintenance Strategy Selection Matrix

Remember, choosing a maintenance strategy needs to be compatible with all the pertinent characteristics of the factory. We can go through a maintenance selection matrix and have it indicate that a preventive strategy is the proper choice. If all company cultures were the same this would be the end of story and we would do just that. Put in place a preventive strategy. But what if we had a company culture that tried to react to every sales opportunity regardless of whether it was feasible to satisfy the customer in a manner the customer needed? What I'm implying is a sales driven company that constantly changes schedule to meet the latest customer crisis regardless of expenses and disruptions. This is a company that has no internal disciplines, which wreaks havoc on all schedules. Every time an opportunity presents itself, this type of company would prefer to abandon all plans in an effort to secure the sale, even if it meant scuttling longstanding and prepaid preventive maintenance shut down plans. Can this company support a preventive maintenance strategy? Most likely not. The choice becomes a reactive strategy and hope for the best, even though logically a preventive strategy would be a cheaper way to go. So, compatibility with company culture is a must in selecting a strategy and often this is the most important factor in the selection process.

Now that we have reviewed the four basic tasks and styles of management choices for the maintenance engineering unit, let us see how the unit is organized.

ORGANIZATION FOR EFFECTIVE MANAGEMENT OF THE MAINTENANCE ENGINEERING UNIT

This unit consists of engineers, specialists, and all the skilled tradesmen necessary to maintain the company's physical equipment. Physical equipment means virtually everything the company owns, from buildings to typewriters to machine tools. In order to organize this vast territory, the unit is usually broken down into the categories shown in Fig. 11-4, with the leader of each reporting directly to the manager of maintenance engineering.

Preventive Maintenance and Repair

The maintenance foremen usually head this subunit, where the plural is used because the responsibilities are usually divided by shift and perhaps by plant geography if the plant is large. The foremen direct the activities of skilled tradesmen such as machine repairmen, electricians, plumbers and pipe fitters, structural steel workers, and carpenters. In addition, a number of laborers are assigned to do the nonskilled portion of the work. No engineers are assigned to this subunit.

Toolroom

The toolroom consists of the skilled toolmakers, senior machinists, and other equipment-related tradesmen. In addition, most toolrooms have draftsmen and perhaps a CNC programmer assigned to the staff. Since this activity tends to be large, with a staff of 40 or more, a supervisor reporting to the manager of maintenance engineering usually administers the toolroom. Under the supervisor there would be one or more foremen and a purchasing clerk. The toolroom is set up in virtually the same way as shop operations units. No engineers are assigned to this subunit.

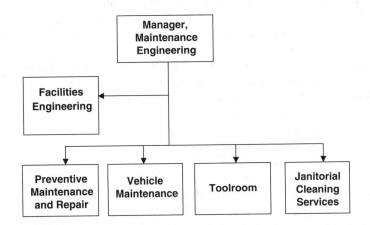

Figure 11.4. Categories of Activities in the Maintenance Engineering Unit

Janitorial and Cleaning Services

This is the cleanup crew for the company and usually consists of low-skilled, entry-level personnel who all have one job description or perhaps two (e.g., cleaners, and floor polishers). The staff varies in size depending on the extent of the company's facility and is managed by a foreman who reports directly to the manager of maintenance engineering. If the activity is large enough, the staff will also include a purchasing clerk. The activity operates semiautonomously, and many of its people operate on the night shift. No engineers are assigned to this subunit.

Vehicle Maintenance

This activity exists only if the company has an extensive fleet of vehicles, for both outside use and internal material handling. If the vehicle population is small, vehicle maintenance is either handled by the preventive maintenance and repair subunit or contracted to an outside firm. If the maintenance engineering unit does have a vehicle maintenance subunit, it is headed by a foreman who reports directly to the manager of maintenance engineering. Under the foreman are a purchasing clerk and a staff of auto mechanics who are experienced in lift truck and other specialty vehicles as well as ordinary trucks and automobiles. No engineers are assigned to this subunit.

Facilities Engineering

All the engineers assigned to the maintenance engineering unit are in this subunit. The organization may be of two types. For a large organization, a supervisor of facilities engineering is appointed. For a smaller organization, each facilities engineer would report directly to the manager of maintenance engineering. The choice of organization depends primarily on the span of control that the maintenance engineering manager feels able to handle effectively.

The facilities engineering subunit supports all the maintenance engineering subunits. In addition, it must perform all the engineering analysis required of the maintenance engineering unit. Therefore, this subunit serves in both an advisory and a direct-action capacity, hence its special place in the organization chart (Fig. 11-4).

METHODS OF MANAGING

Since the maintenance engineering unit is a service organization, the style of management must be such that it can respond quickly and be able to change priorities just as quickly. Unlike the other manufacturing engineering units, a significant part of the detailed work content of the unit is essentially unknown. The preventive maintenance and janitorial and cleaning activities can all be scheduled. However, the repair work cannot be scheduled except for short periods, that is, a week to perhaps a month. For this reason, although many of the assigned responsibilities of the maintenance engineering unit are managed in the normal manner, the quick-response portion requires unique management techniques to be performance and cost effective.

Three unique management techniques are used by maintenance engineering.

1. Order entry system
2. Shop maintenance information system
3. Preventive maintenance system

Each of these systems can be either manual or computer assisted. Each affects the others but can be bounded and studied separately. Keep in mind these are performance measurements used regardless of the type of maintenance management strategy embraced by the company. Therefore, those measurements discussed earlier in the chapter are complimentary to those I will now explain.

Order Entry System

This system prevents chaos in responding to requests to repair product-producing and other equipment. It consists of a method of recording jobs, determining whether the jobs can be worked on, determining the work load for a particular time period, recording results, and projecting the future work load. Figures 11-5 and 11-6 represent the basic documents of the order entry system.

The job log and job effectiveness forms make it possible to plan the work load and measure the effectiveness of the work force. With this system it is possible to determine a manpower level for the short to medium term and schedule effectively for at least 1 month. It is also possible to do partial work scheduling for approximately 3 months.

Let us review the job log form (Fig. 11-5) and discuss the data to be found there. The form provides for recording all unplanned repair jobs given to the maintenance engineering unit by date of entry, job description, and job number. When a job is entered on the form, the responsible facilities engineer must estimate the number of hours it will take to fix the failed item and select the craft or crafts necessary to do the work. Note that the hours are placed under the respective craft columns. If a repair job requires more than one craft to accomplish, the job will be entered as often as there are crafts required to accomplish the repair, with the estimated hours shown for each phase of the work. It is also necessary to estimate whether or not material is available for making the repair. If not, either the material is ordered or a work order is placed in the toolroom to make it. The job log also shows the available for work and unavailable for work backlogs in hours for each individual job and, for statistical information, the number of weeks that the job has been carried on the report. The report is issued weekly, and all jobs that have not been completed as indicated by the last weekly report are carried forward and entered on the updated version. This ensures that all work not accomplished is carried on the unit's open work status.

The part of the form showing hours available to work and work force hours available establishes the next week's work plan. Here we see that the number of hours available to work represents an accumulation of all jobs that could be worked on that week. This means that all materials and instructions are available and, if manpower is applied, the job can continue on to a successful conclusion. This is an important concept. Maintenance engineering can only work on jobs that are available to work on, and all scheduling is based on being able to apply an effort to successfully complete the job. If this is not possible, for example, if there is no repair material, then the job must be ignored for scheduling purposes and not put back on the schedule until it can be worked on.

The number of hours of work force available is simply the allocation of people per craft converted to hours per week. The bottom line of the form—backlog—is the difference between the work available and the work force available. If the number is negative (indicated by parentheses in Fig. 11-5), there is an excess of work force available. This excess is shown in the comments section as hours available to do preventive maintenance work. If the backlog is a positive number, there is more work to do than work force available to accomplish it. This is an available for work backlog and will be carried over to the next week, usually in the form of a complete job rather than a partial job.

> **EXAMPLE:** The backlog of 2 hours for the carpenter craft, if it is a job estimated to require 2 hours, would be carried as a 2-hour job in the job available column for the next week. If the job is estimated to take 5 hours, it would appear as a 5-hour job for the next week and 3 hours would be subtracted from the work available line of the log.

The job log readily informs management of its capability to do the work that is available to be done. A look at the backlog line shows whether it is possible to accomplish the needed work and whether there is an excess of manpower available. To determine whether excess manpower actually exists, it is necessary to consider the unavailable backlog hours, because they will eventually become available hours. Another factor to be considered is the hours that are delegated to perform preventive maintenance, which are also shown on the form. These are not the total hours available for preventive maintenance, since usually a maintenance engineering unit will reserve certain hours

JOB LOG

week 0649			Estimated work hours by craft						Material Available (yes/no)	Hours Backlog	Hours available for work	Weeks on report	Job complete
Date entered	Job no.	job description	Machine repair	Carpenter	Electrician	Plumber	Steel Work	Other					
10/23	83	Fix roof leak		110					yes		110	8	
10/31	94	Repair motor, lathe 2201			6				yes		6	5	
11/1	95	Repair drill press 9024	16						no	16		4	
11/25	116	Repair leaking sink. Ladies room				2			yes		2	1	
11/25	117	Fix stair well crack					20		yes		20	1	
Total	26 total jobs		273	178	210	85	176	16					
Work force hours available			240	160	200	80	160	80					
Work available			165	162	142	38	113	0					
Backlog	17		−75	2	−58	−42	−47	−80					

Comments

Hours available for preventive maintenance

Machine repair = 75
Electrician = 58
Plumber = 42
Steel work = 47

Figure 11-5. Part of a Job Log Form

	Machine repair	Carpenter	Electrician	Plumber	Steel work	Other
Required hours, staff available for repair work	165	160	142	38	113	0
Hours worked on repair work	194	156	191	68	143	0
Efficiency	85.0%	102.5%	74.3%	55.8%	79.0%	-
Jobs available	6	3	5	2	1	-
Jobs complete	5	3	5	1	1	-

Comments: (1) M.R. job no.85 not started, wrong material received. (2) Plumber job 116, required replacement, sink cracked, pipe thread worn and needed re-threading, original estimate was for washer replacement only.

Figure 11-6. Job Effectiveness Form

for preventive maintenance work, and the excess hours shown on the job log are supplemental to this reserved time.

The job effectiveness form (Fig. 11-6) is used to evaluate the efficiency of the repair operation. The first line of the form is a data input from the job log showing the work available to be performed during the planned week. The second line, hours worked on repair work, indicates the hours during which the scheduled work was actually performed. The efficiency of the operation, shown on line three, is simply the planned hours divided by the actual hours. Here the planned hours are the job estimates made by the facilities engineers and are not as exact as planned hour determinations developed for direct labor activities. It must be recognized that planning how to make a product or part is different from planning how to fix a failed piece of equipment. In the former case all the facts are known, whereas in estimating a repair job the facilities engineer must first diagnose what is wrong and what caused the failure. This diagnosis can be a simple procedure, as implied in the second comment in Fig. 11-6, or it can be very complex, such as evaluating the cause of a malfunction on a CNC unit. Even simple diagnoses can be in error, again as illustrated in the second comment in Fig. 11-6. From these brief illustrations we can see that estimating for repairs is an inexact procedure and cannot be fairly compared with the detailed time standards used for production planning.

Nevertheless, the estimate is needed and can be useful. Machine tools and equipment tend to fail at the stress points of their design. Therefore, failures tend to be repeatable over a period of time, and a similar failure would result in a similar estimate. In other words, series of symptoms tend to indicate particular causes of failure, which leads to somewhat standardized repair procedures. For this reason, the use of planned hours versus actual hours does give a good measurement of maintenance labor efficiency. The reasons for measurements of the work force are the same as those discussed in Chapter 7 on methods and work measurement.

The last two lines on the job effectiveness form indicate the ability of the maintenance engineering unit to complete jobs according to schedule. The hours of the first three lines make up the jobs in total shown on the jobs available and jobs complete lines. It is necessary to know this to properly evaluate the efficiency measurement. In the machine repair column of Fig. 11-6 we see 194 hours worked versus 165 hours available for work (planned), which gives an efficiency of 85%. This is the efficiency value to be compared with other weekly performances. However, note that only five jobs were completed, while the hours planned were based on six jobs. Hence, the efficiency calculation is in error; the 194 hours were performed against somewhat less than the indicated 165 planned hours. Thus the last two lines of the job effectiveness form show how valid the efficiency measurements are.

Shop Maintenance Information System

This is an archival record system used by maintenance engineering in order to have data available for decision making. The data are needed in determining whether a machine has reached the point where a complete rebuild or a replacement is necessary, or in determining work assignments based on machine availabilities. Decisions are also required to effectively determine and control operating costs. The shop maintenance information system has three parts: (1) the machine tool capability and replacement log, (2) the daily operating status, and (3) the current period and cumulative cost status.

The titles of these parts may vary, but the content of the SMIS is relatively consistent. This system is usually handled through a computer-based data input program, where the outputs are always the same general triad. Let us consider the philosophy of the system in order to understand the management techniques involved.

Machine tool capability and replacement log.   The MTCRL is the basic document that shows equipment performance data. It includes such information as the accuracy of current machining compared to original specifications and runoff data, the capability of the machine to operate across the entire range of feeds and speeds, and the number of times the equipment has failed and the reasons for failure. The MTCRL can be called up to give the facilities engineer the history of the specific equipment to aid in diagnostic work.

Assistance in diagnosis is an important reason for having an MTCRL, but an even more vital function is supplying data to the AME and MP&WM units. The AME unit uses the data to determine when equipment is approaching the end of its useful life and must be replaced. This information is essential for planning capital investment spending and is used particularly in the development of Long-Range Forecasts (LRFs). Without this information LRFs would be based on the faulty assumption that all existing equipment is in first-class shape, which would be tantamount to assuming that large expenditures for repairs never occur.

Similarly, planning for product production is based on the ability of equipment to perform in accordance with its design. If planning is based on an original or specification sheet value of machine tool performance, it is probable that the hours calculated would be too optimistic. For this reason it is important for methods engineers to have access to current data pertaining to equipment performance. This MTCRL information allows the MP&WM unit to generate planning that is realistic, hence believable by the operators and foremen.

Daily operating status.   The daily operating status is a report showing what pieces of equipment are not available for use ("down") and what pieces are available for limited use ("limping"). If a lathe can only turn up to 50 RPM compared to its normal capability to run at 100 RPM, it is considered to be limping. Similarly, if the lathe cannot rotate the workpiece at all it is considered to be down.

The "down and limping" report is very important to shop operations, particularly the production control unit, which must constantly decide what product components should be loaded to what workstation. The daily operating status is an important data input in determining how such loading should be done. Although production control would like all equipment to be operational, it is important to identify the machines that are inoperable so that alternative manufacturing arrangements can be made.

The production control manager also has information to decide whether a limping machine can be used with any degree of effectiveness. By knowing what is nonfunctional and relating that to the operations the machine must perform on the part, the manager can decide whether to assign work to the machine. Another useful piece of information is the percentage of nonworking features, as shown in Fig. 11-7. If a high percentage of the capability is unavailable, as shown for workstation 12C20, it would be unlikely for the production controller to assign work to that workstation.

The daily operating status report shows at a glance the status of the repair business. This enables the maintenance engineering manager to decide whether to divert personnel from preventive maintenance work or other nonproduction-oriented work to repair work. The three lines at the bottom of the report summarize the magnitude of the current repair work load. Most maintenance engineering managers are keenly aware of the trends of numbers of machines down and limping as well as the number of days a piece of equipment shows up on the report. In many ways this report reflects the performance of the maintenance engineering unit. There are two particular features to evaluate when looking at this report over a period of time.

First is the length of the list. Increasing numbers may mean that preventive maintenance is not being done properly or that there are inexperienced operators at the workstation. Whatever

Daily Operating Status Report					Date 11/29/06
Workstation No.	Description	Down	Limping and non-working feature	Promise repair date	Days on List
2J32	CNC Lathe		30% auto feed	11/29/06	2
17A12	VBM	X		12/2/06	6
12C20	CNC HBM		70% tool changer	12/1/06	5

Number limping: 6
Number down: 4
% Workstation unavailable: 14%

Figure 11-7. Typical Daily Operating Status Report

the cause, the maintenance engineering manager would wish to investigate the changes that have caused a change in the length of the list.

The second feature is how many days an item remains on the list. A trend toward longer stays on the list could indicate problems with the performance of the repair personnel or problems with the repair parts supply system. It might also indicate that the toolroom is not keeping up with maintenance demands, perhaps because it is favoring prototype production. Distinct changes in the average days on the list warrant investigations for cause.

Current period and cumulative cost status.   This third major output of the shop maintenance information system is a summing of the hours required to repair down or limping equipment converted to dollars, plus the cost of materials and other services required to make the equipment operational again. It is important for any business that must work against a predetermined budget.

The information generated here is useful to AME because it helps to quantify decisions involving appropriations of funds for repairs, rebuilds, or replacements. Without such information, and short of a machine failure, it is difficult to convince management to replace worn-out equipment. The manager of manufacturing engineering usually delegates the collection of maintenance cost data to the manager of maintenance engineering, who is charged with keeping accurate records of costs to repair and notifying AME when the cost to repair any piece of equipment becomes excessive.

The cost-collecting activity is also done to compare actual figures with estimated figures related to repair work. The purpose is to identify variances between actual and forecast dollar amounts so that effective corrective actions can be taken.

The Pragmatic Preventive Maintenance System

So far we have learned that excess hours available after repair work is assigned are allocated to Preventive Maintenance (PM) and that the daily status reporting may indicate whether PM has been effective. We will now discuss PM in more detail and show how the maintenance engineering unit approaches this responsibility. Regardless of the type of overall management strategy selected, for example, reactive, predictive, or preventive, PM is a significant aspect of maintenance engineering and needs to be managed effectively.

In PM we are interested in preventing complete or partial machine failure by inspecting the machine, replacing old parts with new parts, and adding disposables such as oil and grease as required. Therefore, a manufacturing engineering definition of PM would be:

"Preventive Maintenance is the replacement of parts or components prior to their actual failure in order to keep the equipment operating as close as practical to an as-new condition."

The primary difference between PM work and ordinary maintenance work is that the latter involves replacing parts that are worn out, while the former consists of replacing parts that soon will become worn out (old parts) or that have reached the end of their useful existence.

A good example of PM, as I mentioned before, is the electric lighting program used for many large office buildings. The calculated life of fixtures is used to establish a schedule for removing old bulbs and replacing them with new ones. Bulbs are replaced even though they are still functioning properly because from the calculation of hours of life and probabilities and statistics, it is known that failure is imminent. Replacement is based on the hours of utilization, not on failure. The result is that darkened rooms are a rarity.

In practice, manufacturing PM systems are not as pure as the light bulb system. They do specify changing oil and filters based on hours of usage, but they do not require taking out bearings that have reached a theoretical effective lifetime limit. The cost of doing so would not be justifiable because machine component lifetime depends on many factors that are at present difficult to measure. For bearings some of these factors would be load parameter histories including temperature differentials (i.e., hot or cold starts), the quality of the lubricating oil, and the exact line-up of the bearing components relative to the shaft. With so many variables the life expectancy is difficult to quantify, and a slightly different approach is taken in practical PM.

A practical PM program would require changing the cheaper or simple things based on the life expectancy approach. For instance, filters and oil would be changed on the basis of hours of operation. Next the maintenance operator would be required to check machine components such as bearings and gears on the basis of hours of operation. Here a probabilistic approach is used to determine when there is a likelihood that a component will fail. If the inspection shows indications of minute failure or wear, the maintenance operator will replace the part. Otherwise the part will be returned to service. Machines are inspected on a set schedule of operating hours based on probabilities of failure of the least survivable parts or components.

Depending on the complexity of the equipment, several stages of PM may be carried out. For example diesel engines usually have PM checks at 500, 1,000, 1,500, and 2,000 hours. The 500 hour check is the most basic one. The 1,000 hour check would include the items of the 500 hour check plus additional ones, and so forth, until the level is reached where a complete overhaul is required. After the overhaul the cycle begins again.

The managerial decision regarding PMs is to determine what they should be and how to schedule their accomplishment. To determine what the PM items should be, the facilities engineer obtains recommendations from the equipment vendor. These recommendations are evaluated for practicality and risks involved in not accomplishing them, and based on the findings the PM items are developed for the particular equipment.

The question of scheduling is always a sensitive one. Shop operations never wants to shut down a production line for PM, while the manufacturing engineer wants to maintain the schedule to prevent damage to the equipment and to allow for PM manpower planning. The maintenance engineering organization must remain flexible in accomplishing the PM work, and it is often done during the night or early morning shifts and on weekends. This poses a practical scheduling problem, which is partly resolved by a pre-notification procedure employed by maintenance engineering. That is, the maintenance engineering unit informs shop operations when a machine tool is due for PM and asks shop operations to report back when the machine will be made available. These notifications, which are often computer generated, become flags for work tickets to be prepared and issued through a dispatch system to the PM and repair subunit. Based on the replies from shop operations on equipment availability, the PM work load is scheduled. This system works fairly well if the factory has a good load-scheduling capability and is operating on or close to schedule, and if there are no emergencies such as machine breakdowns to disrupt the schedule.

Traditionally, PM work fits in where it can, even though efforts are made to schedule it. Therefore, most maintenance engineering units keep a skeleton staff permanently assigned to PM work and supplement the staff with other personnel as they become available. This manpower allocation system works if management has a good log system showing when equipment requires PM and if the PM work becomes a scheduled item similar to emergency and routine repair work indicated on the job log.

SUMMARY

The maintenance engineering unit is a very diverse unit and in many ways is a business in itself. In this chapter we have tried to develop an understanding of the management problems involved in maintenance engineering including the type of management strategy to employ, a subject not usually covered in maintenance engineering texts.

The primary management problem for the maintenance engineering unit is how to control an unknown and constantly varying work load. The job log and the job effectiveness report are used by management to manage this uncertain load. These methods, coupled with the financial control procedures discussed in Chapter 2, give management the ability to adequately handle the maintenance responsibility.

We have also covered the reporting system used to monitor the status of equipment and create a history log. Status reports and documentation are vital in maintaining a well-run factory. The status of equipment reports are short-range data useful for daily shop loading and measuring the effectiveness of maintenance engineering, while the history documents compel management to recognize that no equipment lasts forever and that eventually the company will need to spend money for replacements.

Finally, we have discussed the philosophy of preventive maintenance and seen that it must be tailored for the individual workstations, with the goal of preventing unscheduled outages due to unanticipated malfunctions. The theory of PM is based on calculations of the number of operating cycles to failure, and parts are replaced before the total number of cycles is reached. In practice, modifications must be made to take into account costs of replacement parts and the difficulty in making accurate life-cycle calculations.

REVIEW QUESTIONS

1. What is the basic mission of maintenance engineering? Show how this mission is integrated with the rest of manufacturing engineering.
2. What are the three maintenance management theories and how do they differ?
3. What is the difference in concepts between preventive maintenance and rapid emergency repair? How is this difference reconciled under one management team?
4. Which maintenance management theory fits best for a manufacturer using flow manufacturing exclusively making only one repetitive product, and employing many identical or similar assembly lines?
5. Use the maintenance strategy selection matrix (Fig. 11-3) to recommend a maintenance strategy for manufacturer that has modest lot sizes production of approximately 10,000 units. The products are made on a flow line comprising six dissimilar workstations. Four of the workstations employ state of the art CNC machines requiring moderately skilled operators capable of writing simple CNC programs. The cycle time is 4 days and up time is critical since there are no backup machines. Conformance tolerances are tight but within the range of the equipment.
6. Describe the basic steps in creating a preventive maintenance plan for a machine tool such as a CNC vertical boring mill used to produce tight-tolerance inside diameters.
7. Describe the constraints placed upon the maintenance unit when it is engaged in emergency repairs and how they affect its methods of operation. In particular, include the lack of preplanning time and adequacy of spare parts availability and selection.
8. Discuss how senior management emphasis can influence the selection of emergency repair priorities.
9. In order of priority, outline the tasks assigned to the toolroom. Give reasons for your priority selection.
10. Outline the basic energy management program. Explain why it is a legitimate assignment for maintenance engineering.

11. The basic purpose of the job log (Fig. 11-5) is to identify all jobs and those that are available for work. How would a maintenance engineer decide whether a job is available to work on, and what factors could change it from unavailable to available and vice versa?

12. The job log tells management whether preventive maintenance time is available to be performed by the respective trades. Discuss how an emergency or breakdown repair scheduling system can assign time for non-emergency repair activities.

13. Why is a job effectiveness form (Fig. 11-6) a necessary part of the job log system?

14. The shop maintenance information system exists to provide management with an up-to-date evaluation of the factory's equipment. Describe how the three parts of this system are related and the data that each contains.

15. In reference to the data contained in Fig. 11-7, interpret the meaning of each of the three lines of data shown from the viewpoint of the production control manager and the manager of manufacturing engineering.

16. Discuss the basic tenets of practical PM systems. In particular, what pragmatic compromises must be made, and how does manufacturing engineering strive to comply with the needs of shop operations?

Chapter 12

COMPUTER NUMERICAL CONTROL OF MACHINE TOOLS

The introduction of the computer to control a machine tool or process is the most revolutionary and profound occurrence in industry since the invention of the steam engine ushered in the industrial age. Computer Numerical Control (CNC) allows low-skilled operators to produce craftsmanlike products simply by telling the machine to do it. It can permanently capture the craftsman's skills in the operation of the machine or process and duplicate, triplicate, etc. those skills throughout the entire factory. It opens up new opportunities for mass distribution of complex parts that were previously the sole domain of the skilled machinist or toolmaker. No longer is it necessary to merge a creative mind with an artisan's skills to produce complex shapes; CNC only requires the creative mind to tell the computer what steps are needed to create the complex part. This is a different type of creativity, one that understands how to mathematically model physical shapes and contours, and then devise a sequence a machine tool will follow to create those shapes.

But CNC is not without its costs. Since its introduction in factories in the 1960s, CNC has led to profound changes in manufacturing, the most important being a shift in required manufacturing skills. The skilled operator population has steadily declined, while the skilled CNC programmer population has grown. Factory maintenance personnel have changed from master electricians to electronics technicians and in some cases electronics engineers. Machines and processes have increased enormously in complexity, while the skill levels necessary to use the machines have decreased. The supporting structure, namely manufacturing engineering, has grown considerably to plan and service the equipment. This chapter concerns the philosophical management problems that must be dealt with to succeed in the CNC revolution.

DESCRIPTION OF A CNC MACHINE

A machine tool is a device that can do one or more of the basic metalworking operations:

1. Drilling and boring—making holes
2. Turning—making externally round pieces
3. Milling—making flat pieces with a rotary tool
4. Planning, shaping, and broaching—making flat or concave pieces with a stationary tool
5. Grinding—shaping either curved or flat surfaces with an abrasive
6. Forming—bending, forging, drawing, and shearing

Each operation requires a degree of intelligence to guide the action. With an ordinary machine tool, this is provided by the operator, who tells the machine what to do typically by setting the feed (depth of cut of the cutting edge, e.g., 0.010 or 0.025 in.) and speed (rate at which the cutting tool traverses the work piece, e.g., revolutions per minute or surface feet per minute). Some machine tools require more than two settings. For instance, a lathe or a boring mill requires three settings: (1) depth of cut, the amount of material removed per pass; (2) in feed, the translation rate of the tool into a work piece; and (3) part speed, the rotational speed of the work piece.

With manual control machine tools the operator physically sets these parameters by controlling knobs, levers, and switches directly connected to transmissions, motors, and other devices for activating the machine. The machine continues to operate as set until the operator physically makes a change in the settings, so it is possible to machine off too little or too much material. The operator changes the settings when he has machined off just enough material, has the proper finish, and has achieved the drawing dimension—that is, the operator is intelligently directing the activities of the machine. This is the way industry has operated since the industrial revolution. It requires intimate knowledge of the machine tool's capabilities, skill in operation, and a relatively long apprenticeship to become a master machinist.

Suppose that the operator is absent. Unless we can find another operator, we cannot get the machine tool to work. But if we put a timer on the controls with a mechanical linkage that turned the dials and pushed the levers a measured amount, we could set the timer so that at period 1 the machine runs at feed 1 and speed 1 for X minutes and at period 2 it runs at feed 2 and speed 2 for Y minutes and then stops. Then all we would have to do is push the start button and walk away. We would have created an outside source of intelligence, albeit crude, to control the sequence of operation of the machine.

Now if we had a great many of these mechanical trip lever times, each set slightly differently to give different amounts of mechanical dial twisting and lever pushing, and if we set these timers up in the proper sequence, we could conceivably automate the machine to make many different parts without directly controlling the machine. This would be accomplished by rearranging the mechanical trip levers to suit the particular need at the time, that is, by creating a program. All we would have to do to make parts would be to arrange the timers and trip levers in a programmed order and push the start button. This, essentially, is hard automation.

We have created a programmed sequence and mechanically hooked the controls to an assembly of many timers, spring trips, levers, and so forth. If we correctly calculated the time at each step versus the physical movement of the part and tool we should end up with a good part produced through various operations in sequence without operator guidance. This is automated control. If it is easy to rearrange the sequences, then the control is a flexible automated control. If we replace the mechanical trip lever and timer apparatus with electrical signals and can vary those signals at will, then we are approaching logic or computer control. Finally, if we can write instructions in plain (programming) language and have a microprocessor interpret this into binary code and then into electrical signals to which the machine's controls can respond, we have computer numerical control.

Thus CNC is a modern version of the many different mechanical timer trip lever assemblies. A computer is used to provide intelligence to control the variable operations of the machine tool. The specific steps that we want the machine to progress through can be changed at will, and we have a form of control intelligence directing the machine's activities.

CNC PARTS PROGRAMMING

To make a CNC machine tool work, the computer must be programmed, that is, instructed in the sequence of electrical signals to be sent to the controls of the machine in order to activate the transducers, motors, and gears connected to the operating components. This is done by having a programmer create a CNC parts program. The sequence of creating the CNC program is as follows.

1. The programmer studies the workplace drawing looking for a sequence of operations, much as a skilled machinist does. The skilled machinist determines how many cuts or passes of the machine are required and what knobs and dial settings will be used, by visualizing the part being machined. The CNC programmer also looks at how many cuts are required and visualizes the part being machined, but lists the machine motions necessary to machine the part in a sequential order. This becomes the manuscript, and it is like a step-by-step instruction sequence. However, unlike

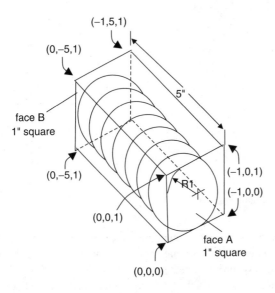

Figure 12-1. Square-faced Block Converted on a Lathe to a Round Dowel

a set of instructions, the manuscript is written in mathematical notation using a space coordinate system. In its most basic form it is a contour tracing system. Imagine visualizing a rectangular block being turned into a circular shaft on a lathe. Figure 12-1 is a simplified before-and-after drawing of a $5 =$ in. $=$ long, square-faced block converted on a lathe to a 1-in.-diameter round dowel 5-in.-long.

We can describe the path the lathe tool will take in cutting the rectangular block, from face A to face B, as the block is rotated. Figure 12-2 shows the trace of the tool as it traverses from right to left and back again and is indexed in along the radius R_1 to the position $R = 0.5$ in.

The manuscript is the tool path written in a mathematical form. The basic manuscript associated with Fig. 12-2 becomes:

1. Location 0,0,1; $R_1 = 0.707$ in.: traverse left to location 0,–5,1.
2. Location 0,–5,1; $R_1 = 0.60$ in.: traverse right to location 0,0,1.
3. Location 0,0,1; $R_1 = 0.50$ in.: traverse left to location 0,–5,1.

In this case R_1 indicates how much the tool must be indexed in along the final part radius from the diagonal of the rectangular block.

This manuscript is not a working manuscript. For one thing, the angle of the tool is not considered. It illustrates the thought process used to prepare manuscripts, not the technically correct procedure. To show the technically correct procedure here would make this example difficult to follow.

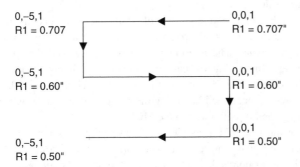

Figure 12-2. Lathe Tool Path

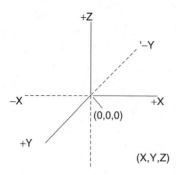

Figure 12-3. Cartesian Coordinate System

2. The manuscript is then converted into something the computer can understand: a punched tape, floppy disk, or cassette, for example. The tape or other medium containing the proper instructions is the work output of the CNC programmer.

3. The information translated to a tape is the CNC program and is now ready to be fed into the machine tool's computer. The trade jargon for this is feeding the Machine Control Unit (MCU). The MCU is a small computer or microprocessor dedicated to a particular machine tool, and the information in it makes the machine tool sequence occur in the desired order. If another sequence is needed for another type of part, another manuscript must be prepared.

Let's now go back and see what is fed into the MCU so that we can better understand what a numerical control program is.

The Society of Manufacturing Engineers defines numerical control as follows:

"Numerical Control is the operation of a machine tool by a series of coded instructions, which are comprised of numbers and other symbols. Webster's definition of a program is 'a logical sequence of operations to be performed.' Coded commands, gathered together and logically organized so they will direct a machine tool in a specific task, comprise an N/C program."*

Thus, a CNC program describes a tool location relative to the work piece or the machine tool's own coordinate system. In addition to describing the locations of the tool, the program contains information dealing with rates and types of motions and auxiliary operations the machine tool must perform, such as changing tools. Tool locations are usually relative to a Cartesian coordinate system, shown in Fig. 12-3.

Cartesian coordinates fit machine tools perfectly, since the tools are constructed on two or three perpendicular planes. All machine tools have a (0,0,0) reference point on the *XYZ* axes. Depending on the number of degrees of freedom of the machine, the location of the tool is based on an *X, Y* or *X, Y, Z* point in the space coordinate system.

This reference point can be at one end of the bed of a lathe, at the tool changer turret location of a horizontal boring mill, at the center of the rotating table of a vertical boring mill, or anywhere the machine tool builder designates. The important thing is that a zero datum point in space is designated so that all positions the cutting tool can reach are described in space in reference to that datum point. The coordinate point *X, Y, Z* shown in Fig. 12-3 is referred to the 0,0,0 datum point. The Cartesian coordinate system is utilized in writing the manuscript and in translating the manuscript into machine language and inputting it into the MCU.

Sometimes contouring must be performed. This can also be programmed into the MCU by using cartesian coordinates, but that requires a considerable number of points to define a curve. As shown in Fig. 12-4, the fewer points defined, the more step like the curve would appear to be.

Numerical Control Fundamentals, Society of Manufacturing Engineers, Dearborn, MI, 1980, p. 56.

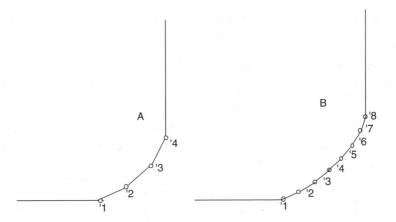

Figure 12-4. Shape of Contour as a Function of the Number of Points Used to Define the Curve

Fortunately, a cutting tool is a finite mass and by its size alone can smooth out most curves as it is translating in space. Therefore curve B can probably be machined adequately well with eight points defining it, while curve A after machining may look like an unsmooth or step function approximation because only four points define the curve.

As one would imagine, calculating the location in space of seven closely adjacent points can be tedious. For this reason CNC programmers define the curve mathematically, as illustrated in Fig. 12-5. Figure 12-5 is a geometric representation of the curved tool path shown in Fig. 12-4B. Note that in this case we only show three points on the curve plus the point designated as R. Point R is located in space and described in the manuscript as any other point would be. However, R is the center of a circle of radius r. When the cutting tool reaches point 2, it is instructed to sweep a circle of radius r; the arc of the circle to be swept will begin at point 2, which is at an angle α away from the vertical perpendicular, and be complete at point 3, which is at an angle β above the horizontal perpendicular. Thus, to make programming easier, arcs of parabolas, circles, and so forth are employed. This is done by locating the center of the arc desired and instructing the translating device of the machine tool to position the cutter there and then trace the arc. Although this is simpler than defining all the points shown in Fig. 12-4B, even it can be a relatively slow process to program.

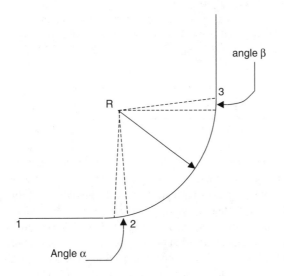

Figure 12-5. Defining a Curved Tool Path

Therefore, most MCUs are designed so that the description of the desired curve, for example, y = ax² + bx + c, can be input directly into the manuscript and none of the points R must be located. The CNC programmer describes a best-fit equation for a curve and the MCU causes the machine tool to perform the necessary translations. By using mathematical descriptions of curves, an infinite number of points can be employed and an exact curve as defined by the design engineer can be created. This was not possible before because a manual machine must be stepped through a curve in a manner similar to that shown in Fig. 12-4.

We can see that the process described above offers many opportunities for error. Therefore, the skill of the CNC programmer, while different from that of the master machinist, is vital to the success of the modern manufacturing concern.

To summarize, we have created a unique set of instructions, called a program, for a specific part. Like any other new program, this program, requires debugging, which is very time consuming. Now let us look at ways to reduce the debugging time so that programs can be supplied to the CNC machines at a significantly higher rate.

COMPUTER-ASSISTED CNC PROGRAMMING

Many programming languages have been developed to help program parts for production on CNC machines. These languages allow the programmer to describe part shape, machine tool motion, and machine functions in English-like commands. The programmer still must prepare a manuscript, but the computer-assisted CNC program does such things as calculation for tool offsets and determination of radius centers. The computer also does all the geometric checking to ensure that a continuous tool path or, if desired, a step function tool path has been input. The program defines the part to be made as a series of points, lines, circles, curves, and planes in three-dimensional space. The output is a series of tool paths that have already been compensated for tool sizes and shapes.

Computer-assisted CNC programming is the only practical way to perform this function. Most companies would find it economically unjustifiable to have a great many programmers on the payroll, even if they could be found and recruited, to do manual programming for each new part. This would be slow and error-prone, and would tend not to derive the full benefit of the CNC machine's capability for productivity improvements. These machines are extremely productive, but only when running, and not when waiting for tapes. For this reason, most companies buy "canned" programs from software companies for specific types of common geometries such as shafts and bolts. These programs, often called family-of-parts programs, are normally written in the common CNC languages such as Apt, Adapt, Compact II, and others that are readily understood by the popular MCUs. These are actually variations of Fortran, so that they are easily taught and generally familiar to technical personnel in manufacturing.

FAMILY-OF-PARTS PROGRAMMING

Creating general CNC programs that can be used for many different parts is the goal of all manufacturing managers. The ideal would be to describe a part to be made, indicate what machine tools should be used, and quickly produce usable CNC tapes. This ideal has not been achieved yet. However, family-of-parts programming begins to approach this ideal. Family-of-parts programming is a subset of computer-assisted programming that is used where parts that are similar in shape will be made, and the variations of instructions to the machine, though large, are not infinite.

For example, Fig. 12-6 shows a generalized outline for producing a rotor shaft. Every dimension and radius is indicated by an appropriate symbol representing a generic variable. A program can be written to perform all radius, length, and diameter machining. That is, all the Rs, Ls, and Ds can be described in general mathematical terms so that tool paths can be described, and

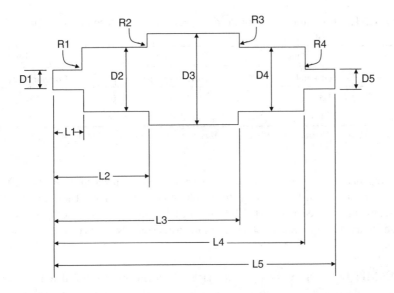

Figure 12-6. Generalized Outline for Producing a Rotor Shaft

substituting specific values for these variables determines the specific geometry for a desired product.

These programs are developed to machine a part of a general shape. It is only necessary to fill in specific values for all the variables, which can be found on the engineering drawing. The manuscript is now prepared and ready to be entered into the MCU, which can be done rapidly. Then, associated with the family-of-parts program, the computer's editing system evaluates the changes (including required omissions for simplified versions of the product) and produces a tape for the specific job. This amounts to matching the generic program to the specific task through the input of specific values for the variables.

USES OF CNC MACHINE TOOLS

Numerical control technology allows the use of general-purpose machines for large lot sizes, that is, mass production. More than one set of operations can be performed on one machine tool, which means more flexible machines, hence less expense incurred for model changeover. With manually controlled machines, the key to mass production has been special-purpose machines, which are very costly and time-consuming to set up but, once running, very economical for producing large numbers of identical parts. With CNC machines, the setup changes are stored in a memory and are essentially electronic rather than mechanical. Special jigs and fixtures are not needed to guide the machine; instead, the proper manuscript must be fed into the MCU.

Numerical control technology also permits small lot sizes to be produced economically with general-purpose machines. Manual control machines may require a cam-type template for the machine to follow (tracer control), requiring the expensive and time-consuming operation of making a template in the toolroom. Small lot sizes are expensive because the cam costs are written off over a small number of pieces, and they take an extended period of time to produce. With CNC machines many of these problems are eliminated. If the shape can be determined mathematically then CNC machines, especially those with six degrees of freedom, can produce the part. This means that many complex parts can now be made quickly and at considerably less cost.

It can certainly be said that CNC technology improves the economics of manufacturing. Some data from a study made by DeGarmo are shown in Table 12-1. These data are representative of time

TABLE 12-1. Times and Costs for Production of a Motor Base by Conventional and N/C Technology

	Setup		Cycle		Tooling	
	Conventional	N/C	Conventional	N/C	Conventional	N/C
Time (hr)	1.3	0.4	0.86	0.65	0.64	0.21
Cost ($)	3140	800	2350	835	1890	260

*From E. DeGarmo, *Material and Processes in Manufacturing Engineering*, 3d ed., Macmillan, New York, 1969, pp. 916–917.

savings reported by many industrial users before N/C evolved into the more productive CNC. Typical savings then range from 1.7:1 to 3.5:1 (the numbers are approaching 7:1 now). Notice that not only is the setup time reduced and practically eliminated, but also the cycle time is significantly reduced, because the machine can operate at a higher rate, particularly when changing from one operation to another.

> **EXAMPLE:** In lathe work a manual machine's operator can proceed at a relatively high rate of metal removal while roughing out the length of a shaft. However, in starting to contour a fit, the operator will probably measure the shaft diameter to make sure it is correct, reposition the cutting tool, and then carefully step-index the cutting tool position in the radial and longitudinal directions. In doing the same job, a CNC lathe would do the roughing to size along the length at about the same speed as a skilled operator. It might do it a little faster because the CNC machine would not have to be concerned about machining the length under size, since its program has instructed it to start traversing at a specified point in space and stop at a second specified point in space. The work piece is perfectly centered in that space and securely fastened in the lathe by the headstock and tailstock, so the exact dimensions will be machined to. After the roughing operation the cutting tool begins to trace the contour tool path, as the CNC program instructed it to. There is no stopping to position the cutting tool as in the manual operation, because the end of the traversing operation, the second point in space, is the beginning of the contouring operation. The contouring operation can be accomplished much more quickly because the radial and longitudinal indexing is preplanned in the program, not done by operator experience. In the manual procedure the operator must take a little metal off in each pass to avoid removing too much, while in the CNC mode the design geometry is input into the MCU by the manuscript and a virtually perfect contour can be achieved rapidly.

With CNC technology the thought process of the design engineer conceiving of the part is virtually duplicated in the MCU of the CNC machine before the part is made. We know exactly the part geometry and the sequenced steps required to achieve the concept within the raw material. Before CNC, manufacturing had only a picture, a drawing with dimensions on it. Skilled operators guided basic cutting machines to try to duplicate the necessary shapes. The MP&WM personnel specified the number of traverses required, but they could never detail in minute steps the very fine contouring step functions required. With CNC it is practical to do so. An exact mathematical match to the engineering drawing is possible and can be produced two to three times faster than in the inherently less accurate manual method.

The lathe is probably the simplest machine tool and the most capable of forming relatively complex shapes in a manual mode. The discussion above shows how even this simple machine is greatly enhanced by the application of CNC. One of the benefits not even imagined when CNC machines were first introduced was that of multiple tool use. Before CNC, only special-purpose, complex machines had the capability of doing more than one operation per work piece setup. A machine of this type has been used to produce internal combustion engine parts for automobiles. In actuality, this is a hard-automation process where parts are fed on a moving rack past milling heads, drills, reamers, boring bars, and so forth. As the work piece passes each tool, that tool performs its work. These machines have the capability to rotate and translate the work piece to present the proper attitude toward the specific cutting tool. They are ideally suited for very long production runs because the cost of initial setup can be prorated over all the parts produced. However, when the design engineer changes the design, the machine must be literally rebuilt to accept the new design. The positioning and tool sequencing must be modified, to an extent determined by the degree of the design change. To make only a few parts on such a machine would be out of the question; the cost would be prohibitive. Therefore, smaller lot sizes had to be made on general-purpose machines, one operation at a time, and economies of scale were nonexistent. With the use of CNC it is possible to construct a general-purpose machine that can do much more than one operation per setup. Economies of scale can be approached with much simpler machines than the hard-automation, special-purpose machines. Let us see how this is done.

The CNC machines discussed above are called multi-tool changer machines or flexible manufacturing machining centers. They are characterized by a track or turret containing different types of cutting tools, such as drills, milling heads, and reamers, that can be called upon and presented to the work piece. More complex machines of this type also have work setup tables that can rotate, translate, and in general present the work piece to the tool at a variety of angles and positions. Virtually all six degrees of freedom of translation and rotation are possible. In essence, the work piece is positionable and the cutting tool is brought to it, all under the control of a computer—the familiar MCU instructed by a properly conceived manuscript. Here we have the capability of planning virtually the entire machining of a part on only one workstation. The program gives an instruction to the machine for what tool should be used for each operation and then an instruction to the machine to position the part for the next operation. The more complex machines are also given instructions for how the work piece should be rotated on the positionable worktable or holder to present the proper aspect for machining.

EXAMPLE: Suppose we are required to machine a gearbox casing, as shown in Fig. 12-7. The casing is a fabricated structure designed to hold the stationary portion of a gear set, with the rotating component contained within the stationary components. The purpose of the casing is to provide a foundation for the gear set. To manufacture the casing, we could mill the faces, drill the numerous holes for securing the gear set, bore the diameters to the proper size, and finally mill the feet. Before the advent of CNC machining centers we would do all the milling on one machine, the drilling on another, and finally the boring on another. With a multipurpose machining center we can do the entire operations on one machine. Let's construct a manuscript summary to show how this is accomplished.

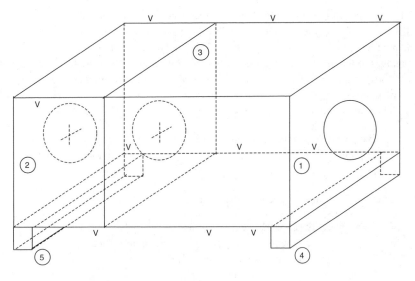

V = clamp positions for supporting the workpiece on the worktable

Figure 12-7. Gearbox Casing

1. Set up work piece in position one. This is the position shown in Fig. 12-7 and is probably a manual operation.
2. Mill face 1. A tool path is created for a milling tool to do the milling work. A milling cutter is selected and its location in the tool turret identified.
3. Drill holes on face 1. A tool path is created for drilling locations to do the drilling. A drill bit is selected and its location in the tool turret is identified.
4. Rotate worktable 180° horizontal. CNC program instructions are given to the worktable to rotate after completion of step 3.
5. Mill face 2. Logic similar to step 2 is created.
6. Drill holes on face 2. Logic similar to step 3 is created.
7. Bore diameters on faces 1, 2, and 3. A tool path is created for a boring tool that has both longitudinal and radial component instructions. A boring tool is selected and its location in the tool turret is identified.
8. Rotate worktable 90° up vertical. Logic similar to step 4 is created.
9. Mill face 4. Logic similar to step 2 is created.
10. Mill face 5. Logic similar to step 2 is created.
11. Drill holes on face 5. Logic similar to step 3 is created.
12. Drill holes on face 4. Logic similar to step 3 is created.
13. Rotate work piece 90° down vertical. Logic similar to step 4 is created.
14. Rotate work piece 180° horizontal. Logic similar to step 4 is created.
15. Off-load work piece.

The list above is a linked series of tool paths and machine instructions designed by the CNC programmer. An entire plan to manufacture a part is created and can be carried out by one machine that is being directed by a computer in accordance with a preconceived logic. Since the computer can only do what is programmed, the sequence will be carried out identically time after time. We have created soft automation and can change it as often as we wish by creating another set of instructions for the computer. Thus CNC creates mass production without the necessity for special templates, cams, or ingenious devices for mechanically instructing a manual machine.

ADVANTAGES AND DISADVANTAGES OF CNC MACHINES

All new technologies introduced in manufacturing can create new types of management problems that must be overcome. To find acceptable solutions to new problems, it is important to understand all aspects of the problem. In the case of CNC that means understanding the advantages and disadvantages to the entire manufacturing organization brought about by this marriage of the computer with production. We will discuss these advantages and disadvantages to develop an understanding of the management options available with CNC.

There are three basic advantages of CNC:

1. The ability to produce parts at two to seven times the rate of manual machines
2. The ability to produce accurate, complex parts at a high rate
3. The absence of a need for skilled machine tool operators

There are also three perceived disadvantages of CNC:

4. A cost roughly twice that of manual machines
5. A maintenance downtime roughly 50% higher than for manual machines
6. The need for a new type of skill, that of the CNC programmer

The six items listed above may be thought of as management opportunities, because any identified situation can be a management opportunity if handled properly. Item 6, the need for CNC programmers, is classified as a disadvantage because it means that a different type of factory skill must be nurtured and people trained in that craft. Scarcity of programmers can create difficulties in using the CNC machines to their maximum capacity. On the other hand, it means that extensive apprenticeship programs to train skilled operators, with their attendant indirect costs, can be pared back, which is a benefit. In addition, a unionized plant may be a disadvantage for management. Skilled operators most often are union members and part of the hourly work force, while CNC programmers are usually part of the technical staff and not union members. Obviously, having the machines directly controlled by the CNC programmer tends to offer management more flexibility. Finally, the effects of the scarcity of CNC programmers can be creatively minimized by the adoption of computer-assisted CNC programs. This, coupled with the fact that CNC programmers can often service several CNC machines while a skilled operator can only operate one machine at a time, makes the advantage/disadvantage trade-off of the need for programmers lean toward being an advantage even though it may originally have been perceived as a disadvantage.

Chapter 11 implied that there are differences in maintenance between CNC and manual machines. At present, CNC machines do require more maintenance than manual machines. However, the two- to sevenfold increased production rate more than offsets the 50% and higher maintenance time increase. Even with a rate of downtime approximately 3 times that of manual machines, CNC machines yield a greater production rate and hence a greater profit.

The same reasoning is involved when a company decides to spend twice as much for a CNC machine as for its manual counterpart. If the machine can produce over twice the volume of a manual machine, that advantage outweighs the disadvantage of higher initial cost and higher maintenance cost. Of course, capital equipment decisions must be based on a case-by-case analysis of the specific situation and not on an industry average. As CNC becomes even more firmly linked to the computer-assisted management techniques the differences in costs between CNC and manual machines will become less and less important. We are rapidly approaching the time when integrated computer control of factories will require only CNC equipment.

One advantage not discussed so far from a management viewpoint is the capability of making complex parts in volume quantities. This enables management to sell tighter quality specifications to its customers, which in many cases is either a marketing advantage or a necessity to stay in business. In addition, by being able to produce complex products in volume and at relatively low cost, it is possible for the company to offer new products or enter businesses that were closed to it before. For instance, consider a company that could only make nuts and bolts on its manual lathes and now can make intricate cams. The company can now enter the machine tool business as a supplier of new equipment. Or it can become competitive in the machine tool rebuild business because it can make complex repair parts quickly and inexpensively.

An analysis of advantages and disadvantages should be looked at as an analysis of different types of opportunities. This is especially true in connection with CNC machines. Here we are dealing with a creative revolution, and all perceived problems associated with that revolution must be looked on as potential opportunities for greater profitability. The examples discussed above show how perceived problems with CNC can be turned into advantages if approached properly. When we speak about the high costs of purchasing CNC equipment, we should think of the even higher costs associated with running uncompetitive factories. Our discussion of advantages and disadvantages should perhaps be rephrased: the advantages of implementing CNC versus the disadvantages of not implementing CNC. Not implementing CNC is stating clearly that a firm intends to withdraw from the marketplace and no longer compete.

MANAGING THE USE OF CNC MACHINES

Effective management of CNC equipment, like that of any other resource, requires an understanding of the resource. In most industrial organizations the responsibility for understanding the CNC equipment and setting guidelines for its use is delegated to manufacturing engineering. Within manufacturing engineering, it is further delegated to the methods, planning, and work measurement unit. Although advanced manufacturing engineering is the procurer and specifier of CNC machines, it is not the responsible unit, because we are concerned with the use of the CNC machines and not their development and implementation. Therefore, the MP&WM unit, not the AME unit, is responsible for interpreting CNC to shop operations.

The MP&WM unit must plan for the use and provide the day-to-day methods support necessary to operate the CNC machines effectively. Let us look at the major considerations involved in operating with CNC that are different from the traditional MP&WM functions.

1. *Establish planning guidelines for selection of parts to be made on CNC machine.* Since most manufacturing concerns find themselves with a mix of CNC and non-CNC machines, it is necessary to route parts to both types of facilities. (However, in the future this will become less of a problem as CNC machines slowly displace all manual machines except for special purpose machines, for example, in the tool room.) The MP&WM unit is charged with establishing guidelines for effectively loading both types of facilities. In doing this, they would want to plan parts to be made on the most efficient machines first and the least efficient as load demands. Since CNC machines tend to be the most efficient, most guidelines require that CNC machines be loaded first, or base loaded, before other equipment is scheduled. Also, since CNC machines are usually more adaptable, they would be favored over manual machines for complex parts manufacturing. This leads us to the second consideration.

2. *Review parts to ensure that CNC application is correct.* We know that CNC machines are ideally suited for making complex parts because all the steps can be preprogrammed. But should this be given preferential scheduling treatment over the ability of CNC machines to make simpler parts significantly faster than manual machines? That is, if CNC machine capacity is limited, should that capacity be utilized for very small lot sizes of complex parts in preference to large lot sizes of simpler parts? This is a consideration MP&WM must deal with constantly. The planner must ensure that the CNC machine is the proper workstation for the part assignment. The decision is based on whether there are technical and economic advantages of making the part on the CNC machine. The overriding consideration must be technical, since it would not benefit the company to run an economic batch on a CNC machine and then struggle unsuccessfully to make a complex part on a manual machine. If the complex part can only be made on the CNC machine, it must be made there, and the batch-run parts must either wait or be scheduled for a manual facility.

3. *Determine machining method and tooling required.* Before a CNC programmer can instruct the CNC machine in performing the required work, a method and tool selection must be made. Unless the CNC programmer is a qualified planner or methods engineer, he is unlikely to know the proper sequence of making the part. Therefore, the planner usually states the method to be followed, including the tooling required, and the CNC programmer then develops a tool path sequence and a manuscript. The difference between this approach and the initiation of manual planning is that in the CNC case, the planner does not have to detail the exact sequence of cuts to be taken. This evolves automatically once the tool path is described mathematically by the programmer. The planner must only describe the tool to be used, which in turn will dictate depths of cuts and feeds and speeds.

4. *Translate drawing dimension into input format.* Since most CNC work is now carried out with generic family-of-parts programs, planners now find themselves entering the geometric data into generic programs on a day-to-day basis. The planner, to generate the required output for the CNC machine tool enters into a family-of-parts program written for a specific CNC machine. The planner uses this output, typically in the form of a punched tape or a direct stored instruction, as part of the planning package dispatched to the workstation. If there is no tape, the planning package simply informs the operator what program should be called up to instruct the CNC machine in making the part.

5. *Develop family programs for new part families.* Usually the CNC programmers are part of the MP&WM organization, and one of their responsibilities is recommending when individual programs should be consolidated into family-of-parts programs. It is MP&WM's responsibility to recognize when a family begins to exist and a productivity improvement could be achieved by allowing nonprogrammers to generate specific parts tapes for CNC machines. A cost trade-off is usually evaluated here, weighing the cost of developing a family-of-parts program against that of continuing to program the parts individually. The decision is based primarily on volume.

6. *Provide training for CNC operators.* Unlike the apprenticeship program for training skilled operators, there is no traditional program for training relatively unskilled people to operate CNC equipment. Contrary to the sales brochures of some CNC vendors, CNC equipment still needs operators—perhaps not the skilled craftsmen needed to get the most out of manual machines, but nevertheless people to monitor and load the CNC machines. These people must instruct the machine, load the raw materials, off-load the finished products, and monitor the machine for proper operation. By instruct and monitor, we mean inputting the properly designed program and determining whether the machine is operating correctly. MP&WM must have the capability of training operators to do this.

Usually the MP&WM unit will have staff responsible for handling the training. These are CNC programmers or methods engineers who are familiar with the machines and usually have the machines as part of their responsibilities for methods and planning. They are the service staff shop operations calls on when there is a need to train new operators.

ROBOTICS

The idea of machinelike imitations of humans has always caught the interest of the public. Therefore, it is hardly surprising that the public was interested when the term robot was used to describe programmable positioning and loading devices. In fact, even the general news magazines have devoted space to discussing this offspring of the CNC revolution. No discussion of CNC can be considered complete if robotics is not included.

A robot may be defined as a mechanical device capable of being programmed to move or manipulate in three-dimensional space and capable of doing work in that space imitating, in a general way, motions capable of being accomplished by a human being. Let us analyze this definition. A vast number of machines can move in space. Fewer machines can manipulate, that is, rotate, translate, and traverse. The ability to imitate human motions is even more severely restricted within the machine population. However, there is still one more important part of the definition that must be fulfilled: the machine must be programmable; its sequence of movements must be capable of being varied by changing instructions to its controlling computer.

The last sentence states the link between a CNC machine and a robot. Manipulators that mimic human motions have been available for a long time. For instance, atomic scientists have used manipulators controlled from a distance to handle radioactive materials behind a radiation barrier. In order to use these manipulators, the scientists had to develop a manual skill, probably by trial and error. These are analogous to manual machines rather than CNC machines. Much greater precision would be possible if these manual manipulators could be driven through three-dimensional space based on a programmable positioning device commanded by a prescribed mathematical sequenced program. This is exactly what a robot is—a manipulator that a CNC programmer has instructed to follow a sequence of commands dictating its spatial motion and when it should grab, start spray painting, start welding, and so forth.

A robot, then, is a specialized CNC machine. The controlling mechanism is the familiar MCU and the intelligence is provided by a manuscript, that is, a mathematical compilation of the specialized tool path. In this case the tool path may be the journey through space to pick up a part and place it somewhere else. The purpose of this tool path is different from that of a CNC lathe's tool path, but the principle is the same. The CNC programmer programs its motions, just as he does for the CNC machine tool's cutting apparatus. The robot's motions usually must be very finely controlled, while the CNC machine can rely on the accuracy of the lead screws and beds. Therefore, the original MCUs are replaced by more powerful mini- and/or microcomputers capable of much finer control.

The rationale for using robots is the same as that for using CNC machines. Robots produce higher levels of quality and work faster than novices and journeymen. They can be programmed to do the work of skilled operators—in this case not machine tool operators but welders, assemblers, and painters. The use of robots to do human manual work is not nearly as advanced as that of CNC machines taking over from manual machines, but the trend is virtually identical. As the MCUs for robots become more powerful and as sensing devices develop from crude pressure sensors to finer sensors for touch, feel, and sight, more human motions, particularly of the fingers, will be mimicked by robots.

Ultimately, this will lead to flexible automated factories where robots do the human work of loading and unloading CNC machines and moving material from workstation to workstation. The technology for control is available now, but the price to accomplish this may still be too high. This is a logical outgrowth of the CNC revolution—the use of a computer to direct a machine based on a program designed by the human controller.

Is this good or bad for society? In my opinion, it is good. It releases humans from performing drudge work on a production line or uncomfortable and possibly dangerous work in foundries and mills. Humans evolve better and better tools so that we can exercise our intellect and have our machines do the drudgery to support our standard of living. This view may seem utopian, but we will never achieve it unless we continue with the CNC revolution. Certainly, there are disruptions in the ability of some to earn a living as these jobs are replaced with robots and CNC machines. We no longer use wheelbarrows and shovels as the main tools for constructing roads; we use machines, and society applauds the improvement. The same should also be true of switching from manual machines and manual work to CNC machines and robots in our factories.

SUMMARY

We have discussed the concept of CNC machine tools by first explaining the meaning of manual control, then hard automation, and finally electronic control via the computer, in order to demonstrate the logic that led to their development.

We have also covered the philosophy of programming for CNC machines. Manuscript development and the relationship to the cartesian coordinate system of points in space were explained. Further simplifications were discussed through the use of equations representing geometric forms to program curves in space.

Management techniques related to CNC machine use have been described. The six steps in planning and implementing work on CNC machines show the impact of CNC on the manufacturing engineering organization.

Finally, the concept of robots has been introduced. Robots are a logical extension of numerical control development. Their programming is similar to that of CNC machines, but they are involved in nonmachining activities such as assembly, welding, and painting. From the discussion of robots it became evident that the flexible automated factory is not only possible but inevitable.

REVIEW QUESTIONS

1. Discuss the differences between the skills required to produce a complex part on a manual lathe and on a CNC lathe.
2. A shaft having several different diameters and a threaded portion is to be machined on a manual lathe. Discuss the judgments the operator must exercise in making the part and what the machine will do in response to these judgments.
3. For the example in question 2, describe how the judgments to be made by the operator can be predetermined so that we have an automated process.
4. Define a programmed sequence with respect to a specific machine tool operation.
5. A manuscript is the CNC version of the detailed sequential planning sheet. Make a manuscript for the machining operations used to manufacture a stud threaded at both ends according to the accompanying sketch. Include a graphical representation of the tool path. Make from carbon steel 1.000-in. diameter rod stock.

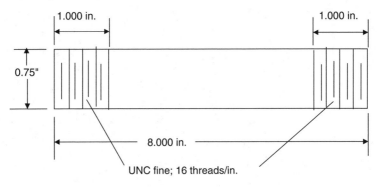

UNC fine; 16 threads/in.

Diameter tolerance = +/– 0.001 in.

Length tolerance = +/– 0.002 in.

6. What is the zero datum point of a machine tool, and how is it related to manuscript preparation?
7. There are two basic ways to define curves in reference to manuscript preparation: the point-in-space method and the radius method. Compare the two, discussing advantages and disadvantages of each.
8. For the shaft shown below, create a set of values for a family-of-parts program as shown in Fig. 12-6. *Hint*: start from the left side; all *L, R,* and *D* values must be used as inputs.

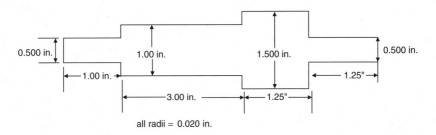

all radii = 0.020 in.

9. Discuss the advantages and disadvantages of special-purpose machine tools and general-purpose CNC machine tools with respect to flow (mass) production of machined components.
10. Why are machine cycle time and setup time significantly reduced with CNC machines versus manual machines?
11. Why can a CNC multi-tool-changer machine (a machining center) accept changes from design engineering more readily and at less expense than a special-purpose multitool machine?
12. Discuss the ways in which CNC machines can open additional marketing opportunities for manufacturing companies.
13. Why are CNC machines always considered to be base-loaded facilities?
14. Under what circumstances should a part not be loaded to a CNC machine?
15. Describe in outline form the similarities between robots and CNC machines.
16. What areas of development are necessary for robots to achieve the same dominance in assembly operations that CNC machines have achieved in machining operations? Give reasons for your answers.
17. Describe the potential benefits of robots over other types of transfer, assembly, and welding methods.

Chapter 13

FUNDAMENTALS OF COMPUTER-INTEGRATED MANUFACTURING

Computerization of the management process is and continues to be the emerging technology in manufacturing, particularly in the areas of design data manipulation, manufacturing data manipulation, and decision making. In this chapter we will explore the fundamental principles involved in defining, developing, and putting into use computer systems that enhance a company's ability to compete. This theme of unifying via computer is a natural follow-on to the introduction of CNC machines and financial accounting systems. In Chapters 14 and 15 we will explore the immense potential for factory data collection systems and computer-aided process planning through group technology. However, it is first necessary to understand the Computer-Integrated Manufacturing (CIM) fundamentals.

CIM is the current title for this unifying concept. Previously it was widely known as Computer-Aided Design/Computer-Aided Manufacturing (CAD/CAM). But that term fell into disuse because it was felt that mentioning only the engineering (design) and the manufacturing functions implied that other business functions, such as marketing and employee relations, need not be integrated for optimum performance. Of course, this is not the case. So another name evolved: Computer-Integrated Manufacturing. This states integration via computers and focuses on the manufacturing nature of most businesses. Also, some experts felt that the slash (/) between CAD and CAM implied a linear nonconcurrent approach that favored islands of automation instead of integration of efforts.

Efforts are still being made to rename this philosophy. Some say it ought to be called Computer-Integrated Business (CIB) because in reality it is more than manufacturing oriented. They argue that the theory is just as applicable to service businesses as it is to those producing physical products. Some say even that does not go far enough. They contend that any organization serving a defined purpose can use this philosophy. Hence they propose the name Computer-Integrated Enterprise (CIE). We will use the acronym CIM. It is still the most widely used name and it suits our purpose because this book is about manufacturing.

INTEGRATION OF FUNCTIONS WITH CIM

Continued progress in productivity improvement is essential for a company to remain competitive. This is another way of saying reducing operating costs, which are defined as direct labor costs, overhead costs, and material costs. A properly conceived CIM system affects all three of these costs categories and, by making possible significant reductions, results in productivity improvements (see Fig. 13-1). No longer can a manager rely on all functions to perform their jobs adequately well but in an isolated manner. With the geometrically rising costs of labor and material, what were once acceptable design cycle times, material procurement times, planning times, queue times, and setup times now lead to losses in market share, lower profits, and, in the extreme, business failure. Methods have been found to significantly improve these factors. These methods involve full integration and they are classified under the title CIM.

CIM is the only way to integrate all job shop functions to minimize the total cost of manufacturing. It is done through the use of common computer databases. This means that all functions can access each other's accumulated and stored information, so that the same information can be used

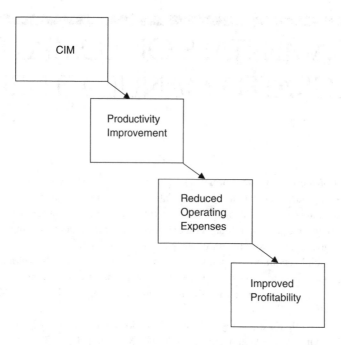

Figure 13-1. Benefits of CIM

in different ways by the different functions. Before common databases were available, functions created their own databases from the information they were capable of gathering, the information passed on to them from others, and the data they created in performing their responsibilities. Each had its own exclusive repository of information, and the only data reasonably sure of being defect-free were the data they inputted themselves.

As an example, design engineering uses the outline of a part or assembly to perform a stress analysis to make sure the part geometry is sufficient for the task. The engineering function is reasonably sure that their work is being done correctly because they are responsible for creating the geometry via the engineering drawing. Eventually the part is released for manufacturing. Manufacturing engineering will then create a CNC tool path manuscript and make the part. This sounds simple, but it can be an error-prone activity. In creating the CNC tool path, manufacturing engineering must recreate the geometry. The engineering drawings produced by design engineering are conceived as instructions for operators to follow in producing parts on manual machines. The final shape is shown along with the final dimensions, but these dimensions are not calculated in three-dimensional space, the cartesian coordinate system. This system is necessary for the MCU input to the CNC machine. Therefore, the manufacturing engineer must create a new database for the MCU input, selecting a zero datum point and then creating points in space relative to the datum point, using the engineering drawing as the source of information. The manufacturing engineer is creating a geometric database by converting the absolute dimensions on the engineering drawing into relative dimensions via the coordinate system, as illustrated in Fig. 13-2. This translation creates opportunities for making mistakes. Data entered incorrectly and misinterpretations of engineering drawing information are common examples. Errors are not only possible, but also probable, which means additional costs and slowdowns in scheduled production to make corrections.

The proper solution to this problem is to have the person who designed the part select the datum point and then show the required dimensions in cartesian coordinates. Design engineering can do this, and does for its own stress analysis needs, for instance, in finite-element mesh generation. However, before the advent of common database generation, there was no practical way to give manufacturing this necessary information, with the result that it had to be recreated. The common database, as required for CIM, allows all functions to have access to primary data,

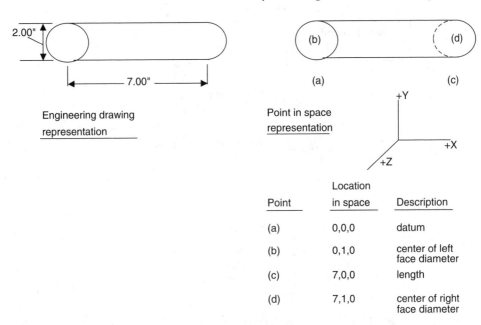

Point	Location in space	Description
(a)	0,0,0	datum
(b)	0,1,0	center of left face diameter
(c)	7,0,0	length
(d)	7,1,0	center of right face diameter

Figure 13-2.   Converting from Engineering Drawing to Database

which eliminates both translation errors and duplication of effort. In our example, the CNC programmer could simply access the database and construct the CNC tool path, without any risk of making interpretation errors. Reduced duplication of effort leads to reduced queue times, setup times, and design times and to an overall improvement in productivity. Thus common databases can optimize the total manufacturing cycle time. We will expand on this theme of the common database throughout the remainder of this chapter and show why it is important and is the heart of the integrated CIM system.

PRODUCTIVITY VIA CIM

CIM is primarily a productivity enhancement system, which is why it is of interest to business. The first thing the manufacturing engineering manager must ask is, "How do I improve productivity with CIM?" If the manager does not know, then obviously no aspects of CIM will be introduced. Similarly, if so-called productivity improvements do not result in lower costs or improved levels of profitability, CIM will not be introduced. CIM must improve the cost picture, and this must be clear to the decision-making manager.

To answer the rhetorical question above, let us look at how job shops specifically, and factories are generally, run. An order is received and is put into the manufacturing master schedule by marketing and agreed to by manufacturing. The specific design is evaluated and produced by design engineering, which translates the design into drawings and instructions for manufacturing. Manufacturing, in turn, takes these drawings and instructions, creates specific task instructions for the various workstations, and orders materials to be used during the fabrication of the product. Manufacturing then schedules the job to be made in the factory. Finally, the job is completed and shipped to the customer, whose technical questions, if any, are answered by marketing or design engineering.

The one common factor in this entire sequence is information (Fig. 13-3). The information needed for all facets of operations is essentially the same: what it is, how it is supposed to work,

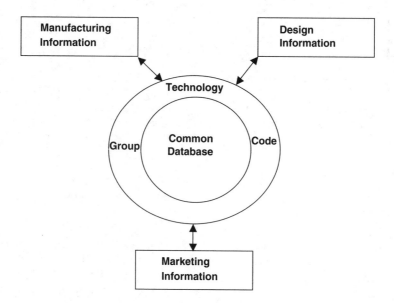

Figure 13-3. Inputs and Outputs of Group Technology Coded Common Database

who has to do the work, when the work has to be done, and why it has to be done in a certain manner. These questions are asked by the various functions of the total company organization in relation to their respective charters. For example, the design engineers are interested in how it works so that they can do the necessary scientific analysis to ensure that it will work and has a high enough margin of safety against failure. The manufacturing engineers are interested in how it works so that they can design the manufacturing process in a compatible manner, that is, not create a situation in which required tolerances cannot be met. Both design engineering and manufacturing engineering are asking the same question in relation to different needs, but based on the same set of facts, the database.

Without CIM, it is difficult to obtain all the facts to answer the question from the viewpoint of each function, and in many cases each function has to create its own library of facts for each individual order. With a CIM system it is possible to have one centrally located library of facts for each job, with each function having access to that library. If manufacturing engineering wants to know how tolerances for a particular job are to be set, it simply visits the library and looks up the information. The library analogy goes even further. Most good libraries are continually expanding, and the same is true for databases. As manufacturing engineering determines how to process a part, this information is fed back into the database. This is important because it allows other functions to know what that specific portion of the plan is so that they do not do further work that may be contradictory.

EXAMPLE: Suppose that manufacturing engineering has decided on a particular form of automated welding because it is suitable for the design created by design engineering. Further, suppose manufacturing engineering has spent company money to purchase the necessary welding equipment. Then, for some reason, design engineering changes the physical nature of the parts, making it virtually impossible to use the recently purchased welding equipment. This is not an impossible occurrence. In fact, this type of "the left hand not knowing what the right hand is doing" goes on every day. With CIM, a database accessible to all could be searched by design engineering before it makes any change, and a situation such as that described above could probably be avoided. Design engineering would know that special-purpose welding equipment had been bought to do the job. A more productive welding technique would be used; there would be no need to further redesign after manufacturing engineering protested;

and there would be no need for each side to prepare its arguments. Hence, many opportunities for overall improved productivity would result from the existence of a common database. Thus CIM amounts to providing a good, reliable communications system.

A second answer to the rhetorical question posed at the beginning of this section can be found by looking at another use of CIM. This is the coupling of the common database with group technology to minimize the need for totally redoing quotes, designs, and plans for each new job offered to the job shop. We may define group technology as a systematic search for sameness in manufacturing techniques, such that larger production runs, commonality of tooling, dedication of machine tools for families of parts, and efficient utilization of designs and planning can be accomplished. The facilitating mechanism is a group technology classification coding system. In Fig. 13-3 we see that the common database is surrounded by the group technology code. This signifies that not only do we define sameness by a code, but we also can enter the database efficiently and quickly using that code. We will cover development and applications of group technology in Chapter 15.

The common database includes large quantities of design, marketing, manufacturing, and financial data pertinent to all contracts worked on by the manufacturing firm. After a relatively short period of time, it is typically found that specific aspects of new jobs are very similar to aspects of previous jobs for which data are stored. Therefore, a great deal of redundant work for each function can be eliminated if we can recall previous work. The group technology classification and coding scheme, used as an index for the common database, allows recall of that information on the basis of coded design characteristics, manufacturing characteristics, and other specific information. That is, we use the group technology scheme to code needs and characteristics of new jobs and then, using the code as an index, search the common database for matches or near matches from previous projects. This recall process is optimized with CIM. For example, when asked to design a new component, the design engineer can search the file quickly and thoroughly for similar designs used previously, and greatly shorten the design cycle by not having to start from scratch. The same is true for manufacturing engineering in determining the planning sequence of the new component.

This shortening of the cycle time means that more useful work can be performed in the same standard workweek. It lowers the indirect costs of manufacturing because work is accomplished faster, but just as thoroughly. This is measurable productivity improvement and enhances the company's competitiveness in the marketplace.

A third answer to the rhetorical question concerns the use of CNC machine tools and processes. General-purpose CNC machine tools can be utilized to produce parts at a higher rate than is possible with manually controlled machine tools. The basic reason for improved productivity with CNC technology is that CNC mimics the performance of the most skilful machinists for all of the CNC machines, whereas the best machinists are only occasionally available to run all the manual machines. The characteristics of CNC machines and manual machines are discussed and compared in Chapter 12.

CNC Machine	Manual Machine
• Skills of best machinists programmed	• Rarely best machinists operating
• All operations sequenced automatically	• Pauses between operations to review next step
• Potential for more parts per shift	• Fewer parts per shift
• Consistent, highly repeatable	• Inconsistent, less repeatable

Another way in which CIM improves productivity is through the productivity/quality measurement capabilities offered by the use of the computer. Before the advent of computers, the only way to keep track of various parts on various workstations was with manually prepared tally sheets prepared by production expediters and foremen. Since machines were slower then, it was possible

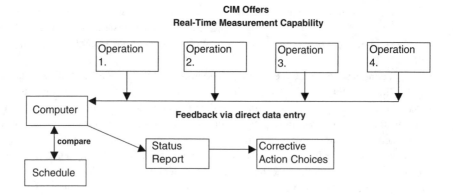

Figure 13-4. Real-time Measurement Capability of CIM

to effectively control operations in this way. With the advent of CNC the pace of manufacturing quickened. Queue bottlenecks occurred as work completed by fast CNC machines reached slow transfer systems. The resulting need to move material more rapidly from station to station effectively eliminated time available to produce manual tally sheets. Human input and manipulation of data was not adequate for the task, and initially good inter-station control was lost. With the application of the computer to manufacturing control, it again became possible to know where parts were in comparison to where they should be according to the schedule. In fact, the schedule became a computer-integrated activity quickly generated from the planning cycle and the delivery dates promised to the customers. With CIM techniques, it is possible to have real-time information on where parts are, where bottlenecks are, at what workstations quality problems are occurring, what workstations are consistently showing poor performance, and other information needed to manage the operation. It is also possible to react to current information to take corrective action. Before CIM job shops frequently reacted to information that was weeks old, after sizable losses had occ-urred, and the problem was aggravated by the implementation of fast CNC machines without the rest of CIM. This capability of CIM is illustrated schematically in Fig. 13-4.

CONTROL OF JOB SHOP OPERATIONS WITH CIM

The use of CIM to control shop operations is becoming one of the most important factors in manufacturing. In its role as the technical resource for the manufacturing function, manufacturing engineering must promote this capability. Let us review this revolution of information management and control made possible by CIM.

The key to success in any endeavour involving many people and many parallel and sequential operations is good communications. It is necessary to let all those involved know what they are supposed to do and when they are supposed to do it. All treatises written about effective management of businesses are, in essence, expounding communications philosophies. Therefore, it is reasonable to say that any device capable of improving our ability to communicate is desirable. CIM computer systems significantly enhance human communications related to design and manufacturing. To understand the revolutionary effects of CIM on these aspects of job shop manufacturing, we must first consider the manual control procedures used before computers, and then see how CIM offers significant improvements.

A typical manual control system would be a gantt chart showing work going across a workstation as a function of time. Figure 13-5 is an example of such a chart. Jobs are plotted in sequential time order across the workstation. In effect, this is a reservation system for jobs at that workstation. If a job shop has only a half-dozen or so different workstations, this manual system of reserving space

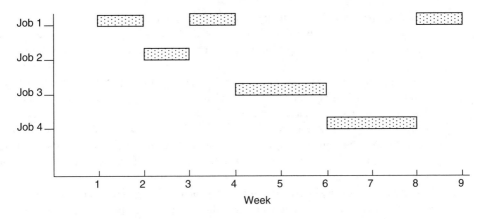

Figure 13-5. Workstation Schedule Chart

on a machine tool or process workstation is adequate and relatively easy to use. To make sure the jobs are completed on time, the manager will compile all the individual workstation schedules into a master production schedule. In addition to the factory portion, the manager must schedule the design engineering function and possibly the assembly of the product at the customer's site. There is, however, a subtle difference in creating a master production schedule from a series of gantt charts for many workstations. Each of the charts applies to a particular workstation with times reserved for various jobs. The master production schedule deals with the chronological sequence of specific steps required to produce each job. It is a compilation of the gantt charts for the workstations, but with the charts dismembered to show the progress of individual jobs as they make their way from raw material and design concept to finished product. Figure 13-6 illustrates the master production schedule.

This difference between individual workstation gantt charts and master production schedules requires a very important mental shift. Shifting and cross referencing are continually going on in the production scheduling of the factory, from workstation primacy to job primacy, and they must complement each other. We cannot have a situation where the master production schedule appears to be balanced but an individual workstation ends up having to accomplish more than one job at the same time.

Let us now consider a typical job shop that has 50 jobs running across 50 different workstations. There could be a matrix of 2,500 workstation-job permutations, resulting in an enormous scheduling task. This can be done manually—and was before CIM—but it requires careful attention to detail. A schedule is made out for each workstation based on the planned or estimated time to

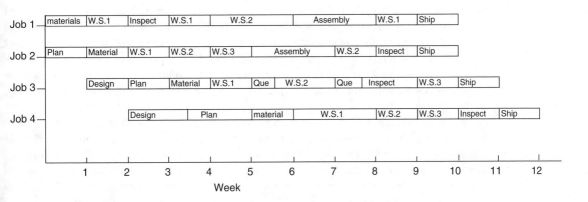

Figure 13-6. Master Production Schedule

perform the work. These time increments are then added up for all the workstations to obtain the total cycle time per job. All the jobs are collated to make the master production schedule, and we have what is called a manufacturing plan. This is a static manufacturing plan in that it is difficult to change.

Obviously, the manager tries to schedule enough time at each manufacturing or design stage to take care of unforeseen events. But, as one would expect with 2,500 possible permutations, it would require a great many planners and production schedulers to anticipate all that could possibly go wrong and have alternative strategies available. This would be economically unjustifiable. Therefore, at best the manager is compelled to tolerate inefficiencies. Jobs are done well before their scheduled due dates and kept waiting until the next workstation becomes available, which increases work-in-process inventory costs. Or additional equipment is purchased to eliminate scheduling bottlenecks. At worst, problems such as delays in receiving materials or unanticipated machine breakdowns stretch out the schedules and delay shipments to customers, with all the cost penalties associated with these failures.

The major problem is the effect of these unknowns on the master production schedule itself. With so many variables to consider, it is very difficult to restructure the schedule when problems occur with one job or a group of jobs. This leads to more than one job being available to work on at a workstation, and requires a decision to favor one of the jobs. The problems multiply as short-range decisions must be made and soon the entire master production schedule is incapable of being followed. The result can be a business out of control, with many emergency production meetings, products scheduled day by day, shipping dates to customers missed, and herculean efforts made at great costs to satisfy preferred customers. When this happens a decision to start all over again is made; the master production schedule is thrown away and the cycle starts once more. The inefficiencies are partially absorbed and partially passed on to the customer, and if neither can be done effectively, the company leaves the business.

This is the classic manufacturing problem: how to create a dynamic schedule, which reacts to change, rather than trying to maintain a static schedule that is too cumbersome to change and hence cannot deal with unforeseen problems. The management of information through CIM is the solution to this problem.

The computer is capable of handling 2,500 permutations and many more. Hence, when unexpected problems occur with a job or a group of jobs, it is possible to reschedule the entire job shop and maintain control. The original master production schedule was a controlling document. If the factory followed the schedule, the product would be produced on time and at an acceptable cost. When the schedule becomes void, the manager may be able to control near-term product production, but will probably lose control of start operations, such as purchasing and planning. So much effort will be required to make the necessary corrections to near-term finishes of production that the manager will not be able to devote the necessary resources to revamping all the starts. A decision to leave the starts alone is not satisfactory, since to do this when production is bottled up will result in an inventory excess at best, which saps the economic vitality of the company. What usually happens is that the delays at the back end of operations lead to relaxed efforts at the front end, so that the starts, too, begin to fall behind. Then as the back end problems are cleared up, the starts are late and cannot support the repaired master production schedule. This is the "accordion effect" that managers strive to avoid. Large amounts of overtime to get back on production schedule and expensive purchasing decisions to get materials in house quickly are two of its many costs. This can be avoided if schedules can be revised quickly and thoroughly, as they can with CIM.

The need to reschedule the entire shop when significant problems occur is evident. A master production schedule is a sequence of different parts going across common facilities in a chronologically arranged order. Therefore, all jobs in the factory are interdependent for scheduling purposes. With the scheduling ability of the computer based on the common database, it is possible to adjust schedules and even make prognoses related to possible manufacturing failures. The human mind may not be capable of inverting a 50×50 matrix efficiently and quickly, but a computer is capable of doing it. Hence, with CIM, constant rescheduling or updating can be done. This means that the manager has control of the design and manufacturing functions, knows what is happening in a specific as well as a general sense, and is in communication with all aspects of the business and can make decisions based on accurate, real-time information.

Advantages of CIM Scheduling

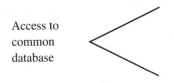

Access to common database

1. Handle large permutations easily
2. Restructuring of workstation loading and high level master production scheduling accomplished quickly
3. "What if" scenarios can be examined
4. Real-time information easily available

CIM can be thought of as a synonym for linking all activities through a common data base. How is this accomplished? In theory, quite simply. All activities involved in manufacturing, engineering, finance, marketing, and employee relations are monitored or controlled by a computer or a series of computers, which in turn are linked together in a communications network. The system is designed so that the data related to each of these activities reside in a common memory, the database. The database may be stored in one computer or distributed among many. That is a system design question left to the computer systems engineers to obtain maximum efficiency.

With CIM, instead of multiple stand alone activities, we have everything tied together. With computerized systems operating independently we have to go outside their domains to transfer information between the various steps of business activities. This results in vulnerability to late and missing information, which leads to errors. The common data base solution, the CIM solution, provides correct up-to-date information that leads to significantly more efficient operations.

For CIM to be used as discussed above, what must be contained within the common data base? Let us examine the index of information most frequently found in the common database and see which types of information affect scheduling.

1. *Design characteristics.* This is the design engineering data on the geometry of the parts and the materials they are made of. This database is usually not consulted when scheduling changes are made.
2. *CNC tool paths.* These generic data, based on the design characteristics database, are fed to the various machine tools to machine the parts. While this database is not consulted in revising schedules, the basic tool path cycle time is included in the planning database.
3. *Planning.* This is the input from the MP&WM time standard and planning activities showing how much time and on what workstation the parts, as defined in the design characteristics database, will be made. Most planning databases give alternative workstations with planned times for parts manufacturing. This is of interest for the master production schedule and rescheduling activities.
4. *Purchasing.* This database contains information on vendors, the material requirements for parts taken from the design database, contracts with vendors, promised dates from vendors, and quality assurance requirements for materials certification. The vendors' promised dates for receipt of materials would be of interest in rescheduling.
5. *Quality requirements.* This database contains the quality plan for the specific parts and becomes a resource input for the planning and purchasing databases. It also includes all quality failures and causes, and therefore becomes a feedback to the design characteristics and CNC tool path databases. There is a direct link to rescheduling activities related to quality failures.
6. *Master production schedule**. This is the key database for planning the strategy of the company. It receives inputs from the planning data base merged with the strategic plan for sales

*Many firms expand the master production schedule to be the Manufacturing Resources Planning (MRP II) database. We define and describe MRP II in Chapter 16, which devoted to that subject.

and income, and keys the purchasing data base sequence as well as the CNC tool path, design characteristics, and planning data bases. It is the controlling database as well as being controlled by the feedbacks from others. It usually includes the dispatching activity. This database is intimately involved with the rescheduling activities.

7. *Data collection.* This is the recording database that informs the master production schedule database of the status of the progress of parts in accordance with the schedule. The reporting back of items accomplished is vital for rescheduling. This reporting back is done by all functions involved in design, procurement, planning, and production. Usually this database is the weak link in a computerized production system. Much work has been done toward automating the reporting back of status and making it a real-time system. Chapter 14 is devoted to this subject.

8. *Cost analysis.* All of the costs from the various databases are collected here, usually in terms of hours for in-house activities and dollars for external activities. This database is essential for determining whether products are being made within budget. It offers opportunities for analysis to show where difficulties have occurred.

The basic databases listed above are typical of what makes up the common database. More databases may exist to meet specific company needs, such as a maintenance database to insert required machine tool downtimes into the master production schedule. Note that the common database is nothing more than the system flow manufacturing production control in Chapter 3, (Fig. 3-6) with the inputs from the various functions and subfunctions rearranged to create a computer-optimized set of programs.

There are many commercial programs called CIM common database systems. They all work after a fashion. However, they are general in nature in order to fit a broad range of products and manufacturing techniques, so that in virtually every case they must be modified to fit specific factory needs, or the specific factory must change its way of doing things to fit the system. Most CIM common database systems tend to be either modifications of commercially available systems or completely locally developed systems. Special management techniques are necessary to implement these types of systems, and will be discussed in a later section of this chapter.

Control of a factory through CIM comprises automated design retrieval, automated planning, control of manufacturing starts and stops by the computerized master production schedule, release of purchase orders for material as it is needed in accordance with a real-time schedule, insertion of quality monitoring and rework requirements into a real-time schedule, and control of machine tools by design-derived tool paths directly fed into the control logics. Such control leads to an order of magnitude improvement in productivity and profitability. By far the most important aspect of this system is the ability to control schedules, which enables the manager to minutely manipulate or "fine tune" the entire sequence. This means that contingencies in scheduling and corresponding cycle times can be significantly reduced, resulting in significant cost reductions. Labor costs are reduced because the work force is not kept waiting as much to be available when "the job breaks." As the cycle time is reduced, material carrying costs are reduced, since the queue time for material is shortened. Design and manufacturing engineering costs are lower because delay times due to unavailability to perform tests and verifications are minimized. The better the schedule is controlled, the less expensive it becomes to produce the company's product.

It is well known that 90 to 95% of all job shop cycle time is nonmachining time. Achieving real-time scheduling through the use of CIM, as summarized in Fig. 13-7, can eliminate a significant portion of the nonmachining time.

A SYSTEMATIC APPROACH TO IMPLEMENTING CIM IN JOB SHOPS

Managers are often fascinated with new technological devices such as CNC machines and machining centers, robots, and flexible automated process lines. However, if these devices are purchased and implemented without a systematic integration plan they will not achieve their full potential. Much

CIM Means:	Most important and overlooked benefit of CIM

1. Automated design and retrieval
2. Automated planning
3. Automated master schedule
  a. Starts and stops
  b. Material order
  c. Inventory control
  d. Real-time production monitoring
4. Automated machine tool CNC tool path generation
	Control of schedule
	Yields: Fine tuning of entire sequence
	1. Contingency time reduced
	2. Cycle times reduced
	3. Less labor cost
	4. Reduced "work in process inventory" costs
	5. Lower engineering costs

Figure 13-7. Reductions in Cost Factors Achievable with CIM

more than the purchase of these new technologies is needed to achieve the productivity improvements possible with CIM.

The "factory of the future" is not so much an automated factory as one where excellent, complete, instantaneous communications are standard, and decisions affecting schedules and production rates are made on a real-time basis. Whether the factory is fully automated, a mixture of computer-controlled and manual tools, or all manual tools is of secondary importance compared to excellence in communications and real-time decision-making. The type of equipment used in a factory depends on the type of products produced. The use of CNC machines in a hydro generator factory, while important, is not nearly as important as in a reciprocating pump factory. The difference is due to the physical size of the products and the typical lot sizes and cycle times. But the communications/decision-making aspects are equally important in these two types of job shops.

During the past three decades we have seen CNC machine tools and processes pervade job shops, which have benefited significantly from their introduction, but not to the fullest extent possible. The reason for this is that we have concentrated too much on only one part of the CAD/CAM triad. Recently, more attention has been focused on the other two parts of the triad.

Figure 13-8 illustrates the CAD/CAM triad: (1) machine/process control, (2) design and planning control, and (3) production and measurement control. All three of these parts are equal in importance, as indicated by the triangular arrangement in Fig. 13-8. All three must be present in order to truly control the factory and optimize the benefits obtainable through CIM. Note that the abbreviation CNC does not appear in Fig. 13-8. This is because all machines, not only CNC ones, can be effectively controlled, or directed, by CIM. For a manual machine this is done by giving the operator real-time instructions, perhaps from a video display terminal. All three parts of the triad derive the necessary information from the common database, and the efficiency of achieving and disseminating this information depends on the communications/decision-making system.

All too often companies make the mistake of thinking of CIM as only the machine/process control part of the triad. This thinking is typical of the company that rushes out and buys CNC equipment and fails to benefit fully from it.

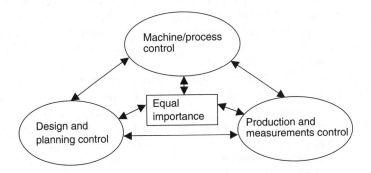

Figure 13-8. The CAD/CAM Triad

EXAMPLE: A CNC lathe versus a manual lathe. The productivity realized with the CNC lathe should be three times that of the manual lathe, but the company only realizes a 1.5-fold improvement in productivity. This company failed to tie the new device into a controllable system. They forgot about the links to design, planning, production scheduling, and measurements. They have not developed a CIM system.

Therefore, designing a CIM system is really the first thing to do in order to implement CIM. This should be done before the first new equipment is purchased. Implementing CIM requires an understanding of the way all product or service producing companies work. To gain this understanding, we will review the concept of the manufacturing system and its seven integrated steps.

The manufacturing system is a methodology that has evolved since the beginning of the industrial revolution. It is simply a logical approach to obtaining information and using that information to manufacture the product in an optimal fashion. Realistically, all companies have to follow this system logic or else chaos occurs. All companies perform the seven steps of the manufacturing system in carrying out their specific process of making products or providing a service. Curiously, the seven steps occur whether or not the company is conscious of them. The seven steps, in chronological order, are:

1. Obtain product specification.
2. Design a method for producing the product, including design and purchase of equipment/processes to produce, if required.
3. Schedule to produce.
4. Purchase raw materials in accordance with the schedule.
5. Produce in the factory.
6. Monitor results for technical compliance and cost control.
7. Ship the completed product to the customer.

The task of applying CIM theory is to integrate the information flow between the seven steps and to automate the operations to minimize random errors in the processes. Information is needed for every step, and much of the information is redundant. Thus we can see the need for a common database. Therefore, the process of implementing CIM becomes one of initiating a common database to link the various operations together. In practice, this means, for example, having CNC machines (the machine/process control segment of the triad) receive their tool path instructions from the design database (the design and planning control segment of the triad) and their sequencing of work order instructions from the production schedule data base (the production and measurement control segment of the triad). Both of these specific databases would be components of the integrated common database (like two books in a library).

Since the entire company is affected by the implementation of CIM, developing a CIM system must be a total company approach, with representatives of all functions being part of the planning process. A cross-functional organization must be established that has the responsibility for developing the criteria for the system and making sure the specifications are defined for use within the company. For lack of a better title, let us call this cross-functional team the systems council. The people on the systems council should be competent in computer usage or potential usage within their home organizations. They should be knowledgeable about the parts processed in their company. They should fully understand how an order from a customer is translated into a design, a producible design, manufacturing instructions, and finally a product shipped to the customer. If they do not have these prerequisites, then the systems council's first order of business is to educate themselves in these areas. Sometimes consultants are hired to help with this education, particularly in computer utilization, but often it is incumbent upon the systems council to train itself. However it is done, this training must be accomplished to avoid costly delays and possibly of failure. The systems council may also find it necessary to train their superiors, the senior members of the company's management. Here consultants may be very worthwhile.

After the preliminary training, the systems council can get on with the task of developing the firm's CIM plan. The first thing to be worked on is the communications/decision-making strategy. The council must decide what kind of manufacturing scheduling system is needed for their product. How is the design engineering data put into the system? How is information on propositions and design transmitted from marketing to design? How is the design information translated to manufacturing information? How should the status of design, proposition activity, and manufacturing progress be reported? When should purchase orders for materials be released? These basic questions must be answered in a manner that is specific to the particular factory and that shows the overall relationships of the communications/decision-making patterns of the company. Once this information is obtained, the systems council must decide whether they like what they have, or whether change is desirable or even necessary to improve efficiency. This decision must be made within the systems council. It cannot be made by vendors of computer equipment or packaged CIM systems or by hired consultants. These sources can help by pointing out similarities to other companies and showing how others have arrived at decisions, but the systems council alone must make the decisions. Once the decisions have been made, the system analysts, programmers, and engineers can convert them into the familiar flowcharts. At this point we are ready to purchase implementing hardware and software. It then becomes the task of the specialists reporting to the systems council, who are usually technical people and often from manufacturing engineering, to specify the various types of equipment and software necessary to implement this tailored communications/decision-making system.

Notice that nothing has been said yet about engineering workstations or CNC or other new technologies. This is not an oversight. While these are very valuable for a manufacturing concern, their full potential cannot be realized unless the communications/decision-making system is in place. After it is operational, the engineering workstations, CNCs, and so forth may be added to further improve productivity. For manufacturing firms that already have engineering workstations and CNCs but do not have a communications/decision-making system, a hold should be put on further expansion of the new equipment until an adequate communications/decision-making system has been implemented.

First Rule of Thumb

Develop and install a communications/decision-making system first for major productivity improvements. Follow up with "computer-automated equipment" for additional productivity improvements.

Second Rule of Thumb

For companies already possessing "computer-automated equipment." Put a hold on further development of such equipment until a communication/decision-making system is developed and installed.

Components of a communications/decision-making system would be a shop floor data collection system, MRP II, or a design retrieval system. These projects would have high priority, while a CNC vertical boring mill would be a medium-priority project that should be approved for purchase only if the cash flow payback period is good, usually less than 2 years. The key here is setting the proper priorities. There must be a framework to build on, otherwise the company will end up with islands of automation that never live up to their potential. This leads to frustration and possibly even abandonment of CIM; in the latter case the company may be unknowingly deciding to leave the business.

In summary, then, the sequence to be followed in implementing CIM is:

1. Recognize the CAD/CAM triad relationships for the company's specific products.
2. Establish a systems council to determine overall strategy and specify the communications/decision-making system.

3. Implement the communications/decision-making system.
4. Add new technologies to further improve productivity.

When implementing a CIM system, it must always be kept in mind that CIM is needed to improve productivity. If an aspect of the program developed by the systems council is vague on this, it must be looked at critically. No projects should be carried out unless the end result is an improvement in productivity, hence profitability.

ARTIFICIAL INTELLIGENCE (AI) IN A CIM WORLD

An often overlooked but an increasingly important application within the CIM environment is the use of Artificial Intelligence (AI). One would logically ask what difference does it make if AI is used via common databases or separate ones. After all, AI is a theoretical approach to help in decision making and not directly in line with the routine operations of a company. This may be so, but there is a subtle but very important difference between AI working with a common database and a more narrowly defined database. I will explain, but first we need to define and demonstrate the potential benefits of AI to make the differences relevant.

We use CIM to smooth out the flow of information in conducting daily activities related to the seven steps of the manufacturing system. Specifically, we use CIM modules to do the clerical and editorial activities, thus freeing ourselves to do the analytical and strategy activities. Even with these higher activities we use the CIM modules to access data and to manipulate data. But the creative thought or idea generating is entirely the domain of men and women, not machines. Now we have AI available to help us even with the creative process. AI can help by giving us logical supposition based on data but not related directly to the specifics of the data. This is the next evolutionary phase of computer use as the industrial revolution merges with the information age.

What is AI? It is not computers with cognizant awareness. Computers do not think in human reference terms. The computer, as we know it, reacts very rigidly to its programmed logic statements. For non-AI, the logic statements are the familiar If/Then routines. This means that if a certain condition exists, then the response (or action) will always be the same. Specific actions always elicit specific reactions. The major innovation AI provides in industry is accomplished by the AI subgroup called expert systems. Expert systems, I must emphasize, cannot make computers thinking machines any more than conventional software does. Expert systems are logical extensions to conventional programs that create an illusion of independent reasoning, more so than we usually associate with computers. The innovation we find useful is that the program appears to learn. This is accomplished by evolving from If/Then logic to If/Then Probable Then. In AI (we will for now refer in this way to expert systems) we use a clever set of programs that allow the database to search for information and then use that information to add to the original logic rules. Hence we say that the computer "learned." This is very useful for many manufacturing activities. Examples of such activities are diagnosis of failures in maintenance activities, process planning when many potential solutions exist, and aspects of statistical process control in which specific results are required; hence "intelligent" monitoring routines are needed. Here, If/Then Probable Then logic expands our ability to find out if a course of action would result in desired results. We shall see how this process works when we review the techniques of expert systems.

Let's delve a little deeper into AI theory so we can build a more complete understanding of why AI capabilities are brought out more fully under a CIM environment. As would be expected, AI programs have terminology specific to them. It will help to explain and demonstrate some of these.

There are three terms to be defined:

Working memory
Knowledge base
Inference engine

The *working memory* is not specifically part of the AI program, but without it the program has no way to learn. It would have no experience base. The working memory consists of the memoirs or knowledge of the people considered experts in the technology we are trying to emulate. Experts are queried in a logical manner to understand how they solve problems that are not necessarily solved using precise causative relationships. This means we are exploring intuition as a method of reaching conclusions about problems. The person doing the debriefing, the knowledge engineer, uses the following sequence to break down instinctive knowledge of the experts.

Problem definition
Symptoms
Root cause
Elimination of root cause (solution)

This is analogous to the traditional scientific method, which may require much iteration to come up with solutions. Thus we can see that knowledge engineers have to be very patient and very logical.

The *knowledge base* is the database of the AI system. One of the advantages of CIM, of course, is that the database can be very large. This holds true for AI databases operating under a CIM system, the only proviso being that the database be structured in a manner that is compatible with the knowledge base needs. This turns out to be an advantage for AI users in a CIM environment because the proviso is easy to comply with. Thus we have a gigantic universe of learning to draw from.

The knowledge engineer structures the information he or she gains from the experts into two categories: facts and rules. Fact are statements of conditions. An example of a fact would be:

"nail length = 1.011 in."

Rules are a set of actions to be taken with sets of facts as directed by the AI program. An example of a rule related to a fact would be:

"if length = 0.75 in. to 1.25 in., then finishing nail"

The *inference engine* is the AI program, similar to the blank format screens of familiar spreadsheet programs. These programs all have logic built into them for spreadsheet uses and we tailor it to our particular needs. So, too, with the inference engine. The inference engine is a generic set of If/Then Probable Then logic statements that can be purchased commercially and is often referred to as the generic shell. The generic shell loaded with the facts and rules of the knowledge base becomes the specific AI program. Shells (the inference engine) have the capability to make an inferred solution based on rules and facts.

What's the difference between AI programs and traditional programs? Traditional programs are numerically related and have only one allowable conclusion based on the facts. Scheduling algorithms are good examples of the traditional approach. AI programs differ in that they deal in and/or logic. And/or logic is not causative mathematical logic but deals more in the realm of probabilities. Hence, an answer is not always correct but close enough to 100% certainty to be valid for most problem solving. This type of reasoning is more like human thought patterns. So we say we have artificial intelligence.

The important factor about AI is its apparent ability to learn. This, of course, is not true in a human sense but it is definitely true in an applications sense. This occurs because the program arrives at its solution through forward or backward chaining. Essentially, with each solution we are adding another link to the chain—hence, the appearance of learning taking place. When the AI program reaches a logic dead end and no answer is apparent through chaining, the program is set up to ask the user (human interface) related questions.

In Fig. 13-9 we see an example of the inference engine content and the schematic of how new rules are added through interaction with the software. Note the list of facts and list of rules contained in the inference engine. From this set of facts and rules, by simple inductive reason, we can

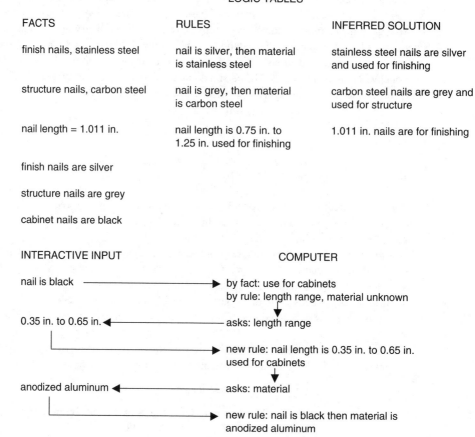

LOGIC TABLES

FACTS	RULES	INFERRED SOLUTION
finish nails, stainless steel	nail is silver, then material is stainless steel	stainless steel nails are silver and used for finishing
structure nails, carbon steel	nail is grey, then material is carbon steel	carbon steel nails are grey and used for structure
nail length = 1.011 in.	nail length is 0.75 in. to 1.25 in. used for finishing	1.011 in. nails are for finishing
finish nails are silver		
structure nails are grey		
cabinet nails are black		

INTERACTIVE INPUT   COMPUTER

nail is black ──────────────► by fact: use for cabinets
                              by rule: length range, material unknown

0.35 in. to 0.65 in. ◄─────── asks: length range

                     ──────► new rule: nail length is 0.35 in. to 0.65 in.
                              used for cabinets

anodized aluminum ◄────────── asks: material

                     ──────► new rule: nail is black then material is
                              anodized aluminum

Figure 13-9.   Learning by Chaining (Based on Inference Engine Logic)

come up with new results or inferred solutions. We can see that the statement "stainless steel nails are silver and used for finishing" comes directly from the fact "finish nails, stainless steel" and the rule "nail is silver, then material is stainless steel." There is no other possible inferred solution. So far, no learning is taking place that is different from non-AI software. Let's look at the interactive action shown in the bottom half of Fig. 13-9.

We see that the user (interactive input) enters the statement "nail is black." The computer, searching the inference engine by looking for relationships involving black nails, finds by fact that black nails are used for cabinets. That's all that exists at this point in time in the database. So the computer states that there are no rules about length or material specifically for black nails used for cabinet making. Therefore the inference engine software has been preprogrammed to ask questions related to other rules it already knows. In this case it asks, "what is the length range?" The human user responds "0.35 in. to 0.65 in." From this information via chaining the computer software creates a new rule: "nail length is 0.35 in. to 0.65 in. used for cabinets." We say the program has learned, and it can be used in future situations.

The importance of the learning ability is that the conclusions the AI program presents always take into account the most current validated information. This is not simply a different numerical to solve the same preprogrammed equation. It is more akin to having the different numerical value and a new equation too. In the real world, not only may data inputs change but also the circumstances affecting those data may change. Humans can accommodate these changes quite readily. Rigid, non-AI programs cannot accommodate changes' at all. It is this limited ability to accommodate change that is the breakthrough expert systems gives us for using computers in a more integrative manner.

AI offers the potential to solve hosts of manufacturing problems. Currently it is most useful for Statistical Quality Control (SQC) programs. Here, the learning capability is beneficial in evaluating whether or not a process is in control. We are also beginning to see AI used in planning programs. Determining how a product could be produced for optimum performance depends on complex logic of CNC availability; types of materials, which have different stress profiles; overall process cycle times; and conceivably the entire manufacturing process knowledge database from forming to painting. Here we can readily see how powerful such a tool could be and how much more effective it would be if we had integrated databases, that is, CIM.

The advent of If/Then Probable Then logic associated with AI opens up a host of opportunities for business. As discussed above, we can actually use the computer as an electronic partner of sorts to solve complex problems with nonrigid algorithms and less than specific related data. This puts a premium on the Probable aspect. Having the wherewithall to find related information for the "learning" to take place is at a premium. This is where CIM, the integration of information in a singular or linked set of databases, really provides an advantage.

CIM has the ability to search for information from a much larger universe than stand-alone data bases. For example, if we were to employ AI for a planning program we would need manufacturing process information, scheduling information, and design geometry information. In a CIM system this information would be contained in several linked databases and be readily accessible. In stand alone systems we would have a design database, a master schedule database, an MRP system database, and a manufacturing engineering database. All would be independent and mute to each other. How would the AI program chain to know the next step in production based on machine availability and customer need data? It would be impossible; hence, AI would not be employed and we would not be able to have a generative planning system. With CIM with data search capabilities, AI can be used, and we would have a planning system with schedule and availability realism built in. Thus CIM expands the potential for AI use, from tightly bounded diagnostic programs using stand alone databases to much more generic scenario and control programs enjoying the resources of CIM linked data bases.

Expert systems while intriguing in their own right are fairly primitive to another type of AI now becoming a mature technology. This system is "fuzzy logic." Here we go far beyond the If/Then Probable Then logic. Let's take a look.

Fuzzy logic is newer than Expert Systems. It was developed primarily for control systems where continuous data flow rather than discrete data points are the rule. For example, how do you define degrees of loudness? You may say by using a logarithmic scale as with decibels but how do you define between 69 and 70 decibels? Can you define 69.5 decibels? What about 69.25, then 69.125, then 69.0625, and then 69.06249, 62.062485, etc.? We can't definitively but we still have to have fairly accurate and repeatable values if we want to have any ability to maintain control. So we need a logic system that allows us to do so. If not we would never have accurate sound recording, or color reproduction, or ability to reach a precise destination, to name a few situations.

Fuzzy logic has uses where precise rules are not practical. This means an Expert System wouldn't work. So we have a system where simplifying assumptions based on rules and facts are essentially not usable. How does this vary from If/Then or If/Then Probable Then logic? The answer is that it is based on a different logic premise. We can call it "If/Then, But Not Precise" logic. It allows a system to handle an uncertainty or a value judgment.

EXAMPLES: The temperature of a room is too warm. The piano is being played softly.

These are uncertainties because we can't precisely define "too warm" or "playing softly." The logic comes from Aristotle's law of the excluded middle. Here is a very brief tutorial on the law of the excluded middle. Also known as set theory. In set theory an object, quantity, or concept either belongs to a set (a classification) or it does not. There is no middle ground. No gray area is allowed.

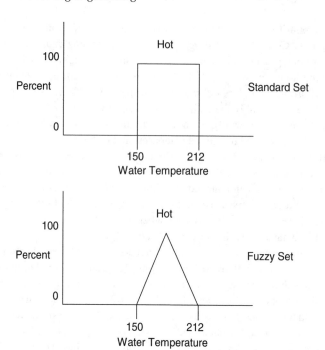

Figure 13-10. Fuzzy Set and Standard Set

EXAMPLES: 5 is a number in the set of odd numbers. 5 is not a member of the complementary set of even numbers.

Set theory does not mimic real life. However, what we call fuzzy set logic does mimic real life.

Fuzzy set theory logic can be described as follows, as compared to set theory logic. We can think of fuzzy set theory being an evolutionary outgrowth of set theory. In fuzzy logic set theory logic an object, quantity, or concept can belong to more than one set at the same time. It can be a major member of one set and a minor member of another, where a special case would be half in one and half in the other.

EXAMPLE: temperature of tap water in a sink, the fuzzy sets would be

Mostly cool (major cool, minor hot)
Just right (half cool, half hot)
Mostly hot (minor cool, major hot)

Fuzzy set (degrees), sometimes called fuzziness, are not probability percentages.

Probability measures whether something will or will not occur. Fuzziness measures the degree something belongs to the set, or which some condition exists. In all set theory unity must be maintained

Standard (Aristotle)
either / or: $(1 + 0 = 1)$, $(0 + 1 = 1)$

Fuzzy: additive
a temperature is cool;
perceived to be 80% comfortable, and 20% uncomfortable;
$(0.8 + 0.2 = 1)$

Now let's look at how the logic of Fuzzy Logic statements would be structured and how used. Recall that fuzzy logic is based on "If/Then, But Not Precise" logic. So statements are written in that manner.

EXAMPLES:

If the music is loud, then set the volume power on low
If the music is quiet, then set the volume power on high

The music sound levels and the power settings are called fuzzy sets. Notice there is nothing precise about these statements. How do you define loud, quiet, low power, high power? You do not. But we hope an expert with a fine and differentially tuned ear can shed some light on the situation and at least create some acceptable ranges. So just as with "Expert Systems," we debrief experts to create data points as to what is the power setting for high or low, also what constitutes loud or quiet vis a vis music. With enough experts debriefed, we can establish a response curve converting the universe of fuzzy degrees to specific actions. The logic being a fuzzy patch of information can lead to chain of information, which in turn leads to the ability to control. And we do this without any precise set points previously known, almost like driving an automobile. We make a turn by visual information, which is translated to rotational motion of the hands on the steering wheel, which controls the motion (path) of the automobile. There is nothing precise about this translation of information. Figure 13-11 shows this same relationship for our power output vs. sound level example.

Computer integrated manufacturing is a systems approach to solving the integration of resources with needs. And we can see throughout our discussion it is certainly possible that there will be multiple opportunities for situations to exist where cause and effect relationships are not precise but we still need to have acceptable solutions to minimize waste. With fuzzy logic we can see it is conceivable to use artificial intelligence as an assist in decision making for

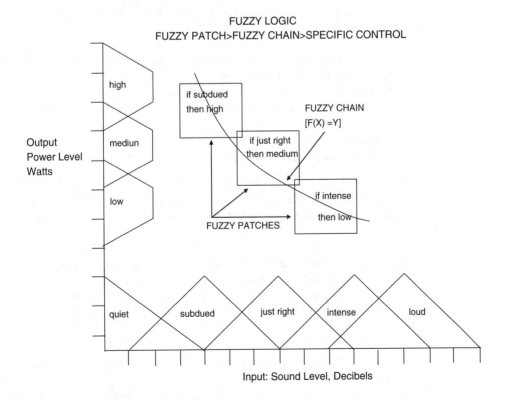

Figure 13-11. Fuzzy Logic Curve

planning where the existence of not precise relationship abound, this especially true concerning work station scheduling, or for that manner any type of project management activity where data are derived from a database.

Expert systems and fuzzy logic should be useful tools for improving project management effectiveness. If AI is suitable for Project Management it must be useful for at least some of the Keys of Project Success. Recall from Chapter 2:

The Keys to Project Success

1. Clearly delineate the project goals
2. Plan the project to meet the goals, including risk assessment.
3. Conduct the work of the project to meet the goals.
4. Measure status and compare to where we are with respect to the plan and the ultimate goals
5. Take corrective actions as necessary to keep the project tracking toward goal achievement.
6. In parallel, communicate results and needs to all interested parties.

There are probably many places where expert systems or fuzzy logic can be employed specifically to defined projects. But are there generic ones? Are there any that apply across the board? I'm confident there are. Here are two examples. Please keep in mind that none are generally applied now, but I think they can be if economically or technically justified.

My examples explore the use of artificial intelligence in project management. Let's first look at an example for use of artificial intelligence via an expert system. Then we will do a similar thing using fuzzy logic. Thereby demonstrating the differences between the two techniques and give indications as to how to chose which one to use. This will be very subjective and depend on the ability to get good data and the type of data to solve the problem.

From the Keys of Project Success: Step 2—Plan the project to meet the goals, including risk assessment.

To be viable the plan has to be dynamic with changing conditions. An expert system can be devised to create rules for project success. This would have:

- An initial state, defining the facts (constraints) the project manager has to contend with.
- A goal state, what the end results of the project are to be.
- A set of rules of how the project is to be conducted.
- The facts of the initial state and the rules that make up the knowledge base.

An example of learning from the rules could be:
Rule: $ value purchases over $5K for high-technology equipment needs manager approval.

Relationships	Rules application	Inferred solution
Automated soldering is wave solder machine	Wave solder machine costs $8.2K, required mgr. approval	Wave soldering machine is high technology

Therefore the program learned that the wave soldering machine is a high technology equipment and by project rules needs to be approved.

In a similar manner, situations applying the AI program during the Work Breakdown Structure development process could help derive rules for managing the entire project.

EXAMPLE:
- A step needing to buy tooling costing over $5K would trigger an AI response to require an additional project step for approval.
- This could have been overlooked if the AI program wasn't monitoring the development of the project.

Now let's look at a slightly different use of artificial intelligence where it is difficult to get rules and facts. Here we can demonstrate the fuzzy logic potential for risk assessment in project management

From the Keys of Project Success: Step 2—Plan the project to meet the goals, including risk assessment.

Plan the project to meet the goals, including risk assessment. To be viable, the plan has to be capable of assessing the risks associated with it. A fuzzy logic program can be devised to test for risks within certain acceptable boundaries. Since levels of risks are not precisely defined we need to write some "imprecise" statements that bound the problem.

Based on: "IF/THEN, BUT NOT PRECISE"

EXAMPLE:

"IF THE RISK IS TOO HIGH THEN ESTABLISH A MORE CONSERVATIVE PROGRAM."
"IF THE RISK IS TO LOW THEN ESTABLISH A MORE AGGRESSIVE PROGRAM."

We would then get expert opinion on the type of project(s) contemplated for our organization as to what constitutes high, medium, and low risk. This may be nothing more than an opinion poll. With enough data points we can establish a "universe of fuzzy degrees to relate to specific actions. Figure 13-12 shows a generic fuzzy logic chain that could be used to aid in risk assessment decisions.

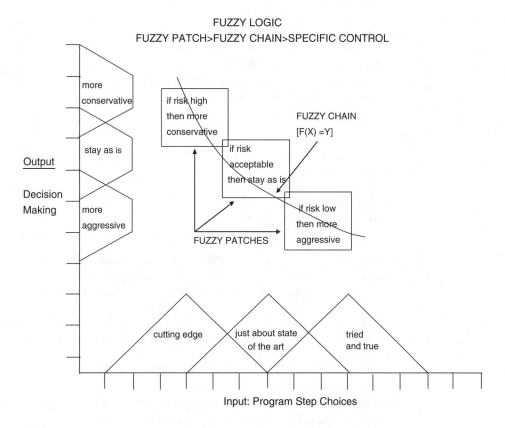

Figure 13-12. Using Fuzzy Chain for Project Management Decision Making

SUMMARY

In this chapter we have defined and developed the fundamentals of CIM. We have seen that CIM is essentially a productivity improvement concept, one so powerful and important that all discrete parts job shops must eventually implement it. Companies that choose not to implement CIM will eventually cease to exist in their business.

The fact that significant productivity improvement is a direct result of CIM implementation has been stressed, and the mechanisms for this improvement have been demonstrated by spelling out in detail why we must use CIM and how it is implemented. We have discussed the common database and learned why this integrating computer information system is necessary to achieve excellence in communications between the functions. We have also learned that CIM means much more than implementing CNC machines in the factory—that, in fact, it means communications/decision-making excellence and CNC machine tools are only a part of it. This led us to the important concept of the CAD/CAM triad: machine/process control, design and planning control, and production and measurements control. Thus, CIM is total control of the business, using real-time feedback systems, and communications excellence is the key factor. The philosophy of implementing CIM via a systems council has been explained. Finally, we looked at the future via applications of AI with CIM.

REVIEW QUESTIONS

1. CIM drives productivity improvement, which in turn drives further improvement in CIM. Explain why this iterative process occurs.
2. Define "common database." Explain why this integrating entity is necessary for productivity improvement.
3. Prior to common databases, why was it impractical for design engineers to create engineering drawings that were suitable for CNC input?
4. Explain how finance, marketing, design engineering, and manufacturing would make use of the following information from a common data base:

   Nuclear valve body,

   Weight—8,200 lb
   outside major diameter—4.25 ft
   valve seat diameter—28.00 in.
   overall length—8.40 ft
   wall thickness at valve seat—4.85 in.
   material: chrome, vanadium, alloy steel
   operating conditions: cyclic; open to close in 0.8 seconds; 1,000 PSI steam, 1000°F

5. Discuss how access to a common database tends to minimize design and process development iterations. Use a hypothetical example to illustrate your answer.
6. Describe the advantages of group technology classification and coding for design and manufacturing engineering activities.
7. One of the early disadvantages of the use of CNC machines was that they caused increased queue times and bottlenecks after the CNC workstation. Explain how CIM mitigated this disadvantage.
8. Explain why excellence in communications is the single most important factor in improving productivity via CIM.
9. Discuss the differences between a static and a dynamic manufacturing plan. Explain the reason for establishing static plans instead of dynamic plans.
10. Explain how CIM can minimize the "accordion effect" caused by unforeseen or unscheduled events within a manufacturing schedule.

11. Why does the modern job shop traditionally generate 90 to 95% nonproductive time for the entire part manufacturing cycle? Why does CIM offer the potential to decrease this nonproductive time?

12. Explain why the purchase of CNC machines is not the highest priority in establishing an effective CIM system.

13. An in-house "systems council" is the preferred mechanism for defining the communications/decision-making system. For each of the following organizations, outline the point of view and expertise each will contribute to the systems council.
    a. Design engineering
    b. Manufacturing engineering
    c. Materials
    d. Shop operations
    e. Finance

14. Discuss the pros and cons of purchasing a communications/decision-making system from a vendor.

15. Explain why all organizations making products or providing services comply with the seven steps of the manufacturing system, even though some organizations may not even be conscious of the existence of the manufacturing system.

16. With reference to AI logic statements, what is the difference between If/ Then, If/Then Probable Then, and If/Then But Not Precise? Which represents AI and which non-AI computer programs?

17. An expert system knowledge base contains the following statements. Which are rules and which are facts?
    a. ball bearings fail at 400°F minimum
    b. roller bearings fail at 300°F minimum
    c. roller bearings fail at lower temperatures than ball bearings
    d. roller bearings are used for loads over 250 lb
    e. if load is over 250 lb then roller bearings
    f. if load is under 250 lb then ball bearings
    g. failed bearing has load of 175 lb and is ball bearing

18. An expert system learning by chaining sequence is outlined below. What is the new rule Inferred by the sequence shown? Hint: Current rules and facts are compatible with those shown in question 17.

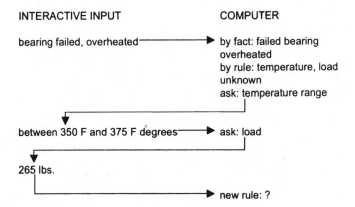

Chapter 14

COMPUTER-AIDED PROCESS PLANNING AND DATA COLLECTION

The subject of this chapter is the CIM activity sometimes known as Computer-Aided Manufacturing (CAM). Chapter 13 dealt with the excellence in communications necessary for achievement of the total CIM potential. In this chapter we will examine what is meant by excellence in communications for the optimal operation of a factory. We will develop the management philosophy behind the integrated factory communications system. The chapters concerning the responsibilities of the various manufacturing engineering units have amply demonstrated the complexity of designing and operating a manufacturing system. We are now ready to explore the use of CAM to significantly improve this communications capability.

HISTORICAL DEVELOPMENT OF COMPUTER-AIDED MANUFACTURING

There have been three very distinct applications of computers in product-producing companies. They are different because they were championed by different organizations in the industrial world.

The first of these applications was scientific computing. Since the computer was developed to solve complex mathematical problems for research scientists, it is natural that the first industrial applications of computers were within the design engineering activity—the closest approximation to the work of research scientists within product-producing companies—and employed modifications of scientific programs to solve design problems. Hence, design engineering had computers before anyone else in industry.

The second application was in the financial function. This again was a natural application. Computers are excellent "number crunchers"; that is, they are good at addition and subtraction and, through programming manipulation, multiplication and division and, of course, other more complex functions. Since finance is basically an operation that evaluates numbers, the attraction of having a superfast method of computing great volumes of figures is understandable. Therefore, finance became a champion of computers for keeping financial accounts accurate and up to date. By the mid-1960s virtually every medium-size and large company and many of the smaller companies had extensive computerized finance sections.

The third application of computers in industry came in the manufacturing area. This was the use of process computers to control machine tools and other manufacturing processing equipment. It was evident that computers could capture the entire method for making a part on a machine, which is an exercise in three-dimensional geometry. If the path of a cutting tool could be mathematically calculated, it would be a relatively simple task to program a computer to direct a controller. The controller activates the drive systems of the machine, and we make a part under computer control. Manufacturing engineering introduced the computer to the factory floor with the advent of numerical control. In Chapter 12 we reviewed the impact of CNC on modern manufacturing.

Until the 1970s computers were used in industry in these three separate, clearly defined ways. Then, as described in Chapter 13, the boundaries became less rigid and we began to consider how the uses of computers could and should be transposed from one function to the others. From finance, manufacturing found that computers could be applied to production control, if production control data could be broken down to numerical values. As we know, these data can be broken down, and we have seen the birth of the computer database that can handle vast quantities of production scheduling and materials purchasing data. This became Manufacturing Resources Planning (MRP II). Manufacturing also borrowed from design engineering, using the interactive graphics tool to define the CNC tool path with the same geometry used to analyze stress and strain. All this borrowing, of course, led to the CIM integration philosophy discussed in Chapter 13.

The one thing that was missing until recently was a way to improve the availability of information characterized by process planning and know what is happening on the factory floor on a real-time basis. These two areas of the manufacturing function's responsibility have no real parallel in the other two major functions that use computers; therefore, they have had to modify techniques borrowed from finance or engineering, or develop new techniques within the manufacturing function. Both of these approaches have been used.

The status of parts completion is of interest to finance so that it can project expenditures of funds and make plans to borrow capital. A time lag in obtaining this information is not a serious concern up to perhaps 1 month. But the basic financial programs already existed within the cost analysis functions. As long as there have been manufacturing firms, the production area has reported completion to the bill paying and collecting area. Therefore, the financial function's computerization of the bill paying and collecting area proved to be of some use to manufacturing. The end result has been a status reporting system developed by finance for finance into which manufacturing inputs data and from which it obtains some limited benefits. The primary benefit is information on how much time it took to complete a part compared to the original planned time, which is used to determine efficiency. For monitoring factory progress, the financial system did not help much.

The only communications programs that were developed entirely by manufacturing were the planning programs. These are an evolutionary development of the scientific time standards based on data related to motion analysis. Motion analysis, or time standards, can be refined to mathematical equations. Therefore, such relationships would be useful in determining how long it should take to make parts and the sequence in which they should be made. If such things can be determined mathematically, computer programs can be written to have the computer do the computations. As we will see, this has been done in computer-aided process planning.

Thus there is a definite historical prerequisite in attempting to automate the communications activities of manufacturing. The desire to automate planning and the desire to know the results of that planning have been evident for many years. With the availability of CIM common databases and excellence in communications this concept is now becoming a mature reality. The computer-aided process planning and real-time data collection systems have been the last major areas within manufacturing to be computerized. Let us see how these cause and effect activities (initiation and reporting of results) so basic to manufacturing success have become useful data bases and what this means to the success of an industrial organization.

EARLY COMPUTER-AIDED PROCESS PLANNING SYSTEMS

Planning is the development of strategy to achieve an aim. It consists of a direct-line approach and contingency plans in case an accident occurs along the way. In manufacturing, this means having a primary set of instructions for shop operations and one or more alternatives if the primary method is not available. Good planners have an excellent knowledge of the parts to be made and the capability of the factory to make those parts.

As one would imagine, contingency planning on a large scale requires either an exceptional memory based on experience or a superb filing system with first-rate indexing. This is especially

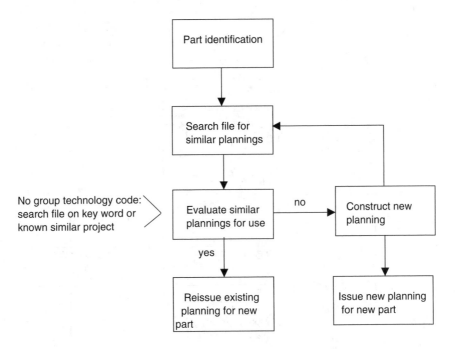

Figure 14-1. Early CAPP Systems Flow

true in a job shop, where part lots being manufactured are seldom identical to previous ones. Therefore, the first idea for Computer-Aided Process Planning (CAPP) was to use the computer as an efficient file cabinet to hold information on the previous experiences for easy recall. No attempt was made to automate planning; the data were simply stored so that the methods engineer could review them and base new planning on similar previous experiences. Figure 14-1 is a flow diagram of the early CAPP systems. It represents a true stand alone system. There is no input from design engineering or any other function, or from any other manufacturing subfunction. The MP&WM unit is entirely on its own. It receives drawings from design engineering and reviews them for information needed to make the plans. The plans are then sent to the master scheduler to fit into the schedule.

Figure 14-1 represents a manual system with computer sorting. A drawing number and/or a key word or phrase identified parts. This meant that in looking for an exact match, the system was limited to finding the exact drawing number issued for an earlier contract, which is not a likely occurrence, or using the key word or phrase to call up the history of drawings with the same nomenclature.

Thus, the methods engineers and planners had a set of parts with similar names that they could review for planning sequences. They could call up each plan and compare it with the new part to be produced. If they had made the part before, it was a simple matter to reissue the plan and the job was done. If they had not made it before, they looked for plans that could be applied to the current part, perhaps with some modification. Finally, if no existing plans fit, the planner had to create new planning to get the job done. In all cases when the job was completed, the hard copy planning was submitted to the master scheduler or production dispatcher to use when requested.

THE CAM CAPP SYSTEM

The system described above is a computer-assisted system but not an integrated system. It uses a computer sort program to find similar plans for further evaluation but the sort is very coarse because there is no way to specify further what is being looked for. The next version of a CAPP system did a much better job of utilizing the computer to do the sorting and evaluating task, but it

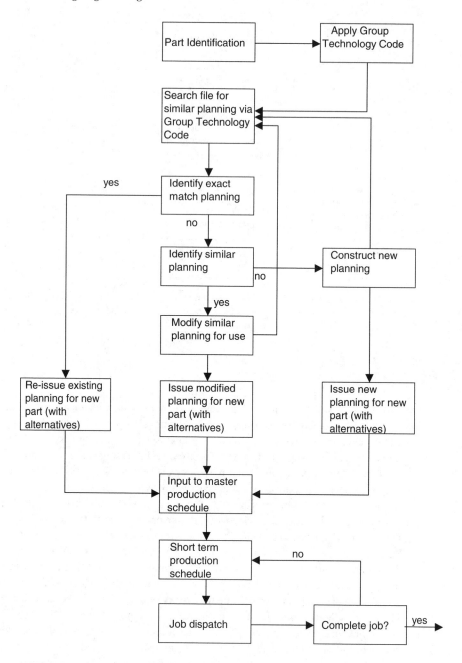

Figure 14-2. CAM CAPP System

was still a stand-alone system. It was still only a CAM system and not CIM. Figure 14-2 represents a typical CAM CAPP system and shows the additional features related to communications and efficiency that it provides.

The CAM CAPP system has three major advances over the early CAPP system:

1. It uses Group Technology (GT) classification coding for finer selection of similar planning.
2. It is integrated with the master production schedule.
3. It provides alternative planning for use in scheduling.

These are significant improvements that we will elaborate on in the following subsections.

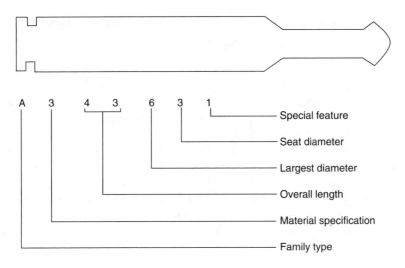

Figure 14-3. Valve Stem Code

Use of Group Technology Classification Coding System

Recall that the early CAPP system used a drawing number to find an exact match and, failing that, used key words or phrases to find similar parts to consider. With a coding system based on the geometry and/or processing sequence, it would be possible to have a prototype planning that could be easily accessed for all similar parts and have the computer select building blocks for the individual characteristics of the new part. Let us review the salient points of a GT classification and coding system. Figure 14-3 shows a very simplified example.

The classification and code number A343631 adequately describes the part called a valve stem. Each digit explains or defines a salient feature of the part. The letter A identifies the family or broadest classification. The first digit specifies the material the part is made of. The second, third, and fourth digits bound its physical size, in this case longest length and largest diameter. The fifth digit identifies the seating diameter. Since A identifies the part as a valve stem of the A family, the next to last digit would specify the seating diameter. If the letter had been a D, the family could have been motor rotor shafts, and the next to last digit could have represented a seal ring groove. Codes can be hierarchical; hence many thousands of parts can be represented via this seven-digit code. (This is a relatively simple code; in practice, some codes have as many as 32 digits.) The final digit in Fig. 14-3 indicates a special feature of the part. In this case the 1 indicates a groove. If the last digit was 0, it might mean that there were no unusual features. Or if it was 6, the part would have a threaded hole off the perpendicular to its longitudinal axis.

Clearly, a code can be a very powerful condensation of significant information about the part. It is certainly more revealing than the key word or phrase search technique used for the early CAPP system. The CAM CAPP system applies the group technology classification code as follows:

1. A part drawing is received from design engineering. The planner, using a code matrix, codes the part.
2. Using the newly created code number, the planner accesses the data base and has it search for a match with other codes. This is started at the most general matching feature, in our example the A for the part family (valve stem), and continued down to the most specific feature, the rightmost digit 1, the special groove. The computer will find the closest match.
3. Once the closest match has been found, the planner again accesses the data base to find the planning for the part or series of parts that are the closest match.
4. Using the closest match as a guide, the planner can make minor adjustments to the previous planning and create new planning. The new planning is then input into the master production

schedule. It is also fed back into the data base as a planning associated with this newly developed group technology code number.

This system has significant advantages over the early CAPP system. It allows a much finer search for similar parts, and it keeps looking until the closest match or matches have been found. This shortens the planner's review of parts considerably, and since the match is as close to an exact match as possible, the system presents as close to what is needed for planning as possible. Hence only fine-tuning modifications will be required. In our example, perhaps the match is good except that the new part is to be 12 in. longer than the closest match from the database. Therefore, by using the planning presented by the system and entering into the machining time formula for the planning, the planner can modify the length of time necessary to do the lathe roughing and finishing cuts.

The early CAPP system required a manual review of all plans presented regardless of how close a match they were. The CAM CAPP system requires a review only when the closest match has been found. The reduction in the time the planner must spend is a significant productivity improvement.

Integration with the Master Production Schedule of the Manufacturing Resources Planning System

The output of the early CAPP system was a hard copy that was mailed or hand delivered to the master scheduler or production control specialist. Usually, a manual system was used to match it with the flow of work orders to the factory floor. When the production control specialist called for that part to be made, he or she would have the dispatcher take the planning from the file and match it with the drawings, then send it to the proper workstation. In the CAM CAPP system, the output from the planning activity is dispatched electronically to the master production schedule, a component of the MRP II system. Essentially, the master production schedule is told that the planning is complete and available. When the schedule says that the part is to be released to the factory floor—that is, material and machine space are available—the computer prints out the planning and associated reporting completion vouchers for shop use. The dispatcher then merges it with the drawings and dispatches them to the starting workstation.

Provision of Alternative Planning for Use in Scheduling

Notice that on the flowchart in Fig. 14-2, the three boxes showing the issuance of planning state in parentheses that alternatives are also issued. This is contingency planning. If the preferred machine is down for maintenance, or a sales schedule need has created a bottleneck on the primary machine, a decision to wait is usually not a viable management choice. Alternative means must be planned to produce the part. An alternative can be planned when it is needed, which tends to be very inefficient, or the plan can exist before any need arises.

With the early CAPP system the capability to do alternative planning was limited. The planners had to spend considerably more time searching files for good matches of existing plans to new parts than with the CAM CAPP system for reasons explained previously. Obviously, before the early CAPP system, the time to do alternative planning was even more severely restricted.

THE INTEGRATED CIM CAPP SYSTEM

A system in which the functions are fully integrated is an integrated CIM system. This system provides the excellence in communications needed to obtain all of the benefits available from CIM implementation. Figure 14-4 defines the flow of the integrated CIM CAPP system.

The Integrated CIM CAPP system is a further improvement over the CAM CAPP system. It realizes the full potential of CAPP for automated information input and retrieval, which is the

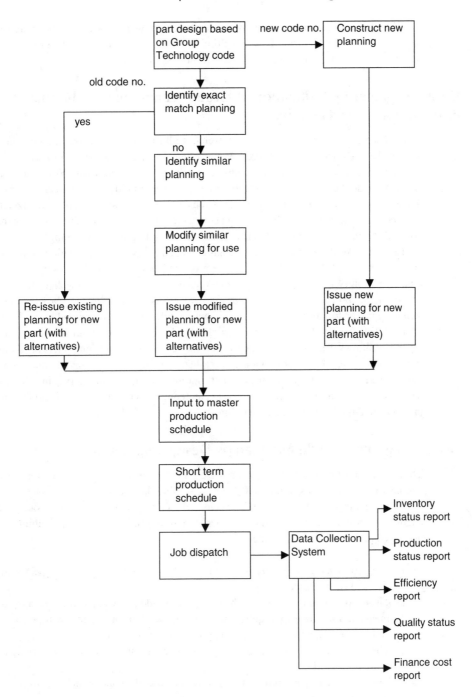

Figure 14-4. Integrated CIM CAPP System

backbone of true CIM. Three significant features available in the integrated CIM CAPP system, which did not exist before but are vital for true CIM, are:

1. Design engineering utilization of group technology to minimize proliferation of parts design.
2. GT code assigned by design engineering.
3. Data collection system used to control factory floor activities.

With these developments the activities of planning how parts are to be made are spread beyond the traditional manufacturing base, resulting in a more integrated manufacturing concern. Let us review in some detail these three new developments.

Design Engineering Utilization of Group Technology to Minimize Proliferation of Parts Design

In our discussions of GT so far we have excluded all functions other than manufacturing engineering. We have shown how the MP&WM personnel apply a code matrix to parts drawings to find existing plans that can be used, or modified to be used, for the new part. Why must manufacturing engineering do the coding? Would there be a greater benefit if design engineering did the classification and coding? The answers to these questions were evident when the potential benefits to design engineering became known. Since design engineering had to code to achieve the benefits, there was no need for manufacturing engineering to repeat the coding for the planning purpose.

Like MP&WM, design engineering has considerable files of information on previous designs, and the number of designs that have to be produced can be reduced by using existing designs in total or as prototypes to support new needs. What has been needed is a mechanism for searching the file to find the proper design to reuse or modify. This is provided by GT classification and coding. Benefits similar to those experienced by manufacturing have been gained by design engineering. Through GT, the absolute number of new designs has been reduced, in some cases by as much as 30%. This allows design engineering to concentrate on existing designs and bring them to a mature state much sooner than was previously possible. This affects all of manufacturing as well, since the reduced volume of designs means that more effort can go into perfecting the manufacturing techniques. In addition, design engineering can reduce its costs because it is producing designs for products at a much higher rate than before.

Group Technology Code Assigned by Design Engineering

This is the unifying aspect of the Integrated CIM CAPP system. By assigning the GT code, design engineering continues its traditional role of designing the product, but also takes on the new role of recommending how the part will be made in the factory. However, we cannot expect the design engineer to become a manufacturing expert. Therefore, this second responsibility is fulfilled as described in the following sequence:

1. A layout of the new project is made by drafting, based on the project specifications detailed by the design engineer.
2. The design engineer evaluates the layout for its adequacy in meeting the customer's requirements and the liability requirement (i.e., checks that the product will perform as intended and will not fail) by performing the proper stress and strain analysis.
3. The layout is reviewed by the design engineers to determine what component parts drawings are needed.
4. The component parts requiring drawings are coded by using the prescribed GT classification code matrix.
5. The GT codes are entered into the database to look for similarities and exact matches for drawings and plans.
6. If exact matches are found, drawing and planning are dispatched electronically to the master production schedule, according to Fig. 14-4.
7. If only similar matches are found, the design engineer evaluates whether they can be used if modified. Plans are referred to manufacturing engineering with the information that modified drawings are being produced that will probably require modified plans.
8. If there is no match, design engineering makes a new design and manufacturing engineering makes a new plan. The results of both functions will be added to the ever-expanding database.

Thus, the basic manufacturing responsibility stays with manufacturing engineering, except that plans that are already good matches need not be referred to MP&WM for further evaluation. One important thing accomplished by such a sharing of databases is that each function gains an appreciation of the other function's work and learns a great deal about that work.

Let us look at the GT codes once more. The code in Fig. 14-3 is very basic. Design engineering or manufacturing engineering can input the code, and the information contained is equally important to both functions. However, there are code portions that are only of interest to one or the other function. Therefore, each function usually retains a capability to input into the code as it sees fit by reserving certain digits for its own use. For example, manufacturing may reserve code space for individual material lots for the purpose of material utilization efficiency, or may use it for special quality control statistical sampling. Similarly, design engineering may wish to use digits of the code to classify the part for different uses of the part, or for allocation to laboratory evaluations. The GT code can be a powerful information gathering and processing tool that far transcends its original use.

Data Collection System Used to Control Factory Floor Activities

With the CAM CAPP system, the electronic portion of the CAPP data information ended with the production scheduling step. Thereafter the system consisted of manual paperwork. The planning was written down step by step, although on a high-speed printer. Along with the planning came a stack of work vouchers that showed the time for each step to be completed and usually left space for the workstation operator to enter the actual time taken.

The computer-aided portion ended with the production of the paperwork and reverted back to pre-computer systems. This was the system put in place by finance to obtain cost data. Data were collected manually, then converted by a keypunch operation into computer input, and finally run in batches once a week to obtain primarily cost data and secondarily manufacturing efficiency data. The problem from the shop manager's viewpoint was that the data were historical in nature and not of much use in correcting current problems.

With the Integrated CIM CAPP system, manufacturing does not depend on the finance system to supply manufacturing data. We can now take CAPP to its logical conclusion and use it to measure the effectiveness of the plan. It is necessary to plan to accomplish anything consistently and necessary to measure to determine whether the plan is working. Prior to the Integrated CIM CAPP system, the information flow back to the master production schedule to measure status based on the plan was severely limited. Therefore, a powerful real-time data collection system was required to make the Integrated CIM CAPP system the communications system needed for the factory of the future. This system had to dispatch information to workstations electronically and receive information feedback in the same way. Such systems have been developed and will be discussed in a later section of this chapter.

MANAGEMENT CONSIDERATIONS, CAPP SYSTEMS

In operating CAPP systems, the manager of MP&WM must first consider CAPP to be different from manual planning, as CNC machine support is different from manual machine support. In the CAPP mode, the common approach would be to modify good planning on file to suit a specific part or assembly. The key word here is "good." If the planning on file is not optimum, we can hardly expect the modified usage of it to be much better. In discussing CNC we made the point that the program fed into the machine's MCU was equivalent to the best, most experienced operator working with a manual machine; therefore, we obtained optimum performance every time the CNC machine was used. We must have the same goal with CAPP. All plans on file must be the best; otherwise, the fact that we have electronically available planning means little.

This best planning is achieved by doing two things: utilizing the most experienced planners in creating the original CAPP database, and establishing standardized ways of performing repeatable operations. By standardized ways we mean a catalogue of methods.

EXAMPLE: If we wanted to paint a garage door, we could immediately start painting with a brush; or we could place drop cloths under the door and in the vicinity, then commence painting with a spray machine; or we could wash the door first, then put down the drop cloths, then tape the door handles and fixtures, then paint with a brush, and so forth. There are many ways of accomplishing the same thing, but usually there are better, less risky ways and inferior ways.

The task in standardization is to identify in the catalogue the overall best ways. In CAPP this is extremely important, because the original data base input can be easily recalled and hence will be used over and over. Therefore, it should be the overall best way of doing the task if the company is to realize optimum productivity. In a manual system a mediocre plan would only affect the specific job it was developed for, but in a CAPP system the same mediocre plan can cause reverberations for a considerable time. For this reason standardization manuals and reference sources are important when developing a CAPP database.

A second problem facing MP&WM management is that of computer paralysis. We find that the senior, more experienced planners in general tend to avoid the computer and therefore do not aggressively pursue CAPP opportunities. This leads management to assign the CAPP data base generation work to the younger, less experienced planners. The result is an even more critical dependence on the standardization manuals and usually some less than optimal planning entering the database. The extent of the problem, of course, depends on the level of training in the MP&WM unit. Computer paralysis can be overcome by judicious use of computer familiarization training and by appealing to the pride of the senior planners. This is psychology, of course, but it is the basis of many facets of good management. It is not necessary to have all senior planners involved in CAPP database development, but there should be enough of them to critically evaluate core plans for practicality and effectiveness. The pragmatic approach is more likely to be taken by the senior planners than the junior planners, and this approach most often results in optimum planning. I believe, however, that computer paralysis as a problem will gradually diminish over time as more people literally grow up with computers, therefore not being in awe of or sceptical of their uses. In fact we are now producing generations of people in the work force who have no recall of times without computers in the work place.

Optimizing techniques such as regression analysis have also been used to ensure that planning is optimal. These techniques are usually cumbersome and the improvements derived are generally not worth the effort. Utilizing the many years of experience of the senior planners is usually the best approach in developing effective CAPP database, ones that are analogous to having captured the techniques of the best manual operator for the CNC machine.

DATA COLLECTION SYSTEMS

With computer-aided process planning we have seen that a unified planning system can be made available to be matched with the master production schedule. The next logical step would be to develop an automated system to report back on how the company is doing in producing products in accordance with the planning. This is what a data collection system does. Let us review the history of such systems, then examine their operations, and finally discuss the management philosophy involved.

As stated earlier, the earliest data collection activities were imposed by finance for cost analysis purposes. Manufacturing told finance when a component part was finished, how many hours it took to complete the task, and what quantities of materials were used. Finance then totaled the cost and compared it to the selling price of the product or component to judge whether a profit had been made. If a loss occurred, manufacturing was asked why. This led to manufacturing asking to see the data collected by finance so that they could analyze the data for cause.

Manufacturing analyzed the data by identifying the high-cost portions of the process and determining the reasons for their cost, after the fact. Bright manufacturing managers then decided that since questions were bound to be asked, they should make use of the data before finance did, or at least concurrently. Hence the planning data were compared with the actual results to determine workplace efficiency and materials usage. As the use of computers grew, so did the availability of data from the finance system. It was inevitable that manufacturing would want real-time data, not batch data at least a week after the fact, in order to optimize operations. This led to the concept of factory data collection.

The goal of the factory data collection system is to develop an ability to determine the current status of all parts going through production. This means that managers should be able to ask a status question at any time and get an up-to-date reply. To satisfy such requests, the system for collecting data must be up to date at all times. The original finance cost accounting system is inefficient for such use. A real-time communications system is required.

In the finance-based data collection system used by manufacturing, paper vouchers for operator recording of status were distributed to the workstation along with the engineering drawings and planning for a particular job. Since a time delay was not considered significantly detrimental, these vouchers were collected at the end of each shift or perhaps weekly, sent to a keypunch operation, and finally fed into the computer 1 week to 10 days after the occurrence. Then the efficiency data were finally made available to the various manufacturing components. Managers knew what the efficiency trends were, and inventory control knew the status of materials usage as of the date of the report. The basic problem that still existed for manufacturing was that this report did not show the progress of each part compared to the schedule. This was partially solved by having many "production chasers" available to follow and report on critical components going through the factory. In fact, this is still the method of control employed in most nonautomated factory operations.

Nothing is wrong with the production chaser method if the factory is making simple components with relatively few parts. Then it is possible to keep track of each part and report status when needed. However, most factories try to fully load their facilities to optimize profitability. Therefore, they have a continuing need to minimize inventory, minimize quality defect problems, and optimize efficiency. They also have too many parts going through at the same time to do all these things manually and do them well. Excellence in communication, which in this case means excellence in handling data, is needed, hence the need for a computerized data collection system as part of the CIM common database.

A modern factory data collection system must overcome the deficiencies of the finance-based data collection format. A way must be developed to record completions and starts of work as they occur and to constantly compare status with the plan. Also, it must be possible to access this constantly updated status database. Therefore, the development of factory data collection systems has been paced by the development of communications terminals suitable for the factory floor.

Terminals have been developed solely as communication devices and they have also been developed as part of CNC MCUs. Regardless of the type, the terminals must be able to easily instruct the operator and to let the operator ask clarifying questions and report status. Along with communications devices, dedicated processing computers had to be developed that were small enough to be efficiently dedicated to the data collection task, but large enough to contain sufficient data. User-friendly software also had to be developed.

In the case of factory data collection, the majority of data inputs via computer terminals are made by people who are not trained in computers. Those making the inputs have virtually no programming skills, nor could the company ever afford to provide such skills. This means that data collection systems must have built into their prime programming all possible contingencies. They must be extremely user friendly. The same characteristic is required for those who access the database to learn current status. The purpose of the data collection system would be thwarted if a computer specialist were required to access the database for the manager of manufacturing.

Now let us look at the flowchart of a generalized factory data collection system so that we can understand the various functions carried out within such systems, and where they interrelate with other communications/decision-making systems.

Figure 14-5 represents a generic factory data collection system. Note the predominance of workstation operator interaction with the computer system. The benefit of this interaction is that

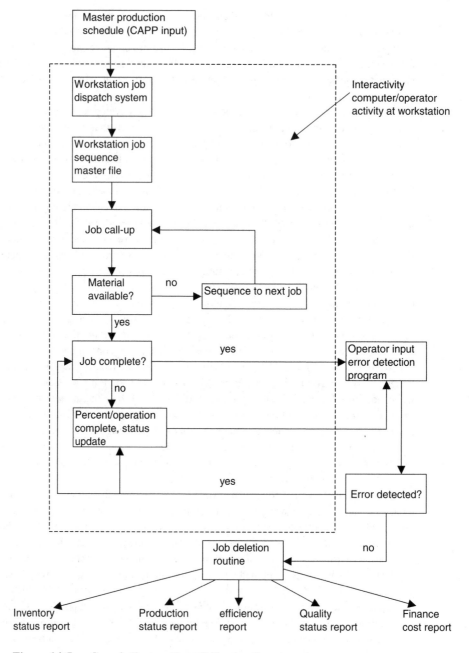

Figure 14-5. Generic Factory Data Collection System

the operator becomes more knowledgeable about the master production schedule and the successes and failures of manufacturing. This familiarity leads to acceptance of greater responsibility, hence improved productivity.

Note that the system allows for errors to be made and corrected by the operator, recognizing that factory floor operators are not computer operators. The error correction routine prevents errors from getting into the database. Typical of errors searched out by this program are:

- Operator identification not on file.
- Start day shown later than finish day.

- Time interval between start and finish unrealistic.
- Start time missing.
- Stop time missing.
- Quality defect data missing, if quality-related input.
- Drawing number missing.
- Part number missing.
- Operation sequence missing.
- Percent complete missing.
- Workstation number missing or incorrect.
- Operation exceeds 100% claimed.
- Labor category missing.

This, of course, is not an all-inclusive listing. A complete listing would depend on the specific factory and types of products being produced. The important fact is that the system is user friendly. If the operator makes an input error, the program states what it is and the operator can easily correct it.

Finally, note the job deletion routine. This is a series of programs that update the status database continuously as input is received. The output is sent to the master production schedule to update the main production control database. The output also goes to a host of information programs, all on a real-time basis. Therefore, all personnel who need the information can access these programs to determine the status of interest to them. The data collection system is a real-time communications system that serves a wide variety of users.

Data collection systems can be very useful for management control of operations. However, there are certain operating considerations that must be complied with. Let us discuss the major ones from a management viewpoint.

A data collection system is only as good as the system that supports it. Therefore, care must be taken that both the hardware and software are up to the rigors of a factory floor environment. If the equipment is not hardened, and continually breaks down because of such things as factory dust, the system will not be in use. If it is not in use a considerable part of the time, then to make its output believable, considerable time will have to be expended in manually feeding in data to keep the system current. A data collection system is worthless if it is not collecting data. Therefore, management must keep the servicing factor in mind when the decision is being made to implement such a system. This should not deter implementation of such systems, but should be recognized as an expense factor in running the system.

Another important factor for management to consider is the increased need for factory discipline to make data collection systems work. By this is meant the way and order in which work is accomplished. Before data collection systems, it was common for foremen to tell the operators to do work out of sequence if it made sense to do so—that is, to do operation 6 before operations 4 and 5 if the earlier operations could not be accomplished because of machine breakdown, unavailability of special tools, and so forth. Of course, these instructions would be given only if a subsequent operation would not unalterably change the ability to do the previous operations. With a data collection system this type of variation becomes more difficult to tolerate. If the system does not know that a step has been skipped it is likely to assume that the step has been completed, thus forever missing the operation and resulting in a quality defect later on. The need for communication discipline is essential. Supervisors must understand that the lack of freedom to deviate from planning is more than compensated for by the overall greater efficiency made possible by the data collection system.

Another aspect of the processing discipline involves the planners feeding the CAPP system. For data collection to be meaningful, the planning sequences must be realistic. The plan cannot be a document that is ignored by the shop floor personnel because it is in serious variance with how things really are accomplished. One of the benefits of CAPP and data collections systems is that they tend to weed out irrelevant and erroneous plans, which might never have been corrected when manual control was the operating mode.

SUMMARY

With CAPP and data collection systems implemented on the factory floor, we have a communication/decision-making system that permits the use of CIM.

In this chapter we have discussed the history of CAPP and data collection systems and have seen the extent of the development that has taken place. Each evolutionary step has brought us closer to the goal of real-time control of the manufacturing process. We have developed an understanding of what is meant by excellence in communication and why it is a central theme in progressing toward the "factory of the future." Each step of this process has been linked to improved productivity—the only reason for implementing CAPP and data collection systems.

We have also discussed the concept of group technology coding and have seen how coding makes possible easy retrieval of historical data. These data can be used to establish new plans. Another important development due to GT code use is that of integrating the design engineering tasks with the responsibilities of the planner. The idea that a function other than manufacturing could do planning work was introduced, and we saw our first practical example of an integrated function. This integrated approach can come about only through the use of CIM, portending significant productivity improvement opportunities for the "factory of the future."

REVIEW QUESTIONS

1. Explain why a time lag in gathering part status data during manufacture is acceptable to the finance function but unacceptable to the manufacturing function.
2. What are the two communications/decision-making computer systems initiated by the manufacturing function? How are their origins alike and how are they different?
3. Referring to Fig. 14-1, discuss the adequacy of the "key word" search function compared to a group technology classification and coding system.
4. What is the difference between a computer-assisted system and a computer-integrated system in relation to process planning?
5. In Fig. 14-3, the overall length is 43 in., the largest diameter is 6 in., and the part has a groove. Suppose we have a similar valve stem with an overall length of 25 in. and the largest diameter is 3 in. This similar valve stem does not have a groove or any other special feature. Using the code described by Fig. 14-3, create the group technology code number for the similar valve stem.
6. Why is it preferable to have alternative planning available when developing a master production schedule?
7. Describe how a computer-automated data collection system improves productivity.
8. List eight uses of GT classification coding: three for design engineering, three for manufacturing, and two for finance.
9. Discuss the differences in management concepts between manual planning and computer-aided process planning.
10. What do "real time" and "batch processing" mean with respect to data collection systems? Which is most beneficial to manufacturing? Explain why.
11. Discuss the interface between the operator and the data collection system. What major items must be dealt with to achieve a successful system?
12. What are the constraints that a data collection system imposes on management? Explain your answers.

Chapter *15*

THE GROUP TECHNOLOGY BASIS FOR PLANT LAYOUT

In Chapter 14 we discussed the use of the classification and coding schemes in design and planning. However, there is much more to Group Technology (GT) than coding for data base retrieval. Group technology concepts can be a powerful tool in plant layout, especially for manufacturers of miscellaneous parts. In this chapter we will describe in pragmatic terms how to justify and proceed with implementing a group technology layout. We will discuss the management concerns that must be successfully and affirmatively answered in order to proceed with such a layout, and we will demonstrate the differences between a traditional layout for job shops and a layout based on the principle of sameness utilized by group technology.

While currently popular terminology such as lean manufacturing will be used rarely in our discussions of this topic. True lean manufacturing the optimization of manufacturing through elimination of waste combined with applications of statistical process control—is virtually impossible without some form of group technology layout principles. This will become quite evident as we go through the subject matter of this chapter.

THE PRINCIPLE OF SAMENESS

Mass production or flow-type manufacturing offers the greatest economies of scale. If we are involved in a business that mass produces its products, and its products are virtually identical, we have the opportunity to employ special-purpose machines optimized to produce the product. However, mass production accounts for only 30% of all manufacturing. The remaining 70% of manufacturing is accomplished in job shops; where it is done on relatively general-purpose machine tools, the cost of optimizing is usually uneconomic, and many unproductive compromises must be made. Therefore, much thought has been given to finding ways to make job shop techniques approach those of flow manufacturing. An approach based on the principle of sameness, the underlying theme of group technology, does just that. All coding, classification, production routing, batching, and layout techniques based on group technology employ the principle of sameness as their theoretical starting point. The principle of sameness refers to the grouping of parts to be manufactured according to similar characteristics, either geometric or processing, so that they can be produced on the same equipment in a batch mode.

Consider the universe of shapes used for an internal combustion engine. There are flat diametric shapes such as piston rings and gaskets; long dowel-like shapes such as piston rods and hold-down studs; cylindrical shapes such as pistons, cylinder liners, and carburetor bowls; flat milled shapes such as head gaskets and flapper carburetor valves; and so forth. Thus, there are many dissimilar parts that share a common geometry. We can also find numerous examples of dissimilar parts that do not have similar shapes, but are consistently manufactured by similar techniques. For example, the crankshaft of an automobile engine requires lathe work followed by grinding; so does a steam valve stem (Fig. 14.3). These parts have dissimilar geometries, but their processing steps are similar and they are probably both constructed of similar steel alloy material.

We have tabulated like characteristics of dissimilar parts because we would like to approach the economies of scale enjoyed in flow manufacturing. For example, if we can find enough parts that require a tapped hole 5 in. in diameter, we can prepare a drill press to do nothing more than make such holes in a flow production manner. Of course, we rarely have products that require only one operation, so our ability to find sameness or commonality will always be constrained. Nevertheless, every time we can make use of a setup more than once, we are saving money and becoming more productive. The principle of sameness encourages us to look for more than one use per setup, and it starts us thinking about other areas where grouping can be beneficial. We will see some of these benefits emerging as we investigate the use of group technology for layout purposes.

GROUP TECHNOLOGY PLANT LAYOUT CONCEPT

According to group technology theory, the plant layout should take advantage of the sameness of parts being produced. This is in conflict with the common practice in job shops of basing the plant layout on machine tools and processes.

Figure 15-1 is a block diagram of a typical job shop layout. With this type of layout we dispatch jobs to each area where work must be performed in the sequence dictated by process planning. Therefore, the number of possible combinations or different sequences going through the factory is 10 factional. In practice, this means that we have a traffic maze and that each part travels a great distance in its journey from raw material to finished product. It means that we need a great effort to load and control each station effectively, and that we will have a larger inventory of work in process. Finally, it means that a new setup is likely for each piece that crosses each workstation. Quite bluntly, the classical job shop workplace layout seems to have been designed for grouped maintenance rather than efficient production. Figure 15-2 shows what this flow is like with only 10 jobs. One can imagine what it would be like with closer to 10 factorial jobs.

A great deal can be done to improve the flows illustrated in Fig. 15-2. We can look for sameness in geometry or production equipment usage and then construct cells of dedicated equipment to produce the similar parts. These are not necessarily parts that are similar in end usage, but ones that are similar in geometry or in how they are manufactured. We are looking for ways to minimize the learning curve. Clearly, if it is necessary to recalibrate, rethink, and retool for each job, it is difficult to achieve expertise in making any of the parts that cross the workstation. On the other hand, if there is a high degree of similarity in operations to make a variety of different parts, we will become expert in making those parts.

If we can find sameness in a sufficient number of parts, we can afford to segregate particular facilities to make those parts. Replace "sameness" by "family of," and we have the common expression for group technology, the family of parts assigned to a common manufacturing cell. The manufacturing cell may consist of one complex CNC machining center, or many manual machines, or combinations of both. The common denominator is that each cell is dedicated to making one family of parts or perhaps a few adjacent families of parts. Figure 15-3 depicts a layout based on the GT cellular manufacturing principle.

Figure 15-3 shows that different facilities are mixed together to form the individual cells. Since the machine tools in each cell are dedicated to making the family of parts assigned to the cell, each

1 Drills	2 Planner Mills	3 Inspection Station	4 Saws	5 Shipping and Receiving
6 Lathes	7 Horizontal Boring Mills	8 Vertical Boring Mills	9 Welding	10 Heat Treat

Figure 15-1. Block Diagram, Job Shop Layout

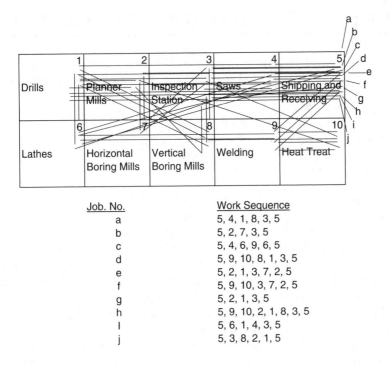

Job. No.	Work Sequence
a	5, 4, 1, 8, 3, 5
b	5, 2, 7, 3, 5
c	5, 4, 6, 9, 6, 5
d	5, 9, 10, 8, 1, 3, 5
e	5, 2, 1, 3, 7, 2, 5
f	5, 9, 10, 3, 7, 2, 5
g	5, 2, 1, 3, 5
h	5, 9, 10, 2, 1, 8, 3, 5
I	5, 6, 1, 4, 3, 5
j	5, 3, 8, 2, 1, 5

Figure 15-2. Job Shop Classical Layout, Showing 10-Job Flow

tool can be optimized to reduce the setup time—in fact, to approach having a universal setup for each machine to match the family of parts it is dedicated to. This, of course, allows significantly reduced throughput times and therefore productivity gains.

The objection is sometimes raised that job shops cannot be set up along group technology manufacturing lines because general-purpose job shops do not know what they will be called upon to produce. However, this is not correct because job shops tend to specialize in certain areas, based on the type of equipment they own. Furthermore, virtually all parts can be broken down by size and shape characteristics, and therefore can be broadly categorized into families of parts even before they exist for job shop production.

Therefore, the question is not whether a group technology cellular layout is beneficial, but how one achieves such a layout. We will review the methodology of doing that, but first we must consider the economics of group technology cellular layouts. This is necessary in order to understand what is theoretically possible; which modified by realities will result in what is practically possible. As manufacturing engineers, we are interested in the practical, not the theoretical.

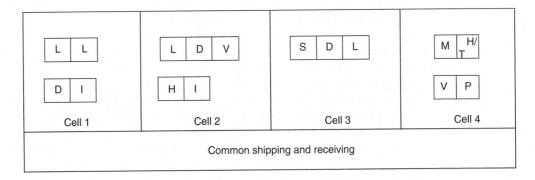

Figure 15-3. Group Technology Cellular Manufacturing Layout

THE ECONOMICS OF GROUP TECHNOLOGY CELLULAR LAYOUT

The benefits of group technology are many and varied. Most practitioners of group technology would list the following as reasons for adoption of the group technology cellular layout philosophy:

1. Reduces setup time
2. Improves direct labor productivity
3. Reduces indirect jobs such as production expediting
4. Reduces manufacturing losses
5. Reduces throughput time

The range of improvements varies, but my experience with such layouts indicates that a 15 to 30% overall improvement is possible. To know exactly what the improvements are, a company needs good historical records against which to measure the accomplishments. Most companies do not maintain sufficient records to obtain precise measurements of improvement. In fact, operations that are in trouble necessitating a change usually have the worst records. This means that changes are often based on experiences of other organizations, which involves some risk. Different factories with different management emphasis may or may not achieve improvements in the range indicated above. In view of this variability, it is mandatory that the feasibility of savings be evaluated early in the project and as pragmatically as possible.

Achieving a group technology cellular layout costs considerable sums of money. Therefore, before venturing into the project, the costs must be carefully evaluated. A cellular layout is not the only means of achieving the benefits of group technology. As a completely cellular layout is approached the potential for gains approaches an optimum. Therefore, every step on the path of implementing group technology will bring incrementally greater benefits. For example, in Chapter 14 we discussed the benefits of coding and classifying but did not mention cellular layouts. These are a good example of the incremental benefits achievable. If we continue and tie coding and classifying to the cellular concept, we can achieve even greater benefits.

In considering a group technology cellular layout, a company must evaluate the feasibility of successfully forming families of parts. Although cellular layouts can be made to work for virtually all job shops, some companies will be able to do this more easily than others. To be able to justify the costs involved to their superiors, those interested in forming a cellular manufacturing layout should be able to make the following statements:

1. The factory is making a wide variety of parts (not necessarily product lines).
2. The parts are repeated over and over again with usually minor and occasionally major variations.

If these two criteria are met, the job shop is well suited to a GT cellular layout. There will be a large enough sample size to look for geometric or processing sameness, and there will be enough repeatability to justify family-of-parts universal fixtures. Whether the parts are presently made in small lot sizes is unimportant in deciding for or against a GT cellular layout.

In addition to the above considerations, there must be a strong economic incentive. If the traditional layout is considered satisfactory in terms of costs and profits by the senior management, they will not be willing to make a change. Since group technology benefits are described in terms of ranges, they are difficult to justify to senior management, especially since larger amounts of money must be spent to convert to the new layout. The fact that savings ranges and averages are difficult to enter into a meaningful cost evaluation means that GT cellular layout projects must often be portrayed as a secondary reason for making a layout change.

Let us examine the group technology benefits and the need for a strong economic reason in terms of a hypothetical miscellaneous parts manufacturing area. We assume that there are 50 major machine tools, lathes, vertical boring mills, horizontal boring mills, shapers, saws, welders, and so forth. Each of these is to be relocated in accordance with a proven group technology principle in order to form manufacturing cells. Suppose that it costs $12,000.00 on average to remove and reinstall one major machine tool. This includes rerouting all services (e.g., air, water, AC and DC electricity, and telephone) and a prorating of rearrangement of supporting areas such as inspection, material storage areas, and project engineering costs. This means that for the hypothetical 50-workstation miscellaneous parts manufacturing area, a cost of $600,000.00 would be incurred. Even assuming a conservative 10% reduction in the need for equipment with the new layout, the facilities move would still cost $540,000.00.

Now let us look at benefits to be achieved by group technology alone. Assume that each workstation can produce four parts per 8-hr shift and that each completed part averages five operations. Four completed parts per set of five workstations, or ten sets of workstations at four completed parts per shift, equals 40 parts per shift for the area. This makes 600 completed parts per week if there are three shifts per day, or 28,000 parts for a nominal 48-week work year. At a nominal $10.00 per hour labor cost, each part costs

$$[(8 \text{ hr/shift})/(4 \text{ parts/shift})] \ (\$10/\text{hr})(5 \text{ operations}) = \$100/\text{part}$$

The cost is thus $2,880,000.00 per year, excluding overhead. The next thing we must do is analyze the possible savings, using the five savings categories attributed to group technology.

1. *Reduces setup time.* We will use a 15 to 30% range of possible savings, as discussed earlier in this section. Average time on each machine for our example is 2.0 hr. Setup time is normally 15% of the total machine cycle time; hence it is 0.30 hr per machine or 1.50 hr per part. Assuming a full 30% reduction, we get 0.45 hr saved, or $4.50 per part. Realistically, one would plan conservatively, so the savings would be at the 15% reduction level, or $2.25 per part. For the remainder of the example I will use the low end of the range to allow for the numerous unknowns and intangibles in a project such as this.

2. *Improves direct labor productivity.* This, of course, must include setup. In order not to double-count, we will subtract the setup time from the total cycle time. Therefore, the theoretical improvement associated with this job enrichment factor is:

$$(5 \text{ operations/part})(2.0 \text{ hr} - 0.3 \text{ hr})(0.15/\text{operations}) = 1.275 \text{ hr/part}$$

At a direct labor cost of $10 per hour this equals $12.75 per part. In my experience, this should be discounted by at least 80%. This productivity improvement comes about because the operator is making similar parts or performing similar operations on dissimilar parts, but parts in the prescribed group technology family. Therefore, the operator is familiar with them and does not have to progress through a learning curve. My observed 15 to 30% savings are for the total group technology process, not for each individual phase. I believe that the group technology benefits for this phase are exaggerated. In the traditional job shop layout, there is considerably more variety of work going through a workstation than in a GT cellular layout. Therefore, the job enrichment often attributed to an operator in a GT cellular layout situation seeing completed parts being made in the operator's cell must be discounted. The learning curve benefit must also be discounted. In practice, most job shops route similar parts to the same workstation, because the foreman usually knows the skill levels of the operators and assigns jobs accordingly. Also to be considered in discounting this particular saving attributed to group technology is the novelty factor. Many theoreticians believe that there is a long-term benefit in manufacturing completed parts in a specific cell. However, my observations indicate that such benefits diminish after the novelty disappears, which happens in approximately 3 months. Therefore, the calculated saving of $12.75 per part will really be in the neighborhood of $2.55 per part for improved direct labor productivity.

3. *Reduces indirect jobs such as production expediting.* Good managers will do this as a matter of course. Successful operations must be as efficient as possible; therefore, there is a constant effort to avoid using excess people. It is difficult to establish a layout as a proven reason for reductions in support personnel. Even though group technology cellular layouts are compatible with personnel reductions, it is difficult to show that this is not a saving that can be achieved through good management. Hence, in practice this saving cannot be proved to exist.

4. *Reduces manufacturing losses.* Making a completed part in a cell where raw stock is converted to a finished part will reduce losses caused by indifference. Some advocates of group technology claim that a 70% reduction of losses will occur. However, I believe that the novelty factor once more comes into play here, and based on a 15 to 30% savings attributable to group technology overall, I would choose a 20% reduction for this part of the calculation. Assuming a traditional 5% manufacturing loss budget, or $5.00 of cost per part associated with losses, we can save 20% of that, or $1.00 per part.

5. *Reduces throughput time.* This appears to be the only aspect of a group technology cellular layout for which a case can be made for tangible savings, the type of savings that manufacturing engineers would be able to place on a plant appropriation request. Therefore, we will assume a 20% reduction. This saving is due to reducing the distance travelled from workstation to workstation by having the succeeding workstations adjacent to each other. This is the core of the cellular concept.

In order not to double-count, we must subtract setup time, machine operating time, and queue time at the workstation from the total cycle and deduct 20% of the remaining time, the transit time, as the saving. Our example analysis shows the setup time to be 0.30 hr. per workstation. The remaining 1.70 hours per part at each workstation is machining and waiting time. Since we lack firm data on the travel time between workstations, which is the case for most factories, we must make an assumption about what part of the 1.70 hours is machining and waiting time and attribute the remainder to travel time. Based on experience, I have chosen 20%.

This assumption of a percentage travel time keeps this category in the intangible area vis-à-vis plant appropriation requests. Some manufacturing engineers have dealt with this problem by commissioning studies of times for moving material from station to station. The studies usually employ timers and reminders to the operators to supply accurate data. Since there is no way to check the accuracy of such data, long periods of data collection are usually substituted for verifiability. There is always some uncertainty about the accuracy of the results.

In our example, assuming that we have a 20% material transport factor and 80% machining and waiting time, the reduction in throughput time per workstation is:

$$[1.70 \text{ hr/part} - (1.70 \text{ hr/part})(0.80)](0.20) = 0.068 \text{ hr/part}$$

Since there are five workstation operations per part, we have a throughput saving of 0.340 hr. per part or $3.40 per part.

This is the most significant factor in the cost reduction calculation, as one would expect. A review of Figs. 15-2 and 15-3 indicates that a positive gain should be made by simplifying the flow path of the parts. The group technology cellular layout allows simple flows, while the traditional job shop layout does not.

Summing the savings attributed to the group technology concept on per-part basis, we have

1. Reduces setup time:		$2.25
2. Improves direct labor productivity:		$2.55
3. Reduces indirect jobs:		$0.00
4. Reduces manufacturing losses:		$1.00
5. Reduces throughput time:		$3.40
	Total	$9.20

At 28,800 parts per year, this corresponds to a $264,960.00 saving in operating expenses less overhead. This is a significant amount of money, equal to 9.2% of the original labor cost for the 28,800 parts.

Would senior management fund a project to rearrange a 50-machine miscellaneous parts area into a group technology cellular layout? Probably not. The reason senior management would give for rejecting such a project is that these are "soft" savings—that is, they are subjective. They are based on average results from other factories, and it is always difficult to determine how close a factory is to the average. Would manufacturing engineering offer such a project for approval? Let us consider the facts.

First, we would be asking for an investment to spend $540,000.00 to save $0.00 (firm) and $264,960.00 (soft or variable).

Second, we could state that savings are on the conservative side and that statistical knowledge tells us that the variable portion would be higher, but only if we do indeed have an average factory.

Third, we must also consider disruption. While the factory is being rearranged it is still necessary to produce products. This will have to be done at a cost greater than the current cost; if it were not, we would always produce by the alternative method. Let us look at these disruption costs in greater detail.

In our example we are contemplating the rearrangement of 50 machines or workstations into a cellular layout. We cannot afford to take all 50 out of service at the same time. Nor can we afford to do only one workstation at a time; using a 2-week rearrangement cycle time per machine, taking out only one machine at a time would result in a nominal 2 years to complete the project. Usually 3 to 4 months is the accepted cycle time for this type of project, and to extend it any longer would create an unacceptable cash flow. We can calculate disruption costs in two ways:

1. Calculate the cycle time per workstation, then add 2 weeks for each machine to be out of production. This is a tedious procedure and most companies cannot afford to support such detailed planning.
2. Use an experience-based factor to calculate disruption costs. My experience shows that for every 3 months of rearrangement project time, there will be 1 month of non-production. This means that the affected factory area will not produce during 1 month for every 3 months the area is undergoing rearrangement.

Since companies normally cannot tolerate a 1-month hiatus in production, the parts must be made somewhere else. This usually means subcontracting to another factory, which in this case means additional costs of about 25% over normal operating costs. Therefore, the disruption costs of our example are:

$$(2,400 \text{ parts/month})(\$100/\text{part})(0.25) = \$60,000.00$$

We can now show the savings and costs of the proposed group technology cellular layout rearrangement project. The costs are:

$540,000.00	facility rearrangement
$ 60,000.00	disruption cost
$600,000.00	total costs

and the savings:

| $0.00 | firm |
| $264,960.00 | subjective (variable) |

Therefore, manufacturing engineering would not present this project in its current form for approval. It would be asking senior management to invest $600,000.00 to save a subjective $264,960.00, and since the savings are subjective, it would be easy to fault the logic for each of the four segments accounting for the savings.

Those who are proponents of group technology, and I count myself as one of those, have wrestled with this approval of project problem for a long time. The classification and coding portion of group technology fits in very well with integrated CIM and therefore can be financially justified with little difficulty. The classification and coding techniques can be used to form part families and lead to the conclusion that cellular layout is the way to proceed. However, cellular layout is very difficult to sell, as shown above. Therefore, group technology cellular layout projects must be piggybacked onto other projects showing firm or hard savings.

What are firm savings associated with rearrangement? Simply stated, firm savings are guaranteed savings. An example is energy savings. If equipment is moved into a tighter layout, less lighting equipment is required; therefore less money is spent on fixture maintenance and the power bill is reduced. We have energy savings. Another example of firm savings involves span of control. By grouping workstations closer together, less supervision (fewer foremen) is required to cover the area. The span of control is over a smaller area. Span-of-control savings always occur when multi-floor layouts are converted to one-floor layouts. A third example is cost of factory space. By consolidating operations into fewer buildings, the costs of maintaining, renting, heating, lighting, and so forth are reduced. In each of these three examples of guaranteed savings, the saving is calculable and can be readily verified. The cost of energy and the number of lights eliminated are known. The span of control is known, so it is possible to calculate the reduction in supervision and the saving in salaries. The cost of occupancy per square foot is known; therefore, space vacated and no longer paid for becomes a firm saving. In all these examples, it is evident that if the company takes the required actions it will gain the stated savings with a high degree of confidence. With subjective savings there are no direct cause-and-effect relations, hence the savings may or may not come about. That is why they are called intangibles and why they are difficult to sell to senior management.

We can see that tying a subjective savings project to a firm savings project is the only way to obtain approval for the subjective project. Of course, the firm savings project must be a good one and its savings must somewhat cover the additional costs of the subjective project. Usually, if a firm project has a payback of 2 years or less, senior management will be willing to extend the payback period up to 25% to accomplish good potential subjective projects. Group technology cellular layout projects should fall into this category. This is especially true if the firm project is a facilities layout and not a new equipment project. Area consolidations, revamping product scopes, and setting up satellite manufacturing locations are examples of firm projects that group technology cellular layout projects can be tied to successfully.

Group technology cellular layout projects will normally not pass a critical financial review. Therefore, to achieve a cellular layout it is necessary to combine it with a firm savings project. Once the cellular layout project has been implemented and is successful, future GT cellular layout projects become easier to sell.

Some proponents of lean manufacturing state that the rearrangement of a factory into cells can be done very quickly. Thus, low cost and hence a nonissue for gaining approval to do so. I wish that were the case. Low cost and quick may be true for factories that have very simple or relatively portable equipment, but it is certainly not true for the majority of job shops who use relatively sophisticated equipment. I have seen advertisements for consulting proposals where lean manufacturing experts claim to be able to re-arrange a factory over a weekend. Be wary of strangers bearing gifts. The truth is complexity breeds confusion and need for specifics oriented project management. Group technology cellular projects are rarely simple to achieve.

TECHNIQUES FOR SELECTING FAMILIES OF PARTS

To form a group technology cellular layout, it is necessary to form families of parts and then determine what processes in what sequence are used to produce the parts in the family. This leads naturally to placement of machines in the cell, location of the inspection facilities, placement of entry and exit to the cell for materials and finished parts, and placement of the support facilities. The actual layout procedure is the same as that used in making a traditional layout: location of the principals and auxiliaries, flow parameters, access parameters, support facilities parameters, and so forth. We have discussed the techniques of making layouts in Chapter 5, and numerous supplemental readings can be found in good engineering libraries. The only new task we must examine is that of forming part families. Part families can be determined by three methods: (1) empirical decision, (2) classification code, or (3) part flow analysis. The method selected will depend on the needs of the individual business. The complexity of the business problem must be analyzed, and this analysis will determine the depth in which the company wants to become involved with group technology.

As stated before, the amount of group technology implemented is directly related to the amount of benefits to be received. Unfortunately, there is another ratio to consider: the more group technology implemented, the higher the implementing costs. Therefore, it is necessary for companies to know their own businesses and the state of the competition. Group technology, like other good cost-saving opportunities, must share in the total dollar pool available for technological improvements. Group technology projects, as demonstrated in the preceding section, are not always sure prospects for investment funds. Therefore, the business analysis must take into account the current performance of the business, the market share, the competition, and, most important, the profit level and potential, before any projects are approved. For example, a company may not have the funds to institute a full-scale group technology project, but may wish to have some hands-on experience so that it can expand into fuller implementation when funds become available. This company would probably choose a limited-area cellular layout and use part flow analysis in place of code to establish the part families. This would certainly be the case if their competition were at the same group technology implementation level. It would be a limited venture into group technology, costing limited funds, but only resulting in limited gains. Each company must evaluate its own needs and make its own decisions concerning group technology implementation. The good thing about this situation is that many levels of implementation can be commissioned, and virtually every implementation has an excellent chance of improving on the current status of operations.

Now let us explore the three methods of forming part families.

Empirical Decision

This is the simplest of the three methods. Experienced planners decide which parts are similar either by shape or by operation sequence and thus create part families. This technique works for factories with minimal frequencies of part types and well-entrenched expertise—for example, long memories for parts made before but not seen in the production schedule for a long time. This technique cannot work for radically new parts, and it depends on the experts to place parts in the proper family.

Classification Code

The classification code is the most complex and most beneficial of the methods. The context of a classification code was described in Chapter 14 for use with a computer-aided process planning and data collection system, where the code is a vital part of a successful data base communications system. Besides forming families of parts for cellular manufacturing, a classification code allows access to many other benefits of group technology such as Manufacturing Resource Planning (MRP II) and those mentioned in Chapter 14: CAPP, design analysis, and data collection. In the case of cellular manufacturing, the classification code yields either geometric similarities or processing similarities, which are easily used to create accurate part families for existing parts, and to assign new parts to existing families or new families.

The classification code method is the most expensive and time-consuming method in terms of personnel and cycle time to implement it. To use the classification code method, a company must first either invest in purchasing an existing code or develop its own code. It must then code all of its parts, or at least the active ones. This requires training analysts and then analyzing each part drawing. The classification code must also be customized to meet the specific needs of the particular factory. For a purchased code this can mean additional expense, depending on how well the commercially available code matches the factory's needs.

A group technology classification code will determine part families to suit the needs of the specific factory very well, thus allowing achievement of the cellular layout. The problem with classification codes is that in many cases they are unnecessarily sophisticated and "overkill," unless management is totally committed to a group technology philosophy and is implementing CAPP and data collection as part of the company's communications system.

An advantage of cellular layouts is that they allow management to derive many benefits of group technology without a major commitment. Of course, the cellular layout must be such that a firm savings can be shown. Requiring that a classification code be used to define part families could easily add another $200,000.00 to the cost in the previous example. This would certainly ensure a negative decision by senior management.

Part Flow Analysis

It is possible to create the part families for a cellular layout without a classification code and with a higher degree of objectivity than in the empirical decision method. Part flow analysis is the middle-of-the-road approach. It develops logical part families based on a planning sequence matrix, and thus has more credibility than a panel of experts making educated guesses. It also helps cellular layouts to work by routing families of parts defined by part flow analysis to the proper manufacturing cells. However, it does not afford the company any further gains in overall productivity because it is not a classification code and therefore does not readily tie into other company systems.

The parts flow analysis technique is simple in concept. We review the planning sequence sheet for a part and list in chronological order the operations necessary to produce the part. After doing enough of these reviews, we can develop a matrix of parts versus operations; from which we derive part families based on similar operation sequences. If this first iteration develops too many families, we then consider as a part family those parts having the same type of operations but not necessarily in the same order. This combining of families is necessary because too many families can result in many underutilized manufacturing cells. Experience shows that a second iteration is usually needed to develop a manageable number of part families.

Considering our example, 28,800 parts in 1 year would require analyzing approximately 120 plans a day just to finish the job in 1 year. Even if the company had the manpower to do this job with its permanent staff, it could not afford to do so. To get the job done, creative management must take over. We must look for a temporary staff that can read a planning sheet and make entries onto a matrix form; then the actual analysis can be done by the permanent MP&WM staff. The solution, then, is to hire temporary clerical staff and quickly train them to review plans and record data on the appropriate matrix forms. However, even with a staff of temporaries to do the work, the task as stated in our example is enormous; 120 plans per day would require about 10 people doing this for an entire year. Something must be done to shorten the cycle, and that is statistical sampling. Most job shops would recognize a repetitious pattern in the work they do. Therefore, they should be able to define a cycle time after which they start to repeat similar jobs. For most miscellaneous parts job shops, where a cellular layout would be beneficial, the cycle turns out to be from 2 to 4 months, with 2 months being the most frequent result. For our example, then, this leads to 4,800 parts to be reviewed. It should be possible to do this in 1 to 2 months.

Once a part flow analysis matrix is completed, we have a manufacturing parts family but not a true parts family according to a classification code. We have identified how these parts are made at present but we have not analyzed whether it is the best way to make the parts. Neither have we developed a capability to load significantly different parts on the same facility based on similar machining characteristics. All we have done is to achieve loading for the makeup of manufacturing cells without determining whether the plans are optimized.

An Example of Applying Parts Flow Analysis:

(Excerpted from my book "*Fundamentals of Shop Operations Management: Work Station Dynamics*, Chapter 5," published by ASME Press and Society of Manufacturing Engineering, 2000.)

In order to gain a fuller understanding of the effort required to create a part flow analysis, I'm including an updated section from an earlier book I wrote, *Fundamentals of Shop Operations Management, Work Station Dynamics*. Here I go into some detail as to how a PFA is achieved. The part

	Simplicity	Handle large no. of parts	Shop knowledge required	Data applicable to other Group Technology concepts	Applicable to new and different parts	Cost	Time to complete
Empirical decision	Easiest	Poor	Yes	No	No	Least	Fastest
Classification code	Most difficult	Excellent	No	Yes	Yes	Most	Slowest
Part flow analysis	Medium	Good	No	Limited (process plng. partially only)	Yes (limited)	Medium	Medium

Figure 15-4. Methods of Forming GT Cellular Layout Part: Families

flow analysis matrix (PFA) in Fig. 15-5 shows a series of part numbers completed by a hypothetical factory involving turned pipe parts of various varieties: pipes, sleeves, washers, disks, and fastener nuts. The matrix shows the method step under each operation for the sequence the part is made. For example the fifth operation for part no. 601B44 is heat treat.

The first thing to observe is that there isn't any pattern to the numbering system and terminology of names is not very specific. For example; examining the matrix shows that the method for making a washer is the same as making a disk. Does this mean they are two names for the same thing? It is entirely possible that this is the case. Which means for determining Family of Parts it is not sufficient to use names only.

For convenience sake the matrix shows workstations by process description. The results of the matrix by process category are:

Incoming Material Inspection:	24 (all)
Cut to Length:	24
Flat Surface Grind:	16
Weld Butt Ends:	7(pipes & Sleeves)
Heat Treat:	7(pipes & Sleeves)
Turn to Dia., Outside:	24
Turn to Dia., Inside:	17
Tap Holes, Radial:	3
Tap Holes, Longitudinal:	6
Thread Dia., Inside:	7
Thread Dia., Outside:	1
Paint:	3
Galvanize:	10

So far we can see that pipes and sleeves appear to be a specific family because they are the only components that require weld and heat treat. But again, we have to be careful jumping to conclusions based on descriptive names only. In Fig. 15-5 we see sleeve part number 653C29 has most of the processing characteristics of a disk or washer than the other sleeves. In fact, other than name it has little if any commonality with the other sleeves. The moral being, that until we look at process steps carefully, we cannot form Family of Parts groups.

So what do we know? We see that there will probably be three Family of Parts groups. The first made up of washer type products. A second group consisting of pipes and sleeves that are manufactured like pipes, and probably require welding. And a third group consisting of fastener nuts. And that's all. We have learned just about as much as we can from the data presented and now its time to look deeper.

| Products | | Processes | | | | | | | | | | | | |
Part No.	Description	Incoming Mat'l Inspection	Cut to Length	Flat Surface Grind	Weld Butt End	Heat Treat	Turn to Dia. Outside	Turn to Dia. Inside	Tap Holes Radial	Tap Holes Longitudinal	Thread Dia. Inside	Thread Dia. Outside	Paint	Galvanize
143A67	washer	1	4	5			2	3						7
153B23	pipe	1	2		4	5	3							
341B12	fastener nut	1	8	7			3	4	6		5			9
560A29	fastener nut	1	8	7			3	4		6	5			9
121C33	washer	1	4	5			2	3					6	
621A17	washer	1	4	5			2	3						
786B54	disk	1	4	5			2	3					6	
601B44	pipe	1	2		4	5	3		6					6
280B43	pipe	1	2		4	5	3							9
785A12	fastener nut	1	8	7			3	4		6	5			6
887C23	disk	1	4	5			2	3						
971C41	disk	1	4	5			2	3						6
101B32	sleeve	1	2		4	5	3							
531A23	washer	1	4	5			2	3						
771B28	fastener nut	1	8	7			3	4		6	5			9
653C29	sleeve	1	6				2	3	5			4		
221A26	washer	1	4	5			2	3						6
681C69	pipe	1	2		4	5	3							
449B67	fastener nut	1	8	7			3	4		6	5			9
532A22	fastener nut	1	8	7			3	4		6	5		9	
223C21	fastener nut	1	8	7			3	4		6	5			9
485B12	sleeve	1	2		4	5	3							
338C34	pipe	1	2		4	5	3							
991A44	washer	1	4	5			2	3						

Figure 15-5. Parts Flow Analysis Matrix
(From Daniel T. Koenig, Fundamentals of shop Operations Management: Work Station Dynamics, ASME Press & SME, 2000.)

The second more in depth analysis comes about by rearranging the data but still using the same work station sequences. Now we can see why numbers in process order were used on the matrices instead of simply check marks. By regrouping we will be able to see if there are any exact matches of different part numbers with identical processing sequences. These would be exact Family of Parts members. Sort of like the nuclear family. We will also see that there are parts that go through the same basic work stations but in a different sequence order. These are near relatives and may be considered to be members of the Family of Parts grouping. This is particularly so if the deviations are caused by specific extra operations for specialized purposes.

Figure 15-6 shows the grouping by common name and we can see that there are some very definite Family of Parts groupings. There are four very distinct Family of Parts groupings.

The first is a washer family made up of washers, disks, and a sleeve. They all have the same manufacturing sequences except for special requirements. For example; the one sleeve in this grouping, part no. 653C29, has two additional manufacturing steps. In this case the threading of the outside diameter and tapping radial holes. This would imply that the part is longitudinally longer than the rest of the parts in the grouping. So special tooling may still be required for this sibling.

The second group is the pipes and sleeves. Again they all require the same manufacturing processes except for additional special features.

Finally, we see the third Family of Parts is the fastener nut family. Here all processes are the same, and again there are additional processes for additional features. We also see that finish requirements are a big variable.

This is about as close as we can get with this technique. For example, we don't know if it's really practical to group all the pipes into one Family of Parts group, because we do not know the size ratio. We could be talking about 5 in. outside diameter pipe mixed with 5 ft outside diameter pipe. Obviously that would be a mismatch. The processes may be the same but the equipment to do so would be vastly different. We also see a little of the limitations of PFA with the special sleeve of the first parts family. We simply cannot tell from the data presented whether the part fits the equipment necessary to manufacture all the siblings. However, in defence of the PFA technique, the personnel involved with manufacturing surely will know the common sizes of their products. So there shouldn't really be too much of a surprise factor once Family of Parts are structured.

We have seen that part families can be formed in several ways. The three primary ways are compared in Fig. 15-4. The choice of method depends primarily on the goals of senior management. If only a cellular layout is desired, then either empirical decision or part family analysis would be sufficient. If the number of the parts to be considered were relatively small, empirical decision would be favored. However, if the management team is going to continue to develop a fully integrated CIM business, then empirical decision and part family analysis are temporary techniques, and will eventually have to be replaced with a classification code. How long it will take to achieve CIM will dictate whether or not a classification code will be employed to achieve a group technology cellular layout. Management must keep in mind that once a method is decided upon, it will be difficult to change at a later date. Management must also decide whether a compromise approach is better than no action at all when contemplating initiation of any group technology-based project.

SUMMARY

The argument for group technology cellular layouts has been presented in this chapter. The compelling reason for such layouts—a 15 to 30% average productivity improvement—has been presented along with detailed examinations of the validity of the claims. We have also examined the strategy of economic analysis of savings and the approach for justifying group technology layout projects.

We saw that the principle of sameness underlies this layout concept and that a method for determining what constitutes a family of parts is essential for achieving any degree of success. Three methods for determining part families were introduced, and their selection is shown to be economically driven. Empirical decision is the simplest method and yields the least benefits, while

Part No.	Description	Incoming Mat'l Inspection	Cut to Length	Flat Surface Grind	Weld Butt End	Heat Treat	Turn to Dia. Outside	Turn to Dia. Inside	Tap Holes Radial	Tap Holes Longitudinal	Thread Dia. Inside	Thread Dia. Outside	Paint	Galvanize
143A67	washer	1	4	5			2	3						7
121C33	washer	1	4	5			2	3					6	
621A17	washer	1	4	5			2	3						
531A23	washer	1	4	5			2	3						
221A26	washer	1	4	5			2	3						6
991A44	washer	1	4	5			2	3						
786B54	disk	1	4	5			2	3						
887C23	disk	1	4	5			2	3					6	
971C41	disk	1	4	5			2	3						6
101B32	sleeve	1	2		4	5	3							
653C29	sleeve	1	6			5	2	3	5			4		
485B12	sleeve	1	2		4	5	3							
153B23	pipe	1	2		4	5	3							
601B44	pipe	1	2		4	5	3		6					
280B43	pipe	1	2		4	5	3							6
681C69	pipe	1	2		4	5	3							
338C34	pipe	1	2		4	5	3							
341B12	fastener nut	1	8	7			3	4	6		5			
560A29	fastener nut	1	8	7			3	4		6	5			9
785A12	fastener nut	1	8	7			3	4		6	5			9
771B28	fastener nut	1	8	7			3	4		6	5			9
449B67	fastener nut	1	8	7			3	4		6	5			9
532A22	fastener nut	1	8	7			3	4		6	5		9	
223C21	fastener nut	1	8	7			3	4		6	5			9

Figure 15-6. Parts Flow Analysis Matrix Grouped by Part Name
(From Daniel T. Koenig, Fundamentals of shop Operations Management: Work Station Dynamics, ASME Press & SME, 2000.)

classification code is the superior method and also the most expensive. The part flow analysis technique has been explained to show how this middle-of-the-road approach can be used. Finally, we saw that for use with an integrated CIM CAPP and data collection system, only the classification code is acceptable.

This chapter has defined the place for group technology layouts in the modern job shop. It is evident that a company must employ the group technology layout philosophy, if not its actuality, if its managers intend to keep it competitive.

REVIEW QUESTIONS

1. Explain how the principle of sameness allows job shops to approach the economies of scale of flow manufacturing.
2. Compare the factory floor layouts shown in Figs. 15-1 and 15-2 with the layout shown in Fig. 15-3. Determine how the flows of jobs 1 through 10 of Fig. 15-2 would fit in the layout of Fig. 15-3. Describe the improvements possible.
3. What is a family of parts and how does it improve the probability of achieving economic production?
4. If a job shop, by definition, makes virtually any type of machined part, what logic is used to justify a group technology approach for its factory layout?
5. There are five reasons for adopting a group technology philosophy. With respect to layouts, state the order of applicability of these reasons. Give reasons for your ratings.
6. It was stated that a present practice of making parts in small lot sizes is not a relevant factor in determining whether a group technology cellular layout is desirable. Explain why.
7. Explain why group technology cellular layout projects are often difficult to have approved by senior management.
8. Describe the differences between subjective savings and firm savings with respect to appropriation requests. Give examples of each.
9. Discuss the reasons a senior manager may have for approving a subjective savings project.
10. Formation of families of parts is called the precursor event in developing a group technology cellular layout. Explain why.
11. In what business situation would the empirical decision method of forming families of parts be chosen?
12. Why is part flow analysis a good compromise method for determining families of parts?

Chapter 16

MANUFACTURING ENGINEERING ASPECTS OF MANUFACTURING RESOURCES PLANNING

The topic of this chapter is manufacturing resources planning, commonly known by the acronym MRP II. It is the scheduling automation technique of choice in modern factories. We will examine this technology as a component of CIM, where it achieves its optimum success. MRP II will be described along with its development. Then we will discuss the manufacturing engineering techniques involved in implementing MRP II. We will also outline the procedures used for implementing and operating MRP II and the ongoing manufacturing engineering responsibilities. This chapter will further enhance our understanding of CIM and how information sharing via the common database is critical for business success.

DEFINITION AND DESCRIPTION OF MATERIALS REQUIREMENTS PLANNING, AND MANUFACTURING RESOURCES PLANNING

Materials Requirements Planning (MRP), sometimes called "little MRP" in deference to its expanded version, Manufacturing Resources Planning (MRP II), is the computer-driven purchasing and inventory management program. It evolved to satisfy the requirements of supplying materials to shop operations. The computer is used to sequence materials inputs in accordance with chronological need. This system optimizes the purchase and distribution of materials to the shop floor. Unfortunately, by focusing on this, it is an island of automation approach and does not consider the entire business process. For this reason it was enhanced and expanded to include the other business processes requiring scheduling in creating a product. This includes shop scheduling of labor and availability of facilities, as well as design engineering due dates.

The evolution of MRP II from MRP is the logical outgrowth of the maturing of the use of computers in manufacturing. We have gone from standalone single-purpose computer applications to integrated and supportive uses. MRP was developed to solve the materials function's needs to cope with ever-increasing complexities of purchasing and inventory control. It was initially developed in the era before integration of all business activities was considered feasible or even desirable. Once awareness of the need for CIM became apparent, the logical extension of MRP into MRP II virtually exploded onto the scene. Let's describe further the power of this development and look at its applications and contributions to business optimization.

MANUFACTURING RESOURCES PLANNING IN A CIM ENVIRONMENT

MRP II is one of the cornerstones of CIM. As we have seen, it derives from the materials function's concept of planning purchasing and allocation of materials in accordance with a computer-driven schedule. It encompasses at least four of the seven steps of the manufacturing system (see

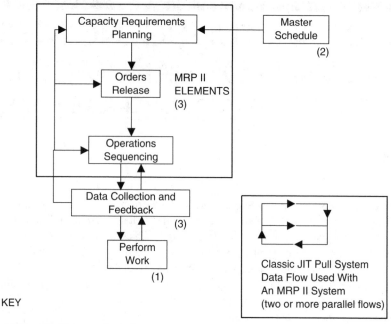

Figure 16-1. MRP II in a CIM World

Chapter 13) directly and the remainder indirectly. MRP II is a major user of the integrated common database and, when used correctly, embodies the philosophy of CIM to its fullest.

Figure 16-1 illustrates how MRP II works. Here we see how it relates to Just In Time (JIT) philosophy and the other modules of CIM. MRP II represents the Production and Measurements Control triad of the CAD/CAM triad. However, it is obvious from Fig. 16-1 that more than production and measurements is involved. In this CIM diagram of MRP II, we can observe the relationships between MRP II and other components of CIM. We see the Design and Planning Control triad represented by the master schedule. The perform work representation can easily be seen to be the Machine/Process Control triad. MRP II is truly a representation of the CIM philosophy of communications excellence applied to real-world activities. The entire essence of CIM is brought to bear in this activity. Let's summarize how MRP II works and then the basic strategy involved in successful implementation of this valuable CIM activity.

MRP II is the major innovation for dynamically allocating resources made possible by the introduction of the computer into manufacturing. It starts with the Bill Of Materials (BOM) prepared by Design Engineering. This lists in assembly hierarchy the component parts of the product. Some parts are purchased. Some are made in the factory. The BOM defines both and also shows assembly sequences. The BOM is the basis of the plan for making the product. Along with the BOM, the MRP II system is also fed by the routing. The routing is the chronological sequence of processing steps and the corresponding workstations where the work of making the specific part or assembly takes place.

MRP II, via computer algorithms, schedules multitudes of operations throughout the factory based on their chronological due dates. However, it still needs one more piece of information— how long it should take to perform each operation. This cycle time information is fed into the MRP II algorithm (and the routings) from the process planning activity. This too can be a CIM module, in which case it is normally called Computer-Aided Process Planning (CAPP). As we discussed in Chapter 14, CAPP systems such as the Integrated CIM CAPP system (Fig. 14-4) are designed to work with other information databases. So we see that MRP II is an enormously large reservation scheduler that performs its function via inputs from other databases. Thus, the benefit of having an integrated common database accessible to all cognizant parties becomes apparent. Also, it becomes obvious that MRP II is only as effective as the information it receives. The more dynamically correct that information is, the more accurate MRP II is. Since CIM is designed to operate in a dynamic situation, we see the power of CIM over stand alone modules.

In stand alone applications of computers, MRP II usage is entirely possible but hardly optimum. We can definitely establish a stand-alone database for all production applications in the traditional "goes into pattern" logic. We can even have shop operations prepare schedules for the machine tools and workstations via MRP II based on a static schedule. And it would work, after a fashion. But what happens when the master schedule changes at the behest of the general manager? How long does it take the new data to be delivered to shop operations? How many no longer required parts are made in the interim? What happens if manufacturing engineering removes a machine tool for maintenance? How does MRP II know not to plan work for that workstation without an integrated database? Obviously, it will eventually when it is manually entered, but at what cost of bottleneck confusion? It becomes obvious that if all this information from other CIM modules were linked via a common database, we would avoid confusion and optimize the utilization of these computer-based systems. A change made in one triad would automatically be reflected in other modules regardless of what triad they happen to be part of.

MRP II is so fundamental to business today that it is hard to imagine any company striving to be a successful world-class competitor surviving without it. Let's look at the steps for implementing MRP II. The plan for implementation is quite straightforward. The basic chronological listing is shown below.

1. Understand how MRP II works and educate all employees.
2. Evaluate the existing Bill Of Materials (BOMs) and routings for accuracy as compared with actual practice.
3. Establish and implement a program to bring the BOM and routing documentation up to a minimum of 98% accuracy as compared to actual practice.
4. Identify and document the cycle time for each step of the routing. This information will be used to establish finite capacity for each workstation.
5. Establish materials ordering, master scheduling, and production planning policies that reflect how your company actually carries out these activities.
6. Investigate commercially available MRP II software packages for compatibility with your company's needs.
7. Implement and debug the selected software in a pilot location.
8. Once the pilot location is operational and debugged, spread one area at a time to the rest of the factory.

The implementation process is indeed straightforward, but there are dangers to be avoided. The primary trap to be aware of is the absolute need to have accurate information. If the BOM is wrong, the algorithms will miss sequence steps needed to build the product. Similarly, the routings must be correct. Otherwise the sequence of producing the product will not coincide with reality and the schedule will be inaccurate. Also, if the cycle times for each operation are wrong, the capacity to do work will be in error. The results will be either too much or not enough cycle time allocated at each workstation. This, again, means the schedule will be in error.

Finally, when MRP II is up and running, I must stress the importance of accurate recording of data pertaining to actual happenings at the various workstations. If this is not done, the system will not know where material resides and in what quantity. Remember, in actuality the MRP II algorithm is a large reservation system. It knows the entire chronological sequence of manufacturing and dispatches materials and labor and reserves facilities accordingly. It modifies the dispatching of resources in accordance with reports on actual accomplishments. If operators report erroneously, the MRP II system cannot order work in the proper manner. Every MRP II system has means for correcting errors. However, the more diligent our engineers and operators are in making correct data inputs, the less likely the MRP II system is to be in error and thus misleading. So one of the critical chores to be completed for a successful implementation and ongoing success is ensuring that data input is easy to do and is done as a routine matter.

One more tip about implementing MRP II. Many well-meaning people suggest that a consultant be engaged to guide the company through the implementing stage. In general, this is sound advice. However, keep in mind that the consultant does not know the product or the proper sequence of fabricating it. This means consultants will be of little help in the critical evaluation of the BOMs, routings, and cycle times. Even in the education process consultants cannot do it alone. They will undoubtedly understand the theory of MRP II and will do justice in that phase of the education of the implementing company's personnel. But consultants cannot be expected to go from the general to the particular of the company. This instruction will require a person familiar with MRP and very familiar with the company's way of doing things. The end result is that a "bootstrap" approach is required for specific instruction needs. The one area in which a consultant is effective, however, is in the software selection arena. Here his or her experience can be very useful and can eliminate some of the "reinventing the wheel" syndrome.

Implementing an MRP II system is an "all hands" evolution. This cannot be overemphasized. The need to educate, educate, and educate some more is significant. People will find that although the benefits of dynamically accurate and timely information are great, there is a much higher need for self-discipline in working the system than in a non–MRP II environment. Understanding the intricacies of the system is very important. MRP II allows companies to cope effectively with the dynamic nature of manufacturing. But there is a price to pay. The laissez-faire approach does not work. Everyone must be constantly aware of the need to input accurate data and resist the temptation to freelance decisions and actions outside the system. Data accuracy is vital for the system to work. However, when it is working, an MRP II–driven factory is a wonder to behold. Promise dates are met. Inventory and production costs are down. Quality is enhanced. All in all, the company functions better and is an improved competitor. The effort to implement MRP II is intense, but the rewards are definitely on par with that effort. No modern manufacturing company can be without an MRP II system.

MANUFACTURING ENGINEERING RESPONSIBILITIES FOR IMPLEMENTING MANUFACTURING RESOURCES PLANNING

Implementing MRP II has a technical content as well as the human factors portion. Manufacturing engineers are responsible for the technical content development, but with such a people-oriented activity the ergonomics must also be considered. MRP II is a well-developed concept in a generic sense. However, since no two companies are identical in products or approaches in manufacturing, the job of the manufacturing engineer along with other user groups is to tailor generic MRP II into specific MRP II. Using Fig. 16-1, let's dissect MRP II to understand the process.

In Fig. 16-1 we see that MRP II consists of three modules—capacity requirements planning, orders release planning, and operations sequencing—plus complementing master scheduling and data collection. We discussed master scheduling and data collection in earlier chapters, and since they are not technically part of MRP II we will ignore them for purposes of this discussion. Also, JIT is discussed in a separate chapter and hence not covered here. "Perform work" in Fig. 16-1 implies value-adding activities at the various workstations in accordance with the production schedule.

Capacity Requirements Planning

Let's start with Capacity Requirements Planning (CRP). This is the highest-order module of MRP II. It receives as its inputs the chronological sequence of work orders from the master production schedule. Typically this would be 6 months to a year's worth of orders, the planning period. This module then evaluates the work to be done versus the capability to do so in the time frame allowed. It shows areas of over and under capacity at various workstations at various times throughout the planning period.

CRP is a refinement procedure for the chronological listing of orders from the master production schedule. The master production schedule has to be correct in a macro sense. That is, there can be no gross errors in capacity and capability made in the master production schedule, but there could conceivably be errors due to estimating during the quest for orders from customers. Since the general manager to fill the company's order books uses the master production schedule, it has to reflect real capability, but only in an overall overview sense. It is the capacity requirements module of MRP II that refines the master production schedule to ensure that the chronological sequence of orders commitments can be met. CRP generally evaluates data from the master production schedule and then recommends what schedule adjustments should be made for capacity with respect to production equipment. This also leads to suggestions for labor loading and for identifying subcontract of work.

It is possible for CRP to refine the master production schedule simply because of the shorter time frame, which allows the vagaries of the company's facilities and labor situation to be better known. The master production schedule usually chronologically loads the factory for 1 or more years, up to a typical maximum of 3 years. The CRP rarely goes over 6 months, so the conditions of the factory are more readily known. Hence the company can take into account more precisely known factors affecting production capabilities. It stands to reason that specifics of labor trends and machine performances will be more easily predicted for 6 months than for 1 to 3 years. So CRP is a definite refining activity of the master production schedule.

CRP is also the module that initiates long-lead-time production materials. Any materials that require a long period of time must be ordered at this juncture. Since we do not want to have excessive materials on hand for just-in-case insurance, one of the primary functions of CRP is to make sure that this category of materials is ordered. To do so means that the engineering specifications generation required for ordering these materials is also scheduled via the CRP. This is one of the important aspects of MRP II. It coordinates the entire business function, not just the manufacturing portion. The output of the CRP module becomes the input of the orders release planning module.

Orders Release Planning

This is the important intermediate level of planning that takes the chronological sequence of orders corrected for capacity and provides details for near-term production. It provides details for planning of subcontract work and in-house work and releases all but long-cycle-time purchase orders. Orders release planning is the logistics module of MRP II for what is called near-term planning. This can be anywhere from 1 week to a full production quarter, 13 weeks. It is the issuance of work orders for either in-house work or purchased services or materials. It is the stage at which strategic planning from the capacity requirements planning module makes a transition to tactical activities to achieve the production schedule.

Orders release planning is the vital function that modifies well-conceived strategies that may be floundering because they do not meet the current tactical situation. It does this by modifying strategies into achievable plans that are compatible with the current reality. It makes adjustments in order release dates if capacity cannot be scheduled economically or unexpected problems occur with production equipment. This module also takes into account unexpected labor situations, such as strikes or influxes of untrained labor into the factory. The module derives its basic sequence of information from the output of the CRP module; thus it depends on the accuracy of CRP to create the details necessary to go to this next, more precise, level.

Operations Sequencing

This module performs the functions commonly known as production control. It controls the queuing sequence of work at every workstation. Thus it provides the instructions about what job is to be done and in what order at the specific workstation. In conjunction with the data collection module it performs the feedback loop so necessary to have a dynamic CIM system.

Production control tries to load workstations to their fullest capacity as measured by the overall optimization of the entire production facility. This module, supported by the CRP and order sequencing planning modules, allows the production controllers to understand fully the integrated needs of the entire factory to optimize output. They accomplish their task by using the computer databases to meet due dates by utilizing available labor and facilities resources, including reserve equipment, overtime, and temporary out sourcing to qualified vendors as required. This correctly implies complex software networking with purchasing, inventory control, and manufacturing engineering.

This is the most dynamic of the MRP II modules. The planning horizon is the work shift out to about 1 week. The time increments are in minutes, not days or even hours. Here we see that the ability to operate in such an environment requires a constant stream of accurate and current information. This we achieve through application of data collection systems, such as described in previous chapters. We see that MRP II requires cohesive integration with the other moduls of CIM, and doing so optimizes the whole of the manufacturing enterprise. The relationship of the operations sequencing module to the rest of the MRP II modules, as seen in Fig. 16-2, demonstrates how the increments of MRP II, and for that matter all of the CIM modules, need each other and work off each other to provide an optimum output of products.

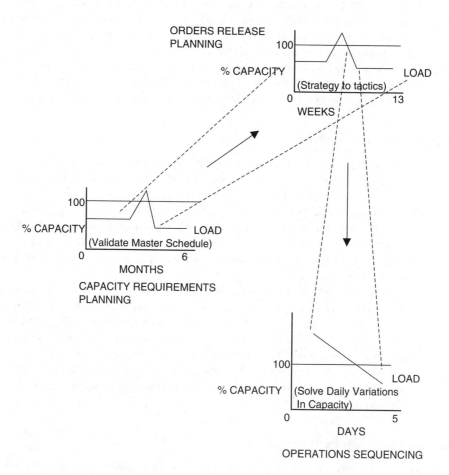

Figure 16-2. MRP II, Cascading or Information from General to specific

TECHNIQUES FOR ENHANCING IMPLEMENTATION OF MRP II

While manufacturing engineers are involved in the entirety of implementing MRP II, there are specific activities in which they have the prime responsibility. Let us look at these prime activities and investigate the manufacturing engineering techniques involved.

Eight generic activities are involved in implementing MRP II, as listed earlier in this chapter. Manufacturing engineers would be expected to take the lead in developing three of these activities. They are:

2. Evaluate the existing BOM and routings for accuracy as compared with actual practice.
3. Establish and implement programs to bring BOM and routing accuracy up to 98%.
4. Identify and document cycle time for each step of the routing, and use this information to establish finite capacity for each workstation.

Two of the three deal with methods and planning activities, and the third deals with factory loading and capacity activities.

Cycle Time, Factory Loading, and Capacity Issues

Factory loading and capacity techniques are discussed in Chapter 3. The requirements for MRP II are the same as for normal considerations. The engineer has to determine the load level that each workstation can handle over specific periods of time. For MRP II purposes this is usually by work shift, a nominal 8 hours. All the interference and nonproductive times are factored in, just as described in Chapter 3. The only additional task the manufacturing engineer has is to input the workstation capacities into the CRP module. This is done at the initiation of the MRP II project and every time there is a significant change in workstation capacity that is considered permanent. By permanent, I mean that the change represents a planned difference that will stay in place for the foreseeable future. For temporary changes in capacity, for example, as a result of equipment failure or inexperienced workers, the engineer will not enter the data into the CRP module. Instead the data will be transmitted to production control for it to be input into the operation sequencing module.

The overall responsibility of manufacturing engineering with respect to MRP II is to implement data to and maintain the CRP module. This means they manage the module. They make sure that data from master scheduling are input in accordance with the policy established by management. These data are then merged with the capacity data and the results are made available to materials management for the logistic planning and scheduling of the orders release planning module. Although this appears to be an ongoing work load of some significance, in reality it is not. Most MRP II systems handle these transactions as a matter of course all within the computer software with little human interaction. The only human interaction is the initial loading of the database and change inputs. Thus, the engineering work is heavy during implementation and start up and then quickly becomes routine.

Bill of Materials and Routings

The two methods and planning activities evolve about the BOM and routings. Let's discuss the BOM first. This document has come into prime use with MRP II, because it is a very organized way to transmit materials and labor needs in chronological order, in a manner easy for systems software to emulate.

The methods and planning activity defines how a part will be fabricated and assembled and how long it should take to do so. The work is based on the engineering drawing and a set of instructions that gives the sequence of assembly. This sequence is called the bill of materials. The bill, as it is called, is sufficient for sequencing operations. It shows how the design is to be made. But it is

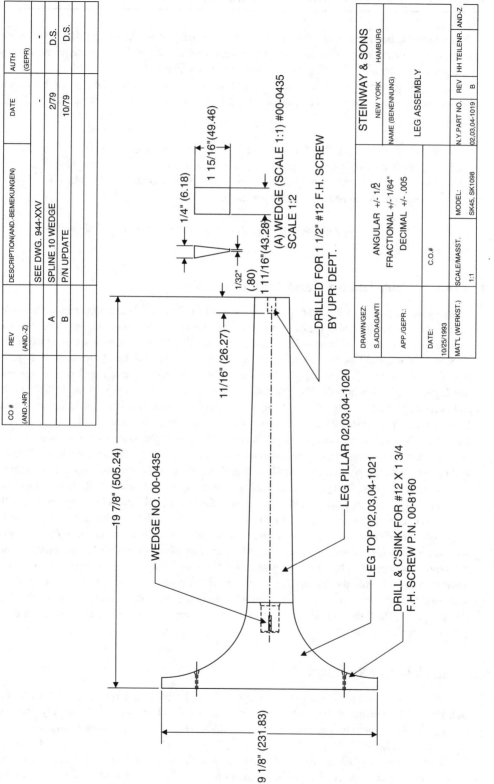

Figure 16-3. Piano Leg Assembly for Demonstration of BOM Principles (Drawing copy courtesy Steinway & Sons, Inc.)

Level	Part Number	CC	Qty Per	UM	LL	Description	Phan
0	021019			EA	3	LEG ASSY 1098 EB	N
1	000435	00	1.0000	EA	4	LEG WEDGE 1098 & K	N
2	119021	00	.0015	BF	8	5/4 BIRCH MLB	Y
3	019021	00	1.3699	BF	9	BIRCH 5/4	N
1	007758	00	.0030	GA	8	WOOD-LOK 40-0272/7045 (5 GAL)	N
1	021020	00	1.0000	EA	4	LEG PILLAR 1098 E	N
2	119024	00	.8680	BF	9	10/4 BIRCH MLB	Y
3	019024	00	2.0000	BF	10	BIRCH 10/4	N
1	021021	00	1.0000	EA	4	LEG TOP 1098 & K EB	N
2	119025	00	.5859	BF	7	12/4 BIRCH MLB	Y
3	019025	00	2.2222	BF	8	BIRCH 12/4	N

Indented Bill of Materials

Review WU   Review Route   Review Bill   Add Struct   Stop   Printed Report   Both   REV INDT BILL EXIT

Part Number 021019

(A) Level   (B) Part Number   (C) CC   (D) Qty Per   (E) UM   (F) LL   (G) Description   (H) Phan

End of retrieval   1016

Figure 16-4. Indented BOM

insufficient for determining how long it should take to do so. Determining the time to do an operation, hence an entire production sequence is the work of manufacturing engineering. This process is described in Chapter 7. This cycle time information is merged with the BOM to become the "routing," sometimes called the planning. Let's now look at the details of the BOM and then discuss how engineers work with the document to ensure the accuracy of the MRP II module.

Figure 16-3 is an example of an assembly drawing of a piano leg.* In order to produce the leg, it is necessary to identify the components that go into the leg assembly and the sequence of fabrication and assembly steps.

The BOM lists these assemblies and parts in two basic formats: the indented bill of materials and the single level bill of materials for a part. These are shown in Figs. 16-4 and 16-5, respectively. The indented bill of materials, commonly known as the indented bill, is the "goes-into" structure, and the single level bill of materials is a part list for making a single part contained in the indented bill. The information in both versions of the BOM refers to information derived from the part drawing, Fig. 16-3. Let's review the information found in both of these documents. Although the details are specific to a piano component part, the headings found in the documents are generic and applicable to any user. For explanation purposes, Figs. 16-4 and 16-5 have their headings labeled A through H and A through J, respectively.

Indented Bill of Materials (Fig. 16-4)

A. *Level*: This shows the "goes-into" sequence of the parts. In this example we see that there are three levels. The construction starts with the lowest level. It then proceeds to the next higher level, where completed components of the previous level are used as material inputs for this level. Note that there is a zero level too. This is the master schedule level, meaning that it is a discrete part recognized at the highest part scheduling module of MRP II—in this case, capacity requirements planning.

B. *Part Number*: This is the drawing number, or reference number for purchased parts.

C. *CC*: This means configuration code. If a number were shown it would mean that a part is being used at more than one level of construction. An example would be glue being used at

*All figures demonstrating drawings, bill of materials, and routings in this chapter are from Steinway & Sons, Inc. and are used with permission.

Single Level Bill of Materials for a Part							REVIEW	BILL
Review WU	Review Alt	Add Struct		Add Part	Stop	Printed Report	Both	EXIT

Part Number 021019

Description	Unit of Measure		Part Class	Part Status	(I)
LEG ASSY 1098 EB	EA		F		

(A) Component	(B) Config. Code	(C) Description	(D) UM	(E) Qty Per	(F) Phantom Flag	(G) Pick Seq	(H) Lead Adj
000435	00	LEG WEDGE 1098 & K	EA	1.0000	N		0
007758	00	WOOD-LOK 40-0272/7045 (5 GAL)	GA	.0030	N		0
021020	00	LEG PILLAR 1098 E	EA	1.0000	N		0
021021	00	LEG TOP 1098 & K EB	EA	1.0000	N		0
End of retrieval							1016

(J)

Figure 16-5. Single level BOM for a Part

subassembly and assembly levels. Another example, in metalworking, would be application of welding at various stages of fabrication.

D. *Qty Per*: This is shorthand for quantity per assembly. It shows the amount of material or parts required to make the next higher level of assembly.

E. *UM*: This means unit of measure and represents how materials or parts are counted for production purposes. In the example we see EA meaning each, BF meaning board feet, and GA meaning gallons. Obviously, these units of measure will be company specific.

F. *LL*: This is the designation lowest level in the database. This shows the parent-component relationship in the entire MRP II BOM. In this example we see Birch 10/4 as 10 under LL but 3 for this particular indented bill of materials. Note that most of the numbers under LL are greater than 3, which indicates that this specific indented BOM is relatively low level within the entire MRP II BOM.

G. *Description*: The description identifies the component parts with enough details so that every cognizant user of the indented BOM understands what is being called for.

H. *Phan*: This is shorthand for phantom flag. This column tells whether or not a component is made or not made (purchased). N means nonphantom, which means it is a component that is scheduled to be made in the factory. Conversely, Y means phantom; hence the part will not be scheduled to be made but has to be identified as internal to the process—for example, so that an inspection can be conducted.

Single level Bill of Material for a Part (Fig. 16-5)

A. *Component*: This is the drawing number identification.

B. *Config Code*: This is shorthand for configuration code and is the same as the CC that appears in the indented BOM, item C.

C. *Description*: This is the same as item G of the indented BOM.

D. *UM*: This is the same as item E of the indented BOM.

E. *Qty Per*: This is the same as item D of the indented BOM.

F. *Phantom Flag*: This is the same as item H of the indented BOM.

G. *Pick Seq*: This is shorthand for pick sequence. This identifies the sequence material needs to be available for making the part. Since no numbers are shown in Fig. 16-5 in this column, we know that all materials are required at the same time.

H. *Lead Adj*: This means material lead-time adjustment for different manufacturing cycle times. Sometimes it is necessary to start a particular operation before others for making the

Part Number: 021020

Description				Common Route ID		(H)
LEG PILLAR 1098 E						
(A)	(B)	(C)	(D)	(E)	(F)	(G)
Route seq No	Wctr ID	Oper Number	Operation Description	Cur Setup Hour	Cur Unit Run Hour	T F
0600	X-CUT	1302100	RIP TO WIDTH CUT TO 20" X 2-1/2"	.000	.0000	Y
0900	MILL	1410400	FACE ONE SIDE- GENERAL NUMBER FACE OPERATION	067	.0040	N
1200	MILL	1411200	CUT PARTS TO SIZE - TENONER CUT LENGTH 19-3/4"	.333	.0060	Y
1500	FRAISE	1635400	SHAPE LEG PILLAR 1098 - K EB TURN TO SHAPE	2.000	.0225	Y
1700	FRAISE	1636400	TURN DOWEL LEG PILLAR 1098 TURN DOWEL 1-1/8"D X 1-5/8"L	2.000	.0116	N
1900	FRAISE	1623800	RIP ON VARIETY SAW - GEN OPER CUT BOTTOM END	.250	.0097	N
2100	FRAISE	1622100	FRAISE TO SHAPE CUT DOWEL TO LENGTH & CUT WEDGE SLOT	.340	.0019	Y
2200	FRAISE	1603200	M SAND LEG PILLAR 1098 K EB SAND 4 FACES MACH SAND 4 FACES ROUGH & FINE AND BREAK CORNERS. CHANGE #3066	.000	.0269	Y

Figure 16-6. Example of a Routing

same part. For example, in working with wood components it may be necessary to kiln dry the wood before it is machined. If this column is blank, there are no lead-time adjustments for this particular single level BOM for a part.

I. *Upper Heading:* The single bill, unlike the indented bill, has a defined upper heading area. This area defines what is called the parent part. We see a description, unit of measure, part classification, and part status fields. Description and unit of measure are the same as described above. Part classification is either F for fabricated (as shown) or P for purchased. Part status is a field reserved for whatever the specific factory requires. It could indicate make for inventory stock. Or it could be used to reserve this particular part to a specific order. Whatever makes sense for the specific factory is an appropriate use for this field.

J. *Components for the Parent Part:* These are all the items A through H of the single level BOM for a part. They form the details supporting item I and give all the necessary information about the material needed to make the specific part.

The BOM is the component part database of MRP II. It has to be very accurate, otherwise the scheduling algorithms at the three MRP II levels will not reflect the true situation. For this reason we strive for 100% accuracy in defining the "goes into" pattern. In fact, we say that the margin of error is very small, only 2%. Thus we need 98% accuracy as a minimum for MRP II to work. It is easy to visualize what kind of havoc would be caused if the BOM were wrong and the database does not reflect the way the factory assembles the product. If this happens, the ordering and dispatching of materials would not occur correctly, and in the extreme case the factory could be idled because of no material at the workstation.

Engineers must be cognizant of these facts and diligent in checking the accuracy of the BOM before these data are entered into the data base. There is no easy way to do this. Usually it requires verification that the planning is being followed on the factory floor. Where it isn't, the reason for the diversion has to be understood and a decision made either to get the factory to follow the plan or to have the plan modified to match the current conditions. The word *current* is chosen carefully. The BOM always has to reflect the current way the product is fabricated and assembled. This means that every time there is a change in how work is performed, the BOM has to reflect that change, because above all MRP II accuracy has to be maintained.

The process of verifying the BOM has to be accomplished before the MRP II module of CIM is implemented. Typically this is done work center by work center. The manufacturing engineer must physically observe the work sequence and resolve all discrepancies before implementation of the system. As a practical matter, this means that we implement MRP II one work area at a time.

The companion database to the BOM is the routing document. This is the end result of the planning activity, as described in Chapters 7 and 14. It shows the sequence of operations, the work center where each operation occurs, and the time it takes to do each operation.

The operations sequencing scheduling algorithm is the overseer database to the routing. Whereas the routing is a set of instructions as to where a part will be made, the operations sequencing is a compendium of the information found in all the routings in the overall chronological sequence. These are compatible databases. Operation sequencing tells the shop what work to do at each workstation and in what chronological order. The routing defines the method to use at the workstation for the particular part or assembly and where the next operation is to be carried out. So, too, does the operations sequencing database. But it also contains the relationships of the particular part or assembly to all the other parts and assemblies. The routing is concerned only with the specific part or assembly.

Figure 16-6 is an example of a routing. Let's review the information contained in a routing. As with the explanations of Figs. 16-4 and 16-5, circled capital letters are used to define the fields for ease of explanation.

Review All Routings for a Part (Fig. 16-6)

A. *Route Seq No:* The route sequence number is from an arbitrary numbering system used to list the manufacturing steps in chronological order. The fact that the sequence starts at 0600 is meaningless. So, too, is the exact numerical distance between steps. The numerical spaces between steps simply allow additional steps to be added later, perhaps as a result of process improvements or design changes. The numbering sequence has only one purpose, to show the chronological order of the operations.

B. *Wctr ID:* This is shorthand for work center identification. It defines where the operation is to be performed. In the example we see department names that are specific for the intended factory.

C. *Oper Number:* The operation number catalogues the operation being performed. This can be a group technology–derived number.

D. *Operation Description:* The operation description defines the work to be accomplished at the particular workstation. The description is normally divided into two levels of information: general and detailed. For example, operation number 1411200 has "cut parts to size—tenoner" on the first line. This is the general information. It identifies the generic procedure to be accomplished and in this case the machine type being used. The second line states the specific length the part is to be cut to. The format of the top line being general information and the bottom line specific information is typical.

E. *Cur Setup Hour:* This column show the time allowed for setting up the part on the workstation value-added equipment. Following the example of operation 1411200, we see that 0.333 hour is required to set up the tenoner. It should be noted that this setup is for one or many identical parts.

F. *Cur Unit Run Hour:* This is the amount of time it takes to perform the value-added work per part. It is the time standard for doing this particular operation. For operation 1411200 the run time per part (unit) is shown as 0.0060 hour.

G. *TF:* This is the abbreviation for track flag. This term tells which operations will be tracked by the MRP II system. The Y or N tells whether the operation is (Y) or is not (N) being monitored. As a practical matter, not all operations need be tracked by the MRP II system. For example, if several operations have very short setup and run times, the practicality of recording the completions may be nonexistent. It makes no sense to measure an operation if the time to do so is virtually the same as the time for the activity itself. We see that operation 1410400 has a combined setup and run time of 0.0674 hour or approximately 4 minutes. We can certainly afford to lose track of the part for that amount of time before it is tracked at the next operation. If we tracked all of these short-term operations we would simply be cluttering the system with data that would have little practical use. The proper choice is to select representative operations that give us good information about where the part is in the production cycle.

H. *Common Route ID:* Common route identification tells us whether this planning is unique. If there were other parts that followed the same routing sequence, an identification number, probably a GT classification code number would be listed here.

As with the BOM, engineers have to ensure that the routing is accurate. The BOM defines the materials that make up the program and the relationship and the sequence of assembly of these sets of materials. The routing tells where the work is to take place and in what sequence. The routing also states how the work is to be performed. For these reasons both the BOM and the routing have to be at least 98% accurate. If not, the MRP II system would not be a useful tool in monitoring and offering suggestions on how to manage the dynamics of manufacturing.

Routing design and authentication are the responsibility of manufacturing engineering. Usually, engineers design the routing as part of the methods and planning activity, as defined and described in Chapters 7 and 14. The method is transcribed into the routing via the CAPP system, which is a companion module of MRP II in the CIM system. The role of the manufacturing engineer is to ensure that the method described is actually the method followed. As is the case with BOM accuracy, it is vital that the routing be accurate, otherwise the MRP II system would end up scheduling a fictitious situation and the factory could not function properly.

SUMMARY

In this chapter we have reviewed the theory of MRP II, and the role of manufacturing engineering in developing and operating the system.

We see that Manufacturing Resources Planning is an expansion of Materials Requirements Planning, the so-called little MRP. MRP planned sequencing of materials purchases and dispatching to the factory. MRP II expands on this idea to handle not only materials but also labor planning and facilities allocations in accordance with a strategic manufacturing plan.

In the study of MRP II, it became evident that the software was necessarily complex so it could handle many iterations of large numbers of discretely scheduled parts. To do this MRP is divided into three distinct but linked scheduling algorithms: capacity requirements planning for long-range planning, orders release planning for intermediate range, and operations sequencing for daily scheduling. The working of these modules was explained.

The process of implementing MRP is presented in the chapter. Key recommendations for successful adoption of the system, in the context of the personnel of the company, were discussed. We see that training and explanation as well as team formation are important for successful implementation.

The chapter concludes with a detailed explanation of the physical manifestations of MRP II, the bill of materials and the routing. These are the outputs of the software, the products of the scheduling algorithms that are used to control manufacturing operations.

REVIEW QUESTIONS

1. Explain why MRP became MRP II. Take into account "islands of automation" versus CIM in your answer.

2. Show which of the seven steps of the manufacturing system are primarily MRP II steps and explain why.

3. Demonstrate the relationship between Computer-Aided Process Planning (CAPP) and MRP II in a CIM environment versus a non-CIM environment.

4. Implementing MRP II is a complex affair involving eight interrelated steps. Step 4 requires the identification and documentation of the cycle time for each step of the routing. Explain how the manufacturing engineer accomplishes this task and how this information is included in the appropriate databases. Also identify those databases with reference to a CIM environment.

5. Explain why step 1, "understand how MRP II works and educate all employees," is vital to the success of MRP II implementation and operation.

6. Capacity requirements planning evaluates for long-term ability to perform to schedule. It evaluates the work center load levels for a sufficient period of time to identify trends. Explain how CRP relates to the BOM and routings to perform this function.

7. Orders release planning is the intermediate-level scheduling algorithm of MRP II. List and defines the differences in approaches to scheduling priority items between it and the other two levels of scheduling.

8. Production control, sometimes known as operations sequencing, is the detailed scheduling activity of MRP II. Discuss what dynamic versus static scheduling means at this level. Consider how MRP II handles changes in schedule at this level in your answer.

9. The BOM is the document used to input materials and sequence information into the MRP II scheduling algorithm. Development of the BOM is a design engineering function but is considered to be insufficient when completed by the design function. Explain why this is so and what the necessary additions are and how they are accomplished.

10. Define the differences between the indented BOM and the single level BOM.

11. The routing gives the MRP II scheduling algorithm the definition of where and how long it should take to make or assemble the product. Discuss how the manufacturing engineer assembles the necessary data and inputs the data into the scheduling data base.

12. Discuss the reason for mandating at least 98% accuracy between actual and planned operations sequencing of the routing.

13. Explain why it is necessary for the routing to have the setup and processing times as part of the information contained in the routing document.

14. Operations sequencing is the lowest level of scheduling algorithm. How does the information presented at this level compare to the information contained in the routing?

Chapter 17

JUST IN TIME AND ITS COROLLARY LEAN MANUFACTURING: A PRAGMATIC APPLICATION OF MANUFACTURING ENGINEERING PHILOSOPHY

Just In Time (JIT), and its new version Lean Manufacturing is fast becoming as familiar and identifiable in the lexicon of manufacturing terminology as mass production. Unfortunately, although the term is familiar to many, its definition remains clouded and not thoroughly understood. The purpose of this chapter is to remove the clouds and to give a pristine and clear explanation of what JIT is, and by extension what Lean Manufacturing is and how it is applied. I will use the term Just In Time (JIT) to mean both JIT and Lean Manufacturing simply for convenience purposes except where Lean Manufacturing may differ in interpretation of understanding for applications (I assure you when all the bells and whistles are removed there is very little of that.).

The most common misunderstanding of JIT is that it means delivery of materials to the factory just when they are needed. This implies reduced inventory levels maintained in the factory. Although there is a modicum of truth in this belief, it is not the true definition of JIT. JIT is one thing and one thing only: an awareness that true optimum manufacturing performance revolves about the dictate to eliminate waste in all of its many manifestations. In this chapter we will examine the various ways we can eliminate waste in accordance with the commonly accepted JIT goals. We will see that these goals are compatible with; in fact identical to, similar ideals we have already seen attributable to manufacturing engineering practice. We will also see that JIT theory, like manufacturing engineering theory, works better with and is very compatible with CIM systems.

JUST IN TIME FROM THE VIEWPOINT OF MATERIALS MANAGEMENT

The first use of JIT appeared in the materials arena, not manufacturing engineering. This may be surprising to some, because JIT is really another name for good engineering application in a factory situation. Considering that non-engineers have popularized JIT, this is understandable. Because of this history we also see that the first claims of manufacturing improvement had a materials management flavor. Let's look at the improvements claimed and how they occur. Some of these claims are more psychological and philosophical than scientific. As we shall see, even though this methodology is based to a high degree on motivational techniques, they can be of use in achieving positive results. But we must be aware that they are not the primary method of achieving results.

First of all, the goal of eliminating waste is very much in play. Anything that does not add value to the product is considered waste. The more visible items that must be analyzed for excesses are the moves and waits that lead to queues, stock levels, setups that could be taking too much time, making more parts than required for sales, and losses due to poor quality. These are but samples of activities that should be evaluated for better methods and policies.

Note that each of the visible items listed for increased scrutiny for waste has to do directly or indirectly with inventory management. In fact, the exhortation is to reduce inventory levels so that we can expose insufficient methodology. The theory is that sloppy practice is being compensated for with large inventory reserves. This is the just in case versus just in time argument, which we will discuss later.

It is interesting to note that non-engineer proponents of JIT always refer to the water level analogy for inventory. Figure 17-1 shows this analogy. We can see that a high water level (analogous to high levels of inventory) floats the boat (the process of manufacturing) high above the rocky bottom (problems retarding manufacturing progress and raising costs). If we lower the water level the boat has problems navigating, and we must remove the rocks (solve the problems). Notice the obsession with inventory levels. The concept is that high inventory levels hide problems. High inventory levels ensure that customer orders can be filled but at the expense of excessive cost. It is much better, the philosophy states, to be more efficient in using materials and thus still meet the customer's needs while doing so at lower cost levels. This makes the company more profitable, hence better able to compete. The end goal is positive, and manufacturing engineering theory certainly supports it.

Managing inventory is a prime facet of materials management; therefore it is not unusual to see this slant given to JIT. Let's look a little closer. It is claimed that we can improve performance if we take the crutches away. This psychological approach parallels the famous Hawthorne experiments in which spotlighting particular activities brought about improvements simply by exposure. These famous experiments showed that placing importance on a technique and how people performed it brought out the best in people, and they did better. This too is the case with "lowering the water." Here we purposely make it more difficult by making it known that there is no or perhaps just a little reserve material; therefore we need to do it right the first time. The proponents say that this

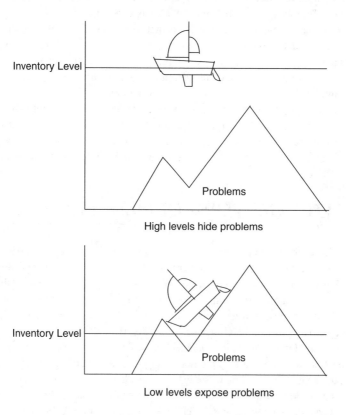

Figure 17-1.   How Lowering Inventory Levels Exposes Problems

allows us to gain significant improvements, necessity being the mother of invention. This is hardly the case. The improvements come about because the need is assigned as a project for improvement, usually requiring manufacturing engineering participation. What we really have is a way of establishing project priority. The goal of eliminating waste is valid. However, putting the company at risk through precipitous lowering of inventory to expose problems should not be done without serious consideration of possible consequences. Just in case is not always inferior to just in time. Let's look at an example of the lowering-the-water technique.

The theory says that we should reduce inventory to expose problems brought about by poor quality of materials and workmanship. Once the inventory is low enough, we cannot depend on high levels of inventory to ameliorate the unsatisfactory quality level. Thus we force the engineers, managers, and operators to take action to fix the problem or else the company will suffer.

One might say this is ridiculous—why create a crisis? Because we want quick action. We want the company to react to a problem that, when solved, will lead to performance improvements. This in turn lets us compete more effectively. The trick is to make sure the artificial crisis created is not too overwhelming. There is no doubt that if the company still demands the production, the pressure is put on the staff to solve the quality problem. The question is, is this the best way to make progress in eliminating waste? Maybe in some situations the answer is yes. However, I contend that rational approaches to problem resolution usually make more sense. Establishing priorities for problem resolution would just as easily get the same results. All we need to do is define the target need date for the project and assign measurable goals and responsibility as described in Chapter 2. I think it is reasonable to assume we can get the same results without the trauma of an artificial crisis.

Eliminating waste is a worthwhile goal and it can be achieved via good manufacturing engineering techniques. The hype and glamour associated with techniques such as lowering the water are really counterproductive and not what JIT is all about.

JUST IN CASE VERSUS JUST IN TIME

The just in case approach versus the JIT approach has been mentioned, and occasionally the former turns out to be a better strategy, as we saw above. But just in case does not emphasize striving for perfection via elimination of all wasted efforts. For this reason JIT is predominantly the correct approach.

Just in case requires extra inventory at strategic locations to be used in certain situations. This could be to take advantage of sudden marketing situations in which the product cycle time is too long to take advantage of unforeseen opportunities. This is a proper application of just in case. Companies can plan to have buffer stocks of materials or partially complete products at strategic locations. They then use this stock when there is an opportunity for very fast delivery times to customers. This sprint capability is an excellent use of just in case theory. This is a properly valid strategy especially in industries where demands for products are cyclic and almost to the point where it is a definable and predictable cycle. Housing for instance is one such cycle which ebbs and flows with monetary interest rates. Any product closely associated with housing then, will follow this pattern.

When just in case is used as a crutch for inferior performance, it ought not be used. Here manufacturing engineering theory agrees with JIT philosophy. Just in case is not for covering up mistakes. It is only for creating opportunity. When we purposely make extra parts to cover deficiencies in people's performance, tooling, equipment capabilities, and vendor deliveries of materials, we are not getting to the root cause of the problem. This is simply perpetuating inefficiencies. We are using resources for purposes other than satisfying customer requirements. We are using resources to cover up problems. The proper procedure would be to establish projects to solve the problems, not cover them up.

The other problem associated with just in case is the propensity to make too much product at a workstation in order to take advantage of complicated setups. Many companies are enamored with economies of scale. They erroneously think that because the setup takes a long time, they must make many more parts with the same setup to lower the cost per piece. By simple arithmetic this is correct, but only for that workstation. If the company does not have any immediate use for the extra parts, the true cost is really higher than the simple arithmetic indicates. We have the cost of storing and caring for excess inventory and the lost opportunity costs of using the capacity to do something else. This is definitely creating waste. A better approach would be to find a way to shorten the time of the setup, thus making it easier to make only the parts that are required to meet current needs. This would eliminate all the extra costs of having excess parts in inventory.

JIT takes an opposite tack from just in case. Here the philosophy is to be miserly with what we produce in the factory. We will make only what is identifiably needed to satisfy requirements and nothing more. By producing only what is needed when it is needed, we preclude the opportunity to produce excess parts. This in turn precludes the opportunity for waste to be generated and passed on to succeeding workstations. Just in case, by putting more material in the production cycle, creates a more difficult to control production schedule because there are more parts and assemblies to account for. JIT does just the opposite. By promoting the miserly approach, we have less parts and assemblies to contend with, hence a more accurate production control system. This accuracy eliminates mistakes and leads to reduction of waste in the manufacturing process.

So far, the case for JIT has been built with pragmatic logic. We have seen that JIT is usually the right approach, but that there can be excessive claims by zealots in its pursuit. Some of these excesses can even lead to damage to the practicing companies if they comply blindly. An example would be lowering of inventory to expose problems. We saw that problems have to be identified and solved, and then the inventory can be lowered. This points out that like any good technique, JIT is not a totally perfect approach. We must understand why JIT becomes an enabling philosophy for accomplishing optimum manufacturing. The techniques for making it happen are tried and true engineering techniques, which have been discussed in depth throughout this book. Let's now examine the true goals of JIT and see how engineering techniques are used go about achieving them.

THE TRUE OVERALL GOAL OF JUST IN TIME

The goal of JIT, also commonly known as lean manufacturing, is entirely compatible with that of manufacturing engineering. In the practice of manufacturing engineering, the focus is to optimize the manufacturing process to the greatest extent possible. The goal is to create the highest level of profit with concurrent highest levels of product quality. JIT has the same goal but those not familiar with manufacturing engineering techniques express it in terms more readily understandable. The JIT goal, simply stated, is to eliminate all waste in the manufacturing process. There is no hidden meaning here. Root out all waste wherever found over the entire manufacturing process from material receipt, to value-added procedures, through quality control, and finally to shipment to satisfied customers.

Some non-engineer proponents of JIT support the goal of elimination of waste in a spirit that would have one believe that JIT is something new under the sun. Actually, it is not. It is good manufacturing practice that appears to have been rediscovered by those perhaps not aware of the manufacturing engineering discipline. This has caused many engineers to belittle JIT as an intruder into their domain. However, this is an improper response. Lean manufacturing (JIT) is, as engineers say, just a repackaging of manufacturing engineering, but it is a very good repackaging. It focuses on what is critical to serving customer needs and does so in a way that is understandable to people not normally involved in manufacturing. For this purpose alone it deserves to be explored, understood, and exploited. JIT is a correct, responsible approach, and when combined with CIM, it is a powerful tool for manufacturing optimization.

The goal of JIT is to eliminate all manufacturing waste. But for this to be useful, we must be more precise. Several authors have taken on the task of doing just this. One author who I believe has succeeded in expanding the concept of eliminating waste into terms that can be properly defined is Robert W. Hall in his book *Zero Inventory*, published by Dow Jones–Irwin. His six points defining elimination of waste are listed below (reproduced with permission of Dow Jones–Irwin).

- Produce the product the customer wants.
- Produce products only at the rate the customer wants them.
- Produce with perfect quality.
- Produce instantly—zero unnecessary lead times.
- Produce with no waste of labor, material, or equipment.
- Produce by methods which allow for development of people.

JIT is called by its promoters a fundamental way of thinking to transform overall manufacturing in the simplest way possible and to generate new and original techniques for doing so. I would dispute that there are any original techniques. However, there is no dispute that the points listed above are certainly attributes we can agree on. JIT theory states we should strive to achieve these goals, and many authors set out to show how this can be done, which in most cases boils down to good manufacturing engineering practice. However, if that's all JIT is, then it is not necessary. Our traditional engineering techniques work well without the hype. The true optimum achievement of the goals, such as those stated above, can be achieved only via the CIM approach.

THE CIM APPROACH TOWARD REALIZATION OF JIT GOALS

The manufacturing engineering precepts espoused by JIT and Lean Manufacturing proponents (There is a difference between the two, primarily based on seniority, The JIT crowd is about two decades older than the Lean Manufacturing group.) can be approached in a traditional manual mode, but there is a much better way. That way is the CIM approach. We will look at the various productivity enhancers available via CIM and show that they are compatible with the six JIT points leading to accomplishment of the overall goal of eliminating waste in the manufacturing process.

Flexible Automation

Flexible automation via the Flexible Manufacturing System (FMS) is the ultimate achievement of computer-controlled machines. These machines are multi-tooled turret types that can do many different types of operations with one setup. Operations such as drilling, milling, cutting, and shaping are commonly done in one setup on these types of machines. Flexible automation is a key goal of a CIM system in order to achieve efficiencies for job shops of the same magnitude as flow manufacturing.

This strategy is definitely compatible with the JIT precept of producing the product the customer wants. Flexible automation allows great customization of production, enhancing a company's opportunity to produce a specific product for a customer without significant increase of cost. These types of machines definitely allow factories to approach the long-standing goal of economic order quantity of unity, usually expressed as

$$EOQ = 1$$

This implies that a company is capable of serving customers with the ultimate in customer satisfaction: the ability to provide customized products at the cost of mass-produced products. This obviously is the JIT goal of producing the product the customer wants.

Flexible automation also means very short setup time because of the many operations that can be performed on the material while it is on the work platform. This is the achievement of the JIT goal of producing instantly—zero unnecessary lead-time. In order to achieve very short setup times manufacturing engineers have devised Computer-Aided Process Planning (CAPP) along with group technology to find and plan families of parts. This makes achievement of short setup times much more likely, and we see the goal of producing instantly much more likely to be achieved. We can see that manufacturing engineering techniques are a very major factor in achieving JIT goals.

Manufacturing Resources Planning

Many proponents of JIT mistakenly think MRP II is a non-compatible strategy. This is not the case. Let us examine how MRP II makes it possible to achieve the JIT goals of "producing only at the rate the customer desires" and "with no waste of labor, material, or equipment."

One of the most widely known terms associated with JIT is *kanban*. Kanban is a Japanese term loosely translated to mean a pull production system. Pioneered by Toyota Motors Corp., it is a system of production control that dispatches work based on the needs of the succeeding workstation. This is in contrast to the more traditional push system. The push system enters the work order at the beginning of the production cycle and depends on sequencing developed by the planning and methods documents to push the product through the factory to conclusion.

Proponents of the pull system claim it to be superior to the push system. They say it is analogous to pulling on a rope, whereas the push system is like pushing on that rope. Obviously, pulling the rope is more effective than pushing it. By pulling along material only when needed at the next workstation we can minimize making unnecessary parts, thus keep work in process inventory down. This would also keep the factory from expending any unnecessary labor. Here we see an example of the JIT goal of eliminating waste being achieved.

The kanban system works by the needing workstation dispatching requirements via an order ticket to the supplier workstation to fill. The supplier would literally be asked to make (or assemble) sufficient quantities for the immediate needs. This order ticket is also an authorization to proceed. It tells the workstation to make only what is required and no more. This is in contrast to the push system, which demands the workstation to make product as long as there is material available to do so, regardless of whether or not the succeeding workstation needs it. Quite often, as the dynamics of the situation change, a succeeding workstation could be glutted with work and not need any more entering the queue. This could be caused by changes in order rates or bottlenecks caused by machinery or quality problems. But the push system cannot easily shut off the flow, and material continues to be made and arrives oblivious to the current situational needs. The pull system has a much better chance of not allowing material to enter the next workstation if it is not needed.

The problem with the kanban system is that it depends on brute force to make it work. It depends on the receiving workstation accurately planning its needs sufficiently on time to keep pulling work at the proper rate. This is very difficult to do for all but the simplest production sequences. When there are more than one or two immediately preceding workstations to contend with, the dynamics of work in process can become very difficult to handle. Not only do we have to know the present needs, we also have to understand the succeeding demands being placed on the entire chain of workstations. We have to be able to predict all the needs on a real-time basis. Only a computerized system can keep track of all the variables with sufficient reaction time, and that system is MRP II. As we saw in Chapter 16, MRP II has the ability to dispatch and control the flow of work through multitudes of workstations and react to changes in situations very quickly.

Figure 16-1 shows how MRP II is designed to operate in a pull mode. Here we see that through data collection the situation of each workstation can be evaluated, and thus we can create the order tickets for each preceding workstation to meet the needs for every succeeding workstation. MRP II develops daily schedules that are, in effect, the pull requirements the kanban system demands.

The pull system cannot work effectively without MRP II. The push system, which is overall less effective than the pull concept, can. All it takes to make a push system work is to enter the orders at the beginning of the production cycle and let it flow. As each step of the process is completed, it will move to the next step, whether or not the next operation is prepared to receive it. Eventually the order will be finished, but there is no guarantee it will occur at the time originally intended. This is why push systems were used before the existence of MRP II. There is no need to make schedule corrections because there is no ability to do so. JIT-inspired pull systems are impractical without MRP II, and the axiom to produce with no waste is only a dream.

Computer-Aided Process Planning

When we think about the JIT principles, one of the most vivid examples portrayed to the public is that of producing with no waste of material. This means striving to convert all the material to useful products. In order to do this effectively, we need a methodology for detailed planning for the material usage. This can be done manually, as we have seen in earlier chapters. But we also know from the chapter on CAPP that we can plan for manufacturing and coordinate sequencing faster and more accurately with computers.

In order to use materials correctly—in this case to use the optimum amount possible per unit of measure—we need to construct the goes-into patterns as precisely as possible and combine them with other parallel uses of the specific material. In this way we use the material to its fullest for all possible products (or parts). Obviously, this can be done manually by use of engineered time standards and methods-driven operation sequencing. But, as we have seen in the chapter devoted to methods, planning, and time standards, this is a tedious and meticulous process. It will be done only if there is sufficient time to do it and if the payback is considered significant enough to justify the effort. This means that most plans are approximations and material usage is seldom optimized. This is a poor strategy because the product cost ends up being higher than it should be.

With CAPP it is entirely possible to apply engineered time standards and methods-driven operations sequencing to the job. This will define the labor and materials needed to perform the job, and because of the link to the common data base the material can be designed to be portioned out to several jobs. It is analogous to saving nuts and bolts with the hope that we will some day find a use for all the excess hardware. As we know, that is hardly the case because we cannot find the right bolt or nut when we want it. But what if the computer was monitoring use and location of excess fasteners? Of course, we would use the excess fasteners instead of buying more. With a computerized system, for example, CAPP, this means more of the material will be consumed in product manufacture and not wasted. The precept of JIT is achieved, whereas it is highly unlikely that it would be in a manual system.

Computer-Aided Statistical Process Control

One of the most prominent of JIT precepts is "produce with perfect quality." A very desirable aid in achieving this precept is Statistical Process Control (SPC). We must be able to evaluate the condition of any process in order to ensure that we can control it and produce parts within desired tolerance ranges. If these evaluations can be made in a pragmatic manner, then we can achieve this precept.

The desire to produce with perfect quality leads to the need to measure process performance and understand the causes of process variance. SPC techniques as described in Chapter 10 are the methodology for measuring process performance. As we have seen, using control chart techniques is an exercise in minutiae and perseverance. It takes collection of data and calculation. Both of these tasks can be done better in a computer-integrated manufacturing environment than by the manual traditional approach.

We saw in the chapter devoted to CAPP that integrated collection of data to update production scheduling dynamically greatly enhances productivity. With computer-aided SPC that same data are entered to perform the control chart calculations, thus achieving virtually real-time SPC monitoring of the process. With this capability, real-time corrections to the process are possible and practical. This is the only way factories can approach zero defects. If we are forced to perform SPC in a manual mode, we never get the chance to make process corrections in real time. We simply cannot collect data and calculate fast enough to do anything but make after-the-fact corrections. The advantage of merging computer technology with SPC is what makes production with perfect quality a practical goal to pursue.

Common Database

Perhaps the most unusual precept of JIT is "produce by methods which allow for development of people." This would imply enrichment programs such as team manufacturing and quality circles. In both of these cases we see empowerment of workers to make production decisions and thus develop their skill sets. Why would the proponents of JIT believe this is important?

The concepts of manufacturing have changed over the years, but all the concepts have had as their core goal to produce at optimum levels to maximize profits. In the early 1900s, it was thought that precisely spelling out details of how each person was to do his or her work was absolutely necessary. Most likely this came about because the engineers did not think the workers could comprehend the requirements the product had to attain. This was probably a logical conclusion at the time, based on the general technical education levels of the work force. However, times have changed, and for the better. The general level of technical education today is superior to that of almost 100 years ago. Unfortunately, the general regard for workers' intelligence for the most part has not changed in 100 years. This has led to stifling of creative initiatives, which means that many potential product and manufacturing methods improvements have not been realized.

The precept that companies should produce through improvement of people is aimed at getting us to realize that more productivity improvement is there for the taking. It is there if we would only make use of the talents of all the people. This means involving the entire work force, not just those who happen to work as supervisors and engineers. Obviously, this means investing resources in training workers in techniques not usually associated with their typical by-rote job responsibilities. Where that has been done the results have been very good. We saw in Chapter 7 that techniques for these participatory programs can be used to advantage. The question is, how do we optimize such programs?

Participatory programs generally require workers to take responsibility for suggesting changes in processes. In some more advanced programs workers do much more. They actually plan, procure materials, and schedule their facilities. It is claimed that these teams having total factory control are very productive and make high profits for their companies. There is certainly evidence that this is the case in specific circumstances. It may also be true for all circumstances to a certain degree. But it cannot be more than a sideshow unless these teams have ways to obtain information efficiently. This is where common databases are useful and, in fact, make this precept a major factor for productivity improvement.

We know that information is necessary for the participatory teams to make effective decisions. We also know that getting correct and up-to-date information can be difficult. Use of common databases in a CIM environment ameliorates this problem considerably. With the CIM axiom of communications excellence, we are continuously generating current information about the status of manufacturing. This means that the teams have access to the most up-to-date information. Any trials or tests they may be employing will benefit, in that results will be obtained faster and be more accurate. The results correlate with much more rapid achievement of gains.

Without common data bases, team projects usually take much longer and consequently the probability of loss of interest and distrust rises. This is the main reason why "produce by methods which allow for development of people" too often is a dormant precept. We need to dispel apathy and to keep the interest level up in participatory groups if we intend to achieve significant results. Keep in mind the people in these groups are predominantly production line employees. Their main responsibility is to perform the methods of transforming raw materials into products. Making improvements in how they do their job is unfortunately a secondary responsibility. Much as we would like it to be otherwise, the nature of production does not allow it. Thus, the best we can hope for is to stimulate the work force to develop improvement ideas and provide some time and resources to assist in this process. We cannot allow this task to become their primary function and take a large proportion of their time. This means that whatever ideas the team has, they need to be implemented and evaluated with as little queue time as possible. The common database usage, which makes communications excellence practical, is the best way to manipulate data for monitoring progress and making evaluations.

We have seen in these few examples that JIT, like manufacturing engineering theory, is advanced toward practical reality through CIM. It became evident that JIT is great in theory but not totally practical without significant help from CIM. CIM is the catalyst necessary to make JIT work.

DESCRIPTION OF THE LEAN MANUFACTURING PROCESS, AN EXTENSION OF THE JIT PHILOSOPHY

Everything I've said so far applies to JIT and lean manufacturing. But there is a reason why over the last ten years JIT has become a lesser used acronym than lean manufacturing (And now this term is even evolving into simply "lean"). Here are some reasons. First JIT has a foreign aura to it (even though it is an American invention) because its most fervent proponents were originally Japanese companies. Second, lean implies tough. We hear terms for rugged individualism or team effectiveness as "lean and mean." Translate this into business acronyms and we get a meaning of being tough on waste and being a "no nonsense" efficient competitor who is successful. A third reason may be because JIT originally was more philosophical in its approach, spelling out reasons for accepting its principles and leaving how to apply it to the individual convert. While lean manufacturing, philosophically identical, gains from its embrace of American pragmatism;, tell people how to do it. So lean manufacturing becomes an applied technology, which is more readily grasped by practitioners, than a higher-level philosophy. So while identical in concept to JIT (it must be because there is no difference in approach), lean manufacturing is thought of as the pragmatic application of JIT theory. As such lean manufacturing is more dogmatic in its approach.

Lean manufacturing encourages improvements to be done in a set manner. Things are to be done in certain step-by-step approach. First goals are set. But even this is constricted. Every firm "doing lean" must create four sets of goals: improve quality, eliminate waste, reduce total cycle time, reduce overall costs. The company will set goals in these categories specific to its unique situation.

The second step is to create a road map of how the company does its seven steps of the manufacturing system. Although the vernacular used is creating the value stream. This is industrial engineering process flow mapping.

The fourth step is to look at the process flow and see how work can be level loaded. This is done by using a variant of the "must/wants" technique. Here the process flow is analyzed for what absolutely needs to be done. All activities are judged on a costs benefits routine as to value added. If a "want" value added is significant it is kept if not done away with. Obviously items that do not add value to the process are very suspect and tend to be eliminated.

The fifth step is simply to implement, trouble shoot, and strive for continuous improvement.

This dogma is taught simply with very little enhancements than I've used to state them, above, or made very complex with lots of bells and whistles for encouragement of the most creative ways to get the most out of the process. A simple matter of judgement tempered by resources and the urgency for need for improvement is required by the implementor. There are literally dozens of subsets of each of the steps. Virtually all of which trace their origins to industrial engineering principles. The thing to keep in mind about a "lean" approach is that there is a ritual to it. Perhaps necessary to motivate the non-engineers into doing the right thing when they haven't had the benefit of an engineering education to understand the "why and where for" of this systematic methodology for eliminating waste in the business process. It is almost religious in its intensity, which can be good or bad. Good because it creates motivated practitioners, and bad because it is a serious deflator if the process goes wrong due to errors along the way. When that happens, those damaged by failure tend to condemn the entire process and will not benefit from further improvements.

Both JIT and lean manufacturing start out stating their reason for being is to eliminate waste, hence enhance profitability. As I said lean does add some additional slants to the techniques/ philosophies for eliminating waste, primarily in the manufacturing floor area. Let me close this discussion by focusing on one. One that is not mentioned at all in JIT discussions but a definite application of the concept of doing work in a manner that grows an individual's understanding. This most prominent one is known as the 5 S's. Let's take a look at this.

The 5 S's is primarily a workstation focused effort to optimize the performance of the work station. In one format or another I've discussed these items independently in previous chapters. Let's look at them together in their synergistic form. My opinion is that this is truly a case of the sum being greater than the addition of all the parts. By looking at this as a system we get better

Various Versions of the 5 S's
(Work Place Organization to Reduce Waste)

The literature reveals many variations of this common theme. Here are the four most common versions

A	B	C	D	Meaning
Sorting	Sort	Sort	Sort	Sort through tooling, supplies, methods instructions at the work station. Keep what is currently needed. Dispose of the rest.
Storage	Straighten	Set in Order	Straighten	Arrange in a logical order so the operator can find items easily when needed.
Shining	Shine	Shine	Scrub	Literally make the work station look like a "show place" by cleaning and eliminating all trash and contaminations.
Standardize	Standardize	Standardize	Stabilize	Assure that everyone is doing all of the "S's" in the same manner, such that a normal routine is maintained.
Sustaining	Sustain	Sustain	Sustain	Maintain training and discipline so that work stations are kept at a certain level of cleanliness and order for optimum production.

Figure 17-2. The Semantics of the 5 S's

results than would be expected by doing them independently at haphazard times and as secondary activities connected to other purposes.

Unfortunately the only thing standard about the 5 S's is that there are five words all starting with the letter S more or less meaning the same thing. I've looked at many sources and here are the four most often used versions, shown in figure 17-2:

As we see from the figure the 5 S's are more or less a pneumonic memory enhancer to get operators to keep clean efficient workstations. With the correct premise being that by doing so we will be better able to reduce waste while producing products at these magnificently maintained and cleaned work stations. As I said the premise is correct. The devil is in the details of constantly admonishing operators to perform cleaning and maintenance operations as part of their on-going responsibilities.

THE MISTAKE PROOFING TECHNIQUE

The JIT philosophy espouses striving to do all work excellently, for example, without mistakes. This is a Japanese contribution to the success of manufacturing processes. The idea that it is possible to do work perfectly, if one focuses efforts to make it impossible to do it wrong. This in a nutshell is "poka yoke" the Japanese phrase for this process. Both JIT and now lean manufacturing jump on this topic "hammer and tong." They embrace it enthusiastically and strive to make it a standard part of the lexicon. Let's briefly review the process.

One of the best ways to gain the highest degree of robustness is through a process called mistake proofing, also known as error proofing. Mistake proofing has been popularized by Shiego Shingo in his book *Zero Quality Control: Source Inspection and the Poka-Yoke System* (Productivity Press, Portland, OR, 1986). Professor Shingo's book is the authoritative reference on this process, but about 3 years ago I came across an excellent pamphlet on this process, which is a great quick to the point tutorial reference for the process It is called *The Basics of Mistake Proofing*, by Michael R. Beauregard, Raymond J. Mikulak, and Robin E. McDermott, Quality Resources, New York, 1997. This little pamphlet is perfect for those who need to have a handy reference manual for the desk drawer, or in their briefcase. In their book Messrs. Beauregard, Mikulak, and McDermott have an excellent definition of mistake(error) proofing;

"mistake-proofing is a technique for making it impossible to make mistakes. Mistake proofing is accomplished by making permanent changes to equipment, operations, or procedures that eliminate opportunities for errors or provide an immediate signal if a mistake occurs"

And that's exactly what its all about. While the technique was developed for manufacturing processes, it has been adopted for engineering processes, therefore it is valid for making project steps robust and is also used extensively in project management.

Mistake proofing depends heavily on the process of problem solving. And as I've stated in Chapter 7, the basis of all problem solving is the classic scientific method; which is reproduced below.

1. Make observations.
2. Develop a hypothesis.
3. Test the hypothesis.
4. Make revisions to the hypothesis based on the test.
5. Test the revised hypothesis.
6. Reach a workable conclusion.

So all variants of process "how to" steps are simply applications specific versions of the scientific method. Here is what I believe to be a most pragmatic series of chronological steps for mistake (error) proofing.

Step 1. Identify Points in the process where problems are most likely to occur.
Step 2. Fully describe the process.
Step 3. Brainstorm potential reasons for failure and identify potential root causes. Use failure mode and effect analysis (FMEA) technique as aid in this process.
Step 4. Prioritize the problems, by frequency of occurrence—use pareto principles
Step 5. Create solution hypotheses for the problems, with the goal of making it impossible to do it wrong.
Step 6. Test the proposed solutions. It is recommended that more than one source be required, corroborate results.
Step 7. Use a costs vs. benefits analysis to validate worthiness of the proposed solution.
Step 8. Implement Solutions, Measure Results, and Iterate as Necessary to Achieve Desired Results.
Step 9. When the first iteration is complete, determine if potential for errors have been eliminated, or at least minimized.
Step 10. Iterate as required.

We can see that Mistake-Proofing is another way of applying the scientific method; only here we have a very focused approach toward catching errors before they occur, or immediately as they happen.

One of the most powerful tools used in Mistake-Proofing is the failure mode and effect analysis (FMEA). Lets briefly review this process.

FMEA Steps
1. Brainstorm list of potential failure modes
2. List all potential causes for a project step to fail
3. Severity—select a consequence level of the failure (scale 1–10). How important is it to the success of the project if the potential failure happens.
4. Occurrence—probability or frequency of the failure happening (scale 1–10)
5. Detection—probability of the failure being discovered (scale 10–1). This is a reverse scale. High detection is a 1, low is a 10.
6. The values for severity, occurrence, and detection are multiplied together to get a Risk Score.
7. The results of the these steps are transferred to a table (as shown in Fig. 17-3) for comparison purposes.

Typical Failure Mode and Effect Analysis Work Sheet

item no.	Potential Problem	Effects	Severity	Frequency of Occurrence	Detection	Risk Score
1						
2						
3						
4						
5						
6						
7						
8						
9						
10						

Figure 17-3. Typical FMEA Work Sheet

The highest score would have the most impact on the project, if it were to occur, therefore the highest level of resources would be applied to prevent it from happening.

Figure 17-3 is a typical work sheet for tabulating various potential problems scores. Please keep in mind; this is a tool for helping to compare the negative worth, so to speak of a problem and helping to select the priority for working to solve potential problems. Never be miss-led into believing that it is an objective activity. It is simply a sorting activity based on biased selections of worth. Therefore it is only as good as the bias demonstrated by those making the selections is correct. In fact if the FMEA results are significantly out of phase with what was previously perceived to be the proper order, then it must be looked at very suspiciously to make sure all prevailing assumptions are acceptable.

SUMMARY

Just in time is a popularized version of good manufacturing engineering practice. And lean manufacturing is pragmatic "how to do it" version of JIT. They both promote the philosophy of doing all phases of manufacturing in a mutually supportive optimum manner. We see that elimination of all waste is the goal of just in time and lean manufacturing, which leads to practices of examining how we do things to look for improvement opportunities. This is not a new theory of engineering; it is simply a repackaging in a popular fashion, especially in a fashion that appeals to non-engineers.

We also examined how to bring JIT principles into practical reality. Here we saw that only through CIM techniques can JIT become anything more than a good attempt to improve manufacturing productivity. We saw the reasons why teaming with CIM philosophy and technology were the only way to achieve tangible results. This showed once more that good manufacturing theory requires computer-based procedures to be truly optimized.

JIT and lean manufacturing offers us the opportunity to popularize sound manufacturing engineering practices. It is not important for non-engineers to recognize the source of the theory. It is important, however, that the principles be practiced in manufacturing operations. To do so makes it more probable that the process can approach optimization.

REVIEW QUESTIONS

1. Explain why JIT/lean manufacturing is more than delivery of materials to the required workstations when they are needed.
2. Identify the claims for improvement attributed to JIT and lean manufacturing that are considered psychological and explain why. As part of your answer, evaluate whether a psychological approach can be used to make permanent changes in operations.
3. Discuss how reduction in inventory levels leads to elimination of wastes in the manufacturing processes. Also discuss the dangers of reducing inventory levels in inappropriate ways. Give an example of inappropriate methods.
4. Define just in case and compare it to just in time. Give an example of a proper use of just in case.
5. Explain the push production control system and the pull production control system with respect to just in case and just in time.
6. Why is JIT considered an enabling philosophy for achieving manufacturing optimization?
7. Explain why the precept "produce by methods which allow for the development of people" is difficult to achieve in a manner that creates a permanent change in practice.
8. Show why flexible automation within a CIM system is compatible with JIT philosophy.
9. Explain why kanban will not work in a complex manufacturing operation without being tied to MRP II. Show how the pull system incorporated in MRP II software allows elimination of waste.
10. Discuss the relationship between CAPP and JIT. Show how CAPP is an enabler for JIT.
11. Explain what the precept "produce with perfect quality" means in a CIM environment.
12. What is the relationship of participatory programs to the goal of waste elimination? Show how common databases enhance the ability of participatory programs to succeed.

Chapter *18*

ENVIRONMENTAL CONTROL
AND SAFETY

Awareness of safety and environmental hazards has led to a greater concern by management for impacts on employees and the company. Self-preservation is a human trait. However, the factory is an artificial environment where more than natural instincts are required to maintain a safe habitat. Hence a body of knowledge called environmental control and safety has been developed.

The safety aspects came first. By dealing directly with causes and effects, it was possible to codify how work should be done to minimize the always present danger. When materials are cut, bent, welded, sintered, melted, formed, ground, milled, and so forth, the very nature of the activity involves an almost violent action against the work piece. If that action is misdirected, it can end up being applied to the operator or plant equipment, resulting in some degree of harm. Safety codes were designed to minimize and at best eliminate these causes and effects.

The environmental aspects came much later, when it was learned from circumstantial evidence that some of the processes being used at the workplace could result in harm to the operators and equipment after a long period of exposure. This has led to the continuing development of a body of knowledge concerning environmental hazards.

Accordingly, the topic of safety is divided here into two categories: one dealing with direct cause and effect, usually of an immediate or short-term nature, called "workplace safety," and the second dealing with indirect cause and effect, usually over a medium to long time period, called "workplace environmental control." A significant part of the understanding and corrective actions required in these two divisions involves applications of engineering disciplines. Therefore, as the technical arm of manufacturing, manufacturing engineering has been assigned the responsibility of providing for a safe workplace. This chapter will develop the managerial strategy involved in carrying out this responsibility.

THE NEED FOR AN ENVIRONMENTAL CONTROL AND
SAFETY CONCERN

It would be a mistake to think that safety and environmental programs are wasteful of capital and of little value. Properly conducted, these morale-building programs lead to greater profitability. Factories with good safety records also have good productivity records. Safe operation means that both the operator and the equipment are protected from harm. If this is not accomplished, the machines will operate less of the time, and profits will be correspondingly reduced.

Let us look at the safety aspect as it affects the operator. Operators are employed to control certain functions of the manufacturing procedure so that the purpose of the factory is accomplished. If they do their work effectively the company can make a profit. It is the manufacturing engineer's responsibility to plan how the operator will do his or her task. As we have seen in previous chapters, manufacturing engineering is continuously designing and redesigning the way work is to be performed with the objective of achieving the highest level of productivity. To achieve the productivity goal, the manufacturing engineer needs a good process, one that does not break down. An accident is a breakdown that causes a loss of productivity and, all too often, human suffering. It also causes property damage and an unanticipated additional cost. Machine breakdowns due to components wearing out and requiring replacement are expected and can be planned for to a certain degree; that is, they can be budgeted and staffed for. But accidents are breakdowns that should not happen, and when they do they are typically expensive in damage and in lost profitability. Therefore, it is to the advantage of management to institute programs to minimize all and eliminate most accidents.

Making the workplace safe for operators is the best way to eliminate the productivity deflator that we call accidents. In effect, we are protecting the operator from the machine and vice versa. Machines are designed to perform certain functions. The designer has taken into account the laws of physics and chemistry and has made the assumption that the function will be carried out and controlled as intended. If the function cannot be carried out and controlled as intended, the machine will not perform its function. The best that can happen is that the machine will not work; the worst is that an accident will occur. In that case, the operator or the machine or both will be injured. Therefore, operator safety is achieved by making sure that the operator can only operate the machine in a safe manner. If this is done successfully, then the operator is working in harmony with the goal of the factory, that is, producing a product in accordance with the plan, hence profitably.

The objective of the safety program is therefore to protect the operator from the machine and the machine from the operator. By doing this, we will substantially increase the factory's capability of producing at the planned productivity level. Accidents stop portions of the factory from working, and therefore they are contrary to our productivity goals.

The discussion above gives the economic foundation of the safety program. But the safety program also has a psychological basis. An operator who is afraid of being injured will work at a slower pace than desired, and we will have less than desired output levels. On the other hand, if the factory has a good accident prevention record, the operator will not fear injury and will probably work at the desired pace. There are many reasons for not working at the desired pace, and many programs have been conceived to eliminate these problems. Unfortunately, many otherwise astute managers forget the safety aspect. We must be aware of the human instinct for self-preservation and make it work to our advantage. A safe workstation is more likely to be a productive workstation.

THE CASE FOR ESTABLISHING FORMAL ENVIRONMENTAL AND SAFETY PROGRAMS

Let us look at environmental and safety issues from the viewpoint of the public at large. Goodwill is an intangible factor that cannot be measured in dollars and cents like capital investment projects, but it is real nevertheless. Consider a company that has a good image and one that does not. If both companies are competing for a contract, the one with the good image will have an advantage in the evaluation process. Safety and environmental programs are vital parts of such image building. From senior managers down to the lowest paid workers, people like to be associated with organizations that show concern for their employees.

We see then, there are more than economic benefits to be derived from the environmental and safety program. We can expect to obtain morale benefits that enhance the company's ability to perform as a team. The goodwill obtained is beneficial in dealing with outside agents such as government regulators influencing rulemaking activities. It facilitates public acceptance, which helps in selling the company's products. And it helps to attract competent workers by indicating that the company will be a good place to work. These are but a few of the intangible benefits of an effective environmental and safety program.

We have two powerful reasons for establishing environmental control and safety programs: economics and image building. Both lead to enhanced profits. Manufacturing engineering is responsible for the technical aspects of this program and frequently for the nontechnical portions as well. Some companies place total control of the program under the human resources (historically known as employee relations) function. Some share control between human resources and manufacturing engineering. Other firms place the entire program within manufacturing engineering with only advisory consultation from human resources. Whatever the management concept employed, the program consists of technical and non-technical aspects.

KNOWLEDGE REQUIREMENTS FOR FACTORY ENVIRONMENTAL CONTROL AND SAFETY PROGRAMS

The typical environmental control and safety program is divided into awareness (publicity) activities and planned action activities. But before we delve into these particulars, let us review the knowledge necessary to effectively carry out such programs.

In most manufacturing engineering organizations, advanced manufacturing engineering will be assigned responsibility for keeping informed about regulatory policies related to workplace safety and environmental control. The specific items covered by such policies are described in the following subsections.

Fire Codes

Provisions for building fire safety vary from jurisdiction to jurisdiction. So too do codes regulating electric wiring and controls for machines and equipment. Hence it is necessary to know the prevailing codes. Some companies also have restrictive codes that they impose on themselves in order to receive the lowest possible insurance rates. Manufacturing engineering must be cognizant of all these rules and regulations and make sure they are applied to all capital-related projects.

There are aids to achieving knowledge in these cases. Most insurance companies belong to underwriters associations, that are willing to document correct safety procedures and assist customers in making site surveys. Manufacturing engineering is expected to take advantage of these sources and make the proper liaisons.

Safety of Structures Above the Work Floor

These structures include ladders and elevated work platforms, both temporary and permanent. Over the years the Occupational Safety and Health Administration (OSHA) has codified standards for such platforms and ladders. It is manufacturing engineering's responsibility to make sure that the factory is in compliance with these standards. Noncompliance is not only unsafe, it is also against the law and can result in criminal as well as civil penalties. Manufacturing engineering must protect the company against this liability.

Also falling in this category would be the placement of apparatus used in a workplace. For example, the placement of machinery, fire extinguishers, and electrical apparatus above the work floor requires guarding and visual identification procedures. Manufacturing engineering must ensure that the appropriate codes are complied with.

Control of Hazardous and Toxic Substances

This is fast becoming one of the most complex and emotional issues facing manufacturing engineering. Toxic and hazardous substances present an immediate and a long-term health problem. Toxic substances are construed as poisons, acids, and other materials that can cause immediate harm. Hazardous substances are materials that, while not having an immediate effect on personnel, represent a statistical health hazard over the long term. Where toxic substances are present, it is necessary to have quick emergency procedures to be followed in case of an accident. Hazardous materials require administrative procedures to ensure that exposures are below the allowable levels mandated by the U.S. Environmental Protection Agency (EPA).

The problem facing manufacturing engineering is in the identification of all toxic and hazardous substances in the factory. Herein lies the complexity of controlling such substances. To ensure that the company does not inadvertently violate an EPA regulation, manufacturing engineering must determine the matrix of all chemicals present. This is a difficult task because most fluids and solids are purchased under trade names and the exact chemical composition is not known.

Therefore manufacturing engineering must solicit information from vendors or conduct analytical tests, probably doing some of both. This is costly and time consuming.

The emotional issue arises when toxic and hazardous substances in the factory are identified, as they almost inevitably will be. Government regulations require notifying the employee of the potential hazard and of the preventive and emergency control procedures to be used to minimize the hazard. This, coupled with the somewhat sensational coverage of previous problems of this nature by the communication media, encourages emotional rather than objective reactions on the part of factory personnel. Therefore, manufacturing engineering must keep abreast of available information on these problems.

Manufacturing engineering is usually the only technical source readily available to interpret regulations and to properly inform management about the potential problems and their solution.

Noise Abatement

A particularly vexing problem whose solution requires considerable knowledge and expertise is that of meeting the factory noise standard of 85 dBA per 8-hour exposure. It has been determined that a consistent noise level above 85 dBA can lead to loss of hearing over a long period of time. Therefore, current OSHA regulations require an engineering solution to reduce the noise level in areas where it is over 85 dBA. Manufacturing engineering must be the source of knowledge for the company's compliance with this regulation.

The fact that an engineering solution is mandated is important. This can be understood most readily by explaining what is *not* an engineering solution. The use of protective equipment for factory personnel is not an engineering solution. OSHA has allowed protective equipment only as an interim solution until an engineering solution can be devised and implemented.

An engineering solution to noise abatement requires a detailed understanding of the physics of sound and vibrations. In the past, this has not been a part of the primary expertise of the manufacturing engineer. But with a requirement to protect factory personnel from the danger of high noise energy, manufacturing engineers must develop expertise in acoustics and related fields.

I have found that searching for engineering solutions to noise problems leads to a greater understanding of the offending tools themselves, and this can have some interesting additional benefits. By understanding the tools better, we can design better applications and hence improve factory productivity. An example of this is an effort I carried out jointly with a nearby university to reduce the noise level of pneumatic grinders for metal polishing from 100 to 85 dBA or lower. In this project, we learned that a noisy tool was an inefficient tool. By using a proper muffler design and making some modifications of the airflow in the tool, we could develop a more efficient tool and one that met the noise standard as well. We were pleasantly surprised to find that the noise abatement project turned out to be a productivity improvement project.

I mention this example to point out the need to avoid thinking of noise abatement projects or any other safety and environmental projects only in terms of compliance. Managers must always encourage the attitude that knowledge applied to these projects should also lead to improvements in producing the company's products. There is no reason to be satisfied with merely complying with regulations. The approach should be one of complying and improving profitability at the same time. This is not always a simple task, but it is a necessary approach in highly competitive industries.

Safety Apparel

Manufacturing engineering is often called upon to specify clothing and protective devices that operators must wear to protect themselves from work hazards. Examples are welding jackets to protect welders from flash and sparks, steel-toed shoes to protect against broken bones, safety eyeglasses with impact-resistant lenses and wire mesh side shields to protect against foreign particles becoming embedded in the eyes, and noise suppression earmuffs.

Knowledge of safety apparel is essential to effective planning of workstation layouts. Manufacturing engineers must not only be competent in designing effective workstations, they must also be knowledgeable about the safety considerations and the proper apparel required.

The designation of proper safety apparel definitely affects the cost of operations. For example, if safety glasses are required, a company must provide prescription as well as plain safety glasses to all employees. This includes new prescription safety glasses if an employee's prescription changes. The fact that safety apparel is a business expense means that manufacturing engineering must consider its choice of such apparel as it would any other device or commodity it specifies.

Reporting Requirements

Understanding government regulations is essential. When government regulations are encountered, they usually require that compliance reports be submitted. This requires time to prepare the reports and an in-depth understanding of the regulations. Manufacturing engineering must be cognizant of the reporting requirements and must issue the reports on time. Many companies have suffered needless embarrassment and fines by failing to report correctly and on time. Management usually assigns this additional compliance responsibility to manufacturing engineering.

These are some of the areas of special knowledge related to safety and environmental control that manufacturing engineering must master. Mastery of these topics leads to the awareness activities and planned action activities that make up the environmental and safety program.

THE ENVIRONMENTAL CONTROL AND SAFETY PROGRAM

This program is divided into awareness activities and planned action activities. The major concept in both parts of the program is a dedication to the principle of "work smart, work safe." We will review a typical two-phase program that exemplifies this commonsense approach and show how such a program improves the safety of the workplace.

Awareness activities and planned action activities are administered by manufacturing engineering or human resources. If the majority of the activities will involve applications of technology to solve safety and environmental problems, then manufacturing engineering is the administrating agency. If the prime thrust is that of maintaining an established safety and environmental program, where the technical problems are minimal, then human recources administers the program. Since we are primarily interested in manufacturing engineering management, we will explore a program administered by manufacturing engineering. This will be an environmental control and safety program skewed in favor of solving the associated technical problems.

The awareness activities in the program are designed to make all employees more aware of the need for safety and environmental care while at work. Awareness is an intangible thing, and those responsible will always have difficulty in determining whether their effort has been successful. Perhaps this can only be known with certainty when an accident occurs and personnel react quickly and correctly. This, however, is counter to the precepts of the program, which is successful only if accidents and mishaps are prevented. Measurement of a successful awareness program must therefore remain essentially intangible. We can measure effectiveness subjectively, however, by noting conditions in the workplace, looking for signs of good housekeeping, lack of safety hazards, properly cared for emergency equipment, and so forth. Awareness programs are, in fact, evaluated by a lack of negative factors being discovered during safety inspections. This gives an indication of the overall awareness level.

The planned action activities are those initiated to correct safety and environmental deficiencies that have been discovered and those required by law to maintain compliance standards. A significant proportion of planned action activities result from awareness activities, that is, subjective audits and inspections.

EXAMPLE: A safety and housekeeping audit may uncover a frayed fire hose, a chance occurrence. This would result in a planned action activity to replace the defective fire hose. This activity is measurable. We can measure the time to place an order for a new hose, monitor for its delivery, and determine whether the hose is installed by the targeted date.

Whereas awareness activities are usually measured in a subjective manner, all planned action activities are measurable in an objective fashion. Whether a safety article is written for the plant newspaper is a measurement of an awareness activity. The measurement is not concerned with whether the article had any impact on its readers in establishing a safer workplace. That is the purpose of the awareness activity, but it is beyond the scope of the measurement. Planned action activities can be measured for their effect. For example, making a sound contour map of the workplace to determine whether it is in compliance with the 85-dBA, 8-hour standard is a planned action activity. The effect of the work gives the answer. After the map is completed, we can easily tell whether the area is in compliance. If it is not, we progress to another measurable planned action activity, that of engineering a solution to lower the noise to an acceptable level.

Both awareness activities and planned action activities are merged to form an overall measurement. This could be a safety standing with reference to a goal set internally by the company or to a compliance level set by a controlling agent, such as an insurance company or government jurisdiction. Such a goal must be very broad-based. It might involve the accident frequency per time period, the accident severity level, or combinations of both. In order to have an environmental control and safety program it is necessary to agree on a goal that can be understood and measured. Figure 18-1 is a flowchart of a typical technically oriented environmental control and safety program. We will review the various components of this program in the next two sections of this chapter. However, the first area to understand is the overall goal designated by the box titled "OSHA Incident Rate Goal." This is the culmination of the awareness activities and the planned action activities. If these activities are carried out effectively, we will see an improvement in the OSHA incident rate. An OSHA incident rate is a combined measure of accident severity and frequency, devised by the Occupational Health and Safety Administration, which measures overall safety improvement over a period of time. The important thing is that it is a measure of the effectiveness of the environmental control and safety program.

AWARENESS ACTIVITIES PROGRAM

Hourly Safety Meetings

Hourly safety meetings are designed to create awareness among the operators. The term *hourly* refers to personnel who are usually paid by the hour and are often members of a recognized labor union. These meetings are conducted by the foreman or other member of management associated with shop operations.

The topic for each meeting and the script to be followed is provided by manufacturing engineering. The AME or MP&WM unit prepares a program for the entire year consisting of the general and specific topics of major concern to their company. Items such as fire safety, first aid, toxic substance control, proper handling and storage of materials, electrical equipment hazards, overhead crane safety, and vehicular safety are covered in a lesson plan format. The format is very explicit in counseling instructors on how the lesson is to be presented, to the level of identifying the key message to be transmitted to the class.

Hourly safety meetings are held during the normal work week to emphasize that safety adherence is an important part of the company's business. Since these meetings interrupt the production of products, there is a tendency on the part of lower level management to be lax in adhering to the schedule. Occasionally meetings are skipped in order to complete an order or to catch up on production. Manufacturing engineering must be very critical of all excuses and must bring pressure to bear to minimize the skipping of meetings or the absence from meetings of segments of the hourly

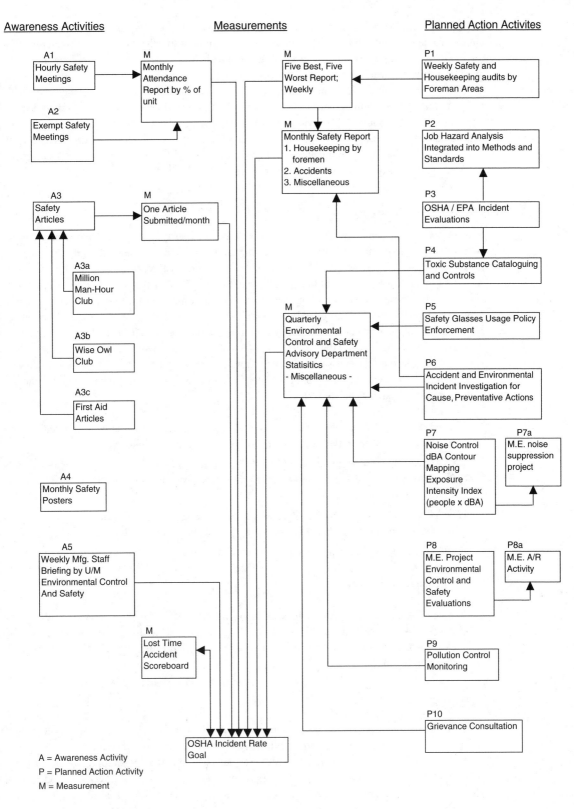

Figure 18-1. Environmental Controls and Safety Programs

work population. Surveys have shown that areas that are lax in safety meetings tend to have a higher accident frequency rate. Over the long term, skipping meetings is counterproductive with respect to overall shop productivity levels. Besides the obvious humanitarian reason, maintaining a good productivity level is a motivation for insisting that these meetings be held.

Manufacturing engineering must also ensure that safety meetings do not become an obstacle to maintaining the rhythm of production. This means that meetings should not be scheduled too frequently or last too long. The frequency of meetings should be a fair compromise between the requirements to maintain production and to keep a high level of awareness of the need to work safely. Therefore, most environmental safety programs schedule hourly paid labor safety meetings to occur on a monthly basis and to last for approximately 30 minutes.

Salaried Safety Meetings

In most respects these meetings are the same as those for hourly personnel. Salaried refers to all employees who do not fall into the hourly classification. The topics for salaried safety meetings are the same as those for the hourly safety meetings. It is good practice to expose office and professional personnel to the same program as the shop floor personnel. This enables the salaried employees to act as unofficial safety monitors during the times when they are in or traversing the factory operations area. Everyone in the entire business organization is expected to report unsafe conditions or practices to manufacturing engineering for corrective actions. The more people who are sensitized to the need for safety, the greater the number of potential dangers that will be reported and corrected.

In addition to the series of safety meeting topics, professional personnel must be made aware of the fact that the results of their work could cause a safety or environmental hazard on the factory floor so that they can guard against this.

EXAMPLE (a): A weld fabrication design that requires a welder to assume a cramped or awkward position to perform the work would expose the welder to a higher accident risk.

EXAMPLE (b): An engineer might decide to use a highly toxic solvent for cleaning parts before assembly. By doing some research the engineer could find a much safer solvent that accomplishes the same purpose.

EXAMPLE (c): A maintenance engineer's desire to save power by turning out lights in areas of the factory might pose a hazard to personnel who have to traverse the area.

EXAMPLE (d): A company personnel policy that limits crane personnel and substitutes operators in making certain lifts does not consider the increased safety hazard incurred by using inexperienced personnel.

The salaried safety program must have as one of its aims increasing the awareness of the safety consequences of the professionals' work. In this way it can reduce the incidence of accident by proxy. An hourly operator who is careless may injure himself. A professional who is careless may cause injury to others.

Safety Articles

This is a companion program to the safety meetings. Again, the purpose is to create a high level of awareness among personnel so that they will carry out their duties in a safe manner.

Many manufacturing organizations have newspapers to impart information of interest to employees. These house organs have a goal of improving morale and thereby improving productivity. They do this by extolling good performance by an employee or groups of employees. They also carry human interest stories that elicit sympathy for the company or a desire on the part of the employees to do their jobs better. Part of this program should be the safety message. Since a good safety goal is a step toward achievement of the productivity goal, it is proper for the plant newspaper to carry safety articles.

Safety articles are usually positive. It would do little good to report an incident that resulted in an injury to an employee as a lesson to other employees to exercise greater care. Safety articles usually report actions through which injury was avoided, or instruct personnel in how to avoid accidents. A third type of safety article is one that commends employees for achieving a recognized goal in minimizing accident frequency and severity. The last type of article usually ends with a challenge to do even better, to reach the next plateau. Let us look at some examples of these types of articles, as outlined in Fig. 18-1.

"Million Man-Hour Club." Articles on the "Million Man-Hour Club" salute the achievement of an area of the shop, or the entire factory, working for a time equal to or better than 1 million man-hours without an accident that causes an injury. They usually compare the area's previous best performance with this milestone and quote workers on the importance of the achievement and how proud they are to work safely and to work with people who care about safety. This is usually followed by a statement of appreciation by the manager of manufacturing with a challenge to the group to continue the accident-free performance and to other groups to emulate this achievement. If the plant newspaper contains photographs it will print one showing the unit manager and a few operators receiving the safety award plaque on behalf of the unit. Usually as many people as practical within the unit will be encouraged to be in the picture.

"Wise Owl Club." The "Wise Owl Club" is a national institution that recognizes individuals for preventing the occurrence of serious eye injury. Articles on this subject commend individuals for avoiding an accident by conducting themselves in accordance with the safety rules of an area. Usually this means that the individual had the good sense to be wearing impact-resistant safety glasses that prevented a missile from entering his or her eye. The article frequently shows a picture of the operator holding the safety glasses with the shattered but intact lens. The text explains the series of events that led to the safety glasses protecting the employee's eyesight. The purpose of this article is to show that adherence to safety rules does, in fact, protect people from harm. Such articles also state that membership in the Wise Owl Club is only given to those with the foresight to work safely.

First-aid articles. These articles are for positive instruction. Their purpose is to inform all personnel on how to react to the need to apply first aid in a particular set of circumstances. The articles are intended not to supplant first-aid manuals or first-aid classes, but also to heighten awareness of the need to apply first aid quickly and correctly if the occasion arises. There are two general types of first-aid articles. The general announcement type states that a first-aid session of a particular kind is to be conducted, gives the time and place, and encourages readers to attend. The second type actually explains a first-aid technique and usually includes instructions on how the reader can obtain further information.

The purpose of all these types of safety articles is to make all personnel more aware of safety. It is easy not to write such articles. Therefore, most managers of manufacturing engineering will take steps to encourage the writing of such articles. They will give a quota of articles to be written to selected personnel and measure their performance in meeting the quotas.

Monthly Safety Posters

Another good way to create safety awareness is through the use of safety posters. Good sources of posters include government agencies, private safety agencies, professional societies, unions, and producers of safety equipment. They produce posters that are witty and eye-catching and convey the specific safety message. It is up to management to use them correctly.

One effective way to use safety posters is to post them prominently and change them periodically, preferably monthly. They should be posted where personnel have to see them but are not distracted by them while working. Good locations are at the time clocks, on the notice bulletin boards, or beside the entrance to the plant cafeteria. They should be changed monthly so that they continue to attract attention. If a poster stays up for too long a period of time, it becomes part of the background or landscape and is no longer read. Once the employees read the poster, its design should get the message across.

Manufacturing Staff Meeting Environmental Control and Safety Briefing

This final component of the awareness activities phase may be the most important one because it is aimed at the senior levels of management. The briefing consists of a manufacturing engineer presenting a summary of the previous period's environmental control and safety activities and occurrences. The engineer also informs the staff of upcoming compliance requirements and what must be done by management to meet them.

The briefing draws attention to the safety and environmental issues facing the manufacturing function. It also reminds senior managers to carry out their environmental control and safety program duties. The manager of manufacturing uses these briefings as a mechanism for showing support for the entire program.

The briefing usually lasts 5–10 minutes and is followed by a question-and-answer session on the briefing or any related subject the managers may bring up. Frequently, discussions occur that require the manufacturing engineer to act as the voice of authority. All this aids the awareness program. If senior managers discuss safety and environmental concerns, so will their subordinates. The briefing and discussion may only last 15 minutes, but this brief time is valuable for raising the safety awareness factor.

PLANNED ACTION ACTIVITIES PROGRAM

Weekly Safety and Housekeeping Audits

This is often considered an awareness activity because the action of taking an audit creates a heightened awareness of proper safety and environmental procedures. However, it is discussed here under planned action activities because it creates action plans to correct discrepancies.

Audits are common activities in manufacturing. We audit for adherence to quality plans, for adherence to procedures, for the status of parts in the production cycle, for adequacy of efficiency measurements, and so forth. so it is natural to also audit for compliance with safety and environmental control rules.

The audit for housekeeping and safety is a detailed examination of things in a specific area to uncover environmental and safety deficiencies. It is usually carried out by a team consisting of the manufacturing engineer, the area foreman, and a shop manager from another area, the latter being included to instill a sense of competition. Managers auditing each other's areas are more likely to uncover deficiencies than if they were auditing their own areas. They will also strive to find more deficiencies than their counterpart finds. To enhance the sense of competition, manufacturing engineering will also publish a best and worst report. Areas exhibiting the best housekeeping will be publicly proclaimed, along with those exhibiting the worst performance.

The result of the safety and housekeeping audit is a planned action report showing the discrepancies found, who is responsible for the corrective action, and when the action will be completed. Figure 18-2 is a sample of such a report. The report is compiled by the manufacturing engineer assigned to the tour, who is also responsible for updating it as actions are completed or new items are added.

The results of the weekly safety and housekeeping planned action report and the weekly best and worst report are used to prepare the monthly safety report. This report is a detailed summary of the planned action activities of the month and a compilation of the safety statistics. The statistics are used to evaluate progress in achieving a safer working environment and to prepare the compliance reports required by government agencies and occasionally insurance companies. Virtually all planned action activities at some time or another create inputs to this report and its quarterly summary, which is known as the "quarterly environmental control and safety advisory department statistics summary." In essence, these reports reflect the effectiveness of planned action activities.

The weekly safety and housekeeping audits are the only specifically scheduled items among the planned action activities. All other planned action activities are responses to a specific need, and after they are accomplished they will not be carried out again until another need arises. A need is usually revealed by the weekly safety and housekeeping audit, but is sometimes generated by an accident.

Item No.	Item Description	Responsible for Completion	Date Entered	Promised Completion	Actual Completion
27	Frayed fire hose, column A-12E	J. Jones	1/26	1/27	1/27
28	Pallets in aisle near lathe 2604	F. Smith	1/29	1/30	1/30
29	Job hazard analysis not posted, VBM 2608	J. Jones	1/29	1/30	
48	D.C. circuit box door won't close, column A-17W	J. Jones	1/29	1/30	

Total items added this report:	32
Total items remaining open since last report	6
Accidents in area since last report	0
Total accidents YTD	1

Figure 18-2. Weekly Safety and Housekeeping Planned Action Report

Item 29 in Fig. 18-2 is an example of the weekly safety and housekeeping audit creating a need to initiate another planned action activity. It notes that a job hazard analysis was not posted on vertical boring mill 2608. This implies that the action is to post the analysis. But a competent engineer would review the current job hazard analysis before reposting it. If it no longer agrees with the methods and standards, it will have to be redone; hence the job hazard analysis will be integrated into a methods and standards planned action activity. Once this activity is completed it will be dormant as far as the environmental control and safety program is concerned, until it is reactivated by a need shown by the weekly safety and housekeeping audit.

Job Hazard Analysis Integrated into Methods and Standards

The job hazard analysis is the safety check on the job design. It is a system for determining whether the job method conceived by MP&WM can be carried out safely and without any medium to long-term environmental hazard to the operator. It is the final check before a job can be safely turned over to shop operations. The procedure for conducting the job hazard analysis is always the same. It starts with the methods sheet and the workplace layout, which the engineer will review for obvious safety hazards.

Let's look at the workplace layout (Chapter 7, Fig. 7-2). Note that it does not show any protective railing to separate the operator from the hazard, perhaps because it is not possible to install one and still have the machine perform correctly. This would be designated as a hazard. By identifying it as a hazard we are creating a heightened awareness of a possibly dangerous condition, but not necessarily an unsafe condition. (Similarly, a railroad station platform is hazardous because we could be hurt if we fall off the platform. Since we know the hazard exists, our awareness is heightened and an unsafe condition is mitigated.)

Now let us look at each methods (Chapter 7, Fig. 7-1) step and ask, "Is the operator being asked to perform an unsafe action?" Item 3 of the methods sheet reads: "Put in table T-slot nuts. Assemble studs to nuts and clamps." This is not a hazardous operation; only normal and reasonable care is required to do it. Item 2 calls for cleaning the table of the machine. This is probably not unsafe, but it introduces a new element to be checked by the engineer making the evaluation. Since we do not designate the solvent to be used in the cleaning operation, the engineer must investigate to see what is being used. If the solvent is known to be a hazardous chemical, the engineer will declare this step hazardous, and have added to the methods sheet the proper precautionary note on how the solvent is to be handled. This step of the method would also appear on the job hazard analysis sheet posted at the workstation.

Continuing our review of the methods sheet, we see that most steps are as simple as item 3 and appear to be non-hazardous. However, we must look a little further and ask ourselves what steps are

Item No.	Operation/Location	Hazard	Preventative Action
1	No railing between operator and workplace.	Operator can come in contact with rotating part.	Observe caution while operating machine.
2	Operation 3, clean machine plate with solvent.	Rash or burns on unprotected hands can occur.	Wear gloves, long sleeve shirt, safety glasses. Observe instructions for solvent use.
3	Operation 18, engage borozon wheel to workpiece.	Fast engage can cause wheel damage, broken wheel, flying missiles.	Engage per instruction. Do not stand in front of wheel. Wear safety glasses with side shields.

Figure 18-3. Job Hazard Analysis

related to activities that have caused accidents in the past, perhaps not at this workstation but at similar workstations. For example, accidents might occur at a point where an operator is addressing a tool to a rotating part. If the tool is not secured properly it could be twisted violently off the tool holder, resulting in damage to the workstation and possibly injury to the operator. Item 18 is a potential hazard by this definition. It instructs the operator to engage the borozon wheel into the sealing ring. Note that it says to do it slowly, so that the methods engineer had in mind possible damage to the machine, work piece, or even the operator if crash engaged. This step would appear on the job hazard analysis sheet.

Figure 18-3 shows part of a job hazard analysis sheet made up of the items discussed above. A job hazard analysis sheet should be posted at each workstation. Also, any change in a method calls for a review of the job hazard analysis sheet to see if it is still valid.

OSHA/EPA Incident Evaluations

This planned action activity occurs only if there is an accident or an OSHA or EPA audit discrepancy. In either case the incident must be investigated for cause and for corrective action taken to remedy the immediate situation and to prevent future occurrences. These investigations are carried out by technically qualified personnel and are reviewed with the manager of manufacturing engineering. Before a report is submitted to the respective agency, the manager of manufacturing, general manager, and perhaps legal counsel will also review it to ensure that the company is not continuing an unsafe and illegal practice and that senior management is aware of the consequences.

Toxic Substance Cataloging and Control

As mentioned earlier in this chapter, the emotional nature of the reaction to possible hazardous chemicals in the workplace must be controlled. Manufacturing engineering has the task of identifying all chemicals and substances used in operations and determining whether they are hazardous to health. This planned action activity is divided into two parts: identifying the chemicals and substances used, then determining whether they are harmful.

The identification step consists of reviewing all purchase invoices and determining what chemicals are contained in products purchased under trade names. This can be done either by laboratory analysis or by requiring vendors to specify the contents of their products. Asking the vendors for such information is preferable to laboratory analysis from a cost viewpoint. However, vendors may not want to disclose what is in their products, because they may be selling something that could be purchased more cheaply under the generic name, or because they feel they must protect proprietary information. This type of withholding of information is becoming rare, and in many jurisdictions violations of law, due to hosts of public disclosure laws. In the case where disclosure is not forthcoming, legal action may be required or the company must determine what chemicals are in the products by laboratory analysis.

Once the chemicals have been identified, manufacturing engineering must determine whether they are hazardous or toxic. Fortunately, this is not a subjective evaluation. The Environmental Protection Agency has published and continues to update a list of what are now legally designated hazardous and toxic substances. Therefore, manufacturing engineering consults the EPA list to see whether chemicals are toxic or hazardous.

The EPA list is a prime factor in dealing with chemical substances in the workplace. If a chemical is found to be on the EPA list, the prudent manager will get rid of it and find a safe substitute. If there is no suitable substitute, a procedure must be introduced to control the chemical in the factory.

The classic example of a hazardous material in recent years has been asbestos. Before the late 1960s, this heat-shielding, insulating material was not generally known to be a health hazard; therefore, its use was considerable. Once the hazard was identified, manufacturing engineering organizations began to eliminate use of asbestos where possible and restrict it where that was not feasible. An example of each would be (1) to eliminate asbestos blankets for heat treating and for welding preheat use and (2) to seal asbestos pipe insulation so that powder from decaying insulation would not become airborne. Asbestos is a familiar hazard. Unfortunately, there are thousands more that must be dealt with, and manufacturing engineering resources must be allocated for the task.

Safety Glasses Usage Enforcement

The planned action activity is to ensure that all personnel requiring safety glasses (impact-resistant and/or side shield types) are provided with them. This is done by requiring all employees to receive safety glasses from the designated dispenser and to wear them on the job. This often involves contracting for an optician to provide prescription services.

Accident and Environmental Incident Investigation

This is the same activity as the OSHA/EPA incident evaluation except that it covers non-OSHA/ EPA areas of concern.

Noise Control

This is a vast subject that is covered adequately in the large numbers of textbooks in the field of acoustics and noise engineering. We will concern ourselves here with the management aspects.

Noise levels have been designated by OSHA for various industries. The standard is set at 85 dBA for an 8-hour exposure, and some wish to lower that to 80 dBA. Manufacturing engineering must ensure that its factory is in compliance. To do this, the manager of manufacturing engineering must determine the level of noise in the factory and then take any necessary corrective action.

The preferred way to determine the ongoing or base-level noise is to conduct a noise contour mapping of the factory and the office areas. This is done by setting instruments in place to gather data, and it provides the basis for measuring future engineering solutions. To be acceptable to the government agencies, noise contour mapping must be done by competent personnel using recognized instruments. Therefore, management must either give its personnel proper formal training or rely on competent outside services. Either alternative is costly. The training approach is preferable because with trained in-house personnel it is possible to deal with noise problems as they occur. When we hire consultants we have a solution to a particular problem, but gain little knowledge to assist in solving future problems. So I believe the best course of action is to hire consultants for simultaneous training of in-house personnel and initial contour mapping, usually in a pilot area. This can be done by making the training a combination of a tutorial and learn by doing approach.

Once noise problems are identified they must be solved. The management approach has been to assign advanced manufacturing engineering to achieve the engineering solutions required by government regulations. This means that the operators must be isolated from the noise or the noise eliminated. Wearing protective devices can be a temporary solution, but management must continue to show progress in achieving an engineering solution, otherwise the company is liable for fines.

I have found that by working with company research laboratories and universities it is possible to meet the intent of the law. The experience with the portable grinder tool described earlier in this chapter is an example of an engineering solution that successfully satisfied the requirement.

Manufacturing Engineering Project Environmental Control and Safety Evaluations

This is essentially the same activity as the job hazard analysis except that it deals with capital equipment projects to ensure that they will be safe to use. To facilitate this, the manufacturing engineer responsible for safety and environmental control is often asked to review appropriation requests during the formulation stage to ensure that there are no safety or environmental problems.

Pollution Control Monitoring

Once hazardous chemicals have been identified, monitoring procedures must be established to ensure that they remain under control. Also, firms that discharge chemicals into the atmosphere, ground, or waterways as a by-product of their manufacturing processes must monitor these chemical discharges. Virtually all discharges are included in government regulations and levels of discharge are set along with permit getting requirements. Discharging at levels higher than those set by regulations or without a permit can result in serious civil and/or criminal penalties.

Manufacturing engineering is responsible for installing the necessary pollution control equipment as part of the capital equipment purchasing program. Once the equipment has been installed, manufacturing engineering must monitor it for proper use, report to the appropriate agencies, secure the necessary required permits, and make corrections and repairs when the equipment is not functioning properly. This means that manufacturing engineering must maintain audit logs and incident logs. The audit log shows when inspections have taken place, and the incident log records out-of-tolerance conditions and shows what was done to correct the situation.

Grievance Consultation

Grievance consultation is a service manufacturing engineering provides to human resources for employee grievances with a technical content. A significant number of grievances have to do with safety of the workplace, hence the inclusion of this item in the environmental control and safety program. Manufacturing engineering does not, as a rule, sit in on grievance sessions. It is only required to provide technical facts to human resources. However, manufacturing engineering is responsible for ensuring that human resources personnel understand the facts as presented.

INDOCTRINATION OF NEW EMPLOYEES

The environmental control and safety program as described in this chapter is meant to be an ongoing activity that grows in competency as experience provides. However, it is also extremely important that new employees are brought up to an acceptable awareness and safety performance level as quickly as possible. Past experience has shown that a higher rate of accidents occurs to new employees, those with less than six months service time than longer longevity personnel. And most of it comes down to ignorance about safety procedures.

To eliminate this problem many companies now embark on safety orientation programs. Most companies have orientation programs for new employees that they must attend during the first few weeks of employment. I have discussed some of that in our discussion of work rules, etc. As a primary feature of the orientation, employees are instructed on safety matters and environmental hazards as discussed above. They are introduced to the company's safety rules and taught rudiments of common sense factory safety before they are allowed to perform their hired job function without one-on-one supervision. Some companies even require that new employees pass a safety

Safety Rules

The following is a list rules to be followed for your protection while on the factory floor.
It is not all inclusive

1   Do not operate any equipment until you are fully instructed and familiar with it.

2   Do not operate any equipment that, in your opinion, is not in a safe condition.

3   Never remove guards from equipment.

4   Safety shoes are required to be worn by all plant, delivery, and installation employees.

5   Safety glasses, goggles, respirators, gloves, hard hats, etc. are required to be worn in all work areas where their use is designated.

6   Do not engage in horseplay anywhere on company property.

7   Avoid distractions - concentrate on your work.

8   When lifting, use approved techniques. Get help for heavy loads.

9   Keep all electric cords and wires off the floors.

10 Do not touch electrical switches, fuse boxes, or other electrical panels that you are not authorized to use.

11 Be familiar with the location and use of all fire extinguishers.

12 Practice good housekeeping at all times.

13 Report immediately any unsafe condition, which you observe.

14 Use the correct tools for the job.

15 Replace any railings and safety chains as required.

16 Keep all trash off the floors. Use trash receptacles provided.

17 Obey all posted safety rules.

18 If you don't know - ASK!

Figure 18-4. Workplace Safety Rules

awareness test before being allowed on the factory floor. The instruction is based on the company's workplace safety rules, whereby each rule is explained and discussed. In some instances, such as fire extinguishers, demonstration and practice is employed as part of the training. Figure 18-4 is an example of a typical set of safety rules used for manufacturing operations.

The desire to have a safe workplace depends on having safety aware employees. As we can see from the work rules, they are quite specific on what constitutes awareness. It is a mixture of dictates such as the requirement to wear safety glass and admonitions to be aware of one's surroundings with regards to being safe, such as knowing where the nearest fire extinguisher is. With training we teach awareness. We also teach that safety is everyone's business. So you could say that the primary focus of safety training is

- Be aware of changes in status quo that could affect you and/or your fellow employees safety.
- Do not do work in an unsafe manner, even if asked to do so.
- Report unsafe conditions to the appropriate supervisor immediately, including requests to do unsafe work.

If we get new employees to abide by these simple conduct rules and have older employees to continue to live by these rules, we will maintain a high level of safety awareness and enjoy a very safe work environment.

SUMMARY

The environmental control and safety responsibilities within manufacturing engineering constitute a very broad and expanding activity that cannot be taken lightly and must be properly managed. Certainly, they are as important as the more traditional responsibilities of manufacturing engineering.

The key to managing this activity is to divide it into awareness and planned action segments. The awareness portion develops a heightened sense of need for proper safety actions, while the planned action activities are specific steps taken to make sure the workplace is as safe as modern technology and management practices can make it. The key planned action activity is the job hazard analysis, a combined AME and MP&WM activity focused on making the workstation as safe as possible.

Throughout this chapter the key emphasis has been on understanding that a safe workplace environment can be a positive factor in enhancing productivity.

REVIEW QUESTIONS

1. Explain why it is good business sense for a factory to have a strong environmental control and safety program.
2. What is meant by direct and indirect causes with respect to environmental and safety programs?
3. Why are accidents called productivity deflators?
4. What is the basic theory on which effective factory safety programs are built?
5. Under what set of circumstances would management of the environmental control and safety program not be assigned to manufacturing engineering?
6. A manufacturing engineer is assigned responsibility for writing the specifications for acquiring a new CNC horizontal boring mill. Make a list of questions the engineer might ask to ensure that the specification complies with the following codes: fire, structures above the floor, hazardous and toxic substances, noise abatement, and safety apparel.
7. Describe how a safety awareness program can be measured for effectiveness.
8. Describe how a safety planned action activity program can be measured for effectiveness.
9. Why is a regularly scheduled safety meeting as important for engineers as it is for workstation operators?
10. What is the purpose of a job hazard analysis?
11. A new cutting fluid is to be implemented for all the lathes of a job shop. Explain how the manufacturing engineer would determine beyond a reasonable doubt that the cutting fluid is not toxic or hazardous.
12. What is a noise contour map and how is it used in a factory?

Chapter 19

THE INTEGRATED PRODUCTIVITY IMPROVEMENT PROGRAM

The preceding chapters covered the many aspects of manufacturing engineering in a relatively isolated sense in order to bring out salient points. However, we all realize that the real world does not allow isolation of subsets of problems that are solved independently of the others. In reality, each problem subset imposes certain constraints on the solution of the others and the whole.

> **EXAMPLE:** A production-level problem of output below expectation may require another workstation for its solution, but this may be economically unfeasible because the funds had to be used to solve a materials problem. The materials problem may be caused by a design that is not producible, requiring more exotic and hence more expensive material. Thus the new facility cannot be purchased because funds are needed to buy material at a greater cost since a producibility engineering problem went unsolved

This may be a round about way of coming to the point, but it shows that real problems are interdependent on many circumstances and events. This leads us to the subject matter for this chapter: the integrated productivity improvement program.

NEED FOR AN INTERACTIVE SOLUTION

By integrated we mean that all the resources the company has will be brought to bear on the problem at hand. We are looking for a solution that transcends manufacturing engineering, and even the manufacturing function in total, requiring the resources of most if not all functions of the company. We wish to discover how to make quantum jumps in improving productivity and thus profitability. Manufacturing engineering skills will be important because they are involved in all technical aspects of the factory operations. But others will also be involved because productivity is a measure of total output divided by total costs to obtain a cost per product value. The productivity improvement problem is to reduce the cost per product value; the lower this becomes, the more profitable the company becomes. It is easy to see that much more than manufacturing costs are involved. Beyond the labor and materials costs we must consider the entire overhead cost structure of all segments of the company. Therefore, the impacts of design engineering, marketing, finance, and employee relations on cost must also be analyzed and reduced to solve the productivity problem.

This overall approach to improving productivity is the only rational one. It makes no sense to have manufacturing drive down factory floor operating expenses if design engineering is adding cost to the product. If we are fortunate, the net effect is close to zero. If both functions are working to reduce costs—manufacturing in operating costs, design engineering in costs of tolerances required and materials selected—then the reductions are additive and the results are significantly greater than zero and beneficial.

Traditionally, each function has looked after its realm of responsibility in relative isolation from the other functions. This method is never optimal, because we are not sure we are spending company resources where we obtain the best results. Each function is competing against the other functions for limited resources and the resources can unknowingly be misapplied.

EXAMPLE: Assume we have a business that costs $60 million annually to run. In attempting to reduce costs, we should be directing our activities where they will have the greatest effect. The major categories of costs are

Direct labor: Cost of hands-on labor to make the products.

Overhead: Costs of all other labor and support activities such as engineers' salaries, heating the building, and so on.

Materials: Cost of supplies directly used to make the products.

The $60 million operating costs can be divided into these three categories. The relative percentages vary from company to company and from product to product. A heavy-apparatus job shop would typically divide the costs as follows:

Direct labor	$10 million
Overhead	$20 million
Materials	$30 million
Total	$60 million

If we have limited resources (i.e., people to effect positive changes in the company's profitability), the most attractive area for reduction would obviously be the materials area. A goal to reduce the cost of materials by 5% would result in a $1.5 million reduction; while the same goal in direct labor productivity would result in only a $500,000 improvement. Clearly, most of the resources should be aimed at reducing materials costs, but the productivity improvement programs set up by most companies give only a small probability of the real target being addressed.

All too often, each function is left to establish a cost reduction plan relatively independently. The result is that the overall picture of opportunity is never seen and we have tactical rather than strategic plans for improvement. For example, within manufacturing a great deal of effort may be expended to reduce direct labor costs by achieving greater workstation effectiveness rather than to reduce the cost of materials, because direct labor productivity is under the control of the manufacturing function. Manufacturing has shop operations to monitor and control work attention time and adherence to methods, manufacturing engineering to design efficient workstations and methods to reduce the time to do the required work, quality assurance to monitor manufacturing losses and feed back corrective action requests, and even materials to batch stock to make effective production runs. Therefore, manufacturing can make an all-out assault on direct labor costs, that is, productivity.

Let us look at the materials situation. Here manufacturing can only affect the purchase price negotiations and perhaps argue successfully for reduced tolerance rigidity so that less scrap is created. But the real saving is in substitutions for materials—using cheaper grades of steel, for example. These substitutions are the domain of design engineering, not manufacturing. Hence the big reduction is not sought, as it should be because it lies in the sphere of responsibility of design engineering and not manufacturing. Why does design engineering not respond to this challenge? There are a few good reasons. First, if left in isolation, design engineering would consider productivity improvement to be the ability to produce designs faster and with fewer people. This affects the design engineering budget in a positive way. Second, design engineers tend to design Cadillacs when stripped-down Chevrolets would be adequate. This is referred to as protecting the design margins. Third, producibility engineering is a manufacturing activity; therefore, there is little reason to expect design engineering to search effectively for the cheapest material if using it requires a design compromise. The result of these productivity improvements in isolation is an extreme under establishment of the materials cost improvement goal.

The solution is to have an integrated attack on the most lucrative area of the cost reduction potential. In the example above, this would mean that manufacturing and design engineering should pool their resources to cut materials costs. This would be done even at the expense of other projects, such as improving workstation performance in manufacturing and reducing drafting time per drawing in design engineering. In an integrated program the producibility and advanced manufacturing engineers would look for ways to make the products with cheaper materials, while the design engineers would evaluate the changes in design needed to allow such materials to be used. In addition, manufacturing engineering would work on methods with the primary objective of saving material and the secondary objective of saving worker time. Purchasing would focus its negotiated price activities on the lower-cost materials that design engineering is striving to use. In this integrated approach the different functions are working toward the same goal. When the functions worked independently and only looked at their portion of the business, this goal was secondary in importance. Thus when the functions are taken out of isolation and given the opportunity to see the area where the greatest overall gain can be made, we can have a quantum jump improvement in productivity.

This is the rationale for the integrated productivity improvement program. Now let us see how an integrated productivity improvement program is established, starting with an explanation of productivity.

TYPES OF PRODUCTIVITY

The basic definition of productivity measurement is output divided by input. To give this useful meaning we divide our productivity measurements among three broad categories selected to show the contributions of the entire organization against established goals. These three categories are (1) productivity, direct labor; (2) productivity, all other; and (3) quality costs.

The first two are positive measurements, while quality costs are expressed in a negative sense. In the first two measurements the higher the absolute value of the ratio of output to input, the better the performance of the organization. The opposite is true for quality costs. The cost to correct quality problems would be a negative factor for profit, while the input is the cost of total work—both direct and all other required to produce both good and bad products.

The productivity measurement must be divided into three segments, otherwise it is too general to be useful. We cannot add direct labor to indirect labor and overhead and come up with a meaningful number. Even though we use dollars as the basic measurement, this does not help management. We would have no way to compare the specific factory with others because we would not know the ratio between productivity, direct labor and productivity, all others. Therefore, we separate them. We can have a direct comparison with all other factories in terms of direct labor productivity. We can also have a comparison of productivity for everything else by using the "all other" category. Because different companies may have different ways of doing the work to support the factory operator, the category productivity, all other is very broad. It measures the productivity of all the support functions and materials and does not try to segregate any component out. Segregating other components, even materials, would result in a very narrow measurement that would be difficult to compare with many other organizations.

This three tier measures of productivity matches well with basic accounting procedures where we have above the line and below the line costs. Above the line being cost of goods sold which is direct labor and the material used to make the product. While below the line is everything else. Notice that materials are segregated out of all others for accounting purposes, which is proper because it is the cost of the material directly related to the product, for example, the board feet of lumber to build a tool shed. However, there is no way to relate material by itself specifically to productivity. It is what it is to make the product. The only productivity involved is the time to select the material by design engineering and the time (ability) for purchasing to source the material and have it delivered. For this reason above the line refers to direct labor productivity while below the line refers to productivity-all other and quality costs, and has these materials and its support items included. We could segregate out true material costs from all other and just leave in the materials support costs, but why bother. It takes extra effort to do so, and won't matter because it is nothing more than a constant with respect to productivity, all other measurements. For example the cost to move material is a variable and effects the value of productivity, all other. The purchase price of the material is a constant, the movement of material does not affect it, but productivity all other is affected by material movement. Inefficient movement techniques result in a lower productivity, all other measurements. Whether the material cost $0.05 or $0.10/lb does not affect the productivity of its usage.

The quality cost measurement is a mitigating factor. It is used to evaluate the purity of the other two productivity measurements. If the two productivity measurements (direct labor and all other) are within targeted goals but the quality costs are high, quality is being sacrificed to meet the output goals. This can have detrimental long-range effects on the business by giving it a reputation for poor quality. Hence, the quality cost measurement is an important modifier in interpreting the results of the other two measurements.

An important fact about productivity measurements is that they are not only for the manufacturing section. Only one of them—productivity, direct labor—is a measurement solely of the manufacturing function, and it applies to a portion of manufacturing, the shop operations sub-function. All other measures of productivity encompass the entire organization.

Productivity, all other measures manufacturing engineering in terms of the effectiveness of utilizing funds to develop tooling and workstations via the cost of equipment purchases and the salaries of the engineers involved. The measurement includes the cost of materials purchased by the materials subfunction, which is affected by the success of negotiations to obtain lower prices. Design engineering is included in the cost of salaries and the cost of design as related to productivity and selection of materials. Finance and the human resources functions are also part of the measurement via the payment of salaries to their personnel and the effectiveness with which they work. Thus, productivity, all other, measures the efficiency of all the functions in all things except direct labor applied to the product.

Since the "all other" category is usually the largest cost of doing business, companies show great interest in improving this aspect of productivity. In fact, it should be of more concern than the category of productivity, direct labor. The value companies place on this measurement is usually shown by the control of people hired for various functions. This is the "head count" control placed on operating components by senior management. The theory is that controlling the number of people will limit salary costs and offer fewer opportunities for those on the payroll to spend money for unnecessary things. By controlling the head count, only the high-priority items will be done; the lower-priority items that contribute less to profitability will not be done. Unfortunately, the proper head count is not always easy to assess. Therefore, this approach to improve productivity, all other is not a simple one to use and can damage a business by allowing necessary activities to be eliminated because of a lack of people to do them.

Quality costs measure all functions because all functions can participate in the creation of defective products. Scrap or repair costs can be due to design engineering specifying defective material, or manufacturing engineering devising ineffective methods that allow operators to make wrong choices in producing a part, or an operator simply making a mistake.

Quality costs can also be attributed to the functions with less direct inputs into producing the products: marketing, finance, and employee relations. These would be classified as indirect causes of quality costs. They would consist of the types of actions that limit the margin of error manufacturing and design engineering should have in the course of their work.

Marketing is responsible for establishing the contracts to which the company works on specific projects. The content of these contracts determines the magnitude of exposure to quality costs.

EXAMPLE: If an agreement is entered into with a customer to warranty a product for 2 years instead of the normal 1 year, there is an additional risk to the company. The margin of error in producing the product is somewhat reduced because it must perform correctly for an additional year. The risk of the customer mistreating the equipment and causing a failure extends over a longer period. In order not to incur the quality cost, the manufacturer will have to prove that the customer did not follow the correct operating procedures, or allowed some other circumstance to cause the failure. Even if the manufacturer is not at fault, there will still be additional costs. The manufacturer will have to use design engineering and manufacturing engineering resources to defend the company. Therefore, there is a build-up of warranty expenses even if the company is not at fault. Warranty costs are typically considered together with scrap and repair costs. Hence, by agreeing to a longer warranty period, marketing can have a detrimental effect on the quality cost measure of productivity.

The impact of human resources on quality costs may be even more distant, but it exists. An example concerns placement of operators on jobs. If human resources allows easy displacements of people from one job to another during an increase or decrease of the work force, a detrimental effect on quality costs can result. Human resources is responsible for negotiating work rules related to seniority, employee "bumping" rights, grouped vacation periods, and so forth. Although in theory human resources only represents management in total during these negotiations with employee groups or unions, in practice they play a considerable role in setting policy, since they tell management what they believe to be obtainable in negotiations and what they believe management will have to concede to achieve or maintain harmony in the labor force. Senior management tends not to vary considerably from the recommendations of its experts in human resources. Therefore, human resources is a powerful indirect force in increasing or decreasing quality costs through its establishment of work rules and seniority rules.

Bumping means that a senior worker can displace a worker with fewer years of service on a job assignment when the work force is being reduced. Similarly, when the work force is being expanded, the most senior members of the hourly ranks have the first opportunity to bid on new jobs as they become available. In either case the displacement causes a chain of personnel moves, thus disrupting the workplace. The more people being moved, the more unfamiliar the work becomes to the work force. This leads to less than optimum performance, in both a direct labor sense and a quality sense. The quality sense is the more problematic since quality errors found outside the factory can cost up to 10 times as much to fix as errors found during manufacture, whereas the loss of direct labor productivity never even approaches twice the cost under optimum conditions. The amount of exposure to excess costs due to bumping is up to human resources in their establishment of the rules. If the provision for bumping is very generous—that is, many people are allowed to displace others because job descriptions are broad and require minimal qualifications—then the exposure to extra costs is great. On the other hand, if human resources narrowly defines jobs and maximizes qualification requirements, bumping becomes a smaller problem. Narrowly defining jobs and strictly interpreting requirements means that only a qualified senior can displace a junior. This is what management strives for. In practice, it means that a senior employee can only move into a job he has held before and performed satisfactorily. If human resources must deviate from this, the amount of deviation is the amount of increased quality cost risk to the business.

The finance function's contribution to quality costs is similar to that of human resources. It is indirect but can have a sizable impact. There are two major ways in which finance can affect quality costs: in the purchase of equipment phase and in the collection of funds phase.

The purchase of equipment phase involves the longest range. Financial analysts exert an influence in evaluating plant appropriation requests. Manufacturing engineering must rely on finance to make the final decision on the validity of the financial data submitted in support of a capital equipment project. Unfortunately, most capital equipment project justifications require assumptions concerning the savings to be achieved. If finance takes a very strict approach to the justification, it may not be possible to acquire the desired equipment, and the company may be forced to make do with lesser equipment or no new equipment at all.

EXAMPLE: A lesser vertical boring mill may have a more difficult time holding required machining tolerances because it is less rigid than the more costly preferred machine. Both machines may be capable of holding the required tolerance, but the cheaper one may require closer attention on the part of the operator, resulting in a higher probability of undesirable quality costs.

In this case, manufacturing engineering may have bought the lesser machine because it was the only one the finance function would endorse. There are many actors and many circumstances involved in capital equipment purchases. However, this is a valid example of the influence of finance causing a greater probability of incurring higher quality costs.

The second way in which finance may cause quality costs is more immediate and direct. One area of quality costs is due to defective material or subcontract work. Theoretically, no costs should be absorbed as a result of these categories, but in practice many such costs are unnoticed or uncollectable. The uncollectable costs are those on which finance has some positive or negative influence. It is the responsibility of the manufacturing function to initiate charges to vendors for defective work or material. This is done through the quality control report process, whereby defects are reported and costs incurred to repair or make the product over again are recorded. If finance has an aggressive policy of tabulating such costs and encouraging the purchasing unit to recover them, the company's exposure to such quality costs is usually minimized. However, if the finance function has poor records of cost because its data collection system is not detailed enough, the costs associated with vendor-caused defects are difficult to collect. Therefore finance has a responsibility for seeing that extra costs due to poor vendor quality are recoverable. Finance can thus have an effect on quality costs.

PRODUCTIVITY MEASUREMENT EQUATIONS

Productivity measurement equations are algebraic notations that describe the existing state of productivity. They describe mathematically what would otherwise take a paragraph or more to communicate.

The productivity equations yield dimensionless numbers that show whether a profit is being made. But it takes a chronological series of these numbers to determine whether productivity is improving, staying the same, or diminishing. Let us look at the equations and review their components.

$$\text{Total Productivity Measurement (TPM)} = \text{output}/\text{input} \qquad (1)$$

This is the overall measurement used to determine whether an operation is profitable. It is also used to detect productivity trends. The output is usually sales dollars and other payments and the input is total cost of operations. A few examples illustrate how the calculated ratio can show the state of the business.

EXAMPLE (a): Sales for company A for 2000 were $180.6 million while costs of operations were $182.4 million. These costs included direct labor, indirect labor, material, and all overhead costs. Therefore, the values input into Eq. (1) are:

$$\text{TPM} = \$180.6 \text{ million}/\$182.4 \text{ million} = 0.99$$

The result, 0.99, is a dimensionless number less than unity. By inspection we can see that company A lost money in 2000. If a loss is suffered the total productivity measurement value will always be less than unity. This is the first important observation concerning productivity formulas; a loss is always less than 1.00, breakeven always equals 1.00, and a profit is always greater than 1.00.

EXAMPLE (b): In 2001 and 2002 company A had sales of $124.3 million and $117.1 million, with respective operating costs of $123.8 million and $114.4 million. We now have data to evaluate the trend of productivity improvement. Including 2000 data, we have

2000: TPM = $180.6 million/$182.4 million = 0.99
2001: TPM = $124.3 million/$123.8 million = 1.004
2002: TPM = $117.1 million/$114.4 million = 1.024

These results show a positive or improving productivity trend. We have gone from a value less than unity, a loss; to a value slightly above breakeven; to a greater than unity value, a profit, in a 3-year period. We can also determine the percent productivity improvement from year to year. This is the second important observation concerning use of the TPM. The percent productivity improvement would be

$$2001/2000: 1.004/0.99 = 1.014; 1.4\%$$
$$2002/2001: 1.024/1.004 = 1.0199; 1.99\%$$

These calculations show that company A is not improving very rapidly. In fact, we would wonder whether any progress is being made at all. This leads us to example (c).

EXAMPLE (c): Company A, not satisfied with progress shown since its loss year of 2000, decides to initiate a bold integrated productivity improvement program. Management has decided that total productivity must improve by 10% in 2003 over 2002. The question is how much they can afford to spend on all operations if the sales forecast is $122.0 million. The TPM equation becomes:

$$TPM = \$122.0 \text{ million}/x = 1.10$$

and x, the total operating costs, equals $110.9 million. This is the third important observation. The TPM permits quick evaluation of the cost reduction that must be achieved to reach desired goals. By neglecting to perform this simple calculation, many managers overestimate goals and consequently fail.

Productivity improvements are an essential part of any realistic business plan. Let's look at the goal set by company A. In order to achieve the stated goal costs must be reduced from $114.4 million to $110.9 million—a 3.15% reduction. This appears to be achievable; therefore, the plan is practical. If the reduction was over 10% we would have a stretched goal. If it was over 15% we would have to say that the goal was not practical.

Productivity, direct labor = sales/direct labor costs   (2)

Here sales are identical to the output part of the TPM equation. Direct labor costs refer to the salaries paid to the factory work force responsible for actually making the product.

This subset of the TPM deals only with the major shop operations portion of the productivity program. The analysis of the numerical values calculated is the same as that for Eq. (1), but there is a much narrower interpretation base. For the productivity, direct labor measurement we are concerned with the practicality of lowering the costs. Of all factory measurements, this is the most analyzed. We calculate the hours required to produce the products and then measure the efficiency of producing the products against the planned time. Productivity, direct labor is capable of being very precisely measured and proposed improvement plans can be evaluated well before they are approved.

EXAMPLE: A program is proposed to reduce overtime by improving the attendance rate of the direct labor work force. By knowing the percentage attendance improvement targeted for, we can calculate the number of additional hours made available to produce products, hence the number of overtime hours that can be reduced. This reduces direct labor costs because overtime labor hours usually cost 50% more than normal labor hours.

Productivity, all other = sales/(indirect labor + material cost + overhead)   (3)

This measures everything but direct labor. Direct labor is very well documented and measured, while the other cost factors are not. The determination of exact times to perform direct labor is a well-established engineering discipline; therefore, the measurement of performance (e.g., productivity, direct labor) is more accurate and specific than the other productivity measurements. We do not want to combine all other labor because only the direct labor portion based on scientific time standards would be accurate, and we would be in jeopardy of skewing the measurement because of the lack of preciseness of the indirect labor portion.

Because direct labor is much more precisely measurable than all other cost factors, we choose to measure it separately. However, because the other cost factors are less precisely controlled, it makes little sense to measure the improvement in performance of these areas separately. Therefore, we choose to combine all the generally measured cost factors into one measurement.

There is another good reason for lumping together all cost factors except direct labor—the advantage of averaging. If we can combine enough general measurement categories, we should, according to statistical theory, be able to say that the positive and negative biases of the individual measurements balance each other. The resulting productivity measurement should then fairly represent the actual situation. This is the case with the productivity, all other measurement, where we combine the costs of indirect labor, materials, and all other overhead factors. If we inadvertently classified as indirect labor a cost that should be charged as a service fee associated with the cost of materials, it would not matter because the value of the denominator would still be the same. This is also the case for the overhead charges. It does not matter whether we are talking about fixed overhead such as the cost of heating the plant with steam, or variable overhead such as the cost of using steam as an energy source for a process. For the purpose of productivity measurement, we do not need to know exactly the account into which the steam costs are placed.

By grouping all other expenses in one productivity measurement category, we can measure the productivity of all the support functions. We can determine whether we have the level of support required for the particular manufacturing entity. The percentage improvements of the factors productivity, direct labor and productivity, all other should be approximately the same. If they are not, there is an imbalance that must be corrected. The imbalance analysis is not precise, but we can use the following guidelines.

1. If productivity, direct labor is growing faster than productivity, all other, and the rate of productivity, direct labor is greater than 5%, then the support functions are probably too ponderous. There is probably excess staff and expenses and lack of standardization. This is indicated by the fact that the direct labor operation is doing well but the staff functions are slowing down. Direct labor performance is outstripping its support, probably because the support is incapable of reacting faster, and reaction time is usually a function of the number of people that must be stopped and pointed in another direction.

2. If productivity, direct labor is growing more slowly than productivity, all other, and the rate of productivity, direct labor is greater than 5%, then the support functions are probably too lean. The support staff appears to be reacting to change very well but does not have the ability to implement change in the way the direct labor work force is instructed. An erroneous decision often made in this situation is to increase the number of foremen. Usually this will not work because the new foremen will be in the same situation as the old foremen, who need new manufacturing instructions. When this is the case, it is essential to have a sufficient manufacturing engineering work force to fully translate the productivity improvements into the direct labor factor. An example would be the development of a new weld technique that reduces the cost of materials and has the potential to shorten the time cycle, but is not translated into lower labor costs because there is insufficient support staff to do so.

3. If productivity, direct labor and productivity, all other shows less than 5% improvement, an analysis is of little value. An improvement of 5% or less is in the scatter band of accuracy of the measurements. Rather than evaluating whether the support functions are too large or too small, it is important to ensure that all programs are investigated to achieve a greater than 5% productivity improvement factor.

The techniques of using Eq. (3) are the same as those for Eq. (1).

$$\text{Quality costs} = (\text{scrap} + \text{repair costs} + \text{warranty costs})/\text{total costs} \qquad (4)$$

This measurement, as discussed previously, is the modifier of all the other measurements. If this measurement is numerically high, we have significant problems. We have processes that are out of control because we are experiencing the high cost of fixing poorly made items, or making them again, or, worse yet, fixing the products in the customer's location and possibly causing the customer loss of income.

The goal, of course, is to have the ratio in Eq. (4) equal to zero. A good quality cost measurement depends on the type of business we are engaged in. In general, an excellent rating would be in the range of less than 0.5%. Fair to good would be in the range of 0.5 to 2.0%. Anything above 2.0% would be considered poor. I must emphasize that these are guidelines and not rules. Each manager must determine what the ratio value should be for a specific operation.

Scrap and repair costs are the two components of the cost factor called manufacturing losses. Scrap refers to the disposal of parts that are defective and either cannot be repaired or are uneconomical to repair. Repair costs are the extra costs incurred to fix defective parts.

It is customary to set an upper-bound budget for manufacturing losses. This is the largest amount of money the company has decided it can spend to fix and remake products. It is determined by past experience and a judgment of what quality improvement programs can do to reduce the incidence of scrap and rework. Therefore, a factor in improving the quality cost ratio is the effectiveness of the manufacturing loss reduction program.

Warranty costs are those associated with fixing deficiencies or replacing whole units after the product's ownership is transferred to the customer. They are the most onerous costs a producer must contend with. Here we are admitting that we were incapable of fixing all problems before the product left the factory.

I have always considered warranty costs to be a measure of the effectiveness of manufacturing engineering. A case can be made that if manufacturing engineering was doing its job correctly, products of poor quality would never have been shipped. Manufacturing engineering is responsible for designing the factory system, training operators, and creating workable designs from proposed designs via producibility engineering. Therefore, having contributions in virtually all phases of the manufacturing process, manufacturing engineering must take a large part of the responsibility for warranty costs.

Manufacturing engineering cannot force operators to work conscientiously, but through process control and MP&WM it has adequate information on the state of operator performance. Thus it should not be surprised if less than adequate quality is being produced. Hence manufacturing engineering must in a sense be the conscience of the organization, seeing to it that shop operations management is cognizant of the current quality status, and taking corrective action as required.

The other major contributing factor in warranty costs is design. No remedial actions within the manufacturing function can correct for deficient design. Therefore, when warranty costs are incurred it is critical to determine whether the cause is design or manufacturing. If we erroneously assume that the cause is a manufacturing inadequacy, many corrective action programs will be initiated that will not fix the problem. This leads to considerable problems beyond money and time wasted. It can create distrust between organizations, which can ultimately result in management paralysis and a business out of control.

The utilization of the quality cost equation is similar to that of the other productivity measurements. We can either work from the desired ratio number and back to the numerator—that is, knowing the total budgeted costs, determine repair cost, scrap cost, and warranty costs. Or we can set the repair, scrap, and warranty budgets and then work out the ratio. In the former case we are taking what is considered an acceptable goal and working out how much repair, scrap, and warranty cost can be tolerated. In the latter case we are comparing current and recent results and determining what can be achieved in the next reporting period. Of course, the latter result may be too conservative and the former too optimistic. Therefore, setting the parameters must involve a subjective blend of both.

IMPLEMENTING THE INTEGRATED PRODUCTIVITY IMPROVEMENT PROGRAM

We will now look at aspects of putting together an integrated program that will produce productivity savings. The process of operating an integrated program is another way of creating an operating plan. Once the plan is put together and approved, it is necessary to implement it and check progress. The basic techniques for creating plans, implementing them, and checking progress have been discussed throughout this text. Our purpose here is to show how these efforts are coordinated and made into an effective integrated productivity improvement program.

Objectives and Goals Input

Chapter 2 is dedicated to this very important management concept. The objective and goals end product is a set of action plans that allow achievement of projects, and hence of their supportive goals.

For an integrated cross-functional plan, we must revisit the objectives and goals exercise to make sure that manufacturing engineering is not working at cross purposes to other functions and sub-functions. We must also be sure that the manufacturing engineering objectives and goals are not so narrowly defined that they miss the intent of an integrated program. It would be a serious error for senior management to allow objectives and goals statements to stand as presented until a review of overall problems is conducted. The objectives and goals statements for each function and subfunction must be reviewed against the perceived needs of the entire organization to ensure that vital concerns of the company are not falling into cracks between the boundaries of responsibility of the respective functions and subfunctions. By doing this we can take the necessary steps to modify various objectives and goals statements so that overall needs take priority over narrower needs of the issuing agency.

EXAMPLE: Manufacturing engineering objectives and goals may include an additional drill press, while the overall needs of the business may dictate that improvements in the appearance of products, even though only cosmetic, is more important in achieving the desired sales volume. The objectives and goals statement of manufacturing engineering may have to be changed to emphasize projects dealing with product appearance rather than a product production capital equipment program.

Here we have an example of a business need taking precedence over a purely technical need. The outcome should be improving the business results and ultimately allowing more funds to be available for technical needs.

The Productivity Seminar

A productivity seminar is an optional step in forming an operating plan, but it is an excellent tool for determining whether the objectives and goals statements have found all the action plans to be pursued. It is also an effective means of creating a cross-functional dialog, and a method by which senior management can review the adequacy of all the objectives and goals statements. Productivity seminars are a technique for obtaining a wide variety of ideas.

A seminar is a call to action to solve a specific set of problems or to teach a particular set of skills. The productivity seminar is an example of the former. It is dedicated to identifying a set of integrated productivity enhancement activities to allow the company to reach a stated goal.

Five broad steps are involved in a successful productivity seminar. These are: (1) announcement of the productivity seminar, (2) selection of attendees, (3) the productivity seminar procedure, (4) incorporation of results into the operating plan, and (5) reporting back of results and plans to attendees.

Announcement of the productivity seminar. This step informs the organization that a productivity seminar is to be held. It states the purpose of the seminar, the goal to be attained, and how the results will be implemented. Usually we point out the need for having such an activity and state why it is important.

> **EXAMPLE:** We have come a long way in 2005 in terms of improving productivity and sales performance of our business. However, we are still not at the point of generating acceptable profit levels and there are new challenges and opportunities facing us in 2006. Therefore, we are scheduling a productivity seminar to take place on December 1, 2005 in order to develop ideas for meeting these challenges.

The statement of purpose tells everyone that a greater effort will be required to meet the generic challenge. Now it is necessary to state what the challenge is. We do this by introducing the goal of the productivity seminar. The goal is defined in the same way as in the objectives and goals systems. It must be a measurable statement of intent bounded by a time period. A productivity seminar goal might be formulated as follows:

> Our segment of the company has been issued a mandate to achieve a $4 million net profit in 2006. To do this we must reduce operating expenses by $1.5 million and all support function costs by $500,000 for this year's operation. We must therefore develop integrated ideas to save $2 million.

Once the purpose and goal have been published, it is time to tell the attendees what will happen to their ideas. This can be done in the following manner:

> In this meeting we will identify specific opportunities available to help us meet our commitment. The ideas generated by consensus in the meeting will be included in our 2006 operating plan.

The targeted audience has been informed about the purpose, goal, and use of the results. We have set the stage for a successful productivity seminar.

Selection of attendees. The announcement of attendees is usually made simultaneously with the seminar announcement. The procedure is to make the announcement in a letter addressed to those who should attend.

The selection of attendees depends on the specific goals of the productivity seminar. For an integrated productivity improvement program, particularly one intended to solve cross-functional productivity problems, the attendees must represent all phases of the operation. Therefore the first rule is:

(1) Select a cross sectional population representing all constituencies.

We want people who can generate ideas for actions to meet the goal. These will be individuals who are primarily knowledgeable in their areas and also have a good understanding of how the business is conducted and how its products are used. Since the latter types of people are not always available, we must be content with the former. The second rule is:

(2) Select people who are knowledgeable in their areas of responsibility.

Another criterion is that of group interaction. We want people who express themselves well and who are creative. Therefore the third rule is:

(3) Select the natural leaders—those who get things done and whom others will follow.

Finally, we want the key managers to participate because they are the person with responsibility for conducting the operating plan. Therefore, the fourth rule is:

(4) Select the key managers as participants.

By following these four selection rules we ensure that the participation is cross-functional and will include the people who are the key contributors to the ongoing success of the business.

The productivity seminar procedure. The procedure is a controlled brainstorming activity. It is made up of four parts:

1. identification of all relevant ideas,
2. evaluation of the ideas,
3. recording of the results of part 2, and
4. ranking the ideas in order of priority.

For a productivity seminar we can divide the activities into three chronological sessions, with the first session covering part 1, the second covering parts 2 and 3, and the third covering part 4. An example of this technique is as follows:

1. Divide the participants into approximately 10-person teams. Give each team the same task. Identify ideas for:
   a. Productivity improvements, direct labor.
   b. Quality improvements.
   c. Productivity improvements, all other.

   Specify:
   a. Expected savings for the current period.
   b. Major measurement milestones.
   c. Time and cost to implement the idea.
   d. Function or subfunction to lead the project.
   e. Category the project falls into (productivity improvement, direct labor; quality improvement; or productivity improvement, all other). Only one category per idea is allowed.

   I have found that six teams is the upper limit of practicality. For each team, select a chairman and a recorder.

2. For period two, divide and shuffle again. This time have only three teams. Assign each team to analyze only one set of ideas:
   a. Team 1: productivity improvement, direct labor.
   b. Team 2: quality improvements.
   c. Team 3: productivity improvement, all other.

   For its assigned area, each team will have the following tasks.
   a. Rank all ideas by importance.
   b. Evaluate feasibility of achievement.
   c. Establish priorities for recommended project ideas.
   d. List ideas deferred with a reason for the deferment.

   As with the first session, select a chairman and recorder, but do not select people who have already served in either of these capacities. This is done to maximize leadership participation in the activity and have the participants develop a sense of personal responsibility for reaching a successful conclusion.

3. Period three is the reporting and final results portion. Here the period two chairpersons report the results of their team's efforts to all the seminar attendees. This allows additional inputs to be made and a consensus to be reached on the results for the three categories.

Such a seminar may take 1 or 2 days, depending on how much time is available. It is important that no distractions interfere with this process. The participants must focus on the task at hand and not be distracted by other responsibilities. Therefore, if a 2-day program is planned, it should take place on 2 consecutive days.

There should also be a sense of urgency—that is, the participants should have a time deadline. Remember that we are looking for ideas, not fully detailed plans. Therefore, we put a time limit on the process in order to minimize the tendency to drift into detailed project planning. Developing of the plans to implement the ideas will take place later.

Finally, someone must be in charge of the productivity seminar and must make sure that the program flows smoothly. This seminar leader must ensure that each team at each session is actually doing the assigned tasks. This is done by briefly visiting each team and observing its performance. The leader participates if asked to do so, but not as a leader. The seminar leader also makes sure that each team is aware of its deadline, encourages teams where necessary, and deftly steers teams away from irrelevant tangents. He must continually challenge the participants to produce and at the same time maintain the schedule. The pressure of the schedule ensures that ideas, and not position papers, are generated.

Incorporation of results into the operating plan.   The ideas generated by the productivity seminar are of no value unless they are acted upon. The mechanism for doing this is the operating plan, which will be discussed later in this chapter. The operating plan is the summary of all ideas and plans that management is supporting for action in the current time period.

The good ideas should be included in the operating plan. But first the productivity seminar ideas must be reviewed to make sure there are no duplications of planned activities already in the objectives and goals and in the present form of the operating plan. It is likely that duplication will exist. The participants in the seminar were probably the same people who had made inputs in the objectives and goals exercise. This would be especially true for the manufacturing engineers and other manufacturing subfunction representatives, because the objectives and goals discipline is very strong within the manufacturing function.

When duplication is found, we must first compare costs and results between the productivity seminar idea and the goal from the objectives and goals statement. This is an opportunity to review the older input (the goal) against possibly fresher or more detailed inputs. If nothing else, the productivity seminar provides an opportunity to clarify the operating plan by updating or corroborating the data. This alone would be a useful result of the seminar.

Once all duplications are culled out we begin the validation and planning cycle for the remaining ideas. First, we must review the ideas to see whether they are still valid. Then we must determine whether the benefits contemplated are accurate. Planning the activity in the same way as the planning described for goals and projects in Chapter 2 does this. We then have a plan and an objectively determined savings. Finally, we must evaluate the productivity seminar idea for cost to implement. If the cost is beyond what the company can afford, the idea must be dropped, possibly to be implemented in future years.

After the productivity seminar idea has passed all these hurdles, it is introduced into the operating plan. This means that the idea has been translated into the objectives and goals system for tracking and implementation.

Reporting back of results and plans to attendees.   This is done to keep up interest in productivity improvement and to show the people who attended that their ideas are indeed shaping the direction the company is taking. It can be done in several different ways. A good approach is to have each subfunction manager report via her/his operating plan and point out the plan segments that trace their inception or at least their verification or updating to the productivity seminar.

Using the operating plans to present results also gives management the opportunity to show the key contributors exactly what the organization is attempting to do and how it will be done. It is important for management to communicate with their people and for everyone to know what they and their company are trying to achieve. With this approach we can show an integrated plan whereby many functions and subfunctions are working on the same problem.

The Operating Plan

An operating plan is a summarized statement of the total objectives and goals statements and other independently obtained ideas. It shows in matrix form all the approved objectives and goals and

Goal	Measurement	Resources	Costs	Risks & Contingency Plans
8. Improve direct labor 10% by: (a) reducing workstation variance 8% (b) Reducing overtime to 7.5% of work-in-process labor	(a) Achieve reduction in measured work station variance from 125% '05 ave. to 115% '06 ave. Save $250K in DL cost. (b) Achieve reduction in WIP overtime from 10% '05 ave. to 7.5% '06 ave. Save $150K in DL cost.	(a)(b) Implement multicode for operators, HR specify 1/2 man-year Improve planning to allow short interval scheduling Mfg. Eng. - 2 man-years Implement attendance and absenteeism follow-up Foremen on-going.	Improved planning will require additional disk storage - $65K Short interval scheduling will cost $35K to develop software.	Risks: Funds not available to add disk storage or software development Contingency: Investigate and implement group incentive pay plan to improve productivity; probably will not achieve goal for one additional year.

Figure 19-1. Section of Operating Plan Matrix

ideas for the current time period. It represents a further sifting of plans at a higher management level, which ensures that the activities finally selected by each function, and its sub-functions are truly integrated to meet the needs of the particular business. By putting the objectives and goals in the matrix, management has a clearer picture of what planned activities will have the greatest beneficial impact on the business. This overview gives management the ability to set priorities and to add to, or subtract from, activities in order to continually optimize the chances of achieving the overall productivity goal. The senior managers can advance, hold, or delay implementation of activities as the current year evolves. The headings to be found on the operating plan matrix are: goal, measurement, resources, costs, and risks and contingency plans (Fig. 19-1).

Goal.   This is the goal derived either from the objectives and goals statement or from another source of ideas such as a productivity seminar. In the objectives and goals format we had a statement and a measurement in one or two sentences. In the operating plan we combine several statements, from several different objectives and goals inputs, but do not continue with the measurement portion, which is found under the next heading.

Measurement.   This is the second half of the goals statement, as indicated above. It represents a summarized condensation of all objectives and goals statements and other ideas that can be combined. The measurements, as explained in Chapter 2, must be bounded by time and be quantifiable.

Resources.   Here we have a summary of the manpower and other resources that will be brought to bear on the particular program to achieve the goal. It is interesting to note that by adding the man-years required to achieve all the goals on the operating plan we can validate its chances for success. If we require more man-years than are available, the operating plan goals will probably not be totally achieved.

Costs.   This is a list of the funds necessary to achieve the goals. Here again, we can use this column to evaluate the probability of achieving all the goals. If the sum of money required to achieve all the goals is more than the sum available, we come to the conclusion that not all the goals will be achieved.

Risks and contingency plans.   In this last column the manager lists the important risks in achieving the goals and then determines what contingency plan should be followed if the risks cannot be overcome. This is a vital portion of the operating plan. It is the pragmatic input that brings overly optimistic plans back to reality and practicality. It is important to give personnel the opportunity to strive for the goals, but it is equally important to have a backup plan that can be quickly implemented if the goals are missed.

In reviewing a section of the operating plan matrix, shown in Fig. 19-1, one very important feature should stand out. All of the goals are "doable." There are no plans to do things that would revolutionize the way the business is conducted. Also note that we are involving cross-functional responsibilities within the matrix. We have human resources responsible for certain aspects of the plan, and manufacturing engineering and shop operations are also involved. This is truly an integrated approach. We wish to achieve a productivity improvement of 10% by having three different organizations carry out segments of a plan. The segments add up to a total, and only the total plan can achieve the goal.

SUMMARY

The operating plan matrix shows the need for integrated planning. Without an integrated approach none of the functions or sub-functions would be interested in pursuing such an activity. Human resources would not on its own aim for a 10% productivity improvement of direct labor. Manufacturing engineering would not try to improve planning to allow shop operations to schedule at short intervals in order to gain greater control unless shop operations was committed to a cooperative effort. Similarly, shop operations would not undertake such an effort without the support of manufacturing engineering and human resources.

If we did not integrate the approach through a tool such as the operating plan, we would end up with each function, and indeed the subfunctions within it, working only on items over which it has complete control. Such a confederation of semiautonomous functions cannot operate as a business team. To achieve their functioning as a business team is the goal of the integrated productivity improvement program. It is done through the use of the operating plan, a thorough understanding of the meaning of the productivity measurements, and continual education of all the key contributors in the organization for the overall needs and goals of the company.

With an integrated program to improve productivity we can use the productivity measurements to set and measure progress toward achieving goals. We can be assured that programs requiring cooperative efforts of the various functions will be coordinated, minimizing the likelihood that the organization will be splintered and the functions will strive toward unrelated goals. Manufacturing engineering plays a very important role in the integrated approach because it is the technical arm of manufacturing and the liaison with marketing, finance, and design engineering. For this reason managers within the manufacturing engineering subfunction must be well versed in the techniques of operating plan implementation and productivity measurement.

REVIEW QUESTIONS

1. Discuss the reasons why the productivity improvement program cannot be solved by manufacturing engineering independently but requires an integrated approach.
2. Describe the three basic categories of operating costs and explain how they relate to a manufacturing company.
3. Explain why a functional approach as compared to an integrated approach to productivity improvement often fails to achieve the desired results.
4. Discuss the reasons for dividing the productivity measurement into productivity, direct labor, productivity, all other; and quality costs.
5. Why is quality costs considered a productivity measurement?
6. What is the rationale behind head count control? How does it affect productivity improvement capabilities?
7. How does bumping affect productivity improvement capabilities?
8. How should manufacturing engineering mitigate possible adverse affects of the finance function on productivity improvements?

9. Construct a productivity trend chart for a company that has recorded the following data. Calculate the year-to-year changes in productivity levels. Dollars are shown in millions.

Year	Sales	Cost
1998	51.6	51.9
1999	50.3	52.8
2000	57.8	53.2
2001	80.6	73.7
2002	76.6	71.4
2003	62.1	63.0
2004	64.2	63.8
2005	42.8	40.1

10. Why is productivity, direct labor segregated out while all other productivity measurements are essentially group measurements?

11. In a manufacturing company productivity, direct labor is growing at a 7.3% annual rate, while productivity, all other is growing at a 4.0% rate. What does this imply about the support staff of the company?

12. Why are warranty costs a direct measure of the effectiveness of manufacturing engineering?

13. What is the purpose of a productivity seminar?

14. Considering an operating plan as only a productivity improvement tool is incorrect. Explain why.

15. If the objectives and goals statements clearly show the company's wishes for the budget year, why is an operating plan required?

16. Explain why it is necessary to include risks and contingency plans in an operating plan.

Chapter 20

USING ISO 9000 AS A MEANS OF BECOMING A "WORLD CLASS" COMPANY

I think everyone has heard about the ISO 9000 quality standard. They might not be aware of what it is all about but they certainly are aware of its existence. Just a drive down any road with factories alongside it will reveal sporadic banners outside proclaiming "We are ISO 9000 certified." Many companies selling products to the general public make claims in their advertisement that they are ISO 9000 certified, thus implying their quality is superior to their noncertified competitors. So ISO 9000 has become part of the manufacturing lexicon. But what is it exactly, and as manufacturing engineers do we need know? The answer to the question is emphatically yes.

In this chapter we will explore the topic of ISO 9000 certification. We will learn where it applies, and perhaps doesn't. Whether or not it is mandatory or even advisable for your company to be certified. We will learn what the standard is and isn't. And most important, how being certified affects quality levels.

WHAT DOES ISO 9000 CERTIFICATION MEAN WITH RESPECT TO COMPETITION?

Is it a means of making your company perform better so it can at least stay on par, but hopefully outpace your competitors? Or is it a lot of hype put together to sell products and services to easily duped customers? More than likely it is the former if taken seriously and implemented correctly.

To perform better means a company has gained better control of its processes and does more things correctly the first time, for example, eliminate waste. Sound familiar? It should. It's the same philosophy as JIT and Lean Manufacturing. But in this case it means getting your administrative act together. Yes, following the precepts of ISO 9000 requirements will make companies perform better because they will become more proficient in understanding the how and why the process dictates a certain methodology of going from order entry to ship to customer.

But keep in mind gaining ISO 9000 certification is a long and sometimes difficult process. The path may be long, but one of the benefits is that as the company travels this journey is continually making incremental improvements in its administrative performance. Constantly tightening controls, creating more focused approaches to problem solving, learning to efficiently document, documenting, and more documenting. Until the point is reached where the company really operates as a well-oiled machine, one virtually impervious to surprises.

So we may say it becomes "intuitively obvious" that a company that truly wants to improve its overall ability to compete, can do so with the help of adopting an ISO 9000 regime. ISO 9000, if implemented properly, will foster a disciplined process for doing the business of the company. It creates a "book" so to speak that says how the company does business. It outlines every administrative, operational, sales and marketing, and financial procedure the company does in bringing its products through the process and to the customers. By creating the book, it gives the company an opportunity to examine its procedures and thus the opportunity to fix and change them in a manner that is compatible and synergistic. By becoming a better-focused company, by eliminating the fuzziness of not quite understanding the ramifications of one action to another, it becomes a company that is faster and more accurate, and thus a much improved competitor.

THE ROLE OF MANUFACTURING ENGINEERING IN THE ISO 9000 SYSTEM

As the technical arm of manufacturing, manufacturing engineering has a lead role in developing and implementing the company's ISO 9000 plan. Let's briefly look at what has to be achieved to be ISO 9000 certified. Remember, ISO 9000 certification is about documentation. In fact a simple definition of ISO 9000 certification could be:

> *"To become ISO 9000 certified:*
> State what you are going to do to produce your product and/or service.
> Document it, and prove you are doing it".

Manufacturing engineering plays a major role in this task.
To become certified:

A. Demonstrate a consistent approach toward quality management, including procedures for implementing and controlling;
   a. A quality manual, stating as minimum; a definition of quality, who is responsible for the quality output, procedures to be followed, and reporting results.
   b. Documentation of how you do your processes.
   c. Controls of records showing you do your processes as documented.
B. Demonstrate that your company has strong and consistent management controls of the business.
   a. Management's commitment to quality is demonstrated through actions.
   b. The company is customer focused to deliver goods and services on time and in accordance with contract specifications.
   c. A quality policy that dictates attention to customer needs
   d. Has consistent planning for;
      i. Quality objectives.
      ii. Implementing and maintaining a quality management system.
C. Has a methodology of defining responsibility and authority that is specified and adhered to
D. A plan to provide sufficient resources, to effectively carry out:
   a. A product realization plan that is documented and consistently carried out, including;
      i. Design and development.
      ii. Purchasing.
      iii. Providing for production and servicing activities.
E. Has the capability and does measure and analyze for continuous improvement.

It is readily apparent, after reading the previous 19 chapters of this book, that manufacturing engineering needs to be a major participant in this activity. This will be so from the scoping of the implementation project for ISO 9000, accomplishing initial certification, and through maintaining the resultant system.

A DESCRIPTION OF THE ISO 9000-2000 STANDARD

ISO is the abbreviation for the International Organization for Standardization (the literal translation from the French language); anglicized, we know it as the International Standards Organization, which is made up of member nations around the world. Its purpose is to promulgate performance and safety criteria, that if applied result in better products and services. As such, while it has no legal authority, its standards produce the best way of manufacturing and producing services for customers, be they commercial or retail. By everyone complying with these standards we have a set base of acceptable quality levels for hosts of different types of products and services.

We have a standard for design to compare products performance against. For example; ISO has standards for film; for example, the speed at which light can change the silver emulsion to register an image on the film. We are familiar with the ISO rating on film packages; ranging from 10 to 8000. We know that the rating will tell us what type of lighting we will need to get a properly exposed photograph. The manufacturers have calibrated their films as compared to a standard. Even with the advent of digital photography, we find these same standards being applied to the equivalency of what the circuits are capable of, for recording images.

We are very used to taking these values for granted. The standard is actually a "voluntary standard" agreed to by many nations, and not dictates of law. However, since ISO standards are agreements as to how we converse on values of performance for international and national commerce, they become like laws. In fact in many cases, these standards are used for contractual agreements between parties involved in commerce. With this as a prelude let's look at what the nations of the world have agreed to as the voluntary standard for defining quality.

The ISO 9000 standard is actually for standards under the same banner; for example, quality. These standards are:

- ISO 9000 (2000)—quality management principles and fundamentals. An index that describes the quality standard with basic definitions.
- ISO 9001 (2000)—requirements for quality management systems to demonstrate capability to meet customer and regulatory requirements.
- ISO 9004 (2000)—guidance for setting up a quality management system that exceeds ISO 9001 requirements in an efficient manner.
- ISO 19011—guidance for executing quality audits.

Only ISO 9001 (2000) requires a certification verifying that your company complies with its provisions. So many people simply ignore the other three. I'd like to point out that ISO 9000 (2000) is important because it contains the definitions that are used through out the other three. ISO 9004 (2000) establishes the links to other standards that are compatible to the ISO 9001 (2000) standard. For example, the automobile industry decided that the "plain vanilla" version standard, ISO 9001 (2000) is not specific enough for their use. So they created an allied standard known commonly as QC 9000. This is based on provisions of ISO 9004 (2000). Most of us, however, will find that ISO 9001 (2000) is sufficient for our industries. The forth standard in the group ISO 19011 (why it doesn't have a 9000 series number I don't know) establishes criteria for doing audits compatible with ISO 9001 (2000) and any derivatives based on ISO 9004 (2000). The bracket (2000) indicates the year the version of the standard was agreed to by the ISO members. The last previous edition was (1994).

Member nations are allowed to modify these standards any way they see fit. But obviously they should not do things that would put their standard at odds with the ISO standard. To do so would put their industries at a disadvantage for international commerce. Usually what is done, if at all, is to put the standard in the popular vernacular of that nation. In the United States we have slightly modified the original standard to be part of the assessment series, officially called "Assessment Standards in the United States; The ANSI/ASQC Q90-Q94 Series."

The American National Standards Institute (ANSI) is the United States' official representative to ISO. It is interesting to note that ANSI is a nongovernmental technical society "owned" by the major volunteer member-based technical societies organized in the United States, for example the American Society of Mechanical Engineers and the Institute of Electrical and Electronics Engineers. The federal government does participate in a liaison role through the Department of Commerce. ASQC is the American Society for Quality Control; recently morphed into simply ASQ, the American Society for Quality. This too is a volunteer technical society. However, regardless of the renaming by ANSI and ASQC, there isn't enough difference between the two versions to even bother to describe. Most of us use the international title, ISO 9000 series.

To give an illustration as to what it takes to become ISO 9001 (2000) certified let's start by looking at what makes up the standard. Keeping in mind that we will have to ultimately prove we can and do comply with these provisions.

There are eight sections that make up ISO 9001 (2000). The first three require no direct action by the part of the company for compliance certification. Here is the official list of the sections and their sub-sections. For a more in depth analysis of the interpretation of the steps with reference to an actual certification project, I refer you to the several references in the bibliography.

1. Scope
2. Reference to associated standards
3. Terms and definitions
4. Quality management system
    4.1 General requirements
    4.2 Documentation requirements
        4.2.1 General
        4.2.2 Quality manual
        4.2.3 Control of documents
        4.2.4 Control of records
5. Management responsibility
    5.1 Management commitment
    5.2 Customer focus
    5.3 Quality policy
    5.4 Planning
        5.4.1 Quality objectives
        5.4.2 Quality management system planning
    5.5 Responsibility, authority, and communications
        5.5.1 Responsibility and authority
        5.5.2 Management representative
        5.5.3 Internal communications
    5.6 Management Review
        5.6.1 General
        5.6.2 Review input
        5.6.3 Review output
6. Resource management
    6.1 Provision of resources
    6.2 Human resources
        6.2.1 General
        6.2.2 Competence, awareness and training
    6.3 Infrastructure
    6.4 Work environment
7. Product realization
    7.1 Planning of product realization
    7.2 Customer related processes
        7.2.1 Determination of requirements related to the product
        7.2.2 Review of requirements related to the product
        7.2.3 Customer communications
    7.3 Design and development
        7.3.1 Design and development planning
        7.3.2 Design and development inputs
        7.3.3 Design and development outputs
        7.3.4 Design and development review

     7.3.5 Design and development verification
     7.3.6 Design and development validation
     7.3.7 Control of design and development changes
  7.4 Purchasing
     7.4.1 Purchasing process
     7.4.2 Purchasing information
     7.4.3 Verification of purchased products
  7.5 Production and service provision
     7.5.1 Control of production and service provision
     7.5.2 Validation of processes for production and service provision
     7.5.3 Identification and traceability
     7.5.4 Customer property
     7.5.5 Preservation of product
  7.6 Control of monitoring and measuring devices
8. Measurement, analysis and improvement
  8.1 General
  8.2 Monitoring and measurement
     8.2.1 Customer satisfaction
     8.2.2 Internal audit
     8.2.3 Monitoring and measurement of processes
     8.2.4 Monitoring and measurement of product
  8.3 Control of nonconforming product
  8.4 Analysis of data
  8.5 Improvement
     8.5.1 Continual improvement
     8.5.2 Corrective action
     8.5.3 Preventive action

What we have here is a list of required action statements. The company must be able to verify that it does everything sections 4 through 8 require. Usually that means the company must have documentation and records of doing these action steps, such that an independent auditor can verify that is the case. So it comes down to convincing the auditor that you actually do what your policies and procedures state you do. These means traceability for all items and implies that the company's employees have been trained to and are doing the dictates of its policies and procedures.

BECOMING ISO 9001 (2000) CERTIFIED

A company that wants to become ISO 9001 (2000) certified must be able to generate a positive assessment for the following items:

- Demonstrate that it is customer focused in that its prime policy is to provide products and services to its customers meeting the customers true needs.
- Management must exhibit leadership in maintaining a viable quality system that allows it to meet customer needs.
- The company must involve all personnel in working the quality system.
- Demonstrate a process approach in achieving quality requirements.
- Uses a synergistic systems approach for identifying potential problems and ultimately solving those that become real problems.
- Demonstrate a continuous improvement philosophy.
- Use a factual approach in decision making

- Have mutually beneficial relationships with vendors.
- Be able to demonstrate continual compliances with sections 4 through 8.

As can be seen it takes a very organized approach to achieve certification. In fact the only practical way to do it is through the project management process as discussed in Chapter 2. I recommend a four-phase approach as described below.

Phase I: Organize for Registration

This is a test to see if senior management has the "fire in the belly" to really go down the path toward certification. Management has to show true commitment, otherwise the project will fail. Management has to be a sincerely dedicated sponsor for the task to succeed.

Management will have to set up a steering committee with true power and authority to do what ever becomes necessary to achieve success. This will include fairly extensive training for the steering committee and others, including senior management, in order to comply with sections 4 through 8. Most often the training is done by outside consults, qualified to do this training.

The steering committee will also start early on doing intensive quality audits. In fact senior management will have to take part in these audits to get a flavor of how far the company must go to achieve certification. Finally, the company early on must select a registrar to do the certification (more on that later).

Phase II: Preparing for Registration.

This is the part where all the grunt work of preparing documents and plans takes place. It is the most intensive and can be the most heartbreaking phase. During this work we really learn how far we are from being a truly competent performer, at least in the eyes of the ISO standard. We see all the foibles of performance we must overcome to be certified.

Also in this phase, the company must learn what the registrar requires for certification. The registrar will willingly give this to the company but is prohibited from acting as a consultant to help implement changes necessary to comply with the requirements. Some companies do hire consultants to help implement required changes. More on that later.

It's easy to become overwhelmed by the complexity and volume of what has to be done if we let it happen. Fortunately, if we have done our project planning work properly, we will have a plan to fall back to and not despair, but simply take it one step at a time. There are hosts of things to do as the following list shows. However, if we take it one step at a time, we will achieve them, and the company will make incremental improvements as each task is achieved.

- Document existing processes with quality plans and methods.
- Determine if all clauses apply to your company. For example a service only company need not comply with all provisions of sections 7 and 8.
- Meticulously identify all areas that need improvement by listing specifics and not generalities.
- Investigate, sort, and adopt and implement improved quality plans and methods.
- Prepare the quality manual based on improved quality plans and methods.
- Apply to the registrar for a preliminary assessment.
- Submit the quality manual to the registrar for review.
- Do a preassessment dress rehearsal audit.
- Submit revisions to the registrar for review.
- Modify the quality manual as per suggestions from the registrar.
- Train all personnel in all aspects of the quality manual.

This list certainly should give an excellent overview of the breath and scope of work required to gain certification.

Phase III: The Registration Audit

Here we are, all the work done, and now nail-biting time begins. The registrar's team arrives and we're ready to begin the audit. First let's backtrack and find out who the registrar is. The registrar is a company that has passed rigorous testing by the national representative of ISO, in the case of the United States—ANSI. They have demonstrated that they fully understand the requirements of all phases of ISO 9000 (2000) series as well as the philosophy of the document. They understand that companies requesting certification must demonstrate the intangibles that the petitioning company truly does have active senior management involvement and above all its program is aimed at customer satisfaction. By doing so the registrar candidate becomes a registrar and is an agent of ANSI or some other national entity of ISO. They are then qualified to examine companies wishing to become ISO 9000 series certified; at a significant sum. Usually certification can cost upward of $25,000. Companies requesting certification can select any registrar they chose. However, the choice usually comes down to a registrar company that is familiar with the industry the client participates in.

The audit process begins. The lead auditor explains the process to the client. Each segment of the company has a team consisting of an auditor and a member of the company's steering committee implementation team familiar with that area. They go through the ISO 9001 (2000) checklist. The company representative demonstrates compliance and the auditor takes notes. At the end of the day the auditors and company team meet to discuss results. The auditors present their findings—pointing out discrepancies. For minor discrepancies, the company is asked to fix them before the next day's auditing activities begin (a typical audit takes 2 to 3 days). Major items are added to a discrepancies list. The audit continues until all items on the certification check list has been covered. If the company has passed all items, they are congratulated and the auditors depart stating they will receive their certification document shortly. And the celebration begins, but don't celebrate too much. Certification is just the beginning. The company will have to be continuously recertified. More on that later.

If the company has discrepancies that do not allow for certification, it has two choices: fix the problems and have a re-audit, or abandon the quest. Either way the registrar company receives its full fee. The majority of companies require some degree of re-audit. After the company fixes the problem(s) that negated certification (90+% of companies do follow through and fix auditor found mistakes) the auditors are called back to verify. But they will not come back on the say so that the problem has been fixed. They will require a description of what has been done and documentation to show it is adequate and working. If they agree that the fix is truly appropriate to eliminate the problem, they will come back to re-audit. If that is the case, then the company will receive its certification. And now the company personnel can have their party. One caveat to keep in mind. When the auditors return, they have the right to look at any other thing they desire to. Here, if they exercise that option, they are looking to see if the company is maintaining the same system as they did at the initial audit. They want to make sure the initial audit wasn't just a show for their benefit.

Phase IV: Maintaining Registration

To gain certification most companies have to change some methodology of how they perform the seven steps of the manufacturing system. New habits have to be ingrained into the operating systems. This is the reason follow-up audits are required to keep certification. The ISO 9001 (2000) system requires that the level of performance measured during the initial audit be maintained. To ensure that this happens, a certified company must maintain a relationship with their registrar. They must notify the registrar of any major changes in their operating systems that occur and the reasons why. The registrar, then, has the option of doing a re-audit (which the client company is contractually required to pay for) to see if the certification premises are still sound. If so, the registrar will note in writing for file added to the client's records.

The company will also undergo a full audit every 6 months to prove it is still complying with the ISO 9001 (2000) requirements. This audit is similar to the initial certifying audit, but typically only takes 1 day. The registrar will try to send the same team that did the initial audit, for familiarity purposes. The company should also try to have the same personnel available to participate. Knowing that there will be follow-up audits ought to be sufficient incentive to maintain the system, and in fact to keep improving the system.

There you have it, the four phases of registration. Perhaps it appears complex to those companies used to doing things in an adhoc informal way. But when you look at it objectively, the only way for a company to become "world class," thus be as efficient as they possibly can be within their environment, is for them to discover the best way to achieve the seven steps of the manufacturing system, document them, train themselves in doing it that way, and not make any changes without a sufficient degree of proof that it is a better way to do so.

SHOULD YOUR COMPANY REALLY CERTIFY OR JUST APPLY THE PRINCIPLES OF ISO 9001 (2000)?

Is it really a help in making your company perform better? Or is it just hype put together to both create business for registrars and gain fees for ISO and its national affiliates? Is it, on the other hand, just something to impress your potential customers with? Perhaps some of the above is true. But I believe adhering to the ISO 9000 process does make companies perform better. And they will do even better if they go through the entire process to become certified. Applying the principles is with out a doubt better than not doing so. However, trying to apply the principles without the threat of constant audit to make sure you are doing so, is not entirely effective. Unless the company is certified, there is no real pressure to really do it right. It is too easy to simply decide "in this case, for the need to be expeditious, we can skip it."

I know from my own experience as a professional engineer, that at times I feel I can sympathize with a client and relent from the letter of the law of building code compliance and make it easier for my client to finish a project on time. But then I know if perhaps in the unlikely event something doesn't work out correctly, I would have to explain my decision to the state licensing board, who I know would not (cannot) be sympathetic. So I don't make it easier, I require the client to do it the code way. Overall everyone is a winner because we've built in a better margin for a successful outcome. The same is true for being certified or not. Being certified holds a company to a higher standard. Not being certified makes it easier for the company to drift from the higher standard, even though it is not the best way to do business in the long run. I believe being required to adhere to the discipline is an excellent insurance policy against being tempted to take a "short cut."

But keep in mind the world is full of good intentions being abandoned. Becoming ISO 9001 (2000) certified is not a simple task, as we've seen in the previous section. Companies should go into this exercise with sufficient understanding of what kinds of changes they must embrace. They must understand that it entails a strong commitment to quality at all levels of the business. They must be willing to fix mistakes as they are discovered and not just say "we'll do better next time." Management cannot disassociate itself from day to day operations. It must recognize it has the ultimate responsibility for good quality, and that at times this might be at odds with short term profits. To give you an example. When I was the vice president of operations for the Steinway and Sons Piano Company we had a policy that a piano not up to quality standards would never be shipped, even if the defect was found at final inspection just before it was scheduled to be placed in a crate for shipment. Our people, all the way up to the president of the company, understood we would rather crush a defective piano with a steam roller rather than ship it. So as a matter of course we fixed all defects discovered, and we looked real hard to find them. Our management and employee team understood that adherence to Steinway and Sons quality levels sometimes were at odds with our desire for profit. But so be it. This philosophy needs to be bought into by the entire company before ISO 9001 (2000) certification is applied for.

THE BENEFITS OF BEING ISO 9001 (2000) CERTIFIED

There is one prime benefit of being ISO 9001 (2000) certified: *higher profit*. This is achieved through a system that minimizes waste (JIT, Lean Mfg.) and affords more opportunities for improvement. Here's how.

The family of ISO 9000 standards are based on best practices. Best practices emanate from knowingly doing the seven steps of the manufacturing system in the most efficient way. The seven steps were presented and explained previously and repeated here.

Seven steps are:

1. Obtain product specification.
2. Design a method for producing the product, including design and purchase of equipment and or processes to produce, if required.
3. Schedule to produce.
4. Purchase raw materials in accordance with the schedule.
5. Produce in the factory.
6. Monitor results for technical compliance and cost control.
7. Ship the completed product to the customer.

In regards to seven steps we should keep in mind

- All companies comply with the seven steps. Those that do so consciously enjoy high profitability.
- The closer a company comes to achieving "best practices" in complying with the seven steps, the higher their profitability will be.
- With the ability to emulate best practices of world class performers readily available, a company cannot afford the status quo of the do nothing option and still survive.

The ISO 9001 (2000) certification process requires proof that the company is applying best practices. Therefore the certification process, itself, teaches a company to perform at a higher level than before. By performing at a higher level, that means the company will automatically enjoy less wastes in its operations, hence more profits. So we can see that those companies that achieve certification, and maintain it, are more likely to continue to be optimal performers than those that do not.

COSTS OF IMPLEMENTATION

Cost of implementing ISO 9001 (2000) are made up of direct and indirect components.

Registrar costs--direct costs
Culture change costs (temporary losses in productivity)-----------------------indirect costs
    Self investigation costs
    People training costs
    Cost of upgrading documentation

While the cost for the registrar is not trivial (upward of $25,000 average as of June 2005), the biggest cost by far is that of changing the company's culture to comply with the ISO 9001 (2000) culture. All companies go through some amount of this culture change while implementing and becoming certified to ISO 9001 (2000) standards. This is true because for virtually every one, but the truly world class companies, they will be going from a looser controlled system to a disciplined one with more checks and balances. The size of the gap of going from a looser to an ISO 9001 (2000) required system sets the magnitude of the required changes, hence costs.

The ISO 9001 (2000) methodology of controlling systems is one of embracing best practices. It is a catalyst that drives companies toward optimal performance through a system that demands continuous improvement. This change puts stress on the company's employees that they are not used to. So it causes slow downs while new traits are being learned. It continuously requires that the company's way of doing business be at an acceptable level in accordance with the eight paragraphs of the standard, especially paragraphs 4 through 8, as described previously. To make these changes management has to "walk the talk" and have continuous presence amongst employees asked to amend their approach to doing their jobs, for the most part thinking and taking responsibilities for doing their jobs properly including documentation. This will require considerable coaching.

Documentation is probably the most challenging change to the status quo. In most companies, employees, especially those on the factory floor were not hired as data keeper, but rather as skilled or semiskilled machine, process, and assembly people. In fact writing and analyzing was a skill that many of these people lacked. Therefore a significant indirect cost will be that of teaching these skills.

Overall, the costs of making these changes will be beneficial. Once this new paradigm of change is achieved, the company will have a set of employees truly trained how to think and look for improved methods, rather than simply being ciphers asked to push buttons and move materials, they will understand why they do what they do and become managers of their workstations.

THE UPGRADE PROCESS LEADING TO CERTIFICATION

It becomes obvious that the ISO 9001 (2000) company needs personnel who have a better understanding of the seven steps of the manufacturing system. They need to be very much aware of the entire business process, what constitutes good and bad quality at every step in that process. This becomes the criteria for a successful implementation. So the process of getting certified becomes one of knowing where the company is with reference to world class companies in working the seven steps of the manufacturing system. This understanding of where we stand in performing these tasks becomes the first task.

By understanding I mean precisely understanding. For example; step 2 of the seven steps is: "Design a method for producing the product, including design and purchase of equipment and or processes to produce, if required." Does the company do this or not? If it does, how? This means we have to ask specific questions to get specific replies. One of the many questions I ask under step 2 is, "How does the company scope the time required to do each operation required to accomplish the manufacture of the product? Do standard times exist for subsets of common tasks? If so, based on estimates or developed time standards?" From this I can get answers that tell me how the company performs this task and measure it against a world-class standard of having a set of reliable time standards to use to scope out factory scheduling via capacity calculations. (Paragraphs 7.1, 7.3.1, 7.5.1, 7.5.2 from the ISO 9001(2000) requirements delve into this topic). Answers I get from various factories are interesting and set the tone for improvements, thus establish the plan to upgrade to ISO 9001 (2000) certification levels. Here is a set of replies I received from one client company and my observations and recommended up grade actions:

"How does the company scope the time required to do each operation required to accomplish the manufacture of the product? Do standard times exist for subsets of common tasks? If so, based on estimates or developed time standards?"

Questionnaire Reply Summary:
- Not done.
- (1) Hr/unit. (2) Yes. (3) Estimates.
- Time estimates do exist for each operation of production. I don't think we use this information to schedule though.
- General time standards exist. No detailed time standards.
- I do not think we did any time studies. Maybe we have a general idea how long an operation takes.

- Our times are consistent.
- Time based on standard finished product; not subroutines, unless very special then involved depts. Give estimate.
- We have a combination of time standards adjusted according to demonstrated output.
- N.A.
- It doesn't. No standard times.

Observations:

A body of data pertaining to time standards is not documented. The amount of work capable of being done in the factory is estimates based on years of experience. The problem with this is that there is no definitive standard as to how much work really could be done over a specified time period. Therefore what was an acceptable rate in the past becomes the norm regardless of whether it is efficient or not. The purpose of time standards is to determine what the objective optimal performance rate for doing a specified job is. This is then used as a planning tool for manufacturing improvement and for sales volume potentials; for example, what can be sold when and in what volume.

Recommendations:

A documented estimate standard is better than none. If individuals possess knowledge about how long it should take to do any specific class of jobs, it ought to be written down and used generically when similar situation arise. In a similar manner time standards can be developed for virtually any operation in the factory. Having documented time standards provides consistency of approach and a way of assessing progress for any job, be it ahead, on-time or behind schedule. The time standard is used to create the master schedule for loading the shop. This will evolve the company into a status of greater capacity without having increased staff or spent significant sums for new capital equipment. This is true productivity improvement that will allow for more sales for delivery in the same period of time.

The process demonstrated above is called a gap analysis. This is the most straight forward way of knowing where a company is with regards to where it should be for ISO 9001 (2000) certification.

THE GAP ANALYSIS TOOL FOR ACHIEVE ACCEPTABLE STANDARDS OF PERFORMANCE

The gap analysis compares the seven steps of the manufacturing system actual conditions to what world class companies achieve. Figure 20-1 shows the how world class companies comply with the seven steps.

A "gap analysis" is a measurement of status from a reference point. It is often confused with "benchmarking," with which it has similarities. The thought process engaged in gap analysis can be expressed as follows: "This is where I want to be. This is where I am. The difference is the gap I want to close." Anyone engaged in an activity that can define how to do the task in the most optimum manner can do a Gap Analysis. The task is to define the optimum end point in a manner that allows it to be measured.

EXAMPLES:

Baseball: pitching a perfect game, 27 batters, 27 outs.
Quality: six sigma performance—3.4 defects/1,000,000 occurrences.

Gap analysis is similar to benchmarking. The whole process of benchmarking is an attempt to set the optimum end points for various business processes by finding others who do some of these processes at those levels, then emulating what they do. Benchmarking differs from gap analysis"

Is Your Company World Class?

Seven steps of the Mfg. System	World Class Companies
1. Obtain Product Specification.	*Use market studies to set new product introduction goals *Use Quality Functional Deployment (QFD) processes. *Use concurrent engineering techniques.

2. Design a method for producing the product, including design & purch. of equipment and or processes to produce if required	*Use three phases of design
	  Concept design
	  Producibility design
	  Mfg. Facilities design
	*Create Bills of Mat'ls & routes
	*Develop Process Methods

3. Schedule to produce	*Use computer aided process planning (CAPP) *Use scientific time standards *Use Mfg. Resources Planning systems (MRP II) *Employ Just In Time philosophy (JIT), Lean Mfg.- principles
4. Purchase raw materials in accordance with the schedule	*Use MRP II purchasing scheduling module *Use supply chain philosophy with vendors *Use only qualified vendors
5. Produce in the Factory	*Use MRP II shop floor scheduling module *Extend schedules to include vendor make Items *Use short interval scheduling techniques
6. Monitor results for technical compliance and cost control	*Use total quality management theory (TQM) *Use Statistical Process Control (SPC) *Employ "zero defects" programs *Are ISO 9000 registered
7. Ship the completed product to the customer	*Use MRP II inventory control system *Inventory records at least 99% accurate *Ship only after 100% of product kitted *Obtain customer concurrence of completed

Figure 20-1. World Class Performance Standards for Complying with the Seven Steps of the Manufacturing System

in that it subjectively decides that a target company does something well. What they do well is not necessarily tied to the definitions of world class embodied in the Seven Steps. For the Seven Steps of the Manufacturing System, we have a set of benchmarks that are designated as world class standards. So bench marking uses a subjective evaluation that a target company does something well and we wish to compare our company against that subjective standard and see how far we differ. Then we follow up with a program to close the differences until we are doing that task as well as the target company. Gap analysis does not use a target company or a subjective evaluation of something done well. Instead we use a set of world class standards to measure against. For ISO 9001 (2000) we use the eight paragraphs of the standard to see if we are in compliance and learn how to come into compliance by comparing our company against the way world class companies perform the seven steps of the manufacturing system.

So the gap analysis becomes a comparison of the ways world class companies go about their business to the way a subject company does. Figure 20-1 shows the world class standard in a macro standard. While it is possible to get a good sense of where a company is by using that as a

**Numerical Gap Analysis - Step 3, Appendix C - Investigation Points;
Seven Steps of the Manufacturing System; World Class Company**

3. Schedule to Produce	Gaps	Magnitude	Score	Histogram
3a. Does a schedule exist showing all open jobs and the required complete dates?	1	5	5	x x x x x
3b. Does the schedule show the total portal to portal cycle time for each job at each work station?	1	10	10	x x x x x x x x x x
3c. Is the cycle time for the schedule derived from step 2, above? What is the link or methodology for accomplishing this task?	0		0	
3d. Is there a capacity document showing the amount of work capable of being completed in a certain time period? Is it used to create the schedule?	2	5 and 6	11	x x x x x x x x x x x
3e. Does the schedule process have an update routine that feeds back actual completions for comparison to plan? If so, is it done daily and is there a policy for revising schedules if the plan becomes too far out of synch. with actual results?	3	4,4,6	14	x x x x x x x x x x x x x x
3f. Is the schedule of sufficient longevity that the supervisors can plan their work load prior to the start of the next work period (day or shift)?	0		0	
3g. Is the schedule broken down by work stations or centers and are workers told the amount of time they have to do the assigned job?	1	3	3	x x x
3h. Are overdue jobs given top priority to complete before new jobs are started?	0		0	
3i. Are there any rules established that give permission to bypass a job in the sequence to start a succeeding job?	1	8	8	x x x x x x x x

Figure 20-2. Portion of Gap Analysis Magnitude Evaluation

standard, for example knowing whether the company has an MRP II system used for scheduling, it is not detailed enough to help prepare a company in developing processes and procedures for ISO 9001 (2000) purposes. For that we need a more detailed and specific comparison. Over the years I have developed such a check list which was first published in my book *The Engineer Entrepreneur* (ASME Press, New York, 2003) and is presented here in Appendix C.

The measurements from the "world class" standard are made via the Investigation Points questionnaires, as shown in Appendix C. The completed the questionnaire gives the company a roadmap for where opportunities exist. Each gap is listed under the step of the seven steps it corresponds to. The importance of each gap should be defined with reference to the company's plans and culture. And it is highly recommended that attempts be made to quantify the magnitude of each gap. Generally, the larger the magnitude, the more serious the gap is to the health of the company. A scale of 1–10 is used to give point values for gaps. A complete absence of ability to do the step per the questionnaire is rated as "10". A minor variance is given a score of "1." A subjective judgment factor is used to score for a situation that is in between full compliance and partial compliance, but it is necessary to score all gaps in between. Figure 20-2 shows a typical section of a gap analysis magnitude evaluation based on the 1–10 point system.

It is a histogram of the number of gaps for step 3 of the seven steps of the manufacturing system as further divided in the investigation points. It is an excellent way of quantifying and visualizing which of the Seven Steps is farthest from the best practice "world class" standard. A Seven Step "step" with lots of small magnitude gaps is a serious situation. As can be seen the scoring is cumulative. When there are many gaps it usually means the reasoning behind the step is not well understood and methods are being followed by rote, not based on knowledge of the reason for the step. Without understanding, problems will occur frequently and quality will suffer. The histogram, then, points out the magnitude of the gap and leads to a corrective action program.

I recommend that the steering committee analyze the gaps. Then put plans in place, usually in project management format to make the corrections. They should first fix those items that are easy to do. So to speak, the steering committee should "clear the decks" before tackling the difficult issues. Some people refer to this as picking the low-hanging fruit off the tree. It is really important to fix easy things first. To not do so deprives the company of easy gain improvements that can have longer lasting positive effects than leaving them for later accomplishments. Why leave easy gains dormant when it doesn't have to be? Remember that fixing more difficult discrepancies may sap the energy of the staff and take a longer than expected time to fix, thus further delaying the easier items from being accomplished.

Keep in mind that all gaps will eventually have to be overcome. Some are necessary immediately to obtain ISO 9001 (2000) registration. Some to achieve optimum performance, thus truly be a world class performer. This may appear to be contrary to what has been stated and described in this chapter. Not so. ISO 9001 (2000) registration says a company has a quality plan and system that is traceable, consistent, and has the ability to identify problems for corrective action. And that the results of the corrective actions are also measurable. It does not say the company has reached world-class performance levels. But implies it is capable of achieving those levels because it has systems in place to measure and control its methods of doing business.

THE HIERARCHY OF QUALITY PLANNING FOR ISO 9001 (2000) CERTIFICATION

The process requiring that the company demonstrate an orderly quality management system is made easy by the format of the ISO 9001 (2000) requirements for certification. The eight paragraphs define what is required, and they do so in a manner that almost automatically dictates a hierarchy of information flow. There are four levels of information flow. From the macro (the strategic) to the tactical to the operational to the archival (the micro). Where the macro is the conceptual and micro is the most detailed and complex level of information. So the levels as they relate to quality are

Strategic—quality policy—What are we going to do?
Tactical—quality procedures (system)—How are we doing it?
Operational—task (process) documents
Archival—quality records

With this relationship of information we can adequately support the ISO 9001 (2000) standard's two key features:

Documentation
Effectiveness of Management Systems

Remember even though it takes an effective system for producing the product or service, ISO 9001 (2000) is a quality management system, not a product conformity standard. Compliance with ISO 9001(2000) provisions does not automatically make the product successful in the marketplace. But it does ensure that the complying company can demonstrate that it does what it says it will do.

INTERNAL AUDITING, A TECHNIQUE THAT HAS REACHED MATURITY THROUGH ISO 9000 APPLICATIONS

One of the unique features of ISO 9001 (2000) is the use of internal audits to ensure a company is remaining in compliance with the standard. It is a dual use activity. Remain in compliance—doing and documenting audits as required by the standard. And auditing to find problems and possible problems and fix before any damage done is excessive. The standard doesn't dictate how an audit should be done, just that it is done. However, based on experience the following procedure seems to be optimal for factory quality audits.

The requirement to audit is based on paragraph 8.22 Internal audit. It states that the company will conduct internal audits at planned intervals to determine whether the quality management system

a. conforms to the planned arrangements to the requirements of the standard and to the quality management system requirements established by the organization, and
b. is effectively implemented and maintained.

It also states that an audit program be planned, taking into consideration the status and importance of processes and areas to be audited, as well as the results of previous audits. The responsibilities and requirements for planning and conducting audits, for reporting results and maintaining records are to be defined in a documented procedure.

If defects are found the ISO 9001 (2000) standard has a hierarchy of corrective actions, both mandatory and suggested

a. Nonconformances for internal audits—mandatory
b. Customer complaints—suggestion
c. Returns—suggestion
d. Suggestions from employees—suggestion
e. Planned improvements—suggestion

The Generic Audit Process

Audits are done to measure the organization's ability to carry out the duties of the audited function. An audit looks at the system the function uses to carry out its responsibilities. It also looks at ability of the system to deliver products and services to customer specifications.

The typical audit process consists of:

a. Understanding how the system is supposed to work.
b. Compares actual system workings to the system plan.
c. Determines if the system plan can ensure customer specification compliance.
d. Observe actual system workings to see if customer specifications are being met.
e. Documents deficiencies for all four points, above.
f. Recommend corrective actions.

The Chronological Sequence of an Audit

The sequence consists of preparing, a preaudit meeting, conducting the audit, writing the audit report, and follow-up responsibilities. In outline form, here are the sequence details

1. Prepare for the audit
   a. Determine scope of audit
      Area to be covered

   b. Determine Type of audit
      Product—detailed examination of some in-process or finished products to verify they are
         in compliance with specification
      Process—examination of work processes to see if they are being performed in accor-
         dance with methods and appropriate instructions
      System—an examination of the quality aspects of various segments of overall systematic
         approaches to applicable procedures and instructions
   c. Know the objectives of the audit.
   d. Obtain and review all available information pertinent to the audit, including auditee's
      previous audit results.
   e. Develop a plan for collecting evidence
   f. Develop a check list of items to be covered
   g. Notify the auditee of the pending audit, agree to date and time
2. Hold a pre-audit meeting with the auditee.
   a. State the objectives and purpose of the audit
   b. Solicit areas of interest from the auditee
   c. Develop a detailed schedule for conducting the audit On-site
      Daily review meeting (if extent of audit goes beyond one day)
3. Conduct the audit
   a. Record findings and observations
   b. Hold daily meetings with members of the audit team to discuss next steps, etc.
4. Prepare audit reports
   a. Verbal and written report to auditee, and the responsible management.
      Present list of items for corrective action
      Written report to audit sponsor
5. Post audit responsibilities
   a. Review corrective action reports for compliance
   b. Report correction status to the audit sponsor

Internal Audits

Internal audits are means for management to periodically check for compliance to policies and
procedures. Internal audits should be among the first activities initiated by the company's ISO 9001
(2000) implementation steering committee. They are absolutely critical for the operation of the
quality assurance (QA) system. However internal audits are not just designed to see if an ISO 9001
(2000) compatible system is present, but to see if the company's QA system is working properly
whether it is contemplating certification or not . This type of audit is aimed at exposing QA system
defects, not the efficiency of a department or person, but to find where documents and practices
have changed in an unauthorized manner. It also verifies that authorized planned and documented
changes have occurred and have been done correctly in accordance with policies and procedures.

Internal audits procedures will be heavily audited by the registrar during the certification process. Internal audit records are the most effective way a company has to prove it is complying with
its quality system. A good internal audit track demonstrates that the company does indeed "walk
the talk." Conversely, internal audit nonconformances not corrected found by the registrar are of
great weight and will certainly cause them to lean heavily against granting certification.

Types of Documentation

There are three general classifications of documentation used in auditing

   Type 1—Procedure is mandatory by the standard and by the company
      —Procedure is auditable by registrar and by the company

Type 2—Procedure is not required by the standard, but is mandatory by the company
   —Procedure is not auditable by registrar, but is auditable by the company
Type 3—Procedure is not required by anyone but is a guideline
   —Procedure is not auditable by registrar or by the company. For example, verbal instructions are never auditable (no audit trail).

Auditor Qualifications

Auditing is a skilled activity that requires training. Basically investigative training. Any potential auditor has to be familiar with what he or she will be auditing, but does not necessarily have to have expert knowledge. The auditor needs to know what part of the process or procedure audited is important and what is not. She has to be familiar with those processes and/or procedures and be able to answer the question:

" Is this particular aspect of the process and/or procedure in compliance with the rules set up governing the process and/or procedure?"

The training program for the auditor has to be comprehensive enough, so after completing this instructive indoctrination the rookie auditor can answer the question properly.

Auditors need to be familiar with the topic they are assigned to audit. For example, if the auditor is assigned to audit a quality procedure such as proper use of quality control reports (QCR), then he must understand the rules and regulations for using the QCR report. In addition to being familiar with the topic to be audited, the auditor needs to be objective, such as like an umpire in a baseball game. He has to be able to call them as he see's it and not be swayed by management or even the senior auditor. Above all the auditor's integrity has to be beyond reproach.

Auditor Conduct

An auditor must start off without any preconceived opinions as to whether or not she will find discrepancies. In fact it is best to proceed with the attitude that everything is in accordance with authorized procedures unless shown to be otherwise. The auditor also has the responsibility to report only what is actually observed, not read into any observation that is not there. She can not suppose something to be if it is not observed.

When a discrepancy is observed it is the auditor's responsibility to track it down to root cause. The auditor has to be persistent and tenacious to get to the truth as to why the discrepancy exists. Otherwise it will not be possible to recommend corrective actions. An internal audit is all about weeding out actions that cause discrepancies. It is the only reason for conducting audits.

Sometimes while driving to root cause an auditor has a difficult time in remaining diplomatic and demonstrating good manners. But she must. The auditor represents the principles of good management and leadership. She has to always be objective in pursuing the truth and not trying to place blame for deficiencies. To do otherwise invites future ill will and hinders opportunities for improvement.

Overall Audit Management

The audit manager schedules the audits in descending order per area importance and problems. Audits can be for a single procedure and plantwide, or for a procedure in a single department. After certification it is required that audits be performed for each area at least every 6 months at a minimum for the first 2 years. In order for the audit schedule to be reasonable and not unduly disrupt production activities, the audits should be spaced out and not done during concentrated time periods. The audit manager also has the responsibility of selecting an auditor, who then is responsible for the audit in the assigned area.

Getting the results of an audit to the audited area's manager and employees has to be relatively fast for it to be of use. Delaying an audit results serves no purpose. It simply detracts from the usefulness of the information. And it is the information the audited area needs in order to make improvements. For this reason, auditors are usually given no more than 2 weeks to perform the audit and write the audit report. Then the audit report is reviewed by the audit manager, who may add comments, and is released to the audited area.

The area manager and his team then sets plans in motion to correct nonconformances. When this is complete the audit manager is requested to re-audit to gain concurrence that the deficiencies have been corrected. Again, it must be emphasized that audits are for finding and correcting deficiencies, nothing more. They are not used for disciplinary reasons, not to be used to measure efficiency or effectiveness of the workers in the area.

Remember:

"An audit is an opportunity for the organization to improve."

SUMMARY

As we see ISO 9001 (2000) certification is an all encompassing exercise for any company and should not be taken lightly. To do so surely will create confusion and be a morale depressor. ISO 9001 (2000) certification requires a disciplined approach to achieve, and that alone makes it a worthwhile journey for a company to pursue. Because doing so almost automatically causes a company to perform more precisely, hence more efficient.

I believe the true strength of the ISO 9000 (2001) certification process is that it teaches companies to use audits as a matter of course in doing its' business. Audits are the true way of finding and correcting problems, even hidden problems. By setting up an ongoing audit system a company creates a continuous improvement atmosphere which virtually assures profit improvement.

Throughout this book I've shown multiple techniques manufacturing engineers can use to improve their company's performance. With an ISO 9001 (2000) certification program, the manufacturing engineer has a capstone project to ply her or his profession and employ all those skills and techniques in an effective synergistic fashion. Manufacturing engineers should welcome the opportunity to participate in an ISO 9001 (2000) program. It is an ideal venue for demonstrating the strength of the discipline. Use it to your advantage.

REVIEW QUESTIONS

1. What role does manufacturing engineering play in an ISO 9001 (2000) certification project?
2. Discuss how management can demonstrate its' commitment to quality.
3. What does continuous improvement mean with respect to gaining ISO 9001 (2000) certification?
4. Discuss the reasons for dividing the ISO 9000 standard into four specific standards. Show how they are related.
5. Why do certain industries modify ISO 9001 (2000) to more specific standards?
6. What is the content of the quality manual, reference paragraph 4.2.2, and how is it prepared?
7. How is management commitment evaluated for certification purposes?
8. What is the responsibility of the human resources department in the certification process?
9. How does an MRP II system fit into the certification process?
10. How does a company demonstrate it is customer focused for certification purposes?

11. Discuss the role of the steering committee in achieving certification for a company?
12. What is the role of the registrar? What can a registrar do to help a company gain certification?
13. What criteria should a company use to select a registrar?
14. What is the purpose of re-certification audits and what is the relationship between the company and its' registrar during this process?
15. How does higher profit manifest itself out of ISO 9001 (2000) certification?
16. What constitutes the indirect costs of certification?
17. Describe how gap analysis can be used to prepare a company for certification?
18. What is meant by the hierarchy of quality planning with respect to information flow?
19. What is the primary difference between a registrar audit and an internal audit?
20. Discuss the responsibilities of the internal auditor.
21. When should the steering committee commence internal auditing and why?

EMPLOYEE HANDBOOK

Table of Contents:

SECTION 1: ADMINISTRATION

Definitions of Employees

Employees are divided into three groups for the purposes of pay and benefits. Company policies apply to all employees.

Regular Full-Time

Employees hired full-time on a full work week basis for a continuous and indefinite period of time are considered regular full-time employees for all pay and benefit purposes.

Regular Part Time

Employees whose work schedule is less than full-time on a full work week basis for a continuous and indefinite period are considered regular part-time employees for all pay and benefit purposes. Regular part-time employees may be eligible for some benefits by specific reference only.

Temporary

Employees hired as temporary replacement for regular full-time or part-time employees, or for short periods of employment such as summer months, peak periods and vacations are temporary employees. These employees are eligible for only statutory benefits regardless of the number of hours or weeks worked.

Federal Wage And Hour Law

Nonexempt

Nonexempt hourly wage personnel are paid overtime pay for all hours worked in excess of 40 in a work week. Hourly employees are expected to confine their work to the normal work day and work week, unless overtime is authorized in advance by your supervisor. Scheduled overtime is considered required work time.

Exempt

The hours of exempt salaried employees are often irregular and begin and end outside of a normal work day. Salaried employees are exempt from the overtime provisions of the Wage and Hour Law and do no receive overtime pay.

Benefit Eligibility

Regular full-time employees are entitled to the benefits stated in this booklet, provided they qualify under each individual benefit. Regular part-time employees are entitled to those employee benefits specifically designated. Temporary employees are eligible for only statutory benefits. The term "eligible employee(s)" used in the benefit portion of this booklet means regular full-time employees, unless otherwise designated. You will be told the status of you position when you are hired.

Orientation

Getting off to a good start is very important. After you have accepted employment, your supervisor will discuss your duties, the company's policies and procedures. Read this booklet thoroughly and ask questions of your supervisor. At the end of this booklet you will find one receipt – the receipt is to be signed by you and dated and witnessed by the Personnel Manager. The completed receipt will become part of your personnel records.

Ninety-Day Evaluation Period

During the first 90 days of employment, the ABC Company and new employees are given an opportunity to evaluate whether they should continue the employment relationship. Any time before the end of this ninety-day period, the employee's performance will be evaluated and he or she will be notified of the future employment status by his supervisor. If, during or at the conclusion of the evaluation period, either the ABC Companyor the employee believes that employment should not continue, termination will follow immediately.

Payroll Information

Soon after you are employed, you will be given federal and state tax forms along with insurance forms to complete. All of the forms, your application, and any information regarding starting pay, starting date and other pay or benefit information will be forwarded to the Personnel Manager.

Continuous Service Date

In order to have a record of your benefits as an employee, a continuous service date will be maintained. Your continuous service date starts with your first day of work and continues unbroken as long as you are a regular employee.

Employee Information

Help keep us informed about any major changes which may affect your job status. Important changes to report to the Personnel Manager include:

— Name
— Address
— Home telephone number
— Marital status
— Names and numbers of dependents
— Any important health information
— Emergency telephone numbers and whom to notify in case of emergency
— Change of beneficiary
— Authorized payroll deductions
— Additional education and special training courses

Records

Your completed application form begins your employment records. Important information regarding your work status with the company is added from time to time. The office maintains a personnel record on each active employee. These records are maintained as confidential company property with other company records in locked files.

SECTION 2: HOURS OF WORK AND PAY

Working Hours

Regular working hours are as follows:

Office   8:00 a.m. to 5:00 p.m. daily, Monday through Friday with one (1) hour lunch.

Factory/Service Staff

1st Shift: 7:00 a.m. to 3:30 p.m. daily, Monday through Friday with one-half (1/2) hour for lunch. Saturday work may also be scheduled and attendance is required.

2nd Shift: 3:00 p.m. to 11:30 p.m. daily, Monday through Friday with one-half (1/2) hour for lunch. Saturday work may also be scheduled and attendance is required.

3rd Shift: 11:00 p.m. to 7:30 a.m. daily, Monday through Friday with one-half (1/2) hour for lunch. Saturday work may also be scheduled and attendance is required.

Certain employees or departments may be required to work a different schedule, and if so advised will be expected to comply as required.

Each employee is expected to complete a normal work day and work week, and work whatever reasonable additional hours that may be required to meet company needs when scheduled by your supervisor, including overtime and Saturday work. The company's work week begins at 12:01 a.m. Sunday, and ends at midnight, Saturday.

Overtime

Non-exempt hourly employees will be paid one and one-half (1.5) times their regular rates of pay for time worked over 40 hours in a work week.

Pay Period and Payment

There are 26 pay periods per year, twice per month. They are nominally divided as follows:

- 1st through the 15th of the month, payable on the 25th of the month
- 16th through the 31st of the month, payable on the 10th of the succeeding month

Rest Breaks (Hourly Employees)

Rest periods, nominally two 15 minute intervals will be set in accordance with applicable federal and state laws.

Time Clock and Time Records

Government regulations require that we keep an accurate record of hours worked by non-exempt employees.

Hourly employees are required to punch or record in when they report to work and punch or record out when they finish work starting and concluding at the Company's offices. If employees go directly to work sites, they are required to maintain a time log and turn it in to their supervisor daily.

It is a violation of company policy for one employee to punch or record another employee's time card, or to alter your time card or that of another employee. Violation of this policy will result in disciplinary action that could include termination.

Attendance

Regular and on- time attendance is essential to efficient operations. Excessive absenteeism and tardiness are not only inconvenient but also a costly problem. While it is recognized that occasional illness or extenuating personal reasons cause unavoidable absence or tardiness from work, regular on-time attendance is required for continued employment.

Employees must call in up to 1 hour after the start of their shift to report any absence or lateness. Notify your supervisor the night before if possible.

YOU MUST MAKE THE EFFORT TO NOTIFY THE COMPANY OF ANY ABSENCE OR TARDINESS

Employees who fail to maintain an acceptable attendance record will be subject to disciplinary measures. Unexcused absence or tardiness will affect your promotion and /or raises during performance reviews.

Family Emergency

In the event the office receives word of an emergency related to your family, you will be notified as soon as possible. Should you be at a location away from the office, we will make arrangements to contact you, and if necessary, see to it that you return home immediately.

Emergency Weather Conditions

If there is any question regarding hours of work during emergency weather conditions, employees are responsible for contacting their supervisor or the office regarding opening and closing hours.

Rates of Pay

You were hired at a certain pay rate for a specific job defined by its job description. A copy of the job description and the pay rate plan have been given to you. Pay rates for all jobs have been determined by an industrial accepted pay evaluation system. It is the ABC Company's police to pay all employees fairly in accordance with the pay evaluation system.

Change(s) in Pay

Your performance will be evaluated in accordance with each job description learning period times and then annually by your supervisor. His or her recommendation for promotion, change in duties or pay raise must be approved by the President of the company before any changes take effect. Performance evaluations do not necessarily mean changes in pay or duties. See "Remedial Actions" in Section 4 for information on performance related delays in pay raises.

Deductions

The company will deduct from your gross pay federal and state (where applicable) income taxes, Social Security taxes and back taxes, garnishments for uniforms and for loss of company property, and any other legal requirements. Mandatory child support falls under the garnishment category. Any deductions other than statutory deductions must be authorized by the employee. No other deductions will be made unless you specifically authorize them in writing. All deductions will be itemized on your pay check stub. If you have any questions, ask your supervisor.

Errors in Pay

Every precaution is taken to ensure that you are paid correctly. If you believe there is an error, notify your supervisor. We will make every attempt to adjust the error immediately and no later than your next paycheck.

Garnishments

State laws require the company to honor garnishments of employee wages as a court or other legal judgment may instruct, including child support. The law also provides for an administrative fee to be charged when a garnishment occurs.

Authorized Check Pick-up

If you are absent on a pay day and instruct someone to pick up your pay check, a note signed by you authorizing the person must be provided before the pay check can be released. The person picking up your check must show proper identification and sign for the check. This protects both you and the company.

Pay on Termination

You will be paid for time worked, less deductions, on the next regular pay day according to the applicable federal and state laws.

SECTION 3: BENEFITS

A comprehensive group of employee benefits has been developed for you by the ABC Company. Benefits are carefully designed to provide you with a balance of time off from work without creating undue hardship, help in cases of injury, illness, or retirement and an opportunity to grow with your job. Most of the cost of the benefits are paid by the ABC Company.

Regular full-time employees are eligible for most benefits if they meet specific requirements. The following benefits may change from time to time. Be sure to keep any information regarding the changes in this booklet for ready access. Questions should be directed to your supervisor. Insurance claim information should be directed to the Personnel Manager.

Vacation

The ABC Company recognizes that employees need a scheduled time away from normal work duties for their personal well-being. The company grants annual vacation with pay to regular full-time employees who meet the service requirement.

Eligibility—(Hourly Personnel)
—Five (5) days of paid vacation time will be provided after you have been employed one (1) full year.
—Following two (2) full years of continuous service, and employee is eligible for 10 days of paid vacation.
Eligibility—(Salaried Office and Supervisory Personnel)
—Ten (10) days of paid vacation time will be provided after you have been employed one (1) full year. Fifteen (15) days will be provided after five (5) full years.
Scheduling
—Your vacation request must be made at least three (3) weeks prior to the time you wish to be off. Vacation may be taken at any time during the year after eligibility with the following provisions.
—Employees are expected to take their paid vacation time as a means of rest and diversion for themselves and their families. Only under special circumstances as determined by the company will an employee be allowed to forego earned vacation. In such cases the employee will be paid for vacation time and also the hours worked at his or her normal rate of pay.
—Five (5) days vacation time may be carried over into the next vacation year at the discretion of the employee. However, the employee will not receive additional pay for time worked instead of taking vacation time.
—A paid company holiday that falls during the vacation period will be considered as a paid holiday and not vacation time. This day of vacation may be taken at another time as approved.

—An employee must work the scheduled work days **before and after** the paid vacation period in order to be eligible to receive vacation pay.

—The company will strive to agree to vacation dates requests, but job requirements must always have precedent over vacation schedules.

—Seniority will be considered in the event a conflict of vacation schedules arises.

Vacation Pay

Pay for vacation time will be at the regular rate of pay. Paid vacation time may not be considered as time worked for overtime purposes.

Pay in lieu of unused vacation at any time will be provided only at the convenience of the company when approved in advance by the President, and, upon separation under certain conditions.

Holidays

After 30 days of employment, employees are eligible for holiday pay for the following holidays (part time employees will be paid for actual work time lost due to a holiday).

New Year's Day	Labor Day
Memorial Day	Thanksgiving Day
July 4th	Christmas Day

You must work the regularly scheduled work days before and after the holiday to receive holiday pay. Paid holiday time will not be considered as time worked for the purpose of computing overtime.

Health/Dental Care Insurance

Health and dental insurance is provided by a group plan for all regular full time employees. The company pays ½ of the premiums for employees. Employees may enroll their dependents in the health and dental plan at a special cost provided by the insurance carrier. Joining the health and dental plan is optional. The insurance carrier will provide eligible employees with a detailed summary of the insurance provided.

Continuation of Health Care Insurance

If an employee who has been covered by the company's health care insurance plan is laid off because of temporary lack of work and is rehired within three (3) months of the layoff, the employee will become eligible to participate in the group insurance plan as of the first of the month after 90 days from the date of rehire.

Employees terminated (other than for gross misconduct) may continue their health care insurance coverage for 18 months at their own expense. Employees whose work hours are reduced to a point where they are ineligible for coverage may also continue coverage for 18 months at their own expense.

Dependents who are no longer eligible to be covered under the employee's insurance contract due to employee's death, a divorce or legally separated spouse or a child ceasing to be a dependent, may continue health care insurance for up to 36 months at the expense of the employee or dependents.

Employees must notify the office of their intent to continue within 30 days and pay the premiums according to the premium schedules. Dependents must notify the office within 60 days of their intent to continue coverage. Failure to notify the office or pay premiums is considered to be a notice of cancellation of this option.

Educational Assistance

The company encourages all regular full-time employees to further develop and improve themselves through education. The company feels employee development is an advantage to both the company and the employee.

With prior approval of the President, the company will reimburse eligible employees with three (3) month's continuous service for 50 percent of the cost of the course(s) upon successful completion with a grade of C or better. Courses must be deemed directly job related.

Employees requesting educational assistance must comply with the conditions listed on the following paragraphs.

— Employee must submit a written request for educational assistance to the Personnel Manager listing the name of the school, the description of the course, costs, its schedule time, and whether or not the employee is working towards a degree. Final approval for all educational assistance is given by the President.

— Employee must continue to be employed full time by the company at the time the reimbursement is paid.

— Upon successful completion of the course, the employee must submit all receipts for books, tuition, student fees, etc., along with a copy of the final grade on the course to the Personal Manager for review before reimbursement.

— Reimbursement for educational assistance will not be made if the course is dropped, failed, or in any way not completed, or if the employee ceases to be employed by the company for any reason.

— Reimbursement will not be made by the company if the employee is receiving payment for course(s) by grant or scholarship from other sources, for example, the G.I. bill. Any special cases or situations not listed above will be at the discretion of the President.

Civic Leave

When a full-time employee is called for jury duty or as a witness, time off with pay will be granted as follows, provided that at least a five (5) day notice is given.

— The company will pay the difference between what the court pays and the employee's regular pay rate.

— A document from the court showing the time and amount paid must be provided to the Personnel Manager.

— Employees must notify their supervisor upon receipt of their summons at least five (5) days prior to the duty period.

— If the court dismisses the jury or witness early, the employee is required to return to work as soon as possible and complete an eight-hour shift comprised of civic time and time on the job.

Voting

The company does not provide time off for voting with or without pay. Employees should make arrangements to vote prior to or following normal work hours.

Funeral Leave

In the event of a death in your immediate family, you will be allowed up to two (2) days off with pay in order to assist in arrangements or to attend the funeral. You must notify your supervisor. Vacation days may be used for additional time provided you are eligible for them.

Immediate family is considered: spouse, mother, father, son, daughter, brother, sister, grandchild, and grandparent.

Birth Leave

In the event of a birth in your family, you will be allowed up to two days with pay to assist in arrangements of the birth. One day of the birth leave day benefits must be the day of the birth.

You must notify your supervisor. Vacation days may be used for additional time provided you are eligible for them.

Family Medical Leave (Without Pay)

To be eligible for a Family Medical Leave (FML), you must have worked for ABC Company:

1. for at least 12 months
2. for 1,250 hours over the previous 12 months

The employment and service requirement will be determined at the time the Family Medical Leave is granted.

Leave Provisions

An unpaid Family Medical Leave can be granted for a period of up to 12 weeks during a 12 month period of time. The 12 month period will begin when FML starts. Leaves can be granted for the following reasons:

1. for the birth of a child and to care for such child, or for the recipiant of placement of a child for adoption or foster care;
2. to care for the employee's spouse, son, daughter, or parent, who has a serious health condition; or
3. for a serious health condition that makes the employee unable to perform their job.

An employee taking leave for "medical reasons" may take intermittent leave or leave on a reduced work schedule for the equivalent of the 12 weeks.

We expect all available **paid time off** (i.e. vacation) would be utilized as the first part of the Family Medical Leave.

An employee who is on Family Medical Leave cannot work for another employer without the written authorization from the company.

Advance Notice and Medical Certification

Anyone wanting to request a Family Medical Leave, will be required to provide advance notice of the leave. The employee must provide 30 days advance notice when the leave is "foreseeable." If circumstances prevent a 30 day notice, a minimum of two business days notice will normally be needed.

Medical certification will be required to support a request for leave due to a serious medical condition, and the company may require, at its expense, a second or third medical opinion, if necessary.

When an employee is ready to return to work, a fitness-for-duty report from the doctor will be needed.

Job Benefits and Reinstatement

For the duration of a Family Medical Leave, group insurance benefits will continue on the same employer-employee cost basis.

The ABC Company will expect to be reimbursed for medical insurance premiums paid on an employee who chooses not to return from Family Medical Leave.

An employee should notify their supervisor not less than one week before the end of any leave of their intent to return to work. If an employee plans to return early, they should notify their supervisor at least two (2) weeks prior to the scheduled return.

Upon returning from leave, employees will be returned to their original job or equivalent position with equal pay, benefits and other employment terms. Any employment benefits accrued prior to the start of a leave, will remain intact.

Military Leave

It is company policy to grant a leave of absence without pay to employees who participate in U.S. armed forces reserve or national guard training programs in accordance with the provisions of the Universal Military Training and Service Act. Employees are encouraged to use their vacation time for military training. See the Personnel Manager for information.

All employees who are called to active duty involuntarily as duly authorized by State and/or Federal authorities are eligible for indefinite leaves of absence. See the Personnel manager for information.

401(k) Plan

401(k) Plan provides employees, who are 21 years old or older and who have completed one (1) year of service, with an excellent opportunity to invest in their own future retirement.

Eligible employees can contribute up to 14% of pay into the 401(k) plan. Employee deferrals are exempt from federal withholding taxes. Employee contributions and earnings are not taxed until withdrawn from the plan.

Employees are always 100% "vested" in their contributions and earnings.

Participating employees select how their contributions will be invested. Hardship withdrawals are possible.

You can get a copy of the 401(k) Plan from the Personnel Manager.

Social Security

The cost of Social Security is shared between you and the company. For every dollar you put into Social Security, the ABC Company puts in a dollar. This provides benefits for you and your family as specified by law in the event of you retirement, hospitalization and medical care after age 65 (Medicare), total and permanent disability before age 65, and death at any time. For details, contact your local Social Security office.

Workers' Compensation

If you lose time away from work due to a job-related accident, you are eligible for payments to cover loss of income as well as medical expenses (as required by law). The benefit cost of this is paid entirely by the ABC Company. If you suffer a job-related accident, report it to your supervisor at once. The report of injury must be filed in the company office no later than 24 hours after the injury occurs.

Workers' Compensation benefits begin only after the first week of absence due to a specific injury. After the fourth full week of absence the first week of benefits is paid.

Unemployment Compensation

This is another form of insurance which is paid for entirely by the ABC Company. It helps an employee meet a loss of income resulting from unemployment beyond his control by paying certain benefits while he is out of work.

Employees will not be eligible for unemployment compensation if employment is terminated for any reason during their initial 90 day probationary period.

SECTION 4: TRANSFER, TERMINATION AND LEAVE OF ABSENCE

Transfer

Transfer of employees for the company's convenience may be made to meet personnel requirements.

Termination

Employees may be separated voluntarily or involuntarily by retirement, voluntary resignation, lack of work, release or discharge.

As you join the ABC Company "team," the prospect of ending your employment should be the furthest thing from your mind. We fully expect and hope that you plan a lifelong career with our company. However, as a part of you full understanding of our company, you should be aware of the types of termination that are possible, the normal steps and procedures which will be required, and the possible benefits that may be available at the time of termination.

You may decide to resign from your job with us. If you make such a decision, you should communicate your resignation in writing to your supervisor. In order to help establish a good reference point for the future, you should give reasonable notice of your decision and put forth your best effort during your notice period.

All employees should give at least two (2) weeks' notice. If you are a supervisor or manager, you should provide four (4) weeks' notice.

You will be assumed to have voluntarily resigned (without proper notice) if you have an unreported absence of three (3) or more days.

A release is a permanent termination which is initiated by the company due to 1) permanent lack of work, when alternate assignment is not possible, or 2) efforts to develop satisfactory work habits or performance have been unsuccessful. The remedial action which we may take to work out personal or performance difficulties are summarized below.

Remedial Action

The ABC Company wants every employee with difficulties to solve their work-related problems and regain their status as productive and effective members of our team. Except in extreme or dangerous situations, supervisors will counsel with affected employees in a positive effort to resolve problems and communicate the seriousness of inappropriate actions. The following process will be used, as necessary, to initiate remedial actions:

1. Verbal warning with a signed copy to file
2. Written warning with a signed copy to file
3. Suspension without pay
4. Release

Please note that it is not meant that the entire process be exhausted before an employee is discharged. Extreme or serious situations, gross or willful insubordination, safety violations, and willful repetitious violations of company policies may result in immediate suspension or discharge.

The following list identifies some of the reasons that remedial actions may be initiated. This list is a sample and is NOT all inclusive.

— Excessive absences or tardiness.
— Not following a reasonable order, or failure to perform work assigned, or failure to comply with work and safety rules.
— Negligent damage to company equipment.
— Gaining unauthorized access to company records.
— Speeding or reckless driving or unauthorized use of company vehicle.
— Use of threatening, profane or abusive language.
— Demonstration of lack of courtesy towards other employees, customers or vendors.
— Not completing assignment up to the quality required by the company.
— Failure to report personal injury resulting from an on-the-job work situation.

Extreme or serious situations may result in immediate suspension and/or discharge.

The following list shows some of the reasons when a discharge would be initiated. This list is only a sample and is NOT all inclusive.

— Individual or collective falsification of employment, personnel or other records. This includes but is not limited to, applications, all reports, records and statements under the responsibility of the employee.
— Disclosing confidential information to outsiders.
— Gambling or fighting on job sites or company property.
— Unethical conduct or serious conflicts of interest.
— Concealing defective work.
— Stealing the company's property, a customer's property or the property of any employee; hiding, concealing or mis-appropriation of company property or the property of other employees or customers; sabotage or willful damage to company property, or the property of other employees or customers.
— Unauthorized use or sale of any company-owned property, salvage material or equipment.
— Reporting to work under the influence of alcohol or illegal drugs or chemicals; possession, sale or use of marijuana or illegal drugs or chemicals or consumption of alcohol while working on job sites, in the office, or in company vehicles.
— Gross negligence or willful acts in the performance of duties resulting in damage to company property or injury to others.
— Gross insubordination—willful and deliberate refusal to follow reasonable orders by a supervisor or member of management.
— Willfully disregarding company property.
— Violation of the company's equal opportunity or sexual harassment policies.
— Serious safety violation resulting in injury.

A discharge is a permanent termination which is initiated by the company due to non-conformance with policies or practices or serious deficiencies in conduct. No notice or severance allowance will be granted, and accrued vacation pay will be forfeited.

Exit Interview

All employees who leave our company will be interviewed by their supervisor and/or another management employee. At that time, their thoughts about improvements and concerns will be solicited, company property will be checked in, and the conversion of group insurance will be discussed.

THE PERSONNEL MANAGER MUST BE ADVISED IMMEDIATELY OF THE DATE AND REASON FOR TERMINATION.

ALL COMPANY PROPERTY IN YOUR POSSESSION MUST BE RETURNED TO YOUR SUPERVISOR UPON SEPARATION AND BEFORE THE FINAL PAY CHECK IS RELEASED.

Pay at Time of Termination

The company will determine if the terminating employee has any outstanding debt owed the company and whether the individual has in his possession any company credit cards, uniforms, tools or other company property. After there is a full accounting of the employee's and the company's accounts as determined by the company, the final pay check will be issued to the employee in accordance with state law.

The company will issue a check which is designated as the final payment for all services rendered. The final check will not reflect any time not actually worked.

Upon resignation or termination, all employees should consult the office for possible conversion of their hospital insurance, and to clear up any financial questions. All employees are expected to return any company property in their possession.

Leave of Absence

An employee may ask for a leave of absence without pay from the company. The company cannot guarantee to hold any job for more that 30 days. The request for a leave of absence must be made in writing and be approved, in advance, by the President.

Due to lack of work the company may require an employee to take an unpaid leave of absence. The length of the company initiated leave of absence may vary.

During any leave of absence the employee is responsible for the payment of all insurance premiums for his individual coverage and dependent health care insurance coverage (if applicable). This money should be paid to the ABC Company by the first day of each month that the employee is on leave.

While on an approved leave of absence, the employee will retain his original employment date, showing no break in service. Credit for paid vacation leave cannot be accrued during an approved leave of absence.

If the employee, during the leave of absence, accepts a job with another company, all company benefits will be terminated.

Leaves of absence are without pay.

SECTION 5: WORK POLICIES AND REGULATIONS

Care of Equipment and Facilities

All employees should be concerned for the care of company equipment and facilities as well as the quality of the company products. This also means good housekeeping is equally important.

Personal Appearance/Clothing

Personal appearance, hygiene and clothing are important to our work practices. Our customers gauge the quality of our company by the care we show in our personal attire and appearance.

Each person is to report to work, in the appropriate uniform, wearing clothing that is clean. A neat, groomed appearance is important to yourself, your fellow workers and to our customers. Keep yourself clean and close shaven. We do not object to reasonably long hair if it is groomed and/or mustaches and beards if they are kept trimmed.

Smoking

Due to safety and health reasons, absolutely no smoking is permitted anywhere in the office areas or factory spaces. Smoking is generally not permitted at customer locations, except in areas designated by the customer.

Uniforms

Uniforms are furnished to all employees by the company. The cost of cleaning is borne by the employee. Cleaning is accomplished through the uniform service provider on a set weekly schedule.

All employees, except office personnel, are required to wear these uniforms at all times during working hours.

Safety Shoes

All employees, except office personnel, are required to wear safety shoes during working hours. Shoes are available through a company sponsored program but are paid for by employees.

Safety Equipment

Employees will be provided safety glasses, respirators, gloves and various other safety equipment items. These must be worn on the job as safety requires. This equipment will be signed for by the employee and replaced at his expense if lost or stolen. Replacements will be provided if the equipment is shown to be worn or defective.

Company Tools

All tools furnished to employees will be signed for by the employee and are to be replaced at the employee's expense if lost or stolen. Replacements and repairs due to fair wear and tear will be at company expense.

Energy Preservation and Waste Prevention

Please conserve energy at every opportunity by keeping thermostats in moderate ranges and changing filters. Drive within speed limits. Waste of energy and materials is costly to the company and ultimately results in losses which must be paid for by other cost reduction actions.

Solicitation and Distribution

Solicitation or distribution of materials, goods, or contests, requests for donations or any other solicitation and distribution which interferes with the work of our employees is not permitted.

Security

All equipment, boxes, doors, files, desks, gates and other equipment with locks are to be kept locked securely when not in direct use and at the end of each day. They are also to be checked regularly. Company vehicles should be kept locked at all times. Lost keys will be reported to the office immediately. Any concerns about security should be directed to your supervisor.

Bulletin Board

The company maintains a bulletin board to keep employees informed of current items of general interest. Check the bulletin board regularly. (Posting and removal of notices must have the approval of the Personnel Manager.)

Moonlighting

We make every effort to keep everyone as fully employed as possible and at a good rate of pay. When you are on the job, this means that it requires 100% of your effort. If you choose to work outside of your job, and the outside work competes with what is expected of you as a company employee, opportunities for promotion and growing along with the ABC Company will be reduced by your decision.

If management feels that outside employment prohibits you from fulfilling your obligations to the company, or creates a conflict of interest, you will be asked to resign or to leave your outside employment. All management and supervisory personnel are expected to enforce this policy and, by example, refrain from conflicting outside employment.

Telephone

Company telephones are important to our everyday operation. Employees should restrict their personal calls to emergency matters only.

No Harassment Policy

It is the policy of the ABC Company to provide, have, and maintain a working environment free of harassment of any kind. Furthermore, employees have a right to work in an environment free of harassment whether racial, sexual, or otherwise. Harassment may be verbal or physical.

The ABC Company does not and will not tolerate harassment of our employees. The term "harassment" includes, but is not limited to, slurs, jokes, and other verbal, graphic or physical conduct relating to an individual's race, color, sex, religion, requests for sexual favors, offensive touching, and other verbal, graphic, or physical conduct of a sexual nature.

Violation of this Policy Will Subject an Employee to Disciplinary Action, Up to and Including Immediate Discharge.

If you feel that you are being harassed in any way by a co-worker or by an employee of a customer or vendor, you should notify you supervisor, manager, or the Personnel Manager immediately. The matter will be thoroughly investigated and, where appropriate, disciplinary action will be taken.

All supervisory personnel are and will be reminded of their responsibilities in this area. Supervisors are instructed to take swift, appropriate, remedial action in response to any report or indication of abuse, threats, intimidation or harassment (sexual, physical or otherwise) directed toward any employee.

You should also be aware that no supervisor or other member of management is authorized to make any employment decision based in any way on an employee's submission to or rejection of sexual conduct or advances. No supervisor or other member of management has the authority to suggest to any employee that the employee's continued employment or future advancement will be affected in any way because the employee enters into or refuses to enter into any form of sexual or other personal relationship with the supervisor or member of management. No supervisor or manager may take disciplinary action against an employee because he or she has rejected sexual advances.

If you believe that a supervisor or member of management has acted inconsistently with this policy, if you are not comfortable bringing a complaint regarding harassment to your immediate supervisor or if you believe that your complaint concerning a co-worker or an employee of a customer or vendor has not been handled to your satisfaction, please immediately contact the President.

You Will Not Be Penalized in Any Way for Reporting Improper Conduct.

All reports will be handled in a prompt, appropriate and confidential manner. Discrimination and harassment will result in appropriate disciplinary action which could include termination.

SECTION 6: CONFLICTS OF INTEREST

A conflict of interest can arise in dealings with anyone whom the ABC Company transacts business with: customers, clients, owners, buyers, suppliers, banks, insurance companies, and people in other organizations with whom we contact and make agreements. Conflicts of interest will be avoided and include the following examples:

— working for any of the above groups for personal gain

— engaging in a part-time activity for profit or gain in any field in which the company is engaged; and/or

— borrowing from, or lending money to individuals representing organizations with whom business dealings are conducted.

Personal Conduct

The company expects that all of its employees will conduct themselves with the pride and respect associated with their positions, their fellow employees, clients and the company. Care should always be taken to use good judgment and discretion in carrying out the company's business. The highest standards of ethical conduct should always be used.

Confidentiality

No employee will store information outside of the company (either in written or electronic form) about any matter pertaining to the conduct of the company's business which may compromise the ABC Company to outsiders.

Any employee who compromises information may be subject to termination.

Bribes, Kick-Backs and Other Illegal Payments

Bribes, kick-backs and other illegal payments to or from any individual with whom we do business (in any form and for any purpose) are prohibited.

Certain types of rebates to the company from suppliers (but not to or from an individual employee) are perfectly legitimate to correct commercial inequity, if done within government trade regulations.

Gifts

ABC Company employees shall not accept gifts of value of greater than $10.00 of any type from any company or individual with whom we do business.

Certain business selling and purchasing conduct requires entertaining clients and being entertained by vendors as a part of the negotiation process. Entertainment of a reasonable value, both given and received, is a legitimate business activity and is permissible. If in doubt whether specific incidences of this type of entertainment is permissible, discuss it with the President before hand.

SECTION 7: SAFETY

General

ABC Company is committed to the safety of its employees and the general public. To this end we will utilize our safety program in our daily activities.

It is necessary that the company establish safety rules and regulations to be observed by all employees at all times. Regarding these rules, the following is to be considered standard procedure for all employees:

1. Should a safety regulation be modified so that the employee's safety is something less than it should be, the employee is to inform his supervisor.
2. All questions concerning the "why" for doing something in a certain manner are to be asked of any member of management at any time.

Employees' decisions should always be guided by the company's commitment to safety. Should a hazardous situation or condition exist, and a decision has to be made on safety or performing work, safety should always be the first concern.

It is the responsibility of all supervisors to see that every employee is provided with safe working conditions and that they observe all safety regulations and use good common sense to protect themselves as well as others. The President, or a designated representative, will periodically inspect work conditions and may suspend all work activity until an unsafe condition is corrected.

The most important part of safety is YOU. It is up to you to abide by the safety rules. They are made for your protection. You are expected to report any personal injury IMMEDIATELY, however minor, and all dangerous conditions, substances and practices to your supervisor.

First Aid Materials Are Kept in the Personnel Office, at Supervisor Stations, and Company Vehicles.
The Personnel Manager is Responsible for Maintaining Sufficient Supplies

Safety Rules

The following list indicates examples of safety awareness activities; it NOT all inclusive.

— Do not operate any equipment that, in your opinion is not in a safe condition.
— Never remove guards from equipment.
— Do not operate any equipment until you are fully instructed and familiar with it.
— Safety shoes are required to be worn by all employees, except office personnel.
— Safety goggles, respirators, gloves, hard hats, etc. are to be worn in all work areas where their use is designated.
— Avoid distractions—concentrate on your work.
— When lifting, use approved techniques. Get help for heavy loads.
— Keep all electric cords and wires off floors.
— Do not touch electrical switches, fuse boxes, or other electrical panels that you are not specifically authorized to use.
— Be familiar with the location and the use of all fire extinguishers.
— Report immediately any unsafe condition which you observe.
— Use the correct tools for the job.
— Keep all trash off floors. Use trash receptacles provided.
— If you don't know—ASK!

Hazardous Materials and Wastes

The Environmental Protection Agency has classified certain chemicals and chemical groups into categories which have been classified as toxic. This means that in concentrated forms or by accumulating and combining with other chemicals, or even the air, these chemicals can be hazardous to human health if you are exposed.

From time to time in the normal course of their jobs, employees may handle materials which have been classified as hazardous by the standards of the Occupational Safety and Health Act (OSHA) regulations.

Hazardous materials that are received from our suppliers should have material safety data (MSD) sheets or labels which state the chemical ingredients of the contents, precautions to take and what to do if you are exposed to a toxic level of exposure. Be sure and note these MSD sheets or labels and exercise immediate caution if you either do not understand them or have any questions. Discuss your questions with your supervisor immediately.

As a company, we are committed to not creating or disposing of hazardous wastes which will contaminate our environment. We will not knowingly dump any wastes into the environment at any time.

Reporting Injuries and Accidents

Employees are to advise their supervisor of all accidents, injuries or illnesses that occur while at work. They are to be reported immediately no matter how slight they may appear. The company will provide the proper forms for reporting accidents, injuries and illnesses, and failure to report these occurrences will be cause for disciplinary action.

In case of a vehicular accident, all information should be reported immediately to your supervisor and the office. In no instance should responsibility for the accident be expressed to anyone until the proper persons in the company have been notified and permission given to make statements.

SECTION 8: TRANSPORTATION AND TRAVEL EXPENSES

Company Owned Vehicles

The following are specific policies related to company owned vehicles,

— Company owned vehicles will be driven only as needed for jobs during working hours.
— Seat belts will be worn at all times.
— Only the driver assigned to the vehicle will sign for gasoline, oil, etc. All charge tickets must show the name and address of the vendor, prices, gallons, vehicle ID number, license tag number and mileage.
— No alcoholic beverages or illegal drugs or chemicals will be aboard a company vehicle at any time.
— No driver who has been drinking alcoholic beverages or is under the influence of drugs or chemicals will be allowed to drive a company vehicle.
— Misuse of company vehicles will be cause for termination.
— Vehicles are to be driven only by authorized persons with current, valid driver licenses on file with the company.
— Vehicles are not to be driven home by any employee except salesman or others who have vehicles specifically assigned to them and who are covered accordingly by company insurance.
— At no time are children or other nonemployees allowed as passengers in company vehicles.

Personal Vehicles

Personal vehicles may be used on official company business when prior approval has been given by a supervisor. A mileage rate based on acceptable industry standard will be paid employees who use their vehicles on official company business. Minimum insurance requirements, which are given to us by our insurance carrier, must be in effect at the time the vehicle is used. Employees' personal insurance will be "primary" in the event of an accident while driving on company business.

Travel Expenses

The company will reimburse employees for expenses incurred when they are traveling on company business.

SECTION 9: EMPLOYEE SUGGESTIONS

If you have a suggestion, please share it, in writing with us. Your written suggestions will be reviewed by the President and may result in a suggestion award.

Acknowledgment of Receipt and Understanding

Read and Sign Immediately

I understand that the statements contained in this booklet are intended to serve the general information purposes concerning the ABC Company and its existing policies, procedures, practices of employment and employee benefits. Nothing contained in this booklet is intended to create, or shall be construed as creating, an express or implied contract or guarantee of employment for a definite or indefinite term.

I understand that from time to time the ABC Company may need to add to, change and otherwise amend the information contained in this booklet and that the company will inform me when these changes are made.

I have received, read and understand the information outlined in this booklet, have asked any questions I may have concerning its contents and will comply with all policies and procedures to the best of my ability to do so.

SIGNED_____

DATED_____

LOCATION_____

WITNESSED_____

(Personnel Manager—The ABC Company)

SALES INCENTIVE PROGRAM

PURPOSE

The sale Incentive program is established to gain additional sales above those generated from traditional company methods for increasing sales. It is a supplemental program to complement the traditional Marketing and Sales efforts.

The program is established to allow employees and selected non-employees to participate in the successful growth of the company. By rewarding them for finding additional clients, the company is expressing its appreciation to personnel for efforts beyond the scope of their normal job assignments and relationships.

CATEGORIES

Class A

Salaried Employees

An employee who is compensated at an annual rate of salary. Normally these are persons hired to perform in managerial, supervisory, and professional positions. For labor classification, the incumbents are referred to as Exempt Personnel.

Class B

Hourly Employees

An employee who is paid for hours worked, or at a set rate per job. Normally these are persons hired to perform the work of manufacturing and/or delivering the company's products and services to the customer. Also included in this class are office personnel and inspection personnel who are compensated for hours worked. For labor classification, the incumbents are referred to as Direct Labor, Indirect Labor, and Nonexempt Personnel.

Class C

Selected Non-Employees

Business associates, clients, vendors, etc. who provide leads that result in additional sales fall into this category.

PROCEDURE

The company will compensate individuals who bring leads into the firm that result in sales. All employees and selected non-employees are eligible to participate in some portion of the program.

The Program

A. Individual Compensation for Additional Sales.
   1. Eligibility:
      a. All internal personnel with the exception of Sales personnel.
      b. All external personnel with the exception of persons requesting services for their own personal or business use.
   2. Leads that are brought to the attention of the Sales Dept. by non-sales personnel are eligible for additional compensation.
      a. The lead must include the name of the individual who is interested in purchasing company products or services.
         1). The individual must have expressed an interest in the company providing the product or service and be willing to have a company representative prepare a quote, and discuss the quote.
      b. The information must be presented in written form to the Sales Dept.
      c. The lead must result in a sale within 90 days to be credited as a sale for this program.
   3. Compensation:
      Class A personnel: 10% of the value of the sale. If it is a contract for repeat service, the 10% is for the first billing only.

      Class B personnel: $25.00 for any lead that complies with the criteria of A.2.a. & b., above. $100.00 for a lead that becomes a sale.

      Class C personnel: 10% of the value of the sale. If it is a contract for repeat service, the 10% is for the first billing only.

B. Group Bonus
   1. Eligibility
      a. All internal personnel.
   2. For each set of 50 sales resulting from the program, all company personnel will receive a one time $50.00 bonus.
      a. Payment will be paid the pay period after the 50th sale is consummated.
      b. There is no limit on the number of times company personnel can earn $50.00 bonuses.

         Example: May 1st, company reaches 50 sales from the program. All personnel receive a $50.00 bonus, next pay period. July 22nd, company reaches 100 sales from the program. All personnel receive a $50.00 bonus, next pay period.

C. Award of Excellence Bonus:
   1. Eligibility:
      a. All internal personnel.
   2. In addition to the bonuses outlined above, employees may receive the following prizes.
      a. Employees will receive a voucher for a dinner for two for every 10 sales they generate.
      b. For non-sales personnel.
         1) Class A personnel with the highest number of sales credited for the year will receive a $250.00 bonus.
         2) Class B personnel with the highest number of sales credited for the year will receive a $250.00 bonus.
      c. Additional prizes will be announce from time to time to provide further incentives for the program.

INVESTIGATION POINTS
(PRODUCT COMPANY)

[Please answer all questions. Be specific. Use back of page or extra blank sheets if necessary.]

Company: _____; Date: _____

Completed by: _____

Seven Steps of the Manufacturing System

1. Obtain product specification.

   1a. What are the company's products?

   _____

   _____

   _____

   _____

   1b. Does the company have a standardized design process to design its products and does it have standard designed components that are compatible with its manufacturing capabilities?

   _____

   _____

   _____

   _____

   1c. Is there a business plan setting sales goals and how they're going to achieve them? If so describe main points.

   _____

   _____

   _____

   _____

   1d. Does the company match its abilities with potential customers real and perceived needs?

   _____

   _____

   _____

   _____

1e. How does the company sell its products?

_____

_____

_____

_____

1f. How are orders obtained? What is the customer base?

_____

_____

_____

_____

1g. What is in the quote package or worksheet used to quote a customer? Is there a standard part that applies to any customer? How are specials handled?

_____

_____

_____

_____

1h. Is there a standard form that lists questions to be asked and answered for each order?

_____

_____

_____

_____

1i. How are orders received and booked?

_____

_____

_____

_____

1j. Does the customer confirm the particulars of the order? In writing? Verbal?

_____

_____

_____

_____

2. Design a method for producing the product, including design and purchase of equipment and or processes to produce, if required.
   2a. Does the design of the product take into account the ability of the factory to make the product with ordinary due diligence for a satisfactory yield and cost?

_____

_____

_____

_____

2b. Does the company design actively consider designs for jigs and fixtures as part of the design of the product, either generic or for a specific customer?

_____

_____

_____

_____

2c. How is the content of each job planned for the scope of work the employees need to do at each workstation?

_____

_____

_____

_____

2d. How does the company scope the time required to do each operation required to accomplish the manufacture of the product? Do standard times exist for subsets of common tasks? If so, based on estimates or developed time standards?

_____

_____

_____

_____

2e. How is the scope of work translated into planned hours to accomplish, tools needed to do the job, and materials types and quantities? Are yield factors considered?

_____

_____

_____

_____

2f. Is there a methods sheet provided for each job? If so does, the methods sheet provide sufficient detail for workers to operate independently and know precisely what constitutes a complete job?

_____

_____

_____

2g. Is there a Bill of Materials (BOM) for each job? If so, is it indented to show the proper assembly order and relationships between parts, sub-assemblies and assemblies?

_____

_____

_____

3. Schedule to produce.
   3a. Does a schedule exist showing all open jobs and the required complete dates?

_____

_____

_____

_____

3b. Does the schedule show the total portal to portal cycle time for each job at each work-station?

_____

_____

_____

_____

3c. Is the cycle time for the schedule derived from step 2, above? What is the link or meth-odology for accomplishing this task?

_____

_____

_____

_____

3d. Is there a capacity document showing the amount of work capable of being completed in a certain time period? Is it used to create the schedule?

_____

_____

_____

3e. Does the schedule process have an update routine that feeds back actual completions for comparison to plan? If so, is it done daily and is there a policy for revising schedules if the plan becomes too far out of synch. with actual results?

_____

_____

_____

3f. Is the schedule of sufficient longevity that the supervisors can plan their work load prior to the start of the next work period (day or shift)?

_____

_____

_____

3g. Is the schedule broken down by work stations or centers and are workers told the amount of time they have to do the assigned job?

_____

_____

_____

_____

3h. Are overdue jobs given top priority to complete before new jobs are started?

_____

_____

_____

_____

3i. Are there any rules established that give permission to bypass a job in the sequence to start a succeeding job?

_____

_____

_____

_____

4. Purchase raw materials in accordance with the schedule.
   4a. Does only one person have responsibility for purchasing regardless of how many others have delegated responsibility?

_____

_____

_____

_____

   4b. Is a maximum/minimum inventory control system or equivalent used to ensure there is an adequate supply of materials on hand?

_____

_____

_____

_____

   4c Are special buy purchases segregated and allocated to their designated jobs?

_____

_____

_____

_____

   4d. Is there reservation system based on the schedule for materials to be allocated for each job?

_____

_____

_____

_____

   4e. Are materials uses needs forecasted based on schedule, previous history, or forecast?

_____

_____

_____

_____

   4f. Is an effort made to find the lowest cost supplier and the most dependable suppliers?

_____

_____

_____

_____

4g. Is an effort made to buy in bulk to get volume discount prices and is this coordinated with the schedule and forecast?

_____

_____

_____

_____

4h. Are lead times for delivery of materials taken into account for ordering materials and is it coordinated with the schedule and forecast?

_____

_____

_____

_____

4i. Are contracts consummated with suppliers to effect a JIT relationship based on long-term commitments to buy and supplier commitments to have on hand on an as needed basis?

_____

_____

_____

_____

4j. Have supplies been differentiated based on shelf life parameters to assure proper limits of bulk buy parameters?

_____

_____

_____

_____

4k. Are report prepared periodically (at least on a quarterly basis, preferably a combination of weekly and monthly) showing as a minimum inventory levels, materials orders outstanding, jobs waiting because of late materials, vendor performance measurements, quality issues, cost values for all categories, and trends measurements for all categories?

_____

_____

_____

_____

5. Produce in the factory.
   5a. Are workers trained or given an orientation before being place on the job, and does this consist of job specific instructions and specific safety and environmental hazards they may encounter?

_____

_____

_____

_____

5b. Is work done in accordance with a daily dispatch list?

_____

_____

_____

_____

5c. Do workers know what job to work on before hand and know how many hours are scheduled for the job?

_____

_____

_____

_____

5d. Do worker receive instructions for each assigned job showing all requirements and methods to follow to do the job?

_____

_____

_____

5e. Are workers required to report back completion of assigned jobs before receiving the authorization to proceed to the next job?

_____

_____

_____

_____

5f. Do supervisors check progress and help workers eliminate obstacles/problems that may hinder job completion on time?

_____

_____

_____

_____

5g. Is there a procedure to list obstacles/problems that can not be resolved quickly for a planned resolution activity?

_____

_____

_____

5h. Are obstacles/problems from 5f. above, purposely resolved and the results reported back to the worker?

_____

_____

_____

5i. Are workers given the opportunity to suggest ways to do the job better and are they rewarded for their initiative?

_____

_____

_____

5j. Are records maintained and reports issued (at a minimum on a monthly basis) for performance based on scheduled versus actual cycle times?

_____

_____

_____

6. Monitor results for technical compliance and cost control.
   6a. Is job performance monitored for compliance with job specifications?

_____

_____

_____

6b. Are reports issued showing noncompliance with root causes identified and corrective actions requirements determined and implemented?

_____

_____

_____

6c. Are compliance trends reports issued at least monthly?

_____

_____

_____

6d. Do workers certify that they did the job in accordance with methods instructions and the results comply with job specifications?

_____

_____

_____

6e. Is TQM theory used to encourage continuous improvement? If so are workers trained in the basics of TQM?

_____

_____

_____

6f. Are vendors monitored for compliance with materials and subcontractor quality requirements?

_____

_____

_____

_____

6g. Is work force absenteeism recorded with reasons and associated corrective actions tracked daily or at least weekly?

_____

_____

_____

_____

6h. Are work hours and costs for all active jobs booked daily and are they coordinated with the work schedules of step 3, above?

_____

_____

_____

_____

6i. Are reports generated daily showing hours worked versus schedule and efficiency, cost for labor expended versus planned for jobs completed and in progress, indirect labor costs expended vs. budget, material costs expended versus plan, and the P/L preliminary estimate for each completed job?

_____

_____

_____

_____

6j. Is a Pro-forma P &L issued monthly with explanations of variances?

_____

_____

_____

_____

6k. Are capital expenditures formally approved and expected results tracked? Is there an established hurdle rate policy set for capital expenditure requests?

_____

_____

_____

_____

6l. Are costs for administration, sales, R&D, finance, nondurable tools and supplies, utilities, maintenance, vehicle operations, and general repairs tracked at least quarterly?

_____

_____

_____

7. Ship the completed product to the customer.

7a. Does the worker have a check list that he or she uses to assure and verify that all aspect of the job are completed before turning over to Shipping (or the customer)? Does the worker initial/sign it?

_____

_____

_____

7b. Does the supervisor or worker inform Shipping when the job is done and available for ship to the customer?

_____

_____

_____

7c. Are products kited against a bill of lading to ensure all of the customer's order is consolidated for shipments prior to dispatching to the customer?

_____

_____

_____

7d. Is it a policy to ask customers to verify that the work has been done to their satisfaction? Is the customer asked to sign indicating verification or items not to their satisfaction?

_____

_____

_____

7e. Is it policy to satisfy all discrepancies with the customer "on the spot" during turn over to the customer? If so, does this take priority over all other work?

_____

_____

_____

_____

GLOSSARY

Accountability, theory of—the basis for operator certification for quality assurance or quality control. People will make sure a job is done correctly if they must identify themselves as the performer of the work.

Action plans—with respect to work planning, very specific, detailed items to be accomplished by a specific individual by a certain date.

AME—advanced manufacturing engineering, a unit within the manufacturing engineering function.

Appraisal cycle—an annual set of chronological steps with interim reviews leading to the annual appraisal for performance of an employee.

Appraisal costs—one of four quality costs; those costs associated with doing the actual evaluating processes to detect whether specification are being met or failures are occurring.

Appraisal form—a standardized form for evaluating employee performance as compared to their job description.

Archival data collection document—a history document for equipment.

Area planning—an activity within advanced manufacturing engineering.

Artificial intelligence—a set of computer software that can add data to its data base, and give the appearance of learning.

Audits—a process of evaluating an organization's ability to carry out its assigned tasks by measuring effectiveness against standards, then recommending corrective actions.

Batch—a grouping of parts or orders together for group manufacturing.

Bid—submittal of costs and performance characteristics by vendors offering to sell equipment.

Bill of materials (BOM)—a listing of materials in sequence order of manufacture of a product.

Bottleneck—a choke point in the manufacturing process as a result of line imbalances.

Budget classification—in reference to capital equipment programs, the categories established for identifying project types.

business plan cycle - an annual event that looks at internal and external factors that effects how well the company did in its' previous year, which becomes the basis for revisions in the company's objectives and the basis for the next set of measurable goals.

CAD/CAM—computer-aided design/computer-aided manufacturing, a generic term meaning extensive coordinated use of computers in industry.

CAD/CAM triad—the three basic aspects of CIM: machine/process control, design and planning control, and production and measurements control.

CAM—computer-aided manufacturing, use of computer technology to assist in the management of manufacturing activities.

Capability—the physical limitations of what can be produced in a factory, including size and weight of parts, processes available, and materials that can be worked on.

Capability matrix—a format for tabulating the ranges and permutations of a factory's ability to produce an array of products.

Capacity—the number of items that the factory has the capability to produce in a given period of time.

Capacity requirements planning—the long-range scheduling program of MRP II.

Capital equipment—facilities, machines, etc., that can be depreciated for income tax purposes.

CAPP—computer-aided process planning, the use of computers for creating, storing, cataloging, and retrieving plans.

Cartesian coordinate system—the three-coordinate *(x, y, z)* system for locating a point in space, used to define positions in space for CNC machines.

Cellular layout—a factory layout based on group technology principles, in which a variety of types of equipment are located adjacent to each other to manufacture a family of parts.

Change management—the techniques for managing deviations from plans in project management to minimize adverse affects.

CIM—computer-integrated manufacturing, a generic term meaning achieving communications excellence via computer synergism throughout the business entity.

Classification and coding—a technique for cataloguing parts in accordance with geometric, manufacturing, and material specifications.

CNC—computer numerical control, an advanced type of numerical control.

Common database—electronically stored product data useful to and accessible by multiple functions.

Communications excellence—the ultimate purpose of CIM, the key to enhanced productivity.

Concept design—the rationalization of a design in terms of science.

Concurrent engineering—the process of team approach to do design and manufacturing engineering activities in parallel instead of series format.

Constructive critique—a methodology of doing employee appraisals in an objective manner and aimed at helping employee improve future performance.

Control chart—a format for recording results of statistical sampling measurements.

Cookbook quality plan method—a method of writing quality plans in which steps are outlined in chronological sequence.

CPM—critical path method, a technique for scheduling and monitoring the accomplishment of a complex project.

Daily accounts controls—a technique for controlling specific expense categories within established budgets.

Daily operating status report—a machine tool and equipment status document pertaining to availability for use.

Data collection—the pulling together of information in a systematic way, usually referring to a computer interface.

Datums—with reference to geometric dimensioning and tolerancing; are points, lines, planes, cylinders, axes, etc. by which the relationship of a feature to other features can be mathematically (geometry) defined.

Day work system—a work measurement system based on pay for hours present at the workplace.

Defect analysis report—a technique for analyzing quality control reports for trends.

Design for manufacturability—see Concurrent engineering.

Design of experiments—a special use of statistics, commonly involving ANOVA methods to fix out of control processes in very special circumstances.

Design review—a review of designs by producibility engineering to check for compliance to producible design criteria.

Detectability—with reference to risk, determine how important the detectability of a risk is to the success of a project.

Direct labor—effort applied at workstations to transform raw materials into finished products.

Down equipment—facilities that are in a state of disrepair that precludes their use for production.

Dynamic scheduling—master scheduling capable of being altered almost at will via CIM techniques.

Empirical decision making—the simplest and most restrictive technique for finding families of parts.

Employee handbook—a comprehensive document setting forth and explaining the policies and procedures for working for the company.

Environmental control—a body of knowledge dedicated to identifying and controlling hazardous and toxic substances so that the probability of harm is minimized.

Error detection program—a sub-program of computer data collection systems used to verify the correctness of entries.

Estimated time standard—a time standard based on estimated cycle times.

Evaluation of performance—a job requirements based judgement of how a person is doing her/his job with respect to those requirements.

Expert system—a type of Artificial Intelligence computer program.

External failure costs—one of four quality costs; those costs associated with defective products being delivered to customers in error

Facilities engineering—a sub-unit of maintenance engineering dealing with machine tools and process equipment.

Facilities list—a list of equipment used for the facilities program.

Facilities program—a capital equipment purchase and upgrade plan linked to the objectives and goals.

Family of parts—parts that are similar in geometry or in the manufacturing procedures used to make them; determined by group technology classification coding, parts flow analysis, or empirical decision methods.

Family-of-parts programming—a generic numerical control program for categories of parts where there are limited geometric variables.

Farm out—having production work done in factories other than the company's factory.

Feature control frame—with reference to geometric dimensioning and tolerancing; a symbolic way in a rectangular box of showing a type of control—the geometric control symbol; the tolerance—form, run out, or location; and datum references and modifiers to the datum references.

Fire codes—safety codes pertaining to flammable materials and fire prevention.

Fire fighting—a technique for solving short-term quality problems.

Firm savings—calculable savings associated with plant appropriation requests.

Flexible automated factory—a factory in which CNC machines and robots directed by central and process computers are used to produce manufactured parts.

Flexible machining center—a complex CNC machine capable of performing many different types of machining operations on varied work pieces and positions.

Fuzzy logic—a form of Artificial Intelligence.

Gantt chart—a bar chart used to monitor the progress of a project.

Geometric Dimensioning and Tolerancing—a standardized method of portraying component or assembly dimensions and their related tolerances based on agreed to datum point conventions.

Goal—measurable statement of specific intent bounded by a specific period of time.

Group technology—a philosophy of manufacturing that exploits the principle of sameness, in which parts are grouped according to similarities in geometry and/or manufacturing process in order to approach flow-type manufacturing in job shop environments.

Group technology classification and coding—see Classification and coding.

Hazardous substances—chemicals that can cause biological damage after long-term exposure.

HBM—horizontal boring mill, a classification of machine tools.

HR—human resources, a department responsible for all policies and procedures dealing relationships between employees and the company.

Incentive system—a compensation system based on pay for accomplishments beyond the established rate of performance set for the measurable goals.

Ideal capacity—the number of products a factory is able to produce based on theoretical feeds and speeds that can be achieved with the equipment.

Incentive system—a work measurement system based on pay per number of parts or operations completed per time period.

Indirect labor—work done to support the direct labor activities but does not add value to the material being worked on.

Inference engine—a part of an expert systems database.

Interactive graphics—a computer-controlled device used to create engineering drawings on a video display tube and then perform engineering calculations and evaluations. Sometimes known as an "engineering workstation."

Interference time—non-productive time during which an operator is available to work but cannot work due to outside circumstances.

Internal failure costs—one of four quality costs; those costs necessary to fix bad products or doing them over again from the start operation.

ISO 9000—the International Standards Organization quality assurance standard.

Job description—a listing and explanation of items to be accomplished with respect to a specific job.

Job deletion routine—a subprogram of the computer data collection system that reports completion of work to the master schedule.

Job effectiveness report—the basic efficiency measurement system for maintenance work performance.

Job hazard analysis—an evaluation of methods for possible safety hazards.

Job log—a format for entering and controlling maintenance work orders.

Job rate evaluation—method of comparing the work of one job skill with another in order to set pay scales.

Job shop—batch, intermittent manufacturing.

Job shop layout—a pre-group technology layout in which like machines and equipment are located together.

Just in case—an inventory strategy stressing emergency buffer stock at strategic locations.

Just in time (JIT)—a popularized version of good engineering practice resulting in elimination of waste in the manufacturing process.

Key jobs—jobs defined as critical for company success and used to measure the worth of other jobs for pay purposes by comparison methods.

Knowledge base—the storage of data in an expert system database.

Layout—an engineering procedure for placing equipment on the factory floor to optimize product flow.

Layout check list—a series of action items to complete before doing a layout.

Lean manufacturing—see Just in time.

Least material condition—with reference to geometric dimensioning and tolerancing; that condition of a part feature that contains the least amount of material within the desired limits of size.

Level loading—a technique for balancing production throughput over time.

Limping equipment—facilities that cannot be used to their full potential because of partial malfunctions.

Loss—a negative value of selling price minus the cost to produce.

LRF—long-range forecast, a document spelling out the goals of a business over a period of time, usually 3 to 5 years.

Machine tool capability and replacement log—chronological capability evaluations of equipment used for shop scheduling and replacement.

Maintenance engineering—a unit within the manufacturing engineering function.

Maintenance engineering strategies—three different approaches in minimizing downtime of production equipment: reactive, predictive, and preventive.

Manufacturing facilities design—the specific design for jigs, fixtures, and processes necessary to implement the producibility design.

Manufacturing losses analysis—a systematic way to determine the causes of processing and manufacturing errors.

Manuscript—a sequential listing of steps to be followed by a CNC machine; part of the programming process.

Master schedule—overall sequenced schedule of multiple orders through a job shop.

Maximum material condition—with reference to geometric dimensioning and tolerancing; that condition of a part feature that contains the maximum amount of material within the desired limits of size.

MCU—machine control unit, the microprocessor or minicomputer that controls a CNC machine.

Mean—a measurement of central tendency, an average of many data points.

Methods—specific procedures for performing work on a workstation or series of workstations.

Methods analysis—a technique for developing a method.

Methods sheet—format for specifying a method.

Monthly quality report—a typical format for documenting product quality status.

MP&WM—methods, planning, and work measurements; a unit within the manufacturing engineering function.

MRP—materials requirements planning, the coordinative computer program for dynamic scheduling of purchasing and inventory control activities.

MRP II—manufacturing resources planning, the addition of labor and facilities scheduling to MRP to form the coordinative CIM computer program for dynamic production scheduling.

N/C—numerical control, a generic term for computer control of machine tools.[*]

N/C parts programming—the concept of writing a manuscript for input into a machine control unit to instruct an N/C machine in how to produce a part.[*]

N/C tool path—the connected points in space an N/C machine will process through as instructed by the machine control unit.[*]

NDT—nondestructive test; an evaluation procedure using x-ray, liquid dye penetrant, ultrasonics, and so forth to examine structures to ensure that flaws, cracks, and other abnormalities are not present above design allowances.

Networking chart—a technique for showing relations between steps of a complex project and planning strategies for successful completion. The critical path method format is an example of a networking chart.

NOI—notification of intent; a preliminary indication of a need to purchase capital equipment.

* Note: The term N/C has generally been superceded by CNC to indicate an advanced control logic in the MCU.

Noise control—programs and techniques for keeping factory noise levels below allowed levels.

Normal distribution—a statistical distribution described by the mean and its three sigma standard deviations. A useful tool for quality control analysis to determine whether a manufacturing process is in control.

Objectives—broad-based generalized statements of intent.

Objectives and goals concept—a management technique for controlling a results-oriented organization. Based on the concept of focusing from broad-based statements of policy to measurable statements of intent bounded by time.

Operating plan—an integrated productivity improvement program and business strategy including an objectives and goals statement.

Operations sequencing—the short-range scheduling program of MRP II.

Operator certification—a technique of self-inspection for quality control.

Order entry system—a method of managing maintenance work.

Orders release—the medium-range scheduling program of MRP II.

Overhead—operating costs exclusive of direct labor.

PAR—plant appropriation request; a document outlining the reasons and justifications for spending funds for capital equipment.

Participatory programs—programs to enable all employees to have inputs into factory operation and problem solving.

Parts flow analysis—a technique for finding families of parts for cellular manufacturing.

Payback period—a criterion for evaluating the worth of plant appropriation requests based on dividing the gross costs of the project by the annualized savings. This yields a time period in years.

Pay grades—a system of levels for pay purposes used in pay systems.

Pay system—a logical approach for defining how much an employee will be paid based on a job rate evaluation plan.

PC form—process control form, a document for recording in-process or finalized dimensions and other data; a quality control form.

Performance requirements—with reference to purchase of capital equipment, the detailed specifications stating the parameters of acceptability.

Periodic review of operations—a technique for evaluating the success or failure of operations as compared to the short-range forecast.

Physical capacity—ability to produce to design specification by the due date.

Piece work system—a work measurement system based on pay for number of discrete parts or assemblies made at the workplace.

Practical capacity—capacity values downgraded for unavoidable delays.

Predictive maintenance strategy—based on engineering calculations of mean times to failure, take equipment out of service and replace parts before they fail.

Preliminary layout—a first attempt to place equipment in the factory to optimize product flow.

Present worth method—a financial analysis procedure using the time value of money concept to evaluate the worth of plant appropriation requests.

Preventive maintenance—work done on equipment to minimize future failures and downtime.

Preventive maintenance strategy—based on set hours of operation (not engineering calculations of mean times to failure), take equipment out of service and replace parts before they fail.

Prevention costs—one of four quality costs; those costs associated with identifying and minimizing activities that are detriments to good quality.

Principle of sameness—the basic theoretical concept of group technology. Parts are grouped on the basis of similar geometry or processing and are produced on the same equipment in a batch mode.

Principles of motion economy—an industrial engineering body of knowledge based on the physics of human body motions related to energy expenditure.

Process control—a unit within the manufacturing engineering function.

Producibility design—the process of customizing the design for the production source.

Producibility engineering—a unit within the manufacturing engineering function.

Product flow—the sequence of making a part in a factory, from raw material to finished product.

Production control system—a methodology of organizing to produce products in a factory and monitoring the results.

Process control system—a closed-loop feedback system in which the results are reduced manufacturing losses, minimized production delays, and improved product quality.

Productivity—in general, a measurement of output of products and goods divided by costs for design, labor, and material; a trend measurement of an organization's effectiveness.

Productivity, all other—the measurement of all cost contributors except direct labor.

Productivity chart—a chart used to plot productivity measurements.

Productivity, direct labor—the measurement of direct labor as a cost contributor.

Productivity rate measurement—a specific productivity measurement of quantity per time period.

Productivity seminar—a technique for developing an operating plan.

Profilometer—a device for measuring surface finish or smoothness.

Profit—a positive value of selling price minus cost to produce.

Project—a specific plan with measurable steps that leads to accomplishment of a goal.

Project management—The art of managing a nonrepeatable multistep assignment that has a beginning and an end point with desired results that are measurable and bounded by time. It is the formalization of tasks into achievable and measurable steps in order to reach a set goal.

Project life cycle—A sequence of step that develop in chronological order leading to the completion of a project. Either four or six steps. Four: concept, planning, execution, close out. Six: concept, definition, design, development or construction, application, and post completion.

Project schedule—a merging of the work breakdown structure, resources to accomplish, and times established to accomplish each task

Project scope—the magnitude and time frame for a project.

QCR—quality control report; a format for recording out-of-tolerance conditions and subsequent dispositions.

Quality assurance—a marketing-oriented document verification procedure used to ensure product compliance with specifications.

Quality circles—a type of participatory program whereby ideas for improving productivity are elicited from production workers in formal meetings and thoroughly investigated for implementation by management.

Quality control—a body of knowledge dedicated to ensuring that products manufactured are in accordance with design requirements.

Quality costs—non-value added costs associated with delivering acceptable quality products to customers; consisting of four components: prevention, appraisal, internal failure, and external failure.

Quality costs measurement—a measurement of negative productivity.

Rapid emergency repair—the expeditious repair of down and limping equipment to minimize production disruptions.

Reactive maintenance strategy—minimizing the effects of downtime by fixing the problem as soon as it's discovered.

Real-time measurement—the measurement and reporting of manufacturing activities as they occur.

Risk—a variance from average (expected) value which occurs in random chance patterns. The larger the variance the larger the risk.

Risk assessment—the evaluation of the probability of success of any project work breakdown structure task.

Risk management—the process of planning for and when necessary mitigating the affects of a risk that becomes a reality.

Robot—a mechanical device capable of being programmed to move or manipulate in three-dimensional space and capable of doing work in that space imitating, in a general way, motions that can be accomplished by a human being.

ROI—return on investment; a financial calculation procedure for evaluating the worth of a plant appropriation request.

Routing—a sequence in order of manufacture of workstations used to make a product.

Sequence of operations—the chronological listing of items to be accomplished in producing a product.

Setup—the procedure of placing and adjusting a workpiece on a machine tool or equipment for the purpose of doing work.

Schedule control chart—a format for monitoring the progress of action plans.

Scientific method—the classical method of observation, development and testing of hypotheses, repeated observation, and iteration to a conclusion.

Scientific time standard—a time standard based on the principle of motion economy and other precise data related to machine and human movement.

Severity—with reference to risk, determining how important the risk is to the success of a project.

Seven steps of the manufacturing system—a series of events that all companies that are producing goods and services traverse through for each and every product they produce (see Chapter 10 for an expanded definition and a listing of the steps).

Shop maintenance information system—a system for cataloguing all events pertinent to the functional performance of equipment.

Short-range forecast—a financial forecast of expenses and credits, usually for up to 3 months.

Six Sigma Approach—a branch of TQM using statistics and probability theory to reach a zero defect state in the manufacturing process.

Standard deviation—a statistical measurement of the dispersion of data from the mean, used in statistical tolerancing and quality control analysis.

Static scheduling—master scheduling of a complex nature that is very difficult to change.

Statistical Process Control (SPC)—the use of the mathematics of statistics and probability to evaluate in process quality during various manufacturing phases.

Subjective savings—intangible savings associated with plant appropriation requests.

Systems council—a temporary cross-functional organization for the implementation of CIM in a factory.

Team production—a type of participatory program whereby productivity is improved by allowing groups of workers to decide how they will divide and perform tasks to produce a finished product.

Technical capacity—ability to conceive and design products.

Theory of Accountability—people will perform at a conscious level consistent with high quality work only when they can be held accountable for that work.

Theory of Constraints—a methodology of identifying those limitation in manufacturing brought about by finite amounts of physical capability and capacity.

Time standard—a calculation of the time it should take to perform a specified task.

The Two Knows—the basis for running any manufacturing plant: "know how to make the product" and "know how long it should take to do it."

Throughput time—the portal-to-portal time to complete a job in the factory.

Tolerance—allowable deviations from a design dimension.

Tollgate inspection—a quality control status review conducted on semifinished or finished parts to ensure that all specifications for design and manufacturing have been complied with.

Tool room—a special-purpose manufacturing area used to make jigs and fixtures, parts for repairs, and prototype products.

Total productivity measurement—an overall measurement of a company's productivity.

Total quality management (TQM)—a process for measuring results of manufacturing activities with the purpose of making continuous improvements, thus lowering costs and delivering ever better products to customer.

TQM triangle—a graphic representation of feedback loop communications for continuous improvements.

Ultrasonic test—a form of nondestructive testing utilizing ultrasound energy.

Uncertainty—the degree of unknown caused by errors in forecasting one or more of the factors significant in determining the future values of the variables making up the variances.

Variance control programs—activities designed to minimize the difference between planned times to accomplish tasks and actual times.

VBM—vertical boring mill, a classification of machine tools.

Vendor—a supplier of raw materials and components.

Virtual condition—with reference to geometric dimensioning and tolerancing; the boundary generated by collective effects of maximum material conditions limits of size of any associated geometric tolerances.

Warner diagram—a statistical diagram used in conjunction with setting tolerances, typically by comparing load characteristics with material strength characteristics.

Weekly problem sheet—raw quality data disseminated for resolution.

Work breakdown structure—a decision process for determining what steps to take and their most efficient sequencing for accomplishing a project's goals

Work measurements—the broad concept of developing and measuring compensation systems and production effectiveness.

Work planning—a technique for establishing individual action plans, schedules, and measurements of accomplishments.

Work rules—a set of general dictates as to how an employee is to work within the company.

Workable design—a design that achieves the design engineer's goals while being economical to produce.

Working memory—the memoirs of an expert used to develop a data base for an expert system.

Workplace—the specific location where manufacturing is accomplished; also denotes a factory environment. A workplace can contain many workstations.

Workplace layout—a graphical description of a workplace.

Workstation—a place where contributed value is added to materials in the process of becoming a finished part (or product)

SELECTED RELATED READINGS

Anderson, Mark J. and Whitcomb, Patrick J., *DOE Simplified: Practical Tools for Effective Experimentation*, Productivity, Inc., Portland, OR, 2000

Bajaria, Hans J. (ed.), *Quality Assurance: Methods, Management and Motivation,* Society of Manufacturing Engineers, Dearborn, MI, 1981

Barish, Norman N., and Kaplan, Seymour, *Economic Analysis for Engineering and Managerial Decision Making,* McGraw-Hill, New York, 1978

Bolz, Roger W., *Production Processes, The Productivity Handbook,* Conquest, Winston-Salem, NC, 1977

Boothroyd, Geoffrey, *Fundamentals of Metal Machining and Machine Tools,* Scripta, Washington, DC, 1975

Breyfogle III, Forest W., *Implementing Six Sigma Smarter Solutions Using Statistical Methods*, John Wiley & Sons, New York, 1999

Brigham, Eugene F., *Fundamentals of Financial Management,* Dryden, Hinsdale, IL, 1978

Buffa, Elwood S., *Modern Production Management,* 4th ed., Wiley, New York, 1973

Childs, James J., *Principles of Numerical Control,* 3rd ed., Industrial Press, New York, 1982

Cokonis, T. J. (ed.), *Computers in Engineering 1982,* vol. 1, American Society of Mechanical Engineers, New York, 1982

Craig, Robert J., *The No-Nonsense Guide to Achieving ISO 9000 Registration,* ASME Press, New York, 1994

Del Vecchio, R.J., *Understanding Design of Experiments*, Hanser/Gardner Publications, Inc., Cinncinnati, OH, 1997

DeVries, W.R. (ed.), *Computer Application in Manufacturing Systems,* American Society of Mechanical Engineers, New York, 1980

Diehl, George M., *Machinery Acoustics,* Wiley, New York, 1973

Feigenbaum, A. V., *Total Quality Control,* 3rd ed., McGraw-Hill, New York, 1991

Goldratt, Eliyahu M. and Cox, Jeff, *The Goal, A Process of Ongoing Improvement, Second Revised Edition*, North River Press, Inc. Great Barrington, MA, 1992

Grant, Eugene L., and Grant, Ireson W., *Principles of Engineering Economy,* 5th ed., Ronald, New York, 1970.

Gross, Erwin E., Jr., and Peterson, Arnold P.G., *Handbook of Noise Measurement,* 7th ed., Gen Rad, Inc., Concord, MA, 1974

Griffith, Gary K., *The Quality Technician's Handbook,* 4th ed, Prentice-Hall, Upper Saddle River, NJ, 2000

Grove, Andrew S., *High Output Management,* Random House, New York, 1983

Hajek, Victor G., *Management of Engineering Projects,* McGraw-Hill, New York, 1977

Hall, Robert W., *Zero Inventory,* Dow Jones–Irwin, Homewood, IL, 1983

Ham, I., Hitomi, K., and Yoshida, T., *Group Technology Applications to Production Management,* Kluwer-Nijhoff, Boston, 1985

Arrington, H. James, Hoffherr, Glen D., and Reid, Robert P., Jr., *Statistical Analysis Simplified The Easy-To-Understand Guide to SPC and Data Analysis*, McGraw-Hill, New York, 1998

Harmon, Paul, and King, David, *Expert Systems,* Wiley, New York, 1985.

Hay, Edward J., *The Just–In–Time Breakthrough, Implementing the New Manufacturing Basics*, John Wiley & Sons, New York, 1988

Heldman, Kim, *PMP Project Management Professional Study Guide*, SYBEX, Inc., Alameda, CA, 2002

Hicks, Charles R. and Turner, Kenneth V., *Fundamental Concepts of Design of Experiments*, Oxford University Press, Inc., New York, 1999

Hicks, Philip E., *Introduction to Industrial Engineering and Management Science,* McGraw-Hill, New York, 1977

Hine, Charles R., *Machine Tools and Processes for Engineers* McGraw-Hill, New York, 1971

Kapoor, S. G., and Martinez, M. R. (eds.), *Statistics in Manufacturing,* American Society of Mechanical Engineers, New York, 1983

Kepner, Charles H., and Tregoe, Benjamin B., *The Rational Manager,* McGraw-Hill, New York, 1965

Kirkpatrick, Elwood G., *Quality Control for Managers and Engineers,* Wiley, New York, 1970

Koenig, Daniel T., *Computer Integrated Manufacturing, Theory and Practice,* Hemisphere, a division of the Taylor & Francis Group, New York, 1990

Koenig, Daniel T., *Fundamentals of Shop Operations Management: Work Station Dynamics*, Society of Manufacturing Engineers, Dearborn, MI, ASME Press, New York, 2000

Koenig, Daniel T., *The Engineer Entrepreneur*, ASME Press, New York, 2003

Koenig, Daniel T., *Interrelationships Between Methods Engineering and Productivity Engineering, ASME Paper 81-DE-3*, presented at the Design Engineering Conference and Show, April 27–30, 1981

Koenig, Daniel T., *The Pragmatic Application of CAD/CAM in Non Aerospace Job Shops, ASME Paper 82-DE-19*, presented at the Design Engineering Conference and Show, March 29–April 1, 1982; reprinted as *How CAD/CAM Can Improve Job Shop Productivity, in Design Engineering, Nov. 1982,* Maclean Hunter Ltd., Toronto, Ontario, Canada

Koenig, Daniel T., Gongaware, Terry, and Ham, Inyong, *Application of Group Technology Concept for Plant Layout and Management of a Miscellaneous Parts Shop,* in *Ninth North American Manufacturing Research Conference Proceedings, May 19–22, 1981,* Society of Manufacturing Engineers, Dearborn, MI, 1981

Kops, L. (ed.), *Toward the Factory of the Future,* American Society of Mechanical Engineers, New York, 1980

Koren, Yoram, *Computer Control of Manufacturing Systems,* McGraw-Hill, New York, 1983

Lazarus, Harold, and Tomeshi, Edward A., *People-Oriented Computer Systems, the Computer Crisis,* Van Nostrand Reinhold, New York, 1975

Leibried, K. H. J., and McNair, C. J., *Benchmarking, a Tool for Continuous Improvement,* Harper Business, New York, 1992

Len, M. C., and Martinez, Miguel R. (eds.), *Computer Integrated Manufacturing,* American Society of Mechanical Engineers, New York, 1983

Levine, David M., Ramsey, Patricia P., and Smidt, Robert K, *Applied Statistics for Engineers and Scientists: Using Microsoft Excel and MINITAB/,* Prentice-Hall, Upper Saddle River, NJ, 2001

Lewis, James P., *The Project Manager's Desk Reference,* 2nd ed., McGraw-Hill, New York, 2000

Lewis, James P. and Wong, Louis, *Accelerated Project Management, How to Be the First to Market,* McGraw-Hill, New York, 2005

Lubben, Richard T., *Just In Time, an Aggressive Manufacturing Strategy,* McGraw-Hill, New York, 1988

McMullen, Thomas B., Jr., *Introduction to the Theory of Constraints (TOC) Management System,* St. Lucie Press, Boca Raton, FL, 1998

Martin, Charles C., *Project Management: How to Make It Work,* Amacom, New York, 1976

Mayer, Raymond R., *Production and Operations Management,* 3rd ed., McGraw-Hill, New York, 1975

Moder, Joseph J., and Phillips, Cecil R., *Project Management with CPM and Pert,* 2nd ed., Van Nostrand Reinhold, New York, 1970

Moore, Franklin, G., *Manufacturing Management,* Irwin, Homewood, IL, 1955

Orlicky, Joseph, *Material Requirements Planning,* McGraw-Hill, New York, 1975

Ostwald, Phillip F. (ed.), *Manufacturing Cost Estimating,* Society of Manufacturing Engineers, Dearborn, MI, 1980

Ott, Ellis R., *Process Quality Control, Troubleshooting and Interpretation of Data,* McGraw-Hill, New York, 1975

Park, William R., *Cost Engineering Analysis,* Wiley, New York, 1973

Pollack, Herman W., *Tool Design,* Reston Publishing Co., Reston, VA, 1976

Radford, J.D., and Richardson, D. B., *Production Engineering Technology,* 2nd ed., Crane, Russak, New York, 1974

Raiffa, Howard, *The Art and Science of Negotiation,* Belknap Press of Harvard University Press, Cambridge, MA, 1982

Reintjes, J. Francis, *Numerical Control, Making a New Technology,* Oxford University Press, New York, 1991

Schonberger, Richard J., *Japanese Manufacturing Techniques,* Free Press, New York, 1982

Shamblin, James E., and Stevens, G. T., Jr., *Operations Research, A Fundamental Approach,* McGraw-Hill, New York, 1974

Tobis, Irene and Michael, *Managing Multiple Projects*, McGraw-Hill, New York, 2000

Tricker, Ray, *ISO 9001-2000 for Small Businesses,* 2nd ed., Butterworth Heinemann, Boston, MA, 2001

Vough, Clair F., with Asbell, Bernard, *Tapping the Human Resource, A Strategy for Productivity,* Amacom, New York, 1975

Wealleans, David, *The Quality Audit for ISO 9001:2000 a Practical Guide*, Gower Publishing Limited, Hampshire, England, 2000

Wight, Oliver W., *Production and Inventory Management in the Computer Age,* CBI Publishing Co., Boston, MA, 1974

Index